ADHESIVES - APPLICATIONS AND PROPERTIES

Edited by **Anna Rudawska**

Adhesives - Applications and Properties
http://dx.doi.org/10.5772/62603
Edited by Anna Rudawska

Contributors

Alexandra Vinagre, João Ramos, Chengsheng Gui, Xiaoqing Liu, Jose Luis Rivera-Armenta, Beatriz Adriana Salazar Cruz, Tzeitel Hernandez-Martinez, Maria Yolanda Chavez-Cinco, Maria Leonor Mendez-Hernandez, Ulises Paramo-Garcia, Halil Ozer, Engin Erbayrak, João Cardoso Ferreira, Paulo Melo, Mario Silva, Patricia Pires, Imed Gargouri, Moncef Khadhraoui, Boubaker Elleuch, Anna Borriello, Aldobenedetto Zotti, Simona Zuppolini, Mauro Zarrelli, Onur Ülker, Jon García-Barruetabeña, Fernando Cortés, Giovanni Belingardi, Valentina Brunella, Brunetto Martorana, Raffaele Ciardiello, Andrea Vernengo, Jun Zhong, Antonija Tadin, Nada Galić, Lidia Gavić, Roberto Pinna, Enrica Filigheddu, Stefano Eramo, Egle Milia

Published by InTech

Janeza Trdine 9, 51000 Rijeka, Croatia

Publishing Process Manager Edi Lipovic

Technical Editor SPi Global

Cover InTech Design team

First published November, 2016

Printed in Croatia

Additional hard copies can be obtained from orders@intechopen.com

Adhesives - Applications and Properties, Edited by Anna Rudawska

p. cm.

Print ISBN 978-953-51-2783-3

Online ISBN 978-953-51-2784-0

For EU product safety concerns:
IN TECH d.o.o., Prolaz Marije Krucifikse Kozulić 3, 51000 Rijeka, Croatia,
info@intechopen.com or visit our website at intechopen.com.

Meet the editor

Anna Rudawska, PhD, DSc, Eng, Associate Professor, is the head of the Technological Processes and Systems Design Chair, Faculty of Mechanical Engineering, and since September 1, 2012, she is the Vice Dean of the Faculty of Mechanical Engineering, Lublin University of Technology, Poland. She is the author or co-author of 150 scientific publications in Polish, English, and Slovak; the author or co-author of monographs and chapters; and the author of patents and patent applications in adhesive field. She is a member of lots of international and national organizations and associations like The Polymer Processing Society, Polish Society of Mechanical Engineers and Technicians, The Polish Maintenance Society and Editorial Board of International Journal of Adhesion and Adhesives.

Contents

Preface

This book presents some specific information regarding adhesives which have applications in industry, medicine and dental fields. Adhesives' advantageous properties and relatively uncomplicated processing methods contribute to their increasing application and their growing popularity in the industry. Some adhesives represent properties superior to those of most adhesive materials, due to their excellent adhesion and chemical resistance. A wide variety of adhesives' (also resins and curing agents) considerable flexibility in modification of properties of adhesives allows adjusting the composition to particular applications.

Joining different structural materials with adhesives is becoming increasingly widespread as it offers multiple advantages when compared to traditional joining methods. Adhesive bonding provides an invaluable alternative to other modern methods of joining structural materials. At present bonding technology offers a range of applications to a number of various branches of industry, with the building, the automotive, the aircraft, the machine-building, or the packaging manufacturing industry. Increasing popularity of adhesive bonding as a method for joining metals results from several factors, for example, high joint strength; lack of stresses within the joint, along with low cost per unit, resulting predominantly from the amount of adhesive used to form a single joint; and the lack of electrochemical phenomena usually accompanying other methods of joining metals and joining dissimilar structural materials, frequently of substantial disproportions in geometric dimensions.

One of its numerous advantages is its joint's relatively high adhesive strength. A number of elements contribute to adhesive joint strength, including one technological factor—the type of adhesive. The various properties of adhesives (like physical, chemical and mechanical properties) decide of their application in the various branches of industry, medicine and various fields of life.

The book was divided into two parts: "Adhesives Applications in Medicine and Dentistry" and "Properties of Adhesive." The aim of such a presentation was to present the usage in very different aspects of application of the adhesives. Each of the types of adhesives is characterized by specific properties.

The dental and medicine adhesives should exhibit the excellent adhesion to various kinds of tissues and other very specific properties. This issue is very important due to human health and life. It is interesting that the part of the book which concerns dental and medicine adhesive covers some test of results which are presented from hospitals and clinics.

Meanwhile industry adhesives should present the excellent adhesion to adherends, the higher strength of adhesive joints prepared by using these adhesives, as well as good properties during their (and also adhesive joints) operating time under various climatic conditions and ageing and degradation process, including biodegradation process.

This book collects the new development in the science and usage of adhesives and adhesive properties and also adhesive joints. I would like to express my gratitude to all of the contributors for their high-quality manuscripts. I hope open-access format of this book will help researches and that they will benefit from this collection.

Anna Rudawska
Lublin University of Technology, Faculty of Mechanical Engineering
Department of Production Engineering, Lublin, Poland

Section 1

Adhesives Applications in Medicine and Dentistry

Chapter 1

Etch-and-Rinse and Self-Etch Adhesives Behavior on Dentin

João Cardoso Ferreira, Patrícia Teixeira Pires, Paulo Ribeiro de Melo and Mário Jorge Silva

Additional information is available at the end of the chapter

http://dx.doi.org/10.5772/64856

Abstract

Hybridization is a key phenomenon in bonding resin-based composite restorations to dentin, and results from a molecular-level interaction between the resin and the demineralized collagen fibrils network. Nanoleakage occurs when small molecules or ions infiltrate into the hybrid layer. Our work aims to evaluate if the type of solvent and adhesive system influences the morphology of the hybrid layer and the occurrence of nanoleakage within it. Human molar teeth were distributed into six groups corresponding to adhesives with different solvents (Scotchbond™ 1XT, XP Bond™, Prime&Bond® NT, One Coat Bond®, AdheSE® and Xeno® V). Dentin disks (specimens) were cut from those teeth. The corresponding adhesive systems and a microhybrid composite (Synergy® D6) were applied onto them. The specimens were thermocycled, fixed, cut, polished, decalcified, dried, and, for nanoleakage evaluation, immersed into a tracer solution. The morphology and nanoleakage analyses were performed with a high-resolution scanning electron microscope (field-emission scanning electron microscope —FESEM), and the results were statistically processed. AdheSE® achieved the overall best performance. The type of adhesive, the number of steps, and the solvent seem to play a significant role in hybrid layer morphology and nanoleakage within it. The hybrid layer water absorption can determine the adhesion longevity.

Keywords: etch-and-rinse adhesives, self-etch adhesives, solvents, hybrid layer, nanoleakage

1. Introduction

Most adhesive systems were designed to promote dentin bonding through the interaction of hydrophilic monomers with the collagen fibrils exposed on the dentin surface after etching.

The effectiveness of current dentin adhesives is believed to depend on the leakage of monomers with high hydrophilic affinity into the network of collagen fibers in the etched dentin structure. This entanglement of monomers with collagen fibrils and scarce residual hydroxyapatite crystals forms a hybrid tissue known as resin-dentin interdiffusion zone or hybrid layer [1].

1.1. Hybrid layer

According to Nakabayashi, Kojima, and Masuhara (1982), the hybrid layer is the zone of contact between the restorative material and the solid dentin [2]. Nakabayashi's team was the first to demonstrate that resins could infiltrate the etched dentin to form a new structure composed of resin matrix reinforced by collagen fibers and named this biocomposite "hybrid layer" [3]. In the dentin, the bonding tension of adhesive systems depends mostly on the hybrid layer and then on the resin tags in dentin tubules, and, finally, on the chemical bond [4].

Hybridization is the key phenomenon in the bonding of resin-based composite restorations to the dentin. It results from the molecular interaction between the resin and the demineralized network of collagen fibers, which occurs at a depth of 3–5 μm [5]. This depth decreases to 1–8 μm [4] with etch-and-rinse systems and to 0.5–1 μm with self-etch systems [6–8].

When resin interacts with the network of collagen fibers, a micromechanical bonding to the dentin occurs through the penetration of polymerizable resin monomers into the exposed collagen network, the open tubules, and their lateral branches—the canaliculi, thus originating, respectively, the hybrid layer or interdiffusion zone, the resin tags, and the resin microtags. The resin penetration through the dentin's tubular/canalicular system is important to the mechanical retention as it may increase the adhesion to the whole dentin [9].

Although the hybrid layer thickness is not important in adhesion, its existence is essential and depends on the collagen fibers not collapsing while the dentin is drying after etching [9]. Under ideal conditions, the formation of the hybrid layer should be the greatest bonding mechanism in the superficial dentin while the resin tags are the major contributors for bonding forces in the deep dentin. On the other hand, since the intertubular dentin is limited due to the enlargement and approximation of dentinal tubules, the hybrid layer contributes little to adhesion [10]. The microtags in the canaliculi provide greater retention and better sealing, and also hybridization, which in this case is named "canalicular hybridization" [11].

The main functions of the hybrid layer are bonding, copolymerization with resin-based composites, and simultaneously acting as a protective layer by preventing microorganisms and toxins from reaching the pulp through the dentin [12]. The hybridized dentin reduces the risk of microleakage, the incidence of secondary caries, and postoperative sensitivity [10].

The hybrid layer's role on dentin bonding has constantly been questioned. After etching, resin monomers do not thoroughly diffuse through the collagen network to the depth of the etched dentin. This incomplete resin leakage into the collagen network may produce a porous layer of collagen that is not protected by hydroxyapatite or impregnated with resin, which will lead to nanoleakage of the hybrid layer. This exposed collagen is thus subjected to hydrolysis and degradation, resulting in increased microleakage and failure with time [13]. Recent studies on micropermeability showed that the hybrid layer can absorb and release water, acting as a semipermeable or even permeable membrane [14, 15]. Before the resin's penetration, part of the organic phase of the dentin matrix composed of type I collagen, proteoglycans, and other noncollagen proteins may not be impregnated with the adhesive, which causes posterior water absorption and may lead to the degradation of bonding forces by nanoleakage.

1.2. Nanoleakage

In 1995, Sano described the nanoleakage phenomenon as an important indicator for evaluating a material's sealing ability [16]. While in microleakage cracks may vary between 20 and 50 μm, in nanoleakage, they have approximately 0.02 μm [17]. Nanoleakage consists in the hybrid layer's permeabilization to small ions or molecules, even when there is no detectable development of interface failures [18, 19]. According to Prati et al., and other authors who studied nanoleakage [16, 18, 20], this phenomenon is believed to result from several mechanisms combined, including the incomplete leakage of monomers in the demineralized collagen matrix, the presence of hydrophilic monomers, and the insufficient removal of solvent or water that is trapped in the hybrid layer. It may also originate from the contraction of restorative material during polymerization or from the elution of residual adhesive monomers [3].

With etch-and-rinse adhesive systems, the demineralized dentin area becomes a weak point for adhesion due to the hydrolysis of type I collagen fibers and proteoglycans located there as they should be completely impregnated in adhesive to ensure a good protection throughout time [19, 21]. Therefore, the incomplete infiltration of resinous monomers in the demineralized dentin matrix should be the main reason for decreased adhesive forces. The poor resin leakage facilitates permeability to external fluids (as well as to oral and bacterial enzymes) and to endogenous proteolytic enzymes that are slowly released from the etched dentin after adhesion and impair the performance of the attached interface [19, 22].

Even though self-etch adhesive systems condition and simultaneously leak into the underlying substrate, a discrepancy between the demineralization and resin infiltration depths may occur [23, 24]. Several references prove that nanoleakage in interfaces formed by self-etch adhesives is not exclusively caused by the incomplete leakage of resin in the demineralized dentin. Other reported causes are areas within the adhesive layer from where water is incompletely removed, thus resulting in areas of improper polymerization and/or hydrogel formation, and hydrophilic areas of acidic monomers that are more prone to water caption [23]. Sometimes, in both etch-and-rinse and self-etch adhesive systems, channels filled with water form in the adhesive layer—water trees, and these are manifested by water bubbles in the resin-dentin interface. This phenomenon may contribute directly to resin degradation due to the extraction of nonpolymerized monomers or small oligomers with time [25].

Besides occurring in the hybrid layer, nanoleakage may also occur in the adhesive layer [16, 26]. Moreover, water absorption by hydrophilic resin monomers in the hybrid and adhesive layers may contribute to the degradation of bonding forces over time [27] since the permeable hybrid layer seems to be highly susceptible to slow water hydrolysis.

Ammoniacal silver nitrate ($AgNO_3$) has been used to identify regions filled with water, with leakage failure, and with the deterioration of hydrophilic polymers in the hybrid and adhesive layers [25]. Silver nitrate is an excellent marker as silver ions are very small and have high solubility, thus allowing the preparation of highly concentrated solutions [28]. The flow of silver ions increases in water trees.

1.3. Objectives

This work was aimed to evaluate if the type of solvent and adhesive system influence the morphology of the hybrid layer and the occurrence of nanoleakage within it, based on high-resolution electronic microscopy (FESEM) analysis.

1.4. Null hypothesis

The type of solvent and the adhesive system used in two-step etch-and-rinse adhesive systems do not influence the hybrid layer morphology, quality, and thickness, as well as the occurrence of nanoleakage within it.

2. Material and methods

2.1. Sample preparation

After appraisal and approval by an Ethics Committee, 78 caries-free human molars ($N = 78$), extracted for periodontal or orthodontic reasons, were included in this study. These teeth had been disinfected in 0.5% chloramine and had been stored for up to 6 months in distilled water (according to ISO/TS 11405, 2003). They were cross-cut into 1 mm-thick dentin disks—specimens—using a slow-speed diamond saw (Accuton 2-Struers, Copenhagen, Denmark). A standardized smear layer was created on the occlusal surface of each specimen using 600-grit silicon carbide (SiC) paper (for 1 min under cooling with distilled water).

2.2. Restorative procedures

Specimens were randomly allocated into the following six adhesive/solvent groups (**Table 1**):

- Group A: Adper™ Scotchbond™ 1XT (3 M ESPE, Seefeld, Germany)—ethanol and water;
- Group B: XP Bond™ (Dentsply, Konstanz, Germany)—tert-butanol;
- Group C: Prime&Bond® NT (Dentsply, Konstanz, Germany)—acetone;
- Group D: One Coat Bond® (Coltène Whaledent, Altstätten, Switzerland)—solvent free, 5% water;
- Group E: AdheSE® (Ivoclar Vivadent, Schaan, Liechtenstein)—water;
- Group F: Xeno® V (Dentsply, Konstanz, Germany)—water/tert-butanol.

Two layers of each adhesive system were applied on the dentin disks, following the manufacturers' instructions, and then light cured for 20 s (BluePhase, Ivoclar Vivadent, Schaan, Liechtenstein). After adhesive application, a hybrid composite resin (Synergy® D6, Coltène Whaledent, Altstätten, Switzerland) was applied and light cured in two increments of 2 mm. Light curing was performed at 1200 mW/cm^2 (BluePhase, Ivoclar Vivadent, Schaan, Liechtenstein).

Adhesives	Composition
Adper™ Scotchbond™ 1XT (3M-ESPE)* Lot number: 5 FL	Bis-GMA, HEMA. Dimethacrylates polyalcenoic copolymer. 5 nm diameter 10% of weight silica spherical particles. Solvents: ethanol and water
XP Bond™ (Dentsply) Lot number: 0609000250	Carboxylic acid modified dimethacrylate (TCB resin); phosphoric acid modified acrylate resin (PENTA); urethane dimethacrylate (UDMA); triethyleneglycol dimethacrylate (TEGDMA); 2-hydroxy ethylmethacrylate (HEMA); butylaned benzenediol [stabilizer]; ethyl-4-dimethylamir obenzoate; camphorquinone; functionalized amorphous silica. Solvent: tert-butanol
Prime & Bond™ NT (Dentsply) Lot number: 0508000096	Di- and trimethacrylate resins, PENTA (dipentaerythritol penta acrylate monophosphate), photoinitiators, stabilizers, nanofillers—amorphous silicon dioxide cetylamine hydrofluoride. Solvent: acetone
One Coat Bond™ (Coltene Whaledent) Lot number: 0090783	HEMA, UDMA, HPMA, hidroxypropylmethacrylate, glycerol, methacrylates, polyalkenoate methacrylized, amorphous silica Solvent: 5% water
AdheSE™ (Ivoclar Vivadent, Schaan – Liechtenstein) Batch number: Primer (1)·M02841 Bonding Agent (2): L56767	AdheSE primer: dimethacrylate, phosphoric acid acrylate, initiators and stabilizers in an aqueous solution AdheSE bond: HEMA, dimethacrylate, silicon dioxide, initiators, and stabilizers. Solvent: water
Xeno™ V (Dentsply, Konstanz—Germany) Batch number: 1002000450	Bifunctional acrylate, acidic acrylate functionalized phosphoric acid ester, acrylic acid, water, tertiary butanol, initiator, stabilizer. Solvents: water and tert-butanol

Bis-GMA = bisphenol A glycidyl methacrylate; HEMA = hydroxyethylmethacrylate; UDMA = Urethanedimethacrylate; HPMA = hydroxypropylmethacrylate.
*Adper™ Scotchbond™ 1XT (Europe) is the same as Adper™ Single Bond *Plus* (USA) and Adper™ Single Bond 2 (Latin America, Gulf countries, and Pacific region including countries such as Australia, New Zealand, and Hong-Kong).

Table 1. Composition and batch numbers of the adhesives used in the study.

2.3. Assessment of the hybrid layer morphology

All 48 specimens were stored at 37°C with 100% humidity for 24 h (Hemmet, Schwabach, Germany) and thermal-cycled (500 cycles) in distilled-water baths at 5 and 55°C (Aralab, mod

200E, Cascais, Portugal) with a dwell time of 20 s. After storage under the same conditions for another 24 h, specimens were fixed in glutaraldehyde and rinsed. Then, they were cross-cut in half, creating 96 restored hemidisks ($n = 16$). The hemidisks were polished with a sequence of sandpapers (320, 500, 1000, and 1500) and diamond pastes with 3, 1, and ¼ μm (Kemet® diamond spray, Kapellen, Belgium) using polishing cloths (DP-NAP, Struers, Copenhagen, Denmark). Immediately after, the hybrid layers were revealed by denaturation and decalcification of the specimens. Finally, the specimens were dehydrated in ethanol and hexamethyldisilazane (HMDS) [20, 29].

2.4. Nanoleakage assessment

A total of 30 specimens were placed in a greenhouse (Hemmet, Schwabach, Germany) at 37°C and 100% humidity. Afterward, they were vertically cut with a microtome (Accuton 2, Struers, Copenhagen, Denmark) into 0.9 mm thick slabs, originating 60 restored hemidisks (slabs)—10 specimens per group ($n = 10$). Each slab was then coated with two layers of nail polish, except for 1 mm around the adhesive interface. Specimens were allowed to rehydrate in distilled water for 20 min before being immersed in 50 wt% aqueous ammoniacal silver nitrate (pH 9.5) for 24 h. Specimens were then immersed in a photodeveloper solution for 8 h (Intra Periomat, Dental Durr, Bietigheim, Bissingen, Germany) under a fluorescent light (Phillips, Amsterdam, Holland) to reduce the silver or diamine silver ions into metallic silver grains [27]—and were washed in water for 1 min.

Fixation was performed by immersion in a solution of 2.5% glutaraldehyde buffered in 0.1 M sodium cacodylate (pH 7.4) for 12 h at 4°C. Specimens were then rinsed three times with 20 ml of 0.2 M buffered sodium cacodylate (pH 7.4) for 1 h, followed by distilled water for 1 min. Surface polishing was achieved using decreasing abrasive grits (600, 800, and 1200) of silicon carbide paper, and diamond spray (Kemet diamond spray) of 2 μm and 1 μm on polishing cloths (DP-NAP, Struers, Copenhagen, Denmark), followed by ultrasound bath in 100% ethanol for 10 min, and demineralization in 0.5% silica-free phosphoric acid for 1 min to remove surface debris. Afterward, the specimens were dehydrated in ascending concentrations of ethanol in water (25% for 20 min, 50% for 20 min, 75% for 20 min, 95% for 30 min, and 100% for 60 min) and were dried by immersion in hexamethyldisilazane for 10 min [29].

2.5. Specimens preparation for the FESEM observation

Specimens were placed in electron microscopy stubs and sputter-coated with Au-Pd (JEOL Fine Coat Ion Sputter JFC-1100, Tokyo, Japan). They were observed in field-emission scanning electron microscope (FESEM) (JEOL JSM 6301 F, Tokyo, Japan) using secondary electrons (SE) and an accelerating voltage of 10 kV for the morphology specimens, and backscattered electrons (BE) and an accelerating voltage of 15 kV for the nanoleakage specimens. Silver was detected using the EDS Microanalysis System (Oxford Inca Energy 350®, Oxford Instruments, Oxfordshire, UK).

2.6. Morphology assessment

The dentin-resin interdiffusion zone of the 16 hemidisks from each group was observed under FESEM (JEOL JSM 6301 F, Tokyo, Japan) at 10 kV, using SE. Electronic microphotographs were then taken at different magnifications and scanned with the EDS Microanalysis System (Oxford Inca Energy 350®, Oxford Instruments, Oxfordshire, UK).

For the analysis of the hybrid layer morphology, in order to standardize the measurements, six images were taken with different magnifications (right, center, and left at 400×, 500×, 800×, and 1500×) per specimen. The hybrid layer thickness was measured in the 1500× images using the "offline" mode of the software (Oxford Inca Energy 350®, Oxford Instruments, Oxfordshire, UK). The hybrid layer thickness was observed in the three images at 1500×, per specimen, and was averaged in order to obtain a single value to represent each tooth.

2.7. Nanoleakage assessment

For the analysis of nanoleakage within the hybrid layer, 10 hemidisks from each group were observed under FESEM. Six images were taken with different magnifications (right, center, and left at 400× and 2000×; 1000× was used for water-tree observation) per specimen. The total length of the hybrid layer and the extension of the ammoniacal silver nitrate impregnation were measured in the 400× images using the "offline" mode of the software (Oxford Inca Energy 350®, Oxford Instruments, Oxfordshire, UK). The nanoleakage length within the hybrid layer observed in the three images at 400×, per specimen, was averaged in order to obtain a single value to represent each tooth.

2.8. Statistical analysis

Scanning electron microscopy observations were registered in an Excel file and were analyzed using the IBM SPSS version 19 software. To detect and localize statistically significant differences in the studied variables from the six groups of adhesives, the Kolmogorov–Smirnov, Levene's, and one-way or Brown–Forsythe ANOVA tests were used. Then, multiple comparisons (and correlations) were performed with the Tamhane posthoc and Bonferroni tests for morphology and nanoleakage in the hybrid layer per 305 μm, respectively. In the comparative analyses between the six studied adhesives, 5 and 10% significance levels were applied, with Bonferroni correction, and only differences with $P < 0.0033$ and $P < 0.0066$, respectively, were considered statistically significant.

3. Results

3.1. Morphology assessment

The final number of specimens used for the statistical analysis was the same as the initial number ($N = 96$) as no specimen was lost during the experimental procedure, even though some in the Xeno® V Group presented fractures.

The hybrid layer thickness values obtained with the different adhesive systems were the following (**Table 2** and **Figure 1**):

- Scotchbond™ 1XT, 3.23 μm ± 0.53 μm;
- XP Bond™, 3.13 μm ± 0.73 μm;
- Prime&Bond® NT, 2.53 μm ± 0.50 μm;
- One Coat Bond®, 1.84 μm ± 0.27 μm;
- AdheSE®, 1.96 μm ± 0.14 μm;
- Xeno® V, 1.29 μm ± 0.14 μm.

Hybrid layer thickness		
Adhesives	**Mean (μm)**	**Standard deviation (μm)**
Scotchbond™ 1XT	3.2287	0.53253
XP Bond™	3.1335	0.7315
Prime & Bond® NT	2.5346	2.3617
One Coat Bond®	1.84	0.2744
AdheSE®	1.964	0.13571
Xeno® V	1.291	0.14187

Table 2. Mean hybrid layer thickness and standard deviation of the different adhesive systems.

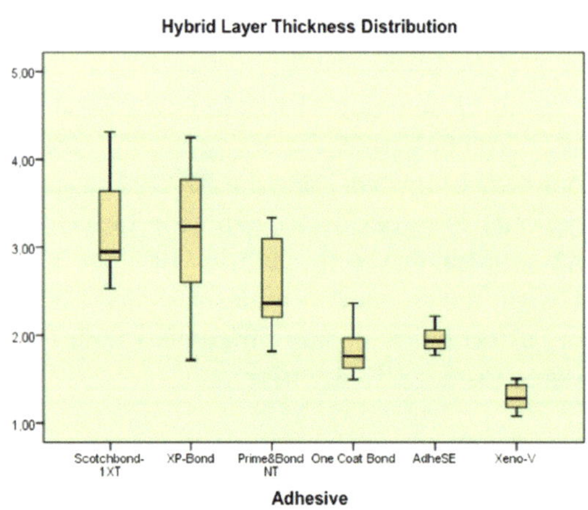

Figure 1. Mean hybrid layer thickness and standard deviation of the different adhesive systems.

Scotchbond™ 1XT showed a greater mean thickness than One Coat Bond®, AdheSE®, and Xeno® V (1.26–1.94 μm, $P < 0.001$). XP Bond™ also showed a greater mean thickness than One Coat Bond®, AdheSE®, and Xeno® V (1.17–1.84 μm, $P < 0.001$). Prime&Bond® NT showed a hybrid layer thicker than One Coat Bond® (0.69 μm; $P < 0.002$) and Xeno® V

(1.24 μm; $P < 0.001$). One Coat Bond® had a hybrid layer thicker than Xeno® V (0.55 μm; $P < 0.001$). AdheSe® had a greater mean hybrid layer thickness than Xeno® V (0.67 μm; $P < 0.001$). Finally, Xeno® V showed the lowest values of all, in a statistically significant way (**Figure 2**).

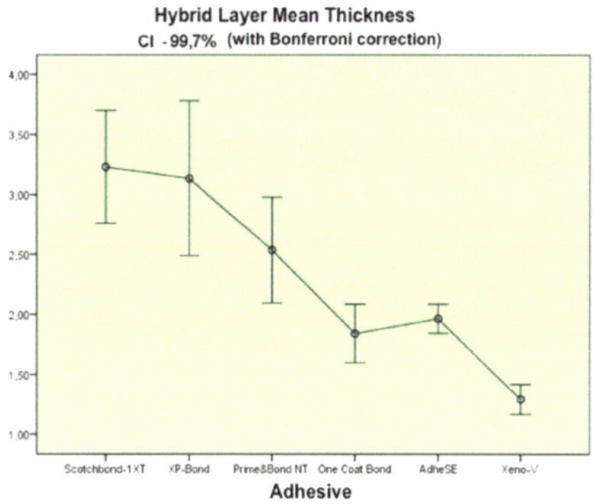

Figure 2. Comparison of the mean hybrid layer thicknesses, with Bonferroni correction, between the six adhesive groups.

3.1.1. FESEM images (morphology)

3.1.1.1. Scotchbond™ 1XT

Figure 3A and **B** shows the adhesive layer (AL), the hybrid layer (HL), the resin tags (T), and dentin (D) with different magnifications (800× and 1500×). In both figures, the hybrid layer of Scotchbond™ 1XT presents good quality.

3.1.1.2. XP Bond™

Figure 4A and **B** shows the adhesive layer (AL), the hybrid layer (HL), and, beneath that, the resin tags (T) at 500× and 1500× magnifications. In both figures, the hybrid layer of XP Bond™ also presents good quality.

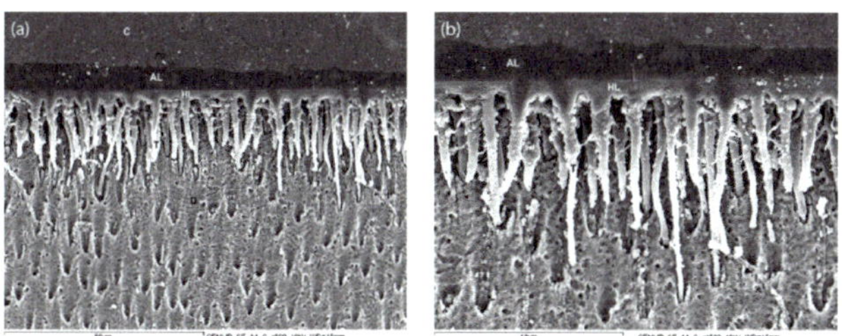

Figure 3. A and B.

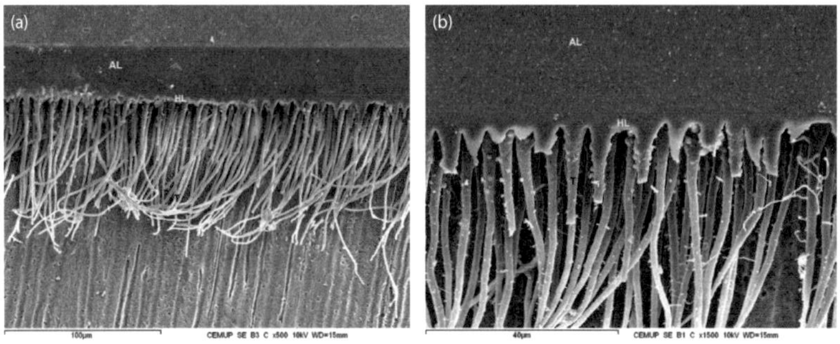

Figure 4. A and B.

3.1.1.3. Prime&Bond® NT

Figure 5A and **B** shows the adhesive layer (AL) and a space in the underlying hybrid layer (HL) marked with arrows that represent a failure in the resin leakage. These figures also show resin tags (T). In this case (Prime&Bond® NT), the hybrid layer is impaired, which may imply greater nanoleakage.

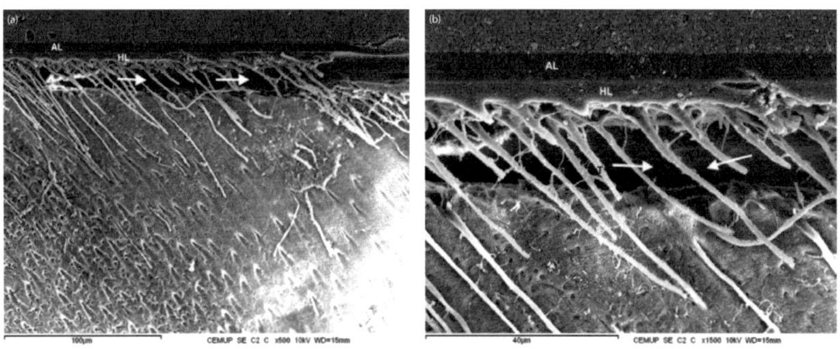

Figure 5. A and B.

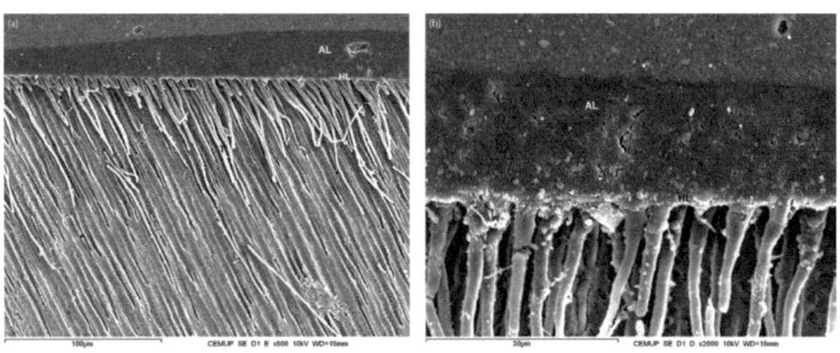

Figure 6. A and B.

3.1.1.4. *One Coat Bond®*

Figure 6A and **B** shows the adhesive layer (AL), the hybrid layer (HL) and the resin tags (T) different magnifications (500× and 2000×). In both figures, the hybrid layer of One Coat Bond® presents good quality.

3.1.1.5. *AdheSE®*

Figure 7A and **B** shows the adhesive layer (AL), the hybrid layer (HL), and, beneath that, the resin tags (T). **Figure 7B** shows the adhesive layer (AL), the hybrid layer (HL), and the underlying resin tags (T) with greater magnification. In both figures, the hybrid layer of AdheSE® presents good quality.

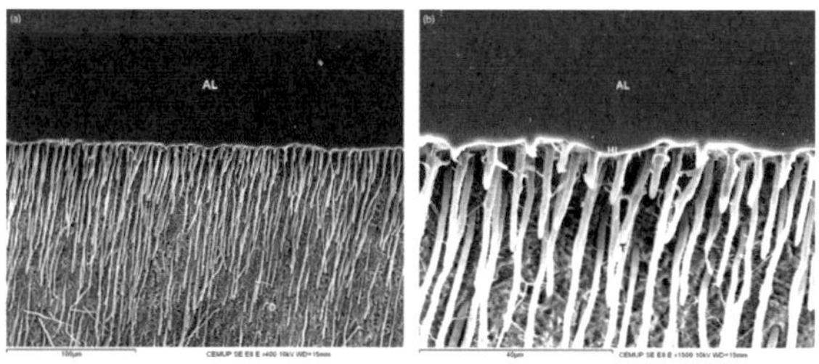

Figure 7. A and B.

3.1.1.6. *Xeno® V*

Figure 8A and **B** shows the adhesive layer (AL) and a space in the underlying hybrid layer (HL) marked with arrows that represent a failure in the resin leakage. That space corresponds to the area where the collagen fibers that were not involved by adhesive resin were located before the specimens' preparation. Those collagen fibers were degraded by either the sodium hypochlorite/hydrochloride acid during preparation or some hydrolysis during thermocycling. These figures also show the dentin (D) and the resin tags (T) of which some are fractured. In this case (Xeno® V), the hybrid layer is impaired, which may imply greater nanoleakage.

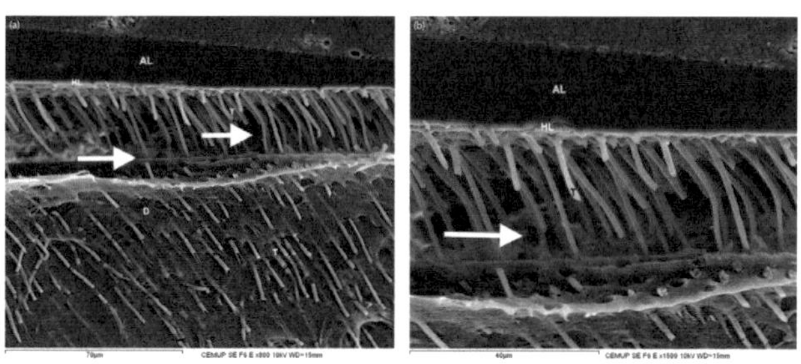

Figure 8. A and B.

3.2. Nanoleakage assessment

The final number of specimens used for the statistical analysis was the same as the initial number ($N = 60$) as no specimen was lost during the experimental procedure, even though some presented fractures, again in the Xeno® V group.

Nanoleakage percentage in the hybrid layer per 305 μm		
Adhesives	**Mean (%)**	**Standard deviation (%)**
Scotchbond™ 1XT	49.43	22.73
XP Bond™	38.81	22.36
Prime&Bond® NT	57.03	31.68
One Coat Bond®	55.29	14.74
AdheSE®	20.94	14.63
Xeno® V	75.68	14.58

Table 3. Mean percentages and standard deviations of the nanoleakage observed in the hybrid layer (per 305 μm) of the different adhesive systems used in this study.

Scotchbond™ 1XT showed a nanoleakage of 49.43 ± 22.73%, XP Bond™ of 38.81 ± 22.36%, Prime&Bond® NT of 57.03 ± 31.68%, One Coat Bond® of 55.29 ± 14.74%, AdheSE® of 20.94 ± 14.63%, and Xeno® V of 75.68 ± 14.58% (**Table 3** and **Figure 9**). Prime&Bond® NT had two outliers, one of which was severe (**Figure 9**). Even so, these outliers were not enough to impede the application of parametrical analysis methods, and their exclusion did not help to detect statistical significances.

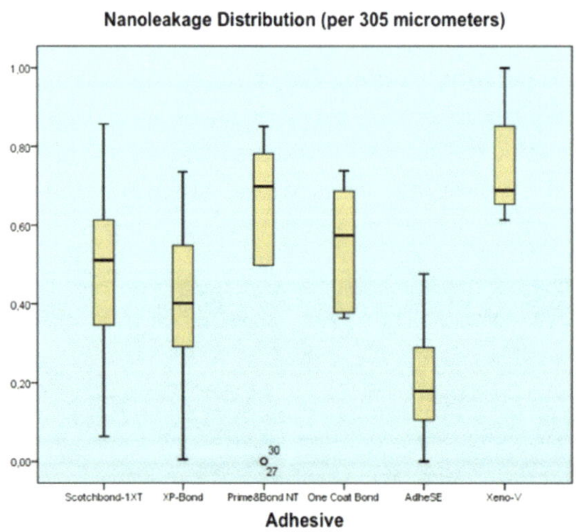

Figure 9. Distribution of the nanoleakage percentage in the hybrid layer (per 305 μm) by the six adhesive system groups (with outliers).

http://dx.doi.org/10.5772/64856

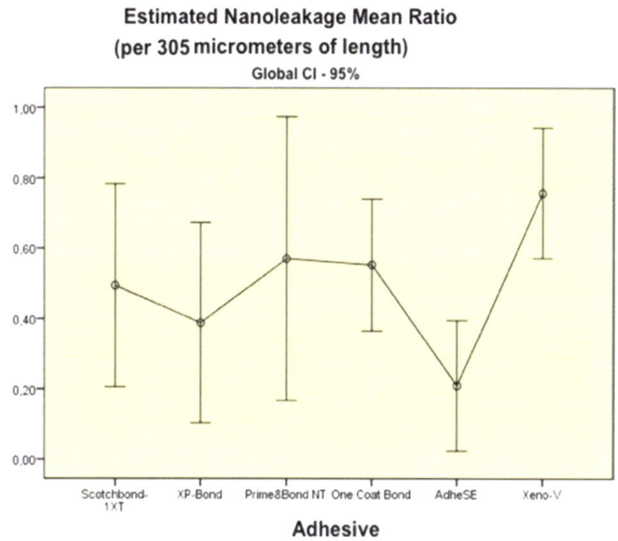

Figure 10. Comparison of the estimated mean nanoleakages for each adhesive system group (with Bonferroni correction).

After outliers' elimination, more analyses were conducted. For a 95% confidence, AdheSE® showed in average less 54.74% nanoleakage than Xeno® V ($P < 0.001$); no statistically significant differences were found in the remaining groups (**Figure 10**).

3.2.1. *FESEM images (nanoleakage)*

3.2.1.1. *Scotchbond™ 1XT*

Figure 11A and **B** illustrates the results of Scotchbond™ 1XT and show the adhesive layer (AL), the underlying hybrid layer (HL), the resin tags (T) and dentin (D) at 400× and 2000× magnifications. The hybrid layer reveals water absorption as it is impregnated with the ammoniacal silver nitrate (lighter shade).

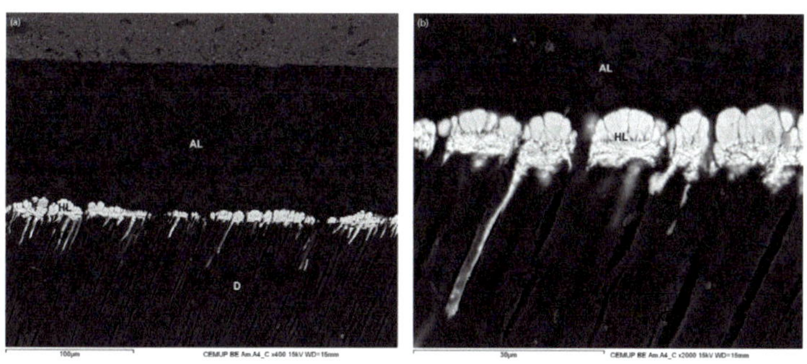

Figure 11. A and B.

3.2.1.2. XP Bond™

Figure 12A and **B** illustrates the results of XP Bond™ and show the adhesive layer (AL), the underlying hybrid layer (HL), and the resin tags (T) at 400× and 2000× magnifications. Figures reveal that the hybrid layer also absorbed some water as it is impregnated with the ammoniacal silver nitrate (lighter shade).

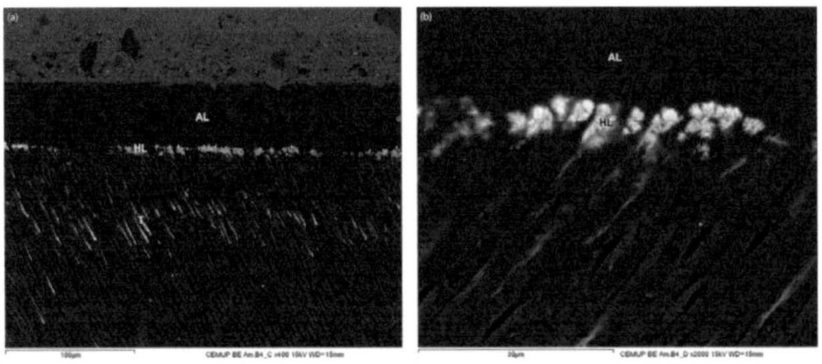

Figure 12. A and B.

3.2.1.3. Prime&Bond® NT

Figure 13A and **B** illustrates the results of Prime&Bond® NT and show the adhesive layer (AL), the underlying hybrid layer (HL), and the resin tags (T) at 400× and 2000× magnifications. It is also possible to identify the resin-based composite (C). The hybrid layer reveals water absorption as it is impregnated with the ammoniacal silver nitrate (lighter shade).

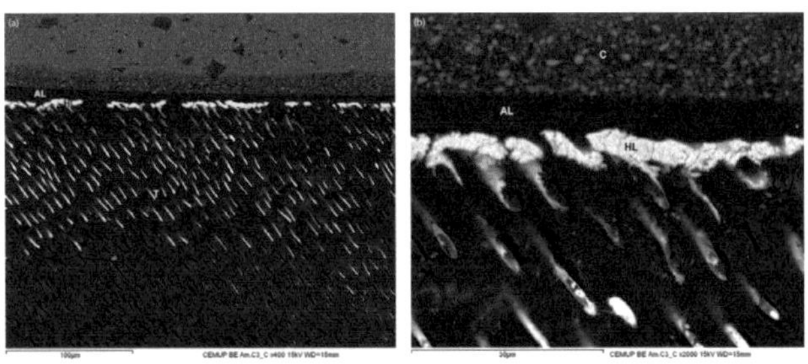

Figure 13. A and B.

3.2.1.4. One Coat Bond®

Figure 14A–C illustrates the results of One Coat Bond® and show the adhesive layer (AL), the underlying hybrid layer (HL), and the resin tags (T) at 400× and 2000× magnifications. The hybrid layer reveals water absorption as it is impregnated with the ammoniacal silver nitrate (lighter shade). In **Figure 14A**, the phenomenon of water trees (WT) formation is clearly visible. It is represented by a branched pattern and reveals nanoleakage in the adhesive layer because the hybrid layer acted as a permeable membrane, thus allowing water to flow through it.

http://dx.doi.org/10.5772/64856

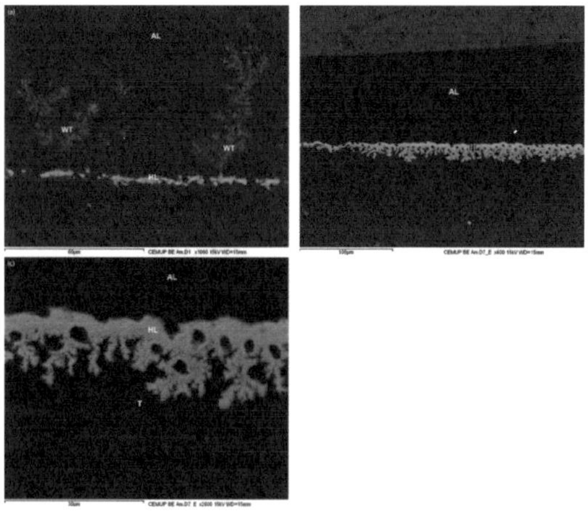

Figure 14. A, B, and C.

3.2.1.5. *AdheSE®*

Figure 15A and **B** illustrates the results of AdheSe® and show the adhesive layer (AL) and the underlying hybrid layer (HL) at 400× and 2000× magnifications. The hybrid layer has absorbed practically no water, and thus the ammoniacal silver nitrate solution was practically not impregnated with this adhesive.

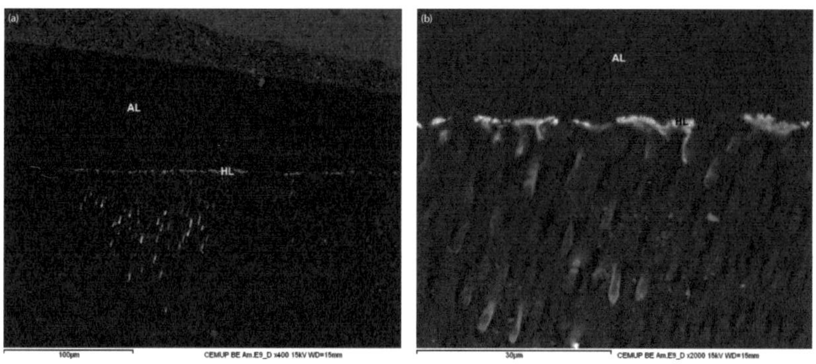

Figure 15. A and B.

3.2.1.6. *Xeno® V*

Figure 16A and **B** illustrates the results of Xeno® V and show the adhesive layer (AL), the underlying hybrid layer (HL), and the resin tags (T) at 400× and 2000× magnifications. The dentin (D) is also identified. The hybrid layer is almost completely impregnated with ammoniacal silver nitrate. The specimens suffered fractures, and this is clearly visible in all figures. Fracture of the resin tags (T) is also visible.

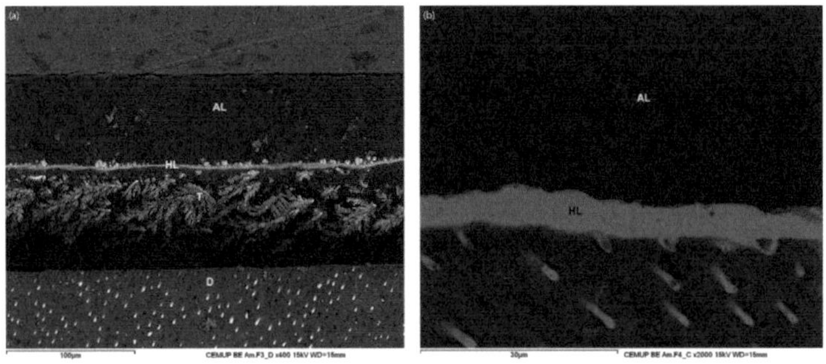

Figure 16. A and B.

4. Discussion

4.1. Hybrid layer morphology

In this study, we evaluated the hybrid layer quality, which may be very important for assessing the bonding process performance and predicting the adhesives durability in the long term in the oral environment. However, that evaluation is somewhat subjective and sometimes may be difficult to perform.

Milia and Santini reported that the identification of a high-quality hybrid layer with strong and stable adhesion is associated with a homogeneous area where monomers leak completely and fill the collagen matrix [30]. When the whole demineralized dentin (with exposed collagen fibers) is protected, i.e., surrounded by adhesive, a high-quality hybrid layer is achieved. When this does not occur, as in the case of etch-and-rinse adhesives, there is a discrepancy between the demineralization and the resin penetration, resulting in a "space" underneath the hybrid layer, which is clear in the electron microscopy images of the Prime&Bond® NT (**Figure 5A** and **B**). Moreover, according to some studies, regardless of the number of application steps, acetone-based etch-and-rinse adhesives have shown a performance worse than the water/ ethanol-based ones. The main reason for their compromised clinical results in the long term is the high technical sensitivity of acetone-based adhesives [31, 32].

In single-step self-etch systems, such as the Xeno® V, the same phenomenon appears in the FESEM images (**Figure 8A** and **B**). In this situation, the cause may lie on the phase separation and poor resin penetration due to an early evaporation of the solvent, or on a poor polymerization of the adhesive due to great quantities of water being absorbed by the hybrid layer of these adhesives [15], which might have lead to adhesive fracture during polymerization. If the solvent does not completely evaporate before light cure, failures may occur, weakening the hybrid layer and causing an early failure of the restoration [33]. Thus, the hybrid layer has poor quality. According to the literature, the good quality of a hybrid layer is essential for adhesion, even more than its thickness. That is why self-etch adhesives often provide greater forces of adhesion to the dentin than etch-and-rinse ones, despite the latter having a larger thickness (up to 5 μm versus 0.5–1.5 μm) [34]. Although these thickness values vary in different studies, they are coherent with the ones obtained in our study. This situation is related to the fact that self-etch adhesives show less discrepancy between dentin demineralization and resin penetration.

Therefore, there is no correlation between the hybrid layer thickness and bonding forces, which suggests that hybrid layer quality is more important than its thickness for the success of bonding forces [35]. The values obtained in this study for the hybrid layer thickness of One Coat Bond® are coherent with the ones obtained in the study by Breschi et al., which varied between 1.4 and 2.1 μm [1]. Xeno® V showed systematically lower values of mean hybrid layer thickness.

For adhesion to be effective, as well as durable, it is essential that the formed hybrid layers do not have defects. Adhesives with poor-quality hybrid layers have short durability in comparison with the ones that form high-quality hybrid layers, which are more qualitatively filled by resin [36]. In poor-quality hybrid layers, collagen fibers or fibrils that are not enveloped by resin will originate a porous interface—nanoleakage, and are thus more prone to hydrolysis (as are all the proteins) and degradation with time; these are areas of adhesion failure [18, 19, 36, 37]. Again, this study's results are coherent with previous studies, as, in the case of Xeno® V, a poor-quality hybrid layer lead to a high percentage of nanoleakage with ammoniacal silver nitrate (75.68%). This situation also occurred with Prime&Bond® NT, showing the highest nanoleakage percentage (57.03%) of all the tested etch-and-rinse systems (and the second highest of all the six groups) (**Table 3** and **Figure 9**).

Collagen fibrils exposed in the resin-dentin interfaces may be digested by matrix metalloproteinases (MMPs) of the host [38]. According to this study, Prime&Bond® NT and Xeno® V may cause bonding problems in the long term due to failures in the filling of nanospaces that were detected in the FESEM images. Therefore, it is possible to extrapolate that the in vivo behavior of these adhesives may involve problems in the mean/long term, considering the precautions that the extrapolation of in vitro results implies. All other adhesive groups achieved an acceptable qualitative formation of the hybrid layer.

4.2. Nanoleakage within the hybrid layer

Nanoleakage is thought to start in the interface between the adhesive layer and the hybrid layer, which is the weakest point in the interior of the dentin-restoration contact area [39, 40] or in the interior of a hybrid layer that has not been perfectly penetrated by adhesive resin [28, 41].

The percentage of nanoleakage with ammoniacal silver nitrate per 305 μm length of the hybrid layer was evaluated in every FESEM microphotograph. It is interesting that the specimens prepared with the two-step etch-and-rinse adhesive systems (Prime&Bond® NT and One Coat Bond®) and with the single-step self-etch adhesive system (Xeno® V) originated a greater rate of penetration of the respective hybrid layers with silver ions (per 305 μm) when compared with the other adhesive systems used in this study (**Table 3** and **Figure 9**). These results may be explained by several factors related specifically to the composition of the adhesive systems.

In fact, several factors influence the development of nanoleakage, such as the type of solvent (water versus acetone), the individual chemical components of the adhesive system (ex: HEMA, PENTA, and Bis-GMA), and the different molecular weights of its components, which vary from 130 Da in HEMA to 513 Da in Bis-GMA, influencing the adhesive's viscosity. Other additives, such as glutaraldehyde, and the application method (dry bonding versus wet bonding) may also influence nanoleakage. With etch-and-rinse adhesive systems, dry bonding increases nanoleakage due to the collapse of collagen fibers, which interferes with the penetration of resin [4].

The reticular mode of the nanoleakage pattern, especially the silver deposits (**Figure 17**), which were perpendicular to the hybrid layer surface, corresponds to the morphologic manifestation of water trees [27]. This well-known phenomenon is clearly identified in the FESEM microphotographs (**Figure 14A**) of the One Coat Bond® and Scotchbond™ 1XT groups (although it appears in a much inferior proportion in the latter).

Given that simplified or two-step etch-and-rinse adhesive systems (such as Scotchbond™ 1XT, XP Bond™, Prime&Bond® NT, and One Coat Bond®) incorporate the primer and adhesive components in a single solution, they originate suboptimal or imperfect hybridization due to having poor ability to penetrate the demineralized dentin substrate. Moreover, their hydrophilic nature makes them more prone to water absorption and, consequently, more susceptible to the effects of hydrolytic degradation. These adhesives have a high percentage of hydrophilic monomers, thus having great permeability after polymerization and facilitating the occurrence of areas filled with water within the hybrid layer [15]. Usually, phosphoric etch adhesives (etch-and-rinse adhesives) tend to cause more nanoleakage than the two-step self-etch ones, as has been shown in some studies [16]; this fact is coherent with our study's results.

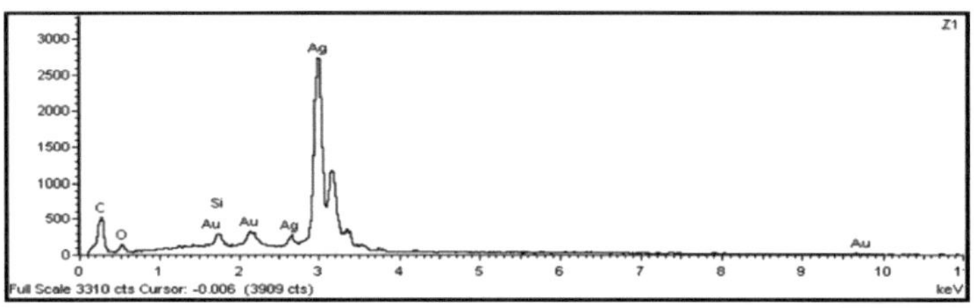

Figure 17. Hybrid layer spectra analysis reveals the presence of silver (Ag).

Furthermore, the solvents in these two-step etch-and-rinse adhesive systems also have an increased difficulty to evaporate, which increases their risk of becoming trapped more often in the adhesive layer after polymerization [6, 42]. This situation may have occurred especially with Scotchbond™ 1XT and One Coat Bond® as water is a cosolvent in the first one and the solvent in the second one. Water has (at 20°C) less volatility and vapor pressure (17.5 mmHg) comparing with organic solvents such as ethanol (43.9 mmHg) or acetone (184 mmHg), and, thus, is more difficult to evaporate, which may have led to the nanoleakage percentages observed in these adhesives. Between both adhesives with water as a solvent, One Coat Bond® had greater nanoleakage than Scotchbond™ 1XT, and this may be explained by the fact that

http://dx.doi.org/10.5772/64856

water is a poor solvent for organic components (such as monomers), which are usually more hydrophobic. Also, adding a secondary solvent, such as ethanol or acetone, helps to overcome this limitation [43]. This fact may explain the better performance of Scotchbond™ 1XT, in terms of nanoleakage percentage, compared with One Coat Bond®, as well as the worst solvent evaporation with the latter, and consequent greater water flow represented by the ammoniacal silver nitrate (**Table 3** and **Figures 11A** and **B**, **14A-C**). Moreover, the high boiling temperature and low vapor pressure of water make this solvent hard to remove from adhesive solutions after these have been applied to the dental substrate. Furthermore, Pashley et al. demonstrated that some monomers, such as HEMA (found in both Scotchbond™ 1XT and One Coat Bond®) lower, even more, the water vapor pressure, which may make difficult the removal of the remaining water [43, 44]. These facts also help explain the greater tendency of One Coat Bond® to form water trees compared with all the other adhesives (**Figure 14A-C**), which may negatively influence the bonding of One Coat Bond® in the long term (despite the good results obtained after 500 thermal cycles).

The slightly higher nanoleakage percentage of the Prome&Bond-NT® is likely to be associated more with the difficult penetration of monomers probably due to the fast evaporation of acetone (which has high vapor pressure and high volatility) or to the inability of rehydration of the collapsed collagen fibers [43]. In this study, we used dentin disks, which may present some level of dehydration of the dentin and absence of positive pulpal pressure, influencing more the performance of an adhesive such as Prime&Bond® NT in the dentin. Many etch-and-rinse adhesive systems usually contain acetone to facilitate water removal [43, 45] as acetone takes the place of the water previously existent in the dentin, removing it after pursuing it. This fact would be optimal if a degree of humidity that allowed acetone to function that way were always present. However, that does not always occur and, especially, in samples with some degree of dehydration, acetone has some difficulty to act, i.e., transport monomers, correctly. Thus, discrepancies may occur due to the improper or incomplete resin filling of open intertubular dentin nanospaces (which vary in size between 15 and 20 nm, according to Fawzy et al.) [46] after etching, and result in nanoleakage (**Figure 13A** and **B**). The PENTA monomer in the Prime&Bond® NT may also have contributed somewhat to the increased nanoleakage because, according to Perdigao et al., this monomer tends to be more hydrophilic [47]. To contradict the propensity of adhesive forms to be more hydrophilic, in the past years, there has been a tendency to reduce the quantity of strongly hydrophilic monomers, such as HEMA, and replace them with UDMA or TEGDMA [43]. XP Bond™, being the most recently released etch-and-rinse adhesive of this study, already has UDMA and TEGDMA, thus compensating the HEMA and PENTA (more hydrophilic) also present in its composition.

Every adhesive evaluated in this study has HEMA in its composition, except for Prime&Bond® NT and Xeno® V. Therefore, the hydrophobic resin component inside the residual water is prone to divide itself into resin globules and water bubbles—the phase separation of resinous materials [48, 49]. In this process, after acetone or tert-butanol has started evaporating, the monomer/solvent balance is broken as water separates from the other components of the adhesive. When the adhesive is light cured, these water bubbles get trapped in the adhesive layer, which may lead to an impaired adhesive effectiveness [6].

Regarding monomers, the AdheSE® also includes *bis*-acrylamide, which is a monomer more hydrolytically stable than the other hydrophilic monomers used in adhesive forms in the dentin [50]. This characteristic may explain why this adhesive had a much lower mean nanoleakage percentage in the hybrid layer (20.94 ± 14.63%) compared with the others. AdheSE® also showed the second lowest standard deviation, and this provides reliability to the results.

In the AdheSE® group's sample, an hydrophobic resin was applied in a stage different than the primer's one, which originated a much lower nanoleakage percentage when compared with the other five groups, even with the elimination of the outliers of Prime&Bond® NT. The values are statistically significant when comparing AdheSE® with Prime&Bond® NT, One Coat Bond®, and Xeno® V (**Table 3** and **Figures 9** and **10**). Consequently, we can predict a great stability of this adhesive with time. Therefore, the existence of two stages (or two steps) in self-etch adhesives may contribute to a better sealing of the adhesive interface [51] and that effect may prevent the occurrence of water trees and promote adhesion durability [27].

A great penetration of the ammoniacal silver nitrate was observed in the hybrid layer of the Xeno® V group (75.68 ± 14.58%) (**Table 3** and **Figures 9**, **16A** and **B**). This group's sample originated the highest mean nanoleakage percentage (75.68 ± 14.58%) in comparison with the other groups, and this was statistically significant when comparing Xeno® V with XP Bond™ and AdheSE® after eliminating the outliers, or even with AdheSE® before outliers were eliminated (**Table 3** and **Figures 12A** and **B**, **15A** and **B**). These results are coherent with several current studies on single-step self-etch adhesive systems that demonstrated the presence of nanoleakage in the hybrid layer due to a poor penetration of resin into the collagen network. Hashimoto et al. found that nanoleakage in these adhesives could be seen in transmission electron microscopy (TEM) [52]. Thus, due to the greater water absorption by single-step self-etch adhesives, these may act as permeable membranes [15, 53]. To contradict that disadvantage, some authors advocate that applying an hydrophobic lining may be essential to improve the adhesion durability [54], as demonstrated by Sillas Duarte Jr. et al. (2009) in a study on etch-and-rinse and self-etch adhesives [55].

A high hydrolysis in the hybrid layer may lead to reduced bonding forces with time, and thus Prime&Bond® NT, One Coat Bond®, and Xeno® V may reveal problems in the long term. Even One Coat Bond® showed water trees, and this water flow through the adhesive layer leads to bonding forces degradation in the long term, which could not be evaluated in this study.

As demonstrated by some studies, a hybrid layer with a high nanoleakage percentage (such as in the case of Prime&Bond® NT and Xeno® V) leads to a greater decline of bonding forces with time due to unprotected collagen fibers present in the resin, which makes them easily degraded by metalloproteinases [19, 22]. Thus, it would be interesting to conduct a complimentary study to evaluate the nanoleakage percentage in the hybrid layer of adhesive systems for a long time (at least one year).

http://dx.doi.org/10.5772/64856

Finally, clinically, adhesive systems should be hydrophilic during their application and, then, they should become hydrophobic to seal the restoration margins thoroughly for a significant amount of time [16].

5. Conclusions

The adhesive with the overall best performance in the dentin was the two-step self-etch system containing water as solvent (AdheSE®), having shown a good morphology with a high-quality hybrid layer and low nanoleakage percentages. These good results appear to be correlated with the application of a hydrophobic resin over the hydrophilic primer in a distinct step. According to our research and other scientific studies, besides other elements that compose the adhesives, the solvent appears to play a significant role in morphology and nanoleakage. The latter is always present in the adhesive interfaces produced by the current commercial adhesives available in the market. The type of adhesive and the number of steps also seems to be important to water absorption by the hybrid layer, which leads to the nanoleakage phenomenon and can determine the adhesion longevity. Thus, according to our research, the null hypothesis was rejected.

Author details

João Cardoso Ferreira*, Patrícia Teixeira Pires, Paulo Ribeiro de Melo and Mário Jorge Silva

*Address all correspondence to: jcferreira@fmd.up.pt

Department of Operative Dentistry, Faculty of Dental Medicine, University of Porto, Porto, Portugal

References

[1] Breschi L, Perdigao J, Lopes MM, Gobbi P, Mazzotti G. Morphological study of resin-dentin bonding with TEM and in-lens FESEM. Am J Dent 2003;16:267–74.

[2] Uno S, Finger WJ. Effect of mode of conditioning treatment on efficacy of dentin bonding. Oper Dent 1996;21:31–5.

[3] Pashley DH, Tay FR, Breschi L, et al. State of the art etch-and-rinse adhesives. Dent Mater 2011;27:1–16.

[4] Ding PG, Matzer AR, Wolff D, et al. Relationship between microtensile bond strength and submicron hiatus at the composite-dentin interface using CLSM visualization technique. Dent Mater 2010;26:257–63.

[5] Cavalheiro A, Vargas MA, Armstrong SR, Dawson DV, Gratton DG. Effect of incorrect primer apllication on dentin permeability. J Adhes Dent 2006;8:393–400

[6] Cardoso MV, de Almeida NA, Mine A, et al. Current aspects on bonding effectiveness and stability in adhesive dentistry. Aust Dent J 2011;56(Suppl. 1):31–44.

[7] Hosoya Y, Tay FR, Ono T, Miyazaki M. Hardness, elasticity and ultrastructure of primary tooth dentin bonded with a self-reinforcing one-step self-etch adhesive. J Dent 2010;38:214–21.

[8] Hosoya Y, Tay FR. Hardness, elasticity, and ultrastructure of bonded sound and caries-affected primary tooth dentin. J Biomed Mater Res B Appl Biomater 2007;81:135–41.

[9] Tyas MJ, Burrow MF. Adhesive restorative materials: a review. Aust Dent J 2004;49:112–21; quiz 54.

[10] Mohan B, Kandaswamy D. A confocal microscopic evaluation of resin-dentin interface using adhesive systems with three different solvents bonded to dry and moist dentinan in vitro study. Quintessence Int 2005;36:511–21.

[11] Van Meerbeek B, Vargas M, Inoue S, et al. Microscopy investigations. Techniques, results, limitations. Am J Dent 2000;13:3D–18D.

[12] Luz MA, Arana-Chavez VE, Netto NG. Scanning electron microscopy examination of 3 different adhesive systems. Quintessence Int 2005;36:687–94.

[13] Montes MA, de Goes MF, Sinhoreti MA. The in vitro morphological effects of some current pre-treatments on dentin surface: a SEM evaluation. Oper Dent 2005;30:201–12.

[14] Breschi L, Prati C, Gobbi P, et al. Immunohistochemical analysis of collagen fibrils within the hybrid layer: a FEISEM study. Oper Dent 2004;29:538–46.

[15] Tay FR, Pashley DH, Suh BI, Carvalho RM, Itthagarun A. Single-step adhesives are permeable membranes. J Dent 2002;30:371–82.

[16] Yuan Y, Shimada Y, Ichinose S, Tagami J. Qualitative analysis of adhesive interface nanoleakage using FE-SEM/EDS. Dent Mater 2007;23:561–9.

[17] Pashley DH, Pashley EL, Carvalho RM, Tay FR. The effects of dentin permeability on restorative dentistry. Dent Clin North Am 2002;46:211–45, v–vi.

[18] Prati C, Chersoni S, Acquaviva GL, et al. Permeability of marginal hybrid layers in composite restorations. Clin Oral Investig 2005;9:1–7.

[19] Suppa P, Breschi L, Ruggeri A, et al. Nanoleakage within the hybrid layer: a correlative FEISEM/TEM investigation. J Biomed Mater Res B Appl Biomater 2005;73:7–14.

[20] Schneider H, Frohlich M, Erler G, Engelke C, Merte K. Interaction patterns between dentin and adhesive on prepared class V cavities in vitro and in vivo. J Biomed Mater Res 2000;53:86–92.

[21] Hashimoto M, Ohno H, Sano H, Kaga M, Oguchi H. Degradation patterns of different adhesives and bonding procedures. J Biomed Mater Res B Appl Biomater 2003;66:324–30.

[22] Carrilho MR, Carvalho RM, Tay FR, Pashley DH. Effects of storage media on mechanical properties of adhesive systems. Am J Dent 2004;17:104–8.

[23] Reis AF, Bedran-Russo AK, Giannini M, Pereira PN. Interfacial ultramorphology of single-step adhesives: nanoleakage as a function of time. J Oral Rehabil 2007;34:213–21.

[24] Oliveira SS, Marshall SJ, Habelitz S, Gansky SA, Wilson RS, Marshall GW, Jr. The effect of a self-etching primer on the continuous demineralization of dentin. Eur J Oral Sci 2004;112:376–83.

[25] Chersoni S, Suppa P, Breschi L, et al. Water movement in the hybrid layer after different dentin treatments. Dent Mater 2004;20:796–803.

[26] Tay FR, Pashley DH, Yiu C, et al. Nanoleakage types and potential implications: evidence from unfilled and filled adhesives with the same resin composition. Am J Dent 2004;17:182–90.

[27] Tay FR, Pashley DH, Yoshiyama M. Two modes of nanoleakage expression in single-step adhesives. J Dent Res 2002;81:472–6.

[28] Sano H, Yoshiyama M, Ebisu S, et al. Comparative SEM and TEM observations of nanoleakage within the hybrid layer. Oper Dent 1995;20:160–7.

[29] Perdigao J, Lambrechts P, Van Meerbeek B, Vanherle G, Lopes AL. Field emission SEM comparison of four postfixation drying techniques for human dentin. J Biomed Mater Res 1995;29:1111–20.

[30] Milia E, Santini A. Ultrastructural transmission electron microscopy (TEM) study of hybrid layers formed beneath a one-bottle adhesive system using the total-etch technique and a self-etching system. Quintessence Int 2003;34:447–52.

[31] Ritter AV, Swift EJ, Jr., Heymann HO, Sturdevant JR, Wilder AD, Jr. An eight-year clinical evaluation of filled and unfilled one-bottle dental adhesives. J Am Dent Assoc 2009;140:28–37; quiz 111–2.

[32] Peumans M, Kanumilli P, De Munck J, Van Landuyt K, Lambrechts P, Van Meerbeek B. Clinical effectiveness of contemporary adhesives: A systematic review of current clinical trials. Dent Mater 2005;21:864–81.

[33] De Munck J, Van Landuyt K, Peumans M, et al. A critical review of the durability of adhesion to tooth tissue: methods and results. J Dent Res 2005;84:118–32.

[34] Pashley DH, Tay FR, Carvalho RM, et al. From dry bonding to water-wet bonding to ethanol-wet bonding. A review of the interactions between dentin matrix and solvated resins using a macromodel of the hybrid layer. Am J Dent 2007;20:7–20.

[35] Cardoso Pde C, Loguercio AD, Vieira LC, Baratieri LN, Reis A. Effect of prolonged application times on resin-dentin bond strengths. J Adhes Dent 2005;7:143–9.

[36] Kato G, Nakabayashi N. Effect of phosphoric acid concentration on wet-bonding to etched dentin. Dent Mater 1996;12:250–5.

[37] Hashimoto M, Tay FR, Ohno H, et al. SEM and TEM analysis of water degradation of human dentinal collagen. J Biomed Mater Res B Appl Biomater 2003;66:287–98.

[38] Reis AF, Giannini M, Pereira PN. Long-term TEM analysis of the nanoleakage patterns in resin-dentin interfaces produced by different bonding strategies. Dent Mater 2007;23:1164–72.

[39] Tay FR, Gwinnett AJ, Pang KM, Wei SH. Variability in microleakage observed in a total-etch wet-bonding technique under different handling conditions. J Dent Res 1995;74:1168–78.

[40] Tagami J, Nakajima M, Shono T, Takatsu T, Hosoda H. Effect of aging on dentin bonding. Am J Dent 1993;6:145–7.

[41] Sano H, Takatsu T, Ciucchi B, Horner JA, Matthews WG, Pashley DH. Nanoleakage: leakage within the hybrid layer. Oper Dent 1995;20:18–25.

[42] Van Meerbeek B, Van Landuyt K, De Munck J, et al. Technique-sensitivity of contemporary adhesives. Dent Mater J 2005;24:1–13.

[43] Van Landuyt KL, Snauwaert J, De Munck J, et al. Systematic review of the chemical composition of contemporary dental adhesives. Biomaterials 2007;28:3757–85.

[44] Pashley EL, Zhang Y, Lockwood PE, Rueggeberg FA, Pashley DH. Effects of HEMA on water evaporation from water-HEMA mixtures. Dent Mater 1998;14:6–10.

[45] Pashley DH, Zhang Y, Agee KA, Rouse CJ, Carvalho RM, Russell CM. Permeability of demineralized dentin to HEMA. Dent Mater 2000;16:7–14.

[46] Fawzy AS, Farghaly AM. Probing nano-scale adhesion force between AFM and acid demineralized intertubular dentin: Moist versus dry dentin. J Dent 2009;37:963–9.

[47] Perdigao J, Ramos JC, Lambrechts P. In vitro interfacial relationship between human dentin and one-bottle dental adhesives. Dent Mater 1997;13:218–27.

[48] Spencer P, Wang Y. Adhesive phase separation at the dentin interface under wet bonding conditions. J Biomed Mater Res 2002;62:447–56.

[49] Tay FR, Gwinnett JA, Wei SH. Micromorphological spectrum from overdrying to overwetting acid-conditioned dentin in water-free acetone-based, single-bottle primer/adhesives. Dent Mater 1996;12:236–44.

[50] Salz U, Zimmermann J, Zeuner F, Moszner N. Hydrolytic stability of self-etching adhesive systems. J Adhes Dent 2005;7:107–16.

[51] Nikaido T, Kunzelmann KH, Ogata M, et al. The in vitro dentin bond strengths of two adhesive systems in class I cavities of human molars. J Adhes Dent 2002;4:31–9.

[52] Hashimoto M, Fujita S, Endo K. Bonding of self-etching adhesives on dehydrated dentin. J Adhes Dent 2011;13:49–54.

[53] Chersoni S, Suppa P, Grandini S, et al. In vivo and in vitro permeability of one-step self-etch adhesives. J Dent Res 2004;83:459–64.

[54] King NM, Tay FR, Pashley DH, et al. Conversion of one-step to two-step self-etch adhesives for improved efficacy and extended application. Am J Dent 2005;18:126–34.

[55] Duarte S, Jr., Phark JH, Varjao FM, Sadan A. Nanoleakage, ultramorphological characteristics, and microtensile bond strengths of a new low-shrinkage composite to dentin after artificial aging. Dent Mater 2009;25:589–600.

Chapter 2

Biocompatibility of Dental Adhesives

Antonija Tadin, Lidia Gavić and Nada Galić

Additional information is available at the end of the chapter

http://dx.doi.org/10.5772/64943

Abstract

The accomplishment of developing a truly adhesive bond between a restorative material and the natural tooth structures is the goal of adhesive dentistry. Dentine adhesive systems come into close contact with dental and oral tissue, especially the pulp and gingival cells. Due to this close and long-term contact, adhesives should exhibit a high degree of biocompatibility. Biocompatibility is one of the most important properties of dental materials, and adhesives are no exception. It has been long demonstrated that different components of adhesives can be released. Numerous *in vitro* investigations have shown that released monomers and other components can cause damage to cultured cells. In addition, many *in vivo* studies have shown that uncured components which reach the pulpal space cause inflammatory response and tissue disorganization. Only a combination of various *in vitro* and *in vivo* tests can provide an overview of the interaction of biomaterials with the host. Therefore, it is necessary on a regular basis to carry out and re-verify the biological compatibility of the increasing number of new dental materials. Adhesives should be biofunctional, protective, and preventive, with health-promoting effects that contribute to a better prognosis for restorative treatments and its biocompatibility.

Keywords: biocompatibility, cytotoxicity, dental adhesives, genotoxicity, *in vitro* assays, *in vivo* assays

1. Introduction

The use of composite dental filling materials along with adhesive techniques has revolutionized today's dental practice. Dental adhesive systems are used to improve contact between restorative material and the walls of the prepared cavity of tooth, in order to increase retention as well as to minimize bacterial leakage. Their purpose is based on a twofold adhesion, adhesion to composite fillings, and bonding to enamel and dentine [1–3].

The introduction of the acid-etch technique by Buonocore in 1955 was the genesis of adhesive dentistry. Adhesive systems are constantly being developed with elaborate and complex chemistry. As a result of technological progress, the large number of new materials on the market has been appeared. Adhesive systems are solutions of resin monomers that make the resin-dental substrate interaction achievable. They are composed of monomers with hydrophilic and hydrophobic groups, organic solvents, initiators, inhibitors or stabilizers, and inorganic fillers. The proportional composition and the chemistry of these ingredients have been different between the different classes of adhesives [1, 4]. Dentine adhesives are currently available as three-step, two-step, and single-step systems, depending on how the three basic steps of etching, priming, and bonding to tooth substrate are accomplished. Modern adhesive strategies depend on how adhesive systems interact with the smear layer dissolving it or making it permeable. Consequently, adhesives can be classified into two types: total-etch (also known as etch-and-rinse) and self-etch adhesive systems [4]. The main difference between them is that the etch-and-rinse adhesives use 37% phosphoric acid for pretreatment of hard dental tissues before application of dental bonding agent, whereas the self-etch adhesives do not require a separate etching step, as they contain acidic monomers that simultaneously condition and prime the dental substrate [2, 5]. Etch and rinse adhesive systems are used either through two or three clinical application steps. First step always involves application of the conditioner or acid etchant on both enamel and dentine. According to the number of clinical application steps, self-etch adhesives are divided into two-step and one-step systems [6].

In order to achieve clinical success, it is necessary for dental adhesives to provide good physical properties. Clinical success of adhesive procedures depends not only on the dental adhesives but also on variables such as substrate condition and composition, clinical technique, restorative material, cavity shape and size, and polymerization quality [6]. However, due to the fact that dental adhesives are in close and extended contact with vital dentine, biocompatibility becomes a property of supreme [5, 7, 8]. The biocompatibility of the adhesive estimated through screening with simple *in vitro* assays, mostly based on cell cultures, is followed by more comprehensive studies on experimental animals and usage tests and ends with clinical studies [9]. The scientific evidence on adhesives is contradictory. Some authors claimed that they are very safe and can be used even in direct contact with the pulp, whereas others believe that they are not suitable for direct pulp capping due to reported associated symptoms of persistent inflammation [10, 11]. Another researchers claim that dental adhesive systems contain certain components that can be released into the oral environment and show genotoxic, allergenic, cytotoxic, estrogenic, and mutagenic activity and that they alter lipid metabolism, glutathione (GSH) concentration, reactive oxygen species (ROS) production, cell cycle, energy metabolism, and mitochondrial activity [2, 5, 7, 12–16]. There are a number of *in vitro* studies that have proven cytotoxic and genotoxic activity of dentine-bonding agents on the cells of human origin (peripheral blood lymphocytes, lung fibroblasts, pulp cells, gingival fibroblasts and oral epithelial cells, pulp cells, dentine, lymphocytes, and leucocytes) [5, 17–20]. Most *in vivo* research has been performed on monkeys and rodents, whereas the rare studies conducted on humans [19].

1.1. Biocompatibility of dental materials

Biomaterial is a substance that is used for a long period within the body with the aim of treating or replacing of tissue, organs, or their functions. Each dental material must be biocompatible or able to function *in vivo* without eliciting an intolerable response in the body either locally or systemically. There are only few materials for which could be said that are biologically inert since most of them contain a potentially harmful or irritating ingredients [9, 21].

Biocompatibility is a measure of body's biological response to a material used in specific application. It is a property of a material to interact with its environment. The biological response to a material could be modified if any change occurs in the host or in the application of the material. Therefore, evaluation of any new material intended for human use requires data from systematic testing to ensure that the benefits provided by the final product will exceed any potential risks produced by device materials [9]. Biocompatibility includes the physical, mechanical, and chemical properties of materials, as well as potential cytotoxic, genotoxic, mutagenic, and allergenic effects. According to EN 1441 (European standard for Risk Analysis, 1996), biocompatible material must not be harmful to the host organism [22]. A characteristic of the material to interact with the biological tissue and thus create a stable connection is essential for biocompatibility. The biocompatibility of material is manifested through a number of parameters: (a) cytotoxicity (systemic and topical), (b) genotoxicity, (c) mutagenicity, (d) carcinogenicity, and (e) the immunogenicity [23].

The toxicity of material is described as the ability of a chemical substance to cause harmful health effects and damage the biological system. It might be local or systemic. Local toxicity occurs at the place of application, as opposed to systemic toxicity where the adverse response occurs distant from its entry point [23, 24]. The term of cytotoxicity is used to describe the cascade of molecular events that cause functional and structural damage to cells. Throughout the years, various assays and protocols were developed to test the cytotoxic effects of biomaterials. The rationale behind doing a cytotoxicity test is to determine how a material sample affects a particular cell type. The primary criteria of these tests are that material must not affect the cell number, cell growth, genetic integrity, membrane integrity, genetic expression, and enzymatic activity of the cells. The cytotoxicity assays are broadly categorized into viability assays, survival assays, metabolic assays, transformation assays, and inflammation assays [25]. The cytotoxicity is related to the damage of the individual cells, for example in cell cultures, and determines the cell viability. There are two modes of cell death, apoptosis (programmed cell death) and necrosis (accidental cell death). They are different significantly in the mechanisms, outcomes, and also in morphology, biochemistry, and biological features. Necrosis occurs as a result of significant damage of the cell, caused by physical or chemical agent. The loss of integrity of the membrane is a hallmark of necrosis. The membrane permeability of damage cell increases, and damage cell cannot control balance of the fluids and ions. It is accompanied by cellular swelling and loses of integrity of organelles. The release of cytoplasmic contents could lead to an inflammatory response in surrounding organs and tissue. Apoptosis could be defined as caspase mediated cell death. Apoptotic cells can be identified based on following morphological features: cytoplasmic and nuclear condensation, chromatin cleavage, formation of apoptotic bodies, maintenance of an intact plasma membrane, exposure of surface molecules targeting cell corpses for phagocytosis and also on the proteolytic activity of certain caspases because these enzymes mediate the process of apoptotic cell death [23, 26, 27]. The primary criteria of these tests are the following: the material must not affect the cell number, cell growth, genetic integrity, membrane integrity, genetic expression, and enzymatic activity of the cells [25].

Genotoxicity is described as adverse effects on the genomic material and may be caused by DNA damage without direct evidence for mutations [23]. Genotoxic effect of a material can induce changes in the genome that disrupt its integrity or function. Depending on the intensity of that effect, the cell could recover, start neoplastic growth, or die [28]. The long-term exposure to low concentrations of the substance may develop neoplasia and death of the whole organism by the effects on the genetic material. Transfer of genetic damage to the next generation of cells can be avoided by programmed cell death (apoptosis) [9, 28].

Transition of genetic injury to the next generation is called mutagenicity. Mutagenicity and carcinogenicity are not the same entities. Carcinogenicity arises from several mutations, which means that all mutagenic events do not lead to tumour development. Mutagenicity serves as an indicator of 'possible' carcinogenicity of substances that damaged DNA. A variety of different methods, mainly *in vitro* assays (Ames test, micronucleus, and hypoxanthine phosphorybosyl transferase (HPRT) test on mouse lymphoma cells), are used in the assessment of the mutagenic potential of materials [9, 23].

The immunogenicity is the ability of a substance to induce an immune response. An allergic reaction to certain substances can be initiated if the organism was sensitized previously. The dosage that could cause the allergic reaction are generally significantly lower than those that could cause toxic reactions. Immunotoxicological screenings are used to identify the influence of substances or materials to the various components of immune system [23].

Biomaterials must meet several criteria before they are put in use, and measuring of the biocompatibility of a material is not simple. It is not possible to biologically characterize a material using a single test. Different characteristics can be explored through both *in vitro* and *in vivo* assays. *In vitro* biocompatibility tests are conducted outside of a living organism with the purpose of simulating the biological response to a material. The influence of the material is determined by measuring the size, growth factors, metabolic functions as well as other functions of a cell treated with a tested material. These tests have lots of advantages; they are standardized, relatively low cost, quick, and easily reproducible. But their clinical relevance is questionable because they provide limited answers due to lack of biological and physiological components [21, 29]. *In vitro* studies are primarily performed to evaluate the cytotoxicity (cell damage) [12, 18, 19, 30] or genotoxicity (specific DNA damage or chromosomal aberrations) [2, 5, 7, 31] of dental materials. *In vivo* biocompatibility testing has provided the influence of tested material in a living organism. Investigations are most commonly conducted on animals. Mostly, tested materials are implanted in the body, followed by monitoring on their influence on a living organism. Thus, it is possible to evaluate the many complex interactions between biological systems and materials, either in the surrounding tissue or at a remote site in an organism [9, 23]. Animal experiments for cytotoxicity tests of dental materials are time-consuming, expensive, and the theme of extensive public discussions [32]. Clinical *in vivo* assays are more reliable, since the tested material is directly applied in the body of volunteers in their ultimate purpose. The results obtained from these studies are of particular interest. However, clinical studies have numerous limitations, are expensive, long term, and difficult to control since many external variables can influence results. Therefore, it is challenging to interpret the obtained results [9, 23].

2. Materials and methods

2.1. Blood sampling

Evaluation of the potential genotoxicity of dental adhesive systems and combination of adhesive systems and composite materials were performed on human leukocytes. The donors were two males and four females with ages ranging from 25 to 32. They were not smokers and had not been exposed to any physical or chemical agents that might have interfered with the results of the genotoxicity testing in the 12-month period prior to blood sampling. The donors were acquainted with the purpose of the study and signed permission for the blood samples to be used for scientific purposes. A peripheral blood sample was collected under sterile conditions by venipuncture into heparinized tubes (Becton Dickenson, Plymouth, UK). The study was approved by the Ethical Committee of the School of Dental Medicine, University of Zagreb, Croatia [3, 5].

2.2. Preparation of materials and cell culture treatment

Three dental adhesives were tested: AdheSE (Ivoclar Vivadent, Schaan, Liechtenstein), G-bond (GC, Tokyo, Japan), and Adper Single Bond (3M ESPE, St. Paul, MN, USA), in their polymerized and unpolymerized form. Also were tested aforementioned dental adhesive in polymerized form in combination with composite material from same manufacturers: Tetric EvoCeram (Ivoclar Vivadent, Schaan, Liechtenstein), Gradia Direct Anterior (GC, Tokyo, Japan), and Filtek Z250 (3M ESPE, St. Paul, MN, USA). Six different combinations of two shades of composite resins and adhesives were examined: Gradia Direct Anterior A1 + G-Bond, Gradia Direct Anterior A3.5 + G-Bond, Filtek Z250 A1 + Adper Single Bond, Filtek Z250 A3.5 + Adper Single Bond, Tetric EvoCeram A1 + AdheSE, and Tetric EvoCeram A3.5 + AdheSE.

In order to test the genotoxicity of non-polymerized materials, the dental adhesives were placed in previously weighed bottles (Sartorius BLG10S, Goettingen, Germany). The mass of each dental adhesive was calculated from the difference in weight of empty and full bottles. For each 0.1 g of dental adhesive, 1 ml of saline solution was added (0.9% NaCl, Sigma, St. Louis, MO, USA) for the purpose of elution.

In order to test the genotoxicity of polymerized dental adhesives, each one was polymerized under aseptic condition in accordance with the manufacturer's instructions using an Elipar TriLight halogen curing unit (3M ESPE) from a 2 mm distance for 40 s. To ensure complete polymerization, only two drops of an adhesive were cured at a time. After polymerization, the dental adhesives were weighted, fragmented, and transferred into bottles. We tended to use the same masses of dental adhesive samples (1 g) regardless of whether they were polymerized or non-polymerized. Each dental adhesive eluate was tested after 1 h, 1 day, and 5 days in two different dilutions of eluate (1: 102 and 1: 104) for each time point.

To assess the genotoxicity of polymerized dental adhesives and their compatible dental composites from same manufacturers, the following was conducted: 20 ± 0.2 μl of each adhesive system was placed on the Mylar strip (Contour™ Strips, Ivoclar Vivadent) and photopolymerized by a halogen curing unit Elipar TriLight in standard mode (800 mW/cm^2) for 20 s. The adhesives were handled exactly by the manufacturers' instructions. Subsequently, 0.025 ± 0.003 g and 0.05 ± 0.002 g of each composite resin were placed on top of the bonding material, covered with another Mylar strip, and mechanically pressed to obtain a 2 mm thick layer. Resin composite samples (n = 2) were for 40 s. The light curing tip was flush pressed onto the Mylar sheet on top of the composite samples. Thereafter, the polymerized composites were separated from the Mylar sheets, and the samples of each material combination were placed in a plastic sterile tube (Greiner Labortechnik Co., Ltd., Frickenhausen, Germany) with saline solution (0.9% NaCl, Sigma, St. Louis, MO, USA) to be eluted for 1 h, 1 day, 7 days, and 30 days. For each 0.1 g of dental composite materials, 1 ml of saline solution was added for the purpose of elution.

Cultures for cytogenetic testing were set up at the end of the elution period. One millilitre of primary leukocyte culture containing 5.6 × 106 cells was introduced into 9 ml of F-10 HAM's medium (Sigma) without the addition of fetal bovine serum or mitogen. Leukocytes were treated with 1 and 100 μl of eluates obtained from each of the tested dental adhesives to, respectively, simulate final mass concentrations of 0.1 and 10 mg of materials/ml. While composite-adhesive elusion solutions were discarded, and the samples were carefully transferred to a new sterile tube (Greiner Labortechnik Co., Ltd., Frickenhausen, Germany), with added 1 ml of primary leukocyte culture (density of 100 cells/μl) and 5 ml of RPMI 1640 (Gibco-Invitrogen, Carlsbad, CA, USA). After that, leukocytes were treated with 0.025 g and 0.05 g of each of the composite/adhesive combinations to, respectively, simulate final mass concentrations of 4.16 and 8.33 mg of sample/ml. Control leukocyte cultures were exposed for 48 h to a clean medium RPMI 1640 (combination of adhesives and composites) and a saline solution (0.9% NaCl, Sigma). Each DNA damage experiment included also positive control, which was hydrogen peroxide, 60 μl, for 15 min on ice. After the treatment period (48 h at 37°C in a 5% CO_2 atmosphere), cultures were centrifuged for 10 min at 70 g, the supernatant was discarded, and cells were transferred into a sterile tube (Nange Nunc International, Naperville, IL, USA). They were resuspended and sampled for vital staining and the comet assay. Each DNA damage experiment included also positive control, which was hydrogen peroxide, 60 μl, for 15 min on ice [3, 5].

2.3. Cytotoxicity testing

Leukocyte viability was tested using the trypan blue exclusion technique [33]. Cell suspension was mixed with 0.4% trypan blue (Sigma) and analysed using an Olympus light microscope (Olympus, Tokyo, Japan) under 100× magnification. For each test group, 1000 leukocytes were analysed by counting unstained (viable) cells. Blue-coloured cells were considered to be nonviable [3, 5].

2.4. Comet assay

The comet assay was carried out under alkaline conditions as described by Singh et al. [34]. All chemicals used to perform the comet assay were obtained from Sigma. Sediment containing leukocytes was suspended in 100 μl of 0.5% low-melting-point agarose. This agarose layer was sandwiched between a layer of 0.6% normal melting point agarose and a top layer of 0.5% low melting point agarose on fully frosted slides. Slides were coded and kept on ice during polymerization of each gel layer. After solidification of the 0.5% agarose layer, slides were immersed in a lysis solution (1% sodium sarcosinate, 2.5 M NaCl, 100 mM Na2EDTA, 10 mM Tris-HCl, 1% Triton X-100, and 10% DMSO) at 4°C. After 1 h, the slides were removed from the lysing solution, drained, and placed in an electrophoresis buffer (0.3 M NaOH and 1 mM Na2EDTA, at pH 13) at 4°C for 20 min to allow the DNA to unwind. Electrophoresis was conducted on a horizontal electrophoresis platform in fresh, chilled electrophoresis buffer for 20 min at 300 mA and 19 V. After electrophoresis, slides were neutralized with Tris-HCl buffer (pH 7.5) three times for 5 min each and stained with ethidium bromide (20 μg/ml) for 10 min. Two slides per material per concentration per polymerization form per time point were analysed using an Ortoplan epifluorescence microscope (Leitz, Wetzlar, Germany) at 250× magnification (**Figure 1**). One hundred comets per slide were analysed by the comet assay II automatic digital analysis system (Perceptive Instruments, Halstead, UK) by measuring the tail length and intensity (% DNA). For the purpose of the analysis, the following were ignored: the edges and eventually damaged parts of the gel as well as debris, superimposed comets, comets of uniform intensity, and comets without a distinct head (i.e., 'clouds', 'hedgehogs', and 'ghost cells') [3, 5].

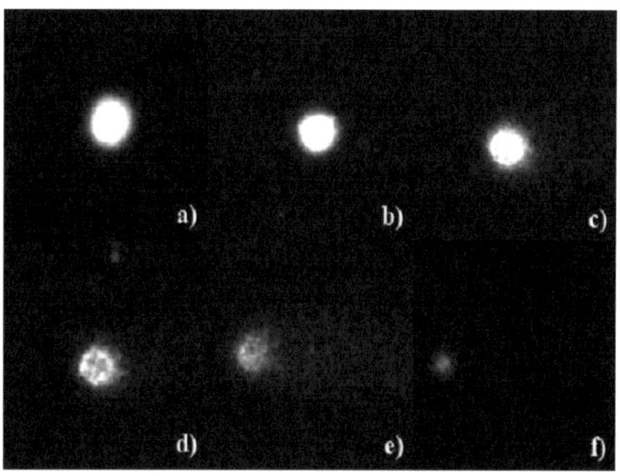

Figure 1. Representative microscopic appearances of comet, taken under an Ortoplan epifluorescence microscope: (a) undamaged cell; (b–e) damaged cells; and (f) apoptotic cell.

2.5. Statistical analysis

The comet test results were tested by the Kruskal-Wallis test to determine the statistical significance. The level of significance was set to 0.05. All calculations were performed using the commercial software, Statistica 7.0 (StatSoft, Tulsa, OK, USA).

3. Results

Before testing, concentrations of the dental adhesives were selected based on the cytotoxicity results. The trypan blue exclusion test showed that the viability was >85% for each material at a dilution of >1:102 and greater than 90% for each material combination at a concentration of 8.33 mg/ml. According to these results, other dilutions and concentrations are considered to be used without any risk. The pH value in the cell culture was always between 7.19 and 7.4, which is, according to the manufacturer, the regular pH value.

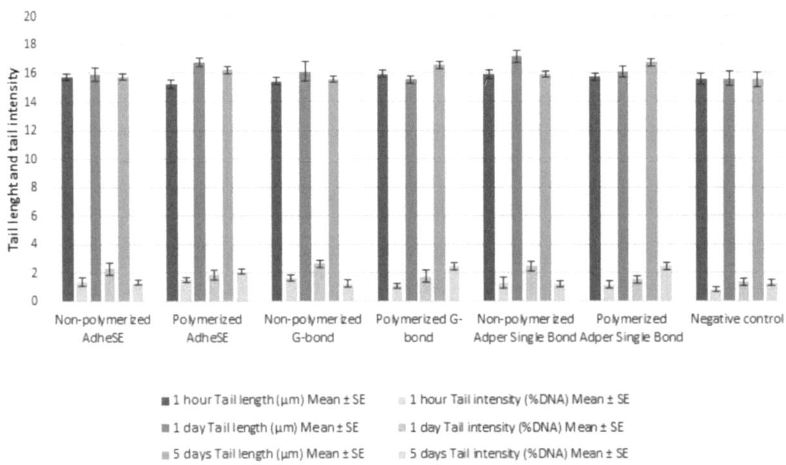

Figure 2. Tail length and tail intensity (mean values and standard deviations) of leucocytes exposed to a 10^{-2} elution of dental adhesives.

The comet test determined the level of primary damage to DNA molecules of leukocytes after treating them with different dilutions of eluates of dental adhesives, depending on the length of time they were rinsed in the saline solution. In the present study for each dilution, elution duration, and polymerization state, 100 comets were analysed. **Figures 2** and **3** showed two basic parameters of the comet tail, length, and the intensity of its fluorescence, of leukocytes exposed to a 10^{-2} and 10^{-4} elution of polymerized and unpolymerized dental adhesives. None of the tested dental adhesives revealed a statistically significant increase in tail length or tail intensity in treated leukocytes, independent of the applied dilution, elution duration, and polymerization form. **Figures 4** and **5** represent the results of comet assay parameters (mean values and standard deviations) in human leukocytes during exposure to tested combinations of adhesive and composite resins in two different shades (A 1 and A 3.5), in concentrations 4.16 mg/ml and 8.33 mg/ml. Obtained results showed statistically higher increase in tail length after 1 day for Tetric EvoCeram in shade A1 for lower concentration (4.16 mg/ml) (15.3 ± 2.52, negative control 14.7 ± 2.85), and after 1 h (15.3 ± 4.70, negative control 14.7 ± 2.85) and 1 day (15.2 ± 9.10, negative control 14.7 ± 2.85) or same material for higher concentration (8.33 mg/ml). Concerning the tail intensity only Gradia Direct in the shade A 3.5 + G-Bond showed statistically higher percentage of DNA in tail in concentration 8.33 mg/ml eluted for 1 day (3.2 ± 5.64, negative control 1.2 ± 3.71) [3, 5].

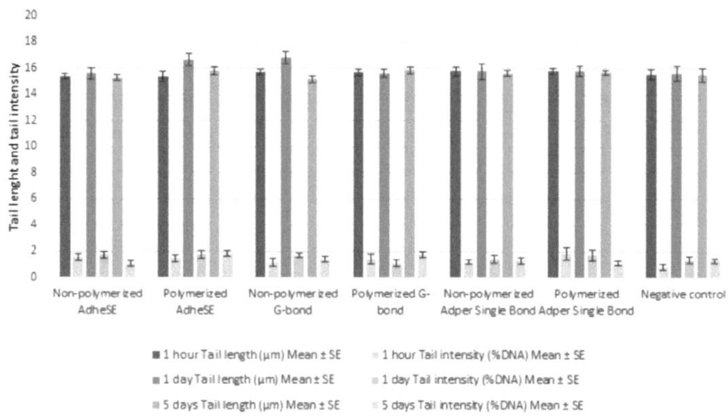

Figure 3. Tail length and tail intensity (mean values and standard deviations) of leukocytes exposed to a 10^{-4} elution of dental adhesives.

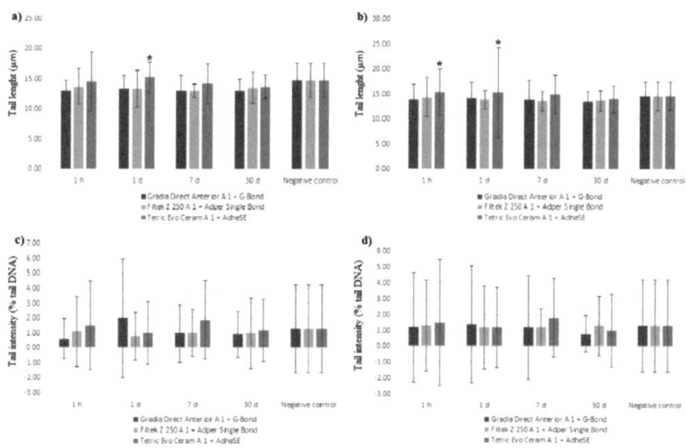

Figure 4. Comet assay parameters (mean values and standard deviations) in human leukocytes during exposure to tested combinations of composite resins A 1 and adhesive: (a) tail length for material combinations in a concentration of 4.16 mg/ml; (b) tail length for material combinations systems in a concentration of 8.33 mg/ml; (c) tail intensity (% tail DNA) for material combinations in a concentration of 4.16 mg/ml; and (d) tail intensity (% tail DNA) for material combinations in a concentration of 8.33 mg/ml. Asterisk denotes material combinations significantly different from the negative control ($p < 0.05$).

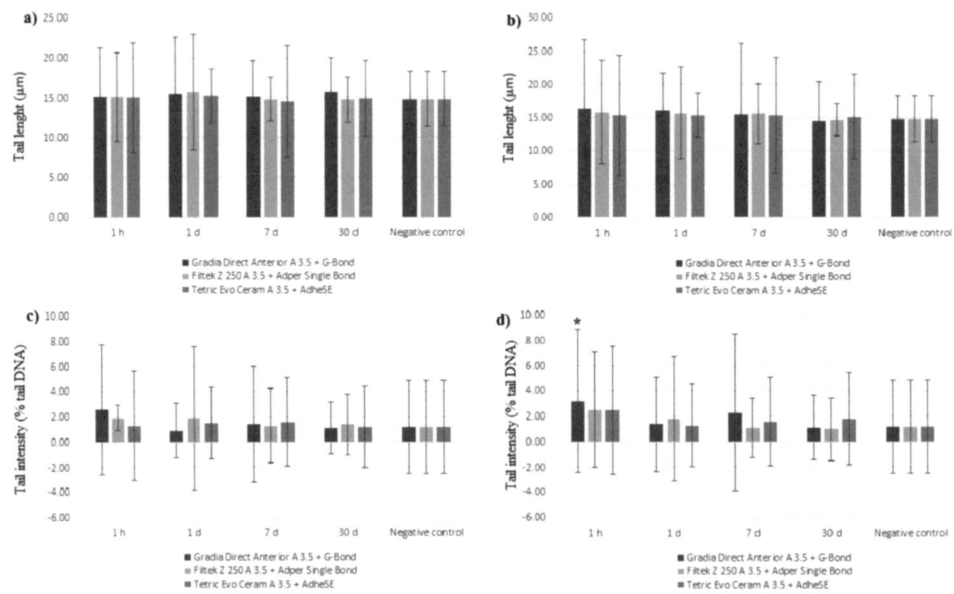

Figure 5. Comet assay parameters (mean values and standard deviations) in human leukocytes during exposure to tested combinations of composite resins A 3.5 and adhesive: (a) Tail length for material combinations in a concentration of 4.16 mg/ml; (b) tail length for material combinations systems in a concentration of 8.33 mg/ml; (c) tail intensity (% tail DNA) for material combinations in a concentration of 4.16 mg/ml; (d) tail intensity (% tail DNA) for material combinations in a concentration of 8.33 mg/ml. Asterisk denotes material combinations significantly different from the negative control ($p < 0.05$).

4. *In vitro* biocompatibility testings of dental adhesives

Contemporary dental adhesive systems are used to improve the contact between composite restorative materials and the walls of the prepared tooth cavity. As these materials come in close and prolonged contact with vital dentine, their influence on pulp cells is critical [31]. Adhesive systems that create a stable relationship with biological tissues and allow both healing and tissue differentiation are considered biocompatible.

The effective toxicity of adhesives and its components is reduced but often not eliminated by the presence of dentine. Adhesive systems are placed on etched dentine (total-etch adhesives) or on cut dentine (self-etch adhesives) that is permeable. Since the available adhesive systems are not able to hermetically seal the deep dentine, in several *in vitro* and *in vivo* studies, it has been demonstrated that after application of bonding agents on the conditioned dentine, residual monomers as well as other components of the adhesive systems may penetrate through the dentinal tubules and reach the dental pulp [18, 35]. The diffusion of chemical products from dentine to the pulp is dependent upon several important variables, including (a) the thickness of the remaining dentine, (b) the surface area of the exposed, (c) the presence or absence of a smear layer, (d) the potency of the microbial products, (e) the rate of pulpal blood flow, and (f) dentine permeability [36, 37]. Smear layers on dentine have been shown to reduce diffusion through dentine by 25–30%, but most current bonding adhesives use primers that remove the smear layer [38]. Dentine permeability increases with the depth of the cavity, reducing the thickness of the remaining dentine and removing the smear layer, whereas decreases with the age due to physiologic sclerosis of dentine. High permeability increases the toxicity of adhesives by allowing increased diffusion of the released components through dentine [39]. Thicker dentine lowers the concentrations of substance that reach the pulp, keeping these concentrations below thresholds that irritate pulp tissue. It has been demonstrated that 0.5 mm of dentine can reduce material toxicity to 75% and that 1 mm of dentine can reduce toxicity by up to 90% of the control value [40]. Also, the presence of dentine, as biological barrier, can also modify toxicity action due to the inherent characteristics of that tissue such as buffer capacity and hydraulic conductance [41].

The toxicity has been attributed to the release of residual monomers from polymerized adhesive systems due to degradation processes or the incomplete polymerization as well. The degree of conversion of dental bonding systems relies mainly on the wavelength and power intensity of the light curing unit [42]. Polymerization can be inhibited by the presence of oxygen, the presence of intrinsic water from dentine, and the presence of residual solvents in the adhesive [1]. The degree of conversion is an important parameter because it specifies the amount of monomers converted in polymer. Even under ideal conditions, the conversion of monomers to polymers is commonly incomplete, and it is about 55–60% [9]. Unconverted free monomers from polymerized adhesive monomers diluted into saliva after curing as result of degradation may be released into the oral environment where they could produce harmful local effects, as chronic inflammatory reactions of human pulp. Also, it has been demonstrated that they are responsible for many cytotoxic and possible endocrine-disruptive impact. The amount of the monomer release ranges from micrograms to milligrams [24].

Conventional adhesive systems require acid etching of the surface of enamel and dentine as a separate step before their applications. Application of acids to dentine surface removes the smear layer, opens the dentinal tubules, and decalcifies the intertubular and peritubular dentine [43]. Despite paste apprehension about potential toxicity of acid penetration in dentine tubules and pulp space, the interaction of etchants with dentine is limited to the superficial 1.9–5.8 μm. It is unlikely that the acid is directly responsible for pulp injury. The effect of acid on the dentine is limited by the action of hydroxyapatite and collagen which acts as a baffler [41]. Total etching is considered safe if the length of the remaining dentine is more or equal to 0.5 mm. Therefore, total-etching adhesives are recommended in shallow cavities, located in superficial or sclerotic dentine where the permeability of the dentine is low and the thickness of the remaining dentine is adequate to prevent any adverse effects from diffusing materials [36, 44]. After acid etching, a significant increase in dentine permeability due to smear layer removal and opening of dentinal tubules can facilitate the permeation of resin monomers towards pulp, mainly in deep dentine [45]. About et al. [46] concluded in their study that dentine-bonding agents do not affect the cytodifferentiation of secondary odontoblasts when the remaining dentine thickness is 0.7 mm and when the materials are properly polymerized. When adhesives are applied in deep cavities, residual monomers might reach the pulp by diffusion, and in etched dentine, this penetration tends to be higher. Odontoblasts are typical pulp cells and are the first cells to be damaged by potentially cytotoxic compounds released from dental materials that diffuse through enamel and dentinal tubules. Injury to the odontoblasts results in loss of the capacity of these cells to secrete reactionary dentine, and secretion of dentine bridge by a new generation of odontoblast-like cells [47]. An evaluation of the cytotoxic effect of total-etch two-step adhesive systems showed that both the acidic and non-acidic components of unpolymerized adhesive resins were responsible for high cytotoxic effects on odontoblast-like cells [48].

The most widely used self-etching adhesive systems involve two application steps: the conditioning of dentine and enamel with a self-etching primer, followed by the application of an adhesive resin. There are also one component self-etching adhesives, which contain etch, prime, and bond functions in a single solution. For these adhesive systems, the adhesive resin infiltrates to the same depth as the acidic primer exposed the collagen in dentine [6]. Therefore, the use of self-etching adhesives systems is indicated for young, deep, permeable dentine [36, 49]. da Silva et al. [8] compared the effectiveness and biological compatibility of different generations of dentine adhesives. Their results showed that the one-step self-etch system had the best bond strength performance and was the least toxic to pulp cells, referring to it as a good alternative for specific cases. These results are in accordance with those given by Hashieh et al. [50] who have shown that one-step dentine-bonding agents were less toxic in cell culture than multistep counterparts. Self-etch adhesives have proved to have less cytotoxicity and better tissue response in histological evaluations than those related to conventional adhesive systems, but at this time they are still harmful to the pulp [51, 52].

Any restorative materials that are placed adjacent to vital pulp can induce biological effects. These effects are controlled by the components that are released from the material and the pulpal response to those components. The composition of the adhesive system plays an important role in the toxic effect it produces, and the choice of adhesive system for clinical use should take into consideration its biocompatibility [17]. The usual composition of bonding agents includes resin monomers, initiators, inhibitors or stabilizers, solvents, and sometimes inorganic fillers [2]. It has been reported that dental adhesives release substances that have biological effects (cytotoxicity, carcinogenicity, mutagenicity, and genotoxicity) and toxic potencies [2, 7]. Acidic and nonacidic components of polymerized and unpolymerized adhesives are considered responsible for the cytotoxic effects on the dentine-pulp system [16, 53, 54]. The resins in the adhesive systems are of interest to many researchers, and its biocompatibility has come under extensive scrutiny [12, 31, 55, 56].

The monomers are found to be cytotoxic in cell cultures and to affect the metabolism of the cells. Bisphenol A-glycidyl methacrylate (Bis-GMA), urethane dimethacrylate (UDMA), hydroxyethyl methacrylate (HEMA), and triethylene glycol dimethacrylate (TEGMA) are considered to be among the most toxic resins [50, 57, 58]. Several *in vitro* studies have indicated that monomers, such as Bis-GMA and UDMA, are strongly cytotoxic to fibroblasts, whereas HEMA and TEGDMA are moderately cytotoxic [55, 59]. Ratanasathien et al. [60] assessed the cytotoxicity of the dentine-bonding components in cell cultures and found that the ranking by toxicity was: Bis-GMA > UDMA > TEGDMA > HEMA after 24 and 72 h of exposure. It is important to emphasize that synergistic interaction between the multiple components may occur resulting in more cytotoxicity than the individual components would have caused by themselves [50, 60]. A bonding agent that contained Bis-GMA was less cytotoxic than the other bonding agents that contained Bis-GMA + UDMA + HEMA and Bis-GMA + HEMA [50]. In addition, it was demonstrated that a combination of Bis-GMA and HEMA was less cytotoxic than a combination of the following three substances: Bis-GMA, UDMA, and HEMA [12]. HEMA is the most frequent hydrophilic monomer present in adhesive systems. It is a low molecular weight monomer, which improves wettability and diffusion, and has penetration properties [60]. HEMA has been shown to diffuse rapidly across the dentine toward the pulp which, in turn, could induce hypersensitivity reactions in susceptible individuals [61]. Furthermore, HEMA may suppress the growth of many cell types and, in particular, induce a delay in primary fibroblast cell cycle progression by increasing the formation of reactive oxygen species (ROS) [56, 62, 63]. Various studies have reported different 50% toxicity (TC 50) values of HEMA. This difference has been suggested to be related to different sensitivities between different cell types [61]. Ratanasathien et al. [60] showed that the concentration of HEMA, which caused 50% inhibition of cellular metabolism after 24 h in contact with mouse fibroblasts, was 3.6 mmol l^{-1}. However, after 72 h, the TC 50 dropped to 1.0 mmol l^{-1}. Extrapolating this *in vitro* result to clinical situations, it could be speculated that the adhesive resin might promote more intense pulp damage over time. The cytotoxic effects of HEMA on human fibroblasts have identified the concentrations that cause TC 50 when compared with controls [64]. A 50% toxicity concentration of HEMA ranging from 10 μmol/l to 10 mmol/l has been demonstrated, mainly by mitochondrial dehydrogenase activity (MTT assay) and lactate dehydrogenase activity (LDH assay). It is well known that HEMA causes cell death by activating apoptosis [36, 55]; it interferes with the cell cycle and DNA synthesis, increases the production of reactive oxygen species, and induces a strong depletion of intracellular glutathione level even after very short times of exposure [24]. Di-methacrylates, Bis-GMA, UDMA, and TEGDMA are the most frequently used cross-linkers in adhesive systems. Some controversy exists about the biocompatibility of these monomers. Bis-GMA is considered cytotoxic with an ability to alter the cell cycle and induce oxidative stress leading to apoptosis and necrosis in a concentration-dependent manner [65]. In study conducted by Chang et al. [65], the concentration of 0.1 mM Bis-GMA reduced the metabolism of pulp cells by 50%. The determinant factor of the death was the monomer concentration, given that low concentrations of Bis-GMA induced apoptosis, whereas high concentrations induced death by necrosis or late apoptosis. Bis-GMA dissolves easily and may cause irritation or a lichenoid reaction of the oral mucosa in close and prolonged contact with gingival tissues, especially after cervical filling procedures [12, 66]. TEGDMA is a copolymer used in major recent biomaterials with resin phase as a diluent. *In vitro* studies have shown that TEGDMA induced toxicity in human fibroblasts associated with early and drastic glutathione depletion with subsequent production of ROS [14, 67], delayed cell proliferation, and mineralization processes. The cellular mechanisms underlying these phenomena remain poorly understood [14]. Experiments with several antioxidants suggested that ROS production is a major cause of TEGDMA-induced cellular cell death. TEGDMA is also known to be more cytotoxic to fibroblasts than HEMA [68]. UDMA provides improved compatibility with water and adhesives, which increases the stability, making it behave similarly in both wet and dry dentins. UDMA released from dental materials was reported to reach a toxic level in human oral cells to induce their necrosis [69]. UDMA also induced morphological changes of pulp cells and decreased cell viability by 29–49% at concentrations of 0.1–0.35 mM [70].

Also, initiators may be released from dental adhesives and have been associated with cytotoxicity, related to their ability to generate free radicals [71]. The cytotoxicity of dentine adhesives may as well be attributed to camphorquinone (CQ), the most commonly used photoactivator [15, 56, 62]. For concentrations higher than 1 mM, CQ caused a significant concentration-dependent increase of intracellular ROS in human pulp fibroblasts (HPF) within 90 min of exposure. The cytotoxicity by CQ can be partly explained by its induction of cell cycle arrest and apoptosis. CQ also inhibited the expression of type I collagen, a major extracellular protein of dental pulp, suggesting the effect of CQ on matrix turnover and pulpal repair. CQ also stimulates prostaglandin E2 (PGE2) and PGF2α production of pulp cells [72]. It has been also documented that the camphorquinone acts not only as a cytotoxic agent but also as a mutagen, and its lixiviation may partly explain why these kinds of resinous products are considered as toxic agents [57]. Another substance known for its toxic, allergenic, and mutagenic effects and present in some dental adhesives is glutaraldehyde, which seems to be harmful by direct contact with mucous tissue, as well as by inhalation of its evaporated form [16].

A further factor directly related to the biological responses to adhesive systems concerns monomer conversion into polymers, obtained by the polymerization technique. It has been reported that the type of light curing unit and light parameters such as light spectrum and intensity affect the cytotoxic properties of dental adhesives, as well as polymerization time and application mode [66, 73]. All these parameters influence directly the degree of monomer conversion and consequently the release of residual monomers. Ye et al. [74] evaluated the effect of light irradiance and source on the photopolymerization of three commercial dental adhesives by monitoring the double bond conversion as function of time during and after irradiation. These authors observed that the time for Single Bond with little solvent to reach the conversion plateau was about 20 s, while the time for OneUp Bond F was about 25 s and for Adper Prompt as long as 40 s, and they indicated that the time required to reach the conversion plateau for adhesive polymerization is a valuable information for the dental clinicians. According to Peutzfeld [75], only 60% of monomers are completely bound in polymerized composite resin. Incomplete conversion results in an increased amount of residual monomer and creates the risk of leaching into the surrounding environment. However, even with sufficient light intensity, a certain amount of unreacted resin monomers may be released, which has a potential impact on the toxicity of the materials [56]. Adhesive/bonding agents cured with light-emitting diode (LED) units demonstrated higher cell survival rates in comparison with those cured with a halogen lamp [73]. Polymerization time can also influence on the cytotoxicity of adhesive systems. For resin-based dental materials, a shorter light-curing duration usually results in a lower degree of conversion, inferior mechanical properties, and a higher cytotoxicity. There is a dramatic difference in the responses of cells to the three conditions of polymerization (light curing for 0, 10, or 40 s) [35].

Cell culture assays provide a convenient, controllable, and repeatable method to assess the biocompatibility of materials. Increasing public concern regarding the use of animals in biocompatibility evaluation of dental materials has made *in vitro* testing more reasonable and ethically more acceptable [76]. In several studies, gingival fibroblasts and dental pulp fibroblasts were chosen as target cells to test the toxic effects of the dental adhesives. This choice was mainly due to the fact that under clinical conditions, gingival fibroblasts and dental pulp fibroblasts seem to be the most exposed cell type to dental adhesives. Pulp fibroblasts may be particularly affected by using dental adhesives on deep tooth cavities or when used as capping material, while the exposition of gingival fibroblasts to the adhesive layer may be more important by treating cavities with margins close to the gingiva, for example in restorations I and V class [77]. When testing is conducted in cell cultures, the monomers appeared to be less toxic in the presence of dentine and adsorption barrier; even the smaller monomer molecules diffuse through the dentine against an externally oriented fluid flow [39, 78]. According to Schmalz et al. [79], the dentine barrier test system for toxicity tests might mimic clinical *in vivo* oral environment and test the ability of the material to diffuse through dentinal tubules, which is better than direct cell material contact *in vitro* methods, and it has the potential to, at least partially, replace animal experimentation. Thus, it is ideal for testing cytotoxic effects of the material on dental pulp cells [25, 80]. One group of investigator show that low pH dentine-bonding agents have no effect on pulp-derived three-dimensional cell cultures when a 0.5 mm dentine barrier was placed between material and cells [81]. Under 0.3 ml/h perfusion conditions, the material with the lowest pH even increased the enzyme activity of the cell cultures. In contrast to them, another group of researchers has shown that components of the dentine-bonding systems may be capable of causing cellular damage, even when an interposing layer of dentine separates the material from the pulp. Therefore, they recommend to clinicians to consider applying a lining agent to the depths of their cavity preparations before applying a resin bonding agent [14]. Studies regarding the cytotoxicity of adhesives revealed potentially harmful effects on the gingival cells and may lead to mild to severe inflammatory reactions, cell changes, and cytotoxicity [77, 82]. Since self-etch adhesives are not fully dispersed during bonding process and cavity restoration with composite resin, there is concern about their contact time with the gingiva. Same authors suggested that contact with uncured primers and adhesives should be minimized. In this regard, long exposure of the acidic part of an aggressive self-etch adhesive without rinsing during its application, along with the chemical composition, might affect gingival fibroblasts [58, 83]. Dentists should follow the rules of adhesive application, precisely dose them, and not allow direct contact with the gums as, even after polymerization, adhesive agents exhibit potential cytotoxic activity [20].

5. *In vivo* biocompatibility testings of dental adhesives

The material is considered biocompatible when the interaction among a host, a material, and an expected function of material is in harmony. Dental material, furthermore, should not contain any toxic, leachable, and diffusible substances that could be absorbed and cause systemic responses such as teratogenic or carcinogenic effects. The ability of a material to even enter into interaction with biological tissue and create a stable connection is essential to verify its biocompatibility [84].

In addition to already mentioned *in vitro* assays for testing biocompatibility of dental adhesives as well as of other dental materials, numerous *in vivo* assays were also used. Non-specific tissue reactions caused by dental materials *in vivo* are usually investigated by histological studies based on implantation of the test material in the different tissues of animals. They are usually used in mammals, but many other kinds of animals have been also used. The tested material can be injected directly or implanted (either directly or in the Teflon, silicone, or polyethylene tubes) in the different tissues determined by the use of the material, such as subcutaneous connective tissue, muscles, or bones [9]. Even the quality and specificity of the data obtained by *in vivo* models are questioned, because the used biological system cannot reproduce the same response as the target organ. However, the results are still valuable and useful. Thus, Santos et al. [85] were evaluating the biocompatibility of orthodontic adhesives in rat subcutaneous tissue. In that study, it has been detected the reduction of the inflammatory response at the end of day 15 compared to day 7 for almost all the tested adhesives.

There are conflicting results concerning direct pulp capping trails with adhesives on animals. Therefore, Watts and Peterson [86] suggested that biocompatibility studies and histological evaluation of the pulp healing after pulp capping need to be performed for each species in particular. So far, investigations were implemented on monkeys, dogs, rats, and cats. Cox et al. [87] showed in the monkey tooth model that most adhesive systems are biologically compatible with pulp, allowing pulpal wound healing and reparative dentine formation. Study of Nowicka et al. [88] analysed the pulpo-dentine response after direct capping with self-etch adhesive system. In the majority of the specimens, it was presented inflammatory pulp response with tissue disorganization and lack of dentinal bridge formation. Contrary to this, the teeth in the control group, capped with $Ca(OH)_2$, showed significantly smaller inflammatory pulp response with tissue disorganization and a considerably higher incidence of reparative dentine formation. These results are also in correspondence with those reported from the study of dogs from Koliniotou-Koumpia and Tziafas [89] and da Silva et al. [8]. They reported a total absence of hard tissue formation after direct pulp capping with dentine adhesive system and presence of chronic inflammatory cell infiltrate, giant cells, and macrophages around the resin globes, morphological cell alteration, and hyalinization of the pulp cells. The pulpal responses to the dentine bonding systems used in the mentioned investigations could be attributed mainly to their acidic nature. *In vivo* studies on primates' teeth showed normal healing of the pulp and also normal differentiation of the pulp cells. One of the reasons for such differentiates in results regarding pulp healing between human and animal teeth might be in different healing capacities among them. Generally, the animals are more resilient and adaptable to various external impacts and noxes around them in their evolutionary path. White et al. [90] in their study on rhesus monkeys in 1994 showed that there were no differences in cytotoxicity when the adhesives were set to dry or wet previously etched dry or wet dentine. Namely, in both cases, it has been observed that there was no inflammation of the pulp and that there has been a creation of reparative dentine beneath the remaining thin layer of dentine.

As opposed to animal studies, there are also clinical studies on humans, which are necessary for determining long-term biocompatibility of dental materials. The findings from above mentored studies on animals were similar to those obtained in study in human teeth of Cui et al. [91]. Namely, more pronounced histological changes in pulpal tissue in teeth treated with self-etching adhesive systems were found than in control group treated with $Ca(OH)_2$. Similar results were obtained also by Preira and associates [10]. In their study, the adhesive systems did not stimulate the dental bridge formation even after over 200 days after the pulp capping. On the contrary, the dental bridge healing was observed after 60 days in control group of pulp capping with $Ca(OH)_2$.

There are lots of opposite opinions of various authors about adhesives. Some authors consider that adhesives are safe, accurately biocompatible, and could be used for direct pulp capping on humans [92–94], whereas other researchers consider that adhesives are not suitable for direct pulp capping because of inducing the constant inflammation that does not heal [95–100]. Gwinnett and Tay [96] observed the pulp response after application of dentine adhesive to etched, deep, and unexposed coronal dentine, and they detected chronic inflammation of the pulp caused by adhesive in dental tubules. These so-called globules of resin in the tubules stimulate an immune reaction to a foreign body and are characterized by the appearance of mononuclear cell infiltration of macrophages or multinuclear giant cells. There has been also observed a degradation and loss of odontoblasts when the remaining dentine thickness was less than 0.3 mm. Furthermore, Hebling et al. [97] have found a hyaline alteration of extracellular matrix associated with oedema and hydropic pulp cell changes. Avles and Sobral [17] evaluated the biocompatibility of an etch-and-rinse adhesive system based on tertiary butanol applied in deep cavity human teeth with approximately 1 mm of remaining dentine by observing histological changes of the pulp tissue of humans at intervals of 1, 7, 14, and 21 days. They observed mild inflammatory infiltrate, preserved pulp tissue morphology, disorganization of the odontoblast layer in most specimens, as well as absence of bacteria at the intervals of 1, 7, 14, and 21 days.

Types of polymerization units also could be affected on the final cytotoxicity of adhesive system. Spagnuolo et al. [98] noticed that after lighting with LED units, there was a greater release of reactive oxygen that decreases cell survival than after lighting with halogen units.

There are a lot of studies that show that failure in the pulp capping is attributed to mistakes during operation, for example inadequate haemorrhage, an incomplete polymerization of the material, or usage of poor material properties. It is also known that a blood clot lyses could launch a series of chemotactic peptides that have a role in inflammation process [99]. Recent studies on human teeth have shown irreversible damage of the pulp and odontoblasts, and also chronic inflammatory response with macrophages and mononuclear giant cell infiltration observed over a period of 300 days following restorative procedure. The appearance of these signs indicates the typical immune response to the foreign substance in the organism. This is a result of interaction of resin and its components with plasma proteins, tissues, and connective tissue cells [100, 101]. In the following studies, it has been shown that non-polymerized resin compounds released from dental adhesives cause chronic pulp inflammatory response [102]. The response of the pulp to the dental materials could be modified by bacteria's and their products, and also affect by carious dentine, cavity preparation, and bleeding. The reparative capacity of damaged pulp in those cases is already weaker, and immunological responses are more drastic; therefore, they might contribute to the cytotoxic effects of dentine-bonding agents [101].

Composite fillings and adhesives contain numerous components that could be released into the environment and cause a biological activity in the organism (cytotoxicity, carcinogenicity, mutagenicity, genotoxicity) [53, 103]. The monomers might be released in the saliva and diffuse into the gingiva, mucosa, salivary glands, and blood where their cytotoxicity or teratogenesis could contribute to the tumour formation which *in vivo* could be detected only after long-term interval. It is well known that high concentrations of UBMA, bis-GMA, TEGDMA, and other similar monomers have an inhibitory effect on the pulp cells. This effect results in chemically induced immunosuppression of the T-cells in the pulp. Thus, the pulp becomes more exposed, vulnerable, and susceptible to the bacterial infection. Acidic and non-acidic components of unpolymerized parts of self-etching dentine-bonding agents are responsible for the cytopathic effects on the odontoblasts and odontoblast-like cells. In an aqueous medium, adhesives dissolve even after polymerization and release components that could cause various effects on organism, such as allergic reactions (contact dermatitis and urticarial), systemic toxicity, cytotoxicity, and mutagenesis [104]. Once again, it is of crucial importance to mention the significance of the remaining dentine thickness [105]. Since the remaining dentine layer is thinner, thus more permeable, the cytotoxicity of applied material increases, and the response of the dentine-pulp complex is more intense [17]. High permeability dentine allows greater penetration of dentine-bonding agent and its components [39]. The highest cytotoxicity was observed in the early period of application, in the first 24–48 h after polymerization [106, 107].

According to these results, there are lots of concerns about pulp therapy with adhesive systems. Also, the results obtained from animal models cannot be extrapolated to human clinical condition.

Most practitioners purchase materials that are commercially available without any concerns about their biocompatibility. Today, despite the fact that clinicians use an increasing number of materials, most relevant studies concentrate on their physical properties with less emphasis on biological compatibility. There is a large gap between the results published by research laboratories and clinical reports. In order to assess the biocompatibility, clinical studies, as well as *in vitro* studies, are needed [108].

Toxicity of adhesive systems is associated primarily with short-term release occurring during setup time of material, due to insufficient polymerization, but also as long-term release of leachable substances generated by degradation over time [80]. Numerous studies showed that the differential toxicity of the adhesives could be attributed to the different ingredients, the interactions between them, and the degree of resin polymerization. Many studies of applied adhesive systems in direct contact with animal pulp tissue have supported the concept that bacteria, rather than dental materials, promote irritation of pulp tissue [87]. On the contrary, in other studies, it has been shown that those resin-based materials do not seem to be appropriate for use as pulp capping material and that no adhesive system should be applied directly on the pulp tissue [10, 35, 101, 109]. Thus, an adequate dentine barrier prevents or even reduces the amount of monomers and other components capable of causing damage to the pulp-dentine complex [6, 110].

6. Conclusion

Biocompatibility is one of the most important properties of dental materials, and adhesives are no exception. Future innovation in adhesive systems should seek novel properties like tissue tolerance. Adhesives should be biofunctional, protective, and preventive, with health-promoting effects that contribute to a better prognosis for restorative treatments and its biocompatibility.

According to results obtained from our previous *in vitro* studies, we can affirm that tested adhesives did not have genotoxic potential. *In vitro* studies are likely to differ from *in vivo* condition because higher concentrations of tested materials are usually used *in vitro* compared to the concentrations that may be used during regular handling with dental materials. Furthermore, *in vivo* studies present numerous factors such as saliva, mucus layer, creatine levels, blood flow, and oral flora that can influence the oral cavity protection. Consequently, we can conclude that commercial available dental adhesives, whether they are used alone or in combination with dental composites, are safe for clinical use.

Author details

Antonija Tadin[1*], Lidia Gavić[1] and Nada Galić[2]

*Address all correspondence to: atadin@mefst.hr

1 Study of Dental Medicine, School of Medicine, University of Split, Split, Croatia

2 School of Dental Medicine, University of Zagreb, Zagreb, Croatia

References

[1] Van Landuyt KL, Snauwaert J, De Munck J, Peumans M, Yoshida Y, Poitevin A, et al. Systematic review of the chemical composition of contemporary dental adhesives. Biomaterials 2007;28:3757–3785. DOI: 10.1016/j.biomaterials.2007.04.044

[2] Prica D, Tadin A, Marovic D, Katunaric M, Prica A, Galic N. Effects of dental adhesives on micronucleus frequency in peripheral blood lymphocytes in vitro. Acta clinica Croatica 2013;52:309–315.

[3] Marovic D, Tadin A, Mladinic M, Juric-Kacunic D, Galic N. In vitro detection of DNA damage in human leukocytes induced by combined effect of resin composites and adhesive systems. American Journal of Dentistry 2014;27:35–41.

[4] Pashley DH, Tay FR, Breschi L, Tjaderhane L, Carvalho RM, Carrilho M, et al. State of the art etch-and-rinse adhesives. Dental Materials: official publication of the Academy of Dental Materials 2011;27:1–16. DOI:10.1016/j.dental.2010.10.016

[5] Tadin A, Galic N, Zeljezic D, Mikelic Vitasovic B, Marovic D, Kovacic I. Ex vivo evaluation of genotoxic effects of four dental adhesives on human leukocytes. Journal of Dental Science 2013;8:37–43. DOI:10.1016/j.jds.2012.12.001

[6] Van Meerbeek B, Perdigao J, Lambrechts P, Vanherle G. The clinical performance of adhesives. Journal of Dentistry 1998;26:1–20. DOI:10.1016/S0300-5712(96)00070-X

[7] Prica D, Galic N, Zeljezic D, Prica A. Genotoxicity evaluation of five different dentin bonding agents by chromosomal aberration analysis. Journal of Oral Rehabilitation 2006;33:462–471. DOI:10.1111/j.1365-2842.2006.01606.x

[8] da Silva JM, Rodrigues JR, Camargo CH, Fernandes VV, Jr., Hiller KA, Schweikl H, et al. Effectiveness and biological compatibility of different generations of dentin adhesives. Clinical Oral Investigations 2014;18:607–613. DOI:10.1007/s00784-013-1000-9

[9] Schmalz G, Arenholt Bindslev D. Biocompatibility of Dental Materials: Springer-Verlag, Berlin, Heidelberg; 2009. DOI 10.1007/978-3-540-77782-3

[10] Pereira JC, Segala AD, Costa CA. Human pulpal response to direct pulp capping with an adhesive system. American Journal of Dentistry 2000;13:139–147.

[11] Dammaschke T, Stratmann U, Fischer RJ, Sagheri D, Schafer E. Proliferation of rat molar pulp cells after direct pulp capping with dentine adhesive and calcium hydroxide. Clinical Oral Investigations 2011;15:577–587. DOI:10.1007/s00784-010-0409-7

[12] Szep S, Kunkel A, Ronge K, Heidemann D. Cytotoxicity of modern dentin adhesives—in vitro testing on gingival fibroblasts. Journal of Biomedical Materials Research 2002;63:53–60. DOI: 10.1002/jbm.10083

[13] Szczepanska J, Poplawski T, Synowiec E, Pawlowska E, Chojnacki CJ, Chojnacki J, et al. 2-Hydroxylethyl methacrylate (HEMA), a tooth restoration component, exerts its genotoxic effects in human gingival fibroblasts trough methacrylic acid, an immediate product of its degradation. Molecular Biology Reports 2012;39:1561–1574. DOI:10.1007/s11033-011-0895-y

[14] Stanislawski L, Lefeuvre M, Bourd K, Soheili-Majd E, Goldberg M, Perianin A. TEGDMA-induced toxicity in human fibroblasts is associated with early and drastic glutathione depletion with subsequent production of oxygen reactive species. Journal of Biomedical Materials Research Part A 2003;66:476–482. DOI:10.1002/jbm.a.10600

[15] Sengun A, Yalcin M, Ulker HE, Ozturk B, Hakki SS. Cytotoxicity evaluation of dentin bonding agents by dentin barrier test on 3-dimensional pulp cells. Oral Surgery, Oral

medicine, Oral Pathology, Oral Radiology, and Endodontics 2011;112:e83–88. DOI: 10.1016/j.tripleo.2011.02.023

[16] Schweikl H, Schmalz G. Glutaraldehyde-containing dentin bonding agents are mutagens in mammalian cells in vitro. Journal of Biomedical Materials Research 1997;36:284–288. DOI: 10.1002/(SICI)1097-4636(19970905)36:3<284::AID-JBM2>3.0.CO;2-A

[17] Alves GC, Sobral AP. Evaluation of biocompatibility of an etch-and-rinse adhesive system based in tertiary butanol applied in deep cavity. The Open Dentistry Journal 2015;9:168–173. DOI:10.2174/1874210601509010168

[18] Koulaouzidou EA, Helvatjoglu-Antoniades M, Palaghias G, Karanika-Kouma A, Antoniades D. Cytotoxicity of dental adhesives in vitro. European journal of dentistry 2009;3:3–9.

[19] Vande Vannet BM, Hanssens JL. Cytotoxicity of two bonding adhesives assessed by three-dimensional cell culture. The Angle Orthodontist 2007;77:716–722. DOI: 10.2319/052706-212.1

[20] Kierklo A, Pawinska M, Tokajuk G, Poplawska B, Bielawska A. Cytotoxicity evaluation of three light-cured dentin adhesive materials on human gingival fibroblasts, ex vivo. Advances in Medical Sciences 2012;57:385–390. DOI:10.2478/v10039-012-0038-2

[21] Moharamzadeh K, Brook IM, Van Noort R. Biocompatibility of resin-based dental materials. Materials 2009;2:514–548. DOI:10.3390/ma2020514

[22] Hauman CH, Love RM. Biocompatibility of dental materials used in contemporary endodontic therapy: a review. Part 1. Intracanal drugs and substances. International Endodontic Journal 2003;36:75–85.

[23] Elshahaw W. Biocompatibility. In: Sikalidis C, editor. Advances in Ceramics—Electric and Magnetic Ceramics, Bioceramics, Ceramics and Environment: InTech, Rijeka; 2011. DOI: 10.5772/18475

[24] Faccioni F, Franceschetti P, Cerpelloni M, Fracasso ME. In vivo study on metal release from fixed orthodontic appliances and DNA damage in oral mucosa cells. American Journal of Orthodontics and Dentofacial Orthopedics: official publication of the American Association of Orthodontists, its constituent societies, and the American Board of Orthodontics 2003;124:687–693. DOI: http://dx.doi.org/10.1016/j.ajodo.2003.09.010

[25] Narvekar A SA, Ramamoorthi M. Cytotoxic effects of restorative biomaterials on dental pulp—a systematized review. World Journal of Pharmacy and Pharmaceutical Sciences 2015;4:239–266.

[26] Bertho AL, Santiago MA, Coutinho SG. Flow cytometry in the study of cell death. Memorias do Instituto Oswaldo Cruz 2000;95:429–433. DOI: 10.1590/S0074-02762000000300020

[27] Fink SL, Cookson BT. Apoptosis, pyroptosis, and necrosis: mechanistic description of dead and dying eukaryotic cells. Infection and Immunity 2005;73:1907–1916. DOI: 10.1128/IAI.73.4.1907-1916.2005

[28] Swift LH, Golsteyn RM. Genotoxic anti-cancer agents and their relationship to DNA damage, mitosis, and checkpoint adaptation in proliferating cancer cells. International Journal of Molecular Sciences 2014;15:3403–3431. DOI:10.3390/ijms15033403

[29] Schmalz G. Use of cell-cultures for toxicity testing of dental materials advantages and limitations. Journal of Dentistry 1994;22:S6–S11. DOI:10.1016/0300-5712(94)90032-9

[30] Porto IC, Oliveira DC, Raele RA, Ribas KH, Montes MA, De Castro CM. Cytotoxicity of current adhesive systems: in vitro testing on cell cultures of primary murine macrophages. Dental Materials: official publication of the Academy of Dental Materials 2011;27:221–228. DOI:10.1016/j.dental.2010.10.006

[31] Demirci M, Hiller KA, Bosl C, Galler K, Schmalz G, Schweikl H. The induction of oxidative stress, cytotoxicity, and genotoxicity by dental adhesives. Dental Materials: official publication of the Academy of Dental Materials 2008;24:362–371. DOI:10.1016/j.dental.2007.06.009

[32] Schuster U, Schmalz G, Thonemann B, Mendel N, Metzl C. Cytotoxicity testing with three-dimensional cultures of transfected pulp-derived cells. Journal of Endodontics 2001;27:259–265. DOI:10.1097/00004770-200104000-00004

[33] Doyle A, Griffiths JB. Cell and Tissue Culture: Laboratory Procedures. London: Wiley; 1998. ISBN: 978-0-471-98255-5

[34] Singh NP, McCoy MT, Tice RR, Schneider EL. A simple technique for quantitation of low levels of DNA damage in individual cells. Experimental Cell Research 1988;175:184–191. DOI:10.1016/0014-4827(88)90265-0

[35] Elias ST, Santos AF, Garcia FC, Pereira PN, Hilgert LA, Fonseca-Bazzo YM, et al. Cytotoxicity of universal, self-etching and etch-and-rinse adhesive systems according to the polymerization time. Brazilian Dental Journal 2015;26:160–168. DOI: 10.1590/0103-6440201300294

[36] Bouillaguet S. Biological risks of resin-based materials to the dentin-pulp complex. Critical Reviews in Oral Biology and Medicine: an official publication of the American Association of Oral Biologists 2004;15:47–60. DOI: 10.1177/154411130401500105

[37] Galler K, Hiller KA, Ettl T, Schmalz G. Selective influence of dentin thickness upon cytotoxicity of dentin contacting materials. Journal of Endodontics 2005;31:396–399. DOI: 10.1097/01.don.0000145428.26880.e5

[38] Pashley DH, Michelich V, Kehl T. Dentin permeability: effects of smear layer removal. The Journal of Prosthetic Dentistry 1981;46:531–537.

[39] Bouillaguet S, Virgillito M, Wataha J, Ciucchi B, Holz J. The influence of dentine permeability on cytotoxicity of four dentine bonding systems, in vitro. Journal of Oral Rehabilitation 1998;25:45–51. DOI: 10.1046/j.1365-2842.1998.00205.x

[40] Meryon S, Jakeman J. Aluminium and dental materials--a study in vitro of its potential release and toxicity. International Endodontic Journal 1987;20:16–19. DOI: 10.1111/j.1365-591.1987.tb00582.x

[41] Bianchi L, Ribeiro AP, de Oliveira Carrilho MR, Pashley DH, de Souza Costa CA, Hebling J. Transdentinal cytotoxicity of experimental adhesive systems of different hydrophilicity applied to ethanol-saturated dentin. Dental Materials: official publication of the Academy of Dental Materials 2013;29:980–990. DOI:10.1016/j.dental.2013.07.006

[42] Rathke A, Alt A, Gambin N, Haller B. Dentin diffusion of HEMA released from etch-and-rinse and self-etch bonding systems. European Journal of Oral Sciences 2007;115:510–516.

[43] Eliades G. Clinical relevance of the formulation and testing of dentine bonding systems. Journal of Dentistry 1994;22:73–81. DOI:10.1111/j.1600-0722.2007.00484.x

[44] Mjor IA, Ferrari M. Pulp-dentin biology in restorative dentistry. Part 6: Reactions to restorative materials, tooth-restoration interfaces, and adhesive techniques. Quintessence international 2002;33:35–63.

[45] Cetinguc A, Olmez S, Vural N. HEMA diffusion from dentin bonding agents in young and old primary molars in vitro. Dental Materials: official publication of the Academy of Dental Materials 2007;23:302–307. DOI:10.1016/j.dental.2005.08.013

[46] About I, Camps J, Mitsiadis TA, Bottero MJ, Butler W, Franquin JC. Influence of resinous monomers on the differentiation in vitro of human pulp cells into odontoblasts. Journal of Biomedical Materials Research 2002;63:418–423. DOI:10.1002/jbm.10253

[47] Goldberg M, Smith AJ. Cells and extracellular matrices of dentin and pulp: a biological basis for repair and tissue engineering. Critical Reviews in Oral Biology and Medicine: an official publication of the American Association of Oral Biologists 2004;15:13–27. DOI: 10.1177/154411130401500103

[48] Costa CA, Vaerten MA, Edwards CA, Hanks CT. Cytotoxic effects of current dental adhesive systems on immortalized odontoblast cell line MDPC-23. Dental Materials: official publication of the Academy of Dental Materials 1999;15:434–441. DOI:10.1016/S0109-5641(99)00071-8

[49] Tay FR, Carvalho R, Sano H, Pashley DH. Effect of smear layers on the bonding of a self-etching primer to dentin. The Journal of Adhesive Dentistry 2000;2:99–116.

[50] Hashieh IA, Cosset A, Franquin JC, Camps J. In vitro cytotoxicity of one-step dentin bonding systems. Journal of Endodontics 1999;25:89–92. DOI:10.1016/S0099-2399(99)80003-X

[51] de Souza Costa CA, Lopes do Nascimento AB, Teixeira HM, Fontana UF. Response of human pulps capped with a self-etching adhesive system. Dental Materials: official publication of the Academy of Dental Materials 2001;17:230–240. DOI:10.1016/S0109-5641(00)00076-2

[52] Medina VO, 3rd, Shinkai K, Shirono M, Tanaka N, Katoh Y. Histopathologic study on pulp response to single-bottle and self-etching adhesive systems. Operative Dentistry 2002;27:330–342.

[53] Schweikl H, Schmalz G. Triethylene glycol dimethacrylate induces large deletions in the hprt gene of V79 cells. Mutation Research 1999;438:71–78. DOI:10.1016/S1383-5718(98)00164-8

[54] Schweikl H, Schmalz G, Spruss T. The induction of micronuclei in vitro by unpolymerized resin monomers. Journal of Dental Research 2001;80:1615–1620. DOI: 10.1177/00220345010800070401

[55] Bouillaguet S, Wataha JC, Hanks CT, Ciucchi B, Holz J. In vitro cytotoxicity and dentin permeability of HEMA. Journal of Endodontics 1996;22:244–248. DOI:10.1016/S0099-2399(06)80141-X

[56] Geurtsen W, Spahl W, Muller K, Leyhausen G. Aqueous extracts from dentin adhesives contain cytotoxic chemicals. Journal of Biomedical Materials Research 1999;48:772–777. DOI: 10.1002/(SICI)1097-4636(1999)48:6<772::AID-JBM2>3.0.CO;2-X

[57] Huang FM, Chang YC. Cytotoxicity of dentine-bonding agents on human pulp cells in vitro. International Endodontic Journal 2002;35:905–909. DOI: 10.1046/j.1365-2591.2002.00589.x

[58] Kaga M, Noda M, Ferracane JL, Nakamura W, Oguchi H, Sano H. The in vitro cytotoxicity of eluates from dentin bonding resins and their effect on tyrosine phosphorylation of L929 cells. Dental Materials: official publication of the Academy of Dental Materials 2001;17:333–339. DOI:10.1016/S0109-5641(00)00091-9

[59] Chen RS, Liu CC, Tseng WY, Jeng JH, Lin CP. Cytotoxicity of three dentin bonding agents on human dental pulp cells. Journal of Dentistry 2003;31:223–239. DOI:10.1016/S0300-5712(02)00088-X

[60] Ratanasathien S, Wataha JC, Hanks CT, Dennison JB. Cytotoxic interactive effects of dentin bonding components on mouse fibroblasts. Journal of Dental Research 1995;74:1602–1606. DOI: 10.1177/00220345950740091601

[61] Falconi M, Teti G, Zago M, Pelotti S, Breschi L, Mazzotti G. Effects of HEMA on type I collagen protein in human gingival fibroblasts. Cell Biology and Toxicology 2007;23:313–322. DOI:10.1007/s10565-006-0148-3

[62] Geurtsen W, Lehmann F, Spahl W, Leyhausen G. Cytotoxicity of 35 dental resin composite monomers/additives in permanent 3T3 and three human primary fibroblast

cultures. Journal of Biomedical Materials Research 1998;41:474–480. DOI: 10.1002/(SICI)1097-4636(19980905)41:3<474::AID-JBM18>3.0.CO;2-I

[63] Imazato S, Horikawa D, Nishida M, Ebisu S. Effects of monomers eluted from dental resin restoratives on osteoblast-like cells. Journal of Biomedical Materials Research Part B, Applied biomaterials 2009;88:378–386. DOI: 10.1002/jbm.b.31067

[64] Yoshii E. Cytotoxic effects of acrylates and methacrylates: relationships of monomer structures and cytotoxicity. Journal of Biomedical Materials Research 1997;37:517–524. DOI: 10.1002/(SICI)1097-4636(19971215)37:4<517::AID-JBM10>3.0.CO;2-5

[65] Chang HH, Chang MC, Lin LD, Lee JJ, Wang TM, Huang CH, et al. The mechanisms of cytotoxicity of urethane dimethacrylate to Chinese hamster ovary cells. Biomaterials 2010;31:6917–6925. DOI:10.1016/j.biomaterials.2010.05.059

[66] Polydorou O, Trittler R, Hellwig E, Kummerer K. Elution of monomers from two conventional dental composite materials. Dental Materials: official publication of the Academy of Dental Materials 2007;23:1535–1541 DOI:10.1016/j.dental.2006.12.011

[67] Lefeuvre M, Amjaad W, Goldberg M, Stanislawski L. TEGDMA induces mitochondrial damage and oxidative stress in human gingival fibroblasts. Biomaterials 2005;26:5130–5137. DOI: 10.1016/j.biomaterials.2005.01.014

[68] Issa Y, Watts DC, Brunton PA, Waters CM, Duxbury AJ. Resin composite monomers alter MTT and LDH activity of human gingival fibroblasts in vitro. Dental Materials: official publication of the Academy of Dental Materials 2004;20:12–20. DOI:10.1016/S0109-5641(03)00053-8

[69] Poplawski T, Loba K, Pawlowska E, Szczepanska J, Blasiak J. Genotoxicity of urethane dimethacrylate, a tooth restoration component. Toxicology in Vitro: an international journal published in association with BIBRA 2010;24:854–862. DOI:10.1016/j.tiv.2009.12.004

[70] Chang HH, Chang MC, Wang HH, Huang GF, Lee YL, Wang YL, et al. Urethane dimethacrylate induces cytotoxicity and regulates cyclooxygenase-2, hemeoxygenase and carboxylesterase expression in human dental pulp cells. Acta Biomaterialia 2014;10:722–731. DOI:10.1016/j.actbio.2013.10.006

[71] Geurtsen W. Substances released from dental resin composites and glass ionomer cements. European Journal of Oral Sciences 1998;106:687–695. DOI: 10.1046/j.0909-8836.1998.eos10602ii04.x

[72] Chang MC, Lin LD, Wu MT, Chan CP, Chang HH, Lee MS, et al. Effects of camphorquinone on cytotoxicity, cell cycle regulation and prostaglandin E2 production of dental pulp cells: role of ROS, ATM/Chk2, MEK/ERK and hemeoxygenase-1. PLoS ONE 2015;10:e0143663. DOI:10.1371/journal.pone.0143663

[73] Ergun G, Egilmez F, Uctasli MB, Yilmaz S. Effect of light curing type on cytotoxicity of dentine-bonding agents. International Endodontic Journal 2007;40:216–223. DOI: 10.1111/j.1365-2591.2007.01225.x

[74] Ye Q, Wang Y, Williams K, Spencer P. Characterization of photopolymerization of dentin adhesives as a function of light source and irradiance. Journal of Biomedical Materials Research Part B, Applied biomaterials 2007;80:440–446. DOI: 10.1002/jbm.b.30615

[75] Peutzfeldt A. Resin composites in dentistry: the monomer systems. European Journal of Oral Sciences 1997;105:97–116. DOI: 10.1111/j.1600-0722.1997.tb00188.x

[76] Imazato S, Tarumi H, Ebi N, Ebisu S. Cytotoxic effects of composite restorations employing self-etching primers or experimental antibacterial primers. Journal of Dentistry 2000;28:61–67. DOI:10.1016/S0300-5712(99)00039-1

[77] Garner AD, Tucci MA, Benghuzzi HA. Functional and structural responses associated with potential restorative dental adhesives using human gingival fibroblasts as a model. Biomedical Sciences Instrumentation 2012;48:141–148.

[78] Gerzina TM, Hume WR. Diffusion of monomers from bonding resin–resin composite combinations through dentine in vitro. Journal of Dentistry 1996;24:125–128.

[79] Schmalz G, Garhammer P, Schweiki H. A commercially available cell culture device modified for dentin barrier tests. Journal of Endodontics 1996;22:249–252. doi:10.1016/S0099-2399(06)80142-1

[80] Goldberg M. In vitro and in vivo studies on the toxicity of dental resin components: a review. Clinical Oral Investigations 2008;12:1–8. DOI:10.1007/s00784-007-0162-8

[81] Schmalz G, Schuster U, Koch A, Schweikl H. Cytotoxicity of low pH dentin-bonding agents in a dentin barrier test in vitro. Journal of Endodontics 2002;28:188–192. DOI: 10.1097/00004770-200203000-00011

[82] Teixeira HM, Do Nascimento AB, Hebling J, De Souza Costa CA. In vivo evaluation of the biocompatibility of three current bonding agents. Journal of Oral Rehabilitation 2006;33:542–550. DOI:10.1111/j.1365-2842.2005.01591.x

[83] Banava S, Najibfard K, Garcia-Godoy F, Saghiri MA, Ghahremani MH, Ostad N. Impact of dilution and polymerization on cytotoxicity of dentin adhesives to human gingival fibroblasts: early exposure time. Journal of Dental Research, Dental Clinics, Dental Prospects 2015;9:151–158. DOI:10.15171/joddd.2015.029

[84] Brune D. Metal release from dental biomaterials. Biomaterials 1986;7:163–175. DOI: 10.1016/0142-9612(86)90097-9

[85] Santos RL, Pithon MM, Fernandes AB, Cabral MG, Ruellas AC. Biocompatibility of orthodontic adhesives in rat subcutaneous tissue. Journal of Applied Oral Science: revista FOB 2010;18:503–508. DOI: 10.1590/S1678-77572010000500013

[86] Watts A, Paterson RC. A comparison of pulp responses to two different materials in the dog and the rat. Oral Surgery, Oral Medicine, and Oral Pathology 1981;52:648–652. DOI: 10.1016/0030-4220(81)90085-2

[87] Cox CF, Keall CL, Keall HJ, Ostro E, Bergenholtz G. Biocompatibility of surface-sealed dental materials against exposed pulps. The Journal Of Prosthetic Dentistry 1987;57:1–8. DOI: 10.1016/0022-3913(87)90104-1

[88] Nowicka A, Parafiniuk M, Lipski M, Lichota D, Buczkowska-Radlinska J. Pulpo-dentin complex response after direct capping with self-etch adhesive systems. Folia Histochemica et Cytobiologica/Polish Academy of Sciences, Polish Histochemical and Cytochemical Society 2012;50:565–573. DOI:10.5603/20325

[89] Koliniotou-Koumpia E, Tziafas D. Pulpal responses following direct pulp capping of healthy dog teeth with dentine adhesive systems. Journal of Dentistry 2005;33:639–647. DOI:10.1016/j.jdent.2004.12.007

[90] White KC, Cox CF, Kanka J, 3rd, Dixon DL, Farmer JB, Snuggs HM. Pulpal response to adhesive resin systems applied to acid-etched vital dentin: damp versus dry primer application. Quintessence International 1994;25:259–268.

[91] Cui C, Zhou X, Chen X, Fan M, Bian Z, Chen Z. The adverse effect of self-etching adhesive systems on dental pulp after direct pulp capping. Quintessence International 2009;40:e26–34.

[92] Fujitani M, Shibata S, Van Meerbeek B, Yoshida Y, Shintani H. Direct adhesive pulp capping: pulpal healing and ultra-morphology of the resin-pulp interface. American Journal of Dentistry 2002;15:395–402.

[93] Gorecka V, Suliborski S, Biskupski T. Direct pulp capping with a dentin adhesive resin system in children s permanent teeth after traumatic injuries: case reports. Quintessence International 2000;31:241–248. DOI: 10.1111/j.1600-9657.2010.00907.x.

[94] Kopel HM. The pulp capping procedure in primary teeth "revisited". ASDC Journal of Dentistry for Children 1997;64:327–333.

[95] Nascimento MM, Gordan VV, Qvist V, Bader JD, Rindal DB, Williams OD, et al. Restoration of noncarious tooth defects by dentists in The Dental Practice-Based Research Network. Journal of the American Dental Association 2011;142:1368–1375. DOI:10.14219/jada.archive.2011.0138

[96] Gwinnett AJ, Tay F. Early and intermediate time response of the dental pulp to an acid etch technique in vivo. American Journal of Dentistry 1998;11 Spec No:S35–44.

[97] Hebling J, Giro EM, Costa CA. Biocompatibility of an adhesive system applied to exposed human dental pulp. Journal of Endodontics 1999;25:676–682. DOI:10.1007/s00784-003-0247-y

[98] Spagnuolo G, Annunziata M, Rengo S. Cytotoxicity and oxidative stress caused by dental adhesive systems cured with halogen and LED lights. Clinical Oral Investigations 2004;8:81–85.

[99] Shoaf HK, Pashley EL, Myers DR, Pashley DH. Quantitation and control of pulpal bleeding. Pediatric Dentistry 1979;1:177–181.

[100] Costa CA, Giro EM, do Nascimento AB, Teixeira HM, Hebling J. Short-term evaluation of the pulpo-dentin complex response to a resin-modified glass-ionomer cement and a bonding agent applied in deep cavities. Dental Materials: official publication of the Academy of Dental Materials 2003;19:739–746. DOI:10.1016/S0109-5641(00)00008-7

[101] Costa CA, Hebling J, Hanks CT. Current status of pulp capping with dentin adhesive systems: a review. Dental Materials: official publication of the Academy of Dental Materials 2000;16:188–197. DOI:10.1016/S0109-5641(00)00008-7

[102] de Souza Costa CA, do Nascimento AB, Teixeira HM. Response of human pulps following acid conditioning and application of a bonding agent in deep cavities. Dental Materials: official publication of the Academy of Dental Materials 2002;18:543–551. DOI:10.1016/S01095641(01)00089-6

[103] Jontell M, Hanks CT, Bratel J, Bergenholtz G. Effects of unpolymerized resin components on the function of accessory cells derived from the rat incisor pulp. Journal of Dental Research 1995;74:1162–1167. DOI: 10.1177/00220345950740050401

[104] Kleinsasser NH, Wallner BC, Harreus UA, Kleinjung T, Folwaczny M, Hickel R, et al. Genotoxicity and cytotoxicity of dental materials in human lymphocytes as assessed by the single cell microgel electrophoresis (comet) assay. Journal of Dentistry 2004;32:229–234. DOI:10.1016/j.jdent.2003.11.002

[105] Jontell M, Okiji T, Dahlgren U, Bergenholtz G. Immune defense mechanisms of the dental pulp. Critical Reviews in Oral Biology and Medicine: an official publication of the American Association of Oral Biologists 1998;9:179–200. DOI: 10.1177/10454411980090020301

[106] Camps J, Tardieu C, Dejou J, Franquin JC, Ladaique P, Rieu R. In vitro cytotoxicity of dental adhesive systems under simulated pulpal pressure. Dental Materials: official publication of the Academy of Dental Materials 1997;13:34–42. DOI: 10.1016/S0109-5641(97)80006-1

[107] Bagis B, Atilla P, Cakar N, Hasanreisoglu U. Immunohistochemical evaluation of endothelial cell adhesion molecules in human dental pulp: effects of tooth preparation and adhesive application. Archives of Oral Biology 2007;52:705–711. DOI:10.1016/j.archoralbio.2007.01.014

[108] Goldberg M, Lasfargues JJ, Legrand JM. Clinical testing of dental materials—histological considerations. Journal of Dentistry 1994;22 Suppl 2:S25–28. DOI: 10.1016/0300-5712(94)90036-1

[109] Costa CA, Mesas AN, Hebling J. Pulp response to direct capping with an adhesive system. American Journal of Dentistry 2000;13:81–87.

[110] Van Meerbeek B, Yoshihara K, Yoshida Y, Mine A, De Munck J, Van Landuyt KL. State of the art of self-etch adhesives. Dental Materials: official publication of the Academy of Dental Materials 2011;27:17–28. DOI: 10.1016/j.dental.2010.10.023

Chapter 3

Adhesion in Restorative Dentistry

Alexandra Vinagre and João Ramos

Additional information is available at the end of the chapter

http://dx.doi.org/10.5772/65605

Abstract

Bonding agents play a crucial role in the effective sealing and retention of resin-based composite restorations, which have been increasingly placed and replaced by dentists in many countries around the world. In fact, direct adhesive restoration with composite resins has become the procedure of choice for the treatment of anterior and posterior teeth. However, long-term durability of those restorations may be compromised due to progressive loss of the integrity of adhesive interfaces. This means that no adhesive strategy is free from technique sensitivity. The specificity and proportion of different constitutive molecules, the interaction between them and substrates can differ greatly from one class of adhesive system to another, which can affect bond quality. Protocol simplification has been an inevitable trend that has boosted the use of self-etching and universal systems in adhesive dentistry. However, there is a lack of randomized clinical trials to prove the effectiveness of these systems. This chapter gives an overview of the most important issues in dental adhesion and adhesive systems, as well, discussing their composition and clinical use.

Keywords: dental bonding systems, enamel, dentin, hybrid layer, durability

1. Introduction

The advent of adhesion to hard tissues of the tooth was a turning point and featured a new era in the dental field. Historically, adhesive systems are classified in 'generations' according to their chronological appearance on the market. Nevertheless, this nomenclature lacks objectivity, as it does not categorize adhesives according to strict scientific-based criteria, leading to confusion and misinterpretation if clinicians assume a hypothetical best performance of the last generations. One of the classifications that best characterizes adhesive systems scientifically is that proposed by Van Meerbeek et al. [1], as it considers their effect on the smear layer and the number of steps involved in their application. Currently, etch-and-rinse, self-etch and universal adhesive systems are the three main strategies to promote adhesion of composite resins to dental substrates [2–4]. Phosphoric acid continues to be the elective approach to optimize enamel surface conditioning, whereas the best approach to bonding on the dentin surface is still a controversial issue, mainly because of structural heterogeneity and higher hydrophilicity of dentin [2].

The quality and stability of adhesive interfaces in enamel and/or dentin play a major role in the long-term clinical success of composite resin restorations. Three-step etch-and-rinse adhesive systems have been shown to perform well in both laboratory and clinical assays and are still seen as the gold standard among bonding systems [1, 5–8]. The trend in bonding in recent years has been towards simplification by combining the primer and bonding agent, the conditioner and primer or even all three components into a single solution. All contemporary adhesive systems contain various specific molecules in their formulation that perform identical functions and therefore share the same basic bonding mechanism. This mechanism relies on the replacement of inorganic tooth material by resin monomers that, upon *in situ* setting, become micromechanically interlocked in the microporosities created in the substrate [1–3]. However, the molecular chemistry, specificity, proportion of the incorporated ingredients and their interaction with different tooth structures vary greatly between adhesive systems, which affects the quality of the bond achieved with each system [5, 7, 9].

Self-etch adhesives were developed to overcome several shortcomings of the etch-and-rinse approach, in particular, the susceptibility to variations in the degree of dentin moisture and the differential between deep demineralization and resin infiltration. Major concerns have been expressed in the literature regarding interfacial ageing due to degradation of the bonded interface [10–13]. Only two-step self-etch adhesives, mainly those containing the 10-methacryloyloxydecyl dihydrogen phosphate (10-MDP) monomer in their composition, have evidenced dentin bonding effectiveness similar to the gold standard three-step etch-and-rinse adhesives. This has been attributed to the chemical interaction potential that contributes to a more durable and stable bonding [14, 15]. Despite that, the quality and durability of adhesion to enamel using self-etch adhesives is still a controversial issue, particularly for those presenting mild acidity [3]. The latest one-step all-in-one self-etch systems have shown an inconsistent and tendentiously weaker adhesive performance that depends more on their specific formulation. These systems are alleged to have several drawbacks, especially regarding the high hydrophilicity, which makes them act as permeable membranes, allowing water movement across the adhesive layer [3, 5, 7, 9, 13, 16]. In spite of the compromised *in vitro* and *in vivo* findings associated with the one-step self-etch adhesives, simpler and faster-use adhesive materials have been developed, taking over a large fraction of the adhesive dentistry market. Besides, they have gained popularity among clinicians, regardless of the scarce proof of clinical efficacy.

A new group or generation of dental adhesives has been more recently introduced and were classified as universal or multi-mode adhesives. They are more versatile, giving the professional the opportunity to decide which adhesive strategy to use: etch-and-rinse or self-etch, since they can be adapted to a specific clinical situation. They are essentially based on the 'all-in-one' concept of the already existing one-step self-etch adhesives and can be used as self-etch systems, in association with previous enamel and dentin phosphoric acid etching or combined with a selective enamel etching mode. The *in vitro* literature seems to suggest that bond strength is improved by the use of multi-mode adhesives with prior acid etching only for enamel; this effect was not evident for dentin with the use of mild universal adhesives [4]. More research is needed to study and compare those adhesives with those currently regarded as gold standards, as there is still very little clinical evidence available along with short follow-up periods.

The main challenge for current dental adhesives is to provide a bond that is equally effective for a range of dental substrates of different natures, such as enamel, sound, caries-affected or sclerotic dentin. From the clinical point of view, when bonding to both enamel and dentin, selective etching of enamel followed by the application of a mild 10-MDP-based two-step self-etch adhesive to both etched enamel and non-etched dentin would be the best option for an effective and more durable bond to the dental substrate.

2. Adhesion

The advent of adhesion to hard tissues of the tooth defined a turning point and featured a new era in dentistry. The substrates more frequently available for adhesion in restorative dentistry are enamel and dentin. These mineralized substrates present distinct ultra-morphology and composition. Regardless of depth or where it is located, enamel has a crystal, homogeneous structure mostly formed by hydroxyapatite (HAp) (96%wt/86%vol HAp; 3%wt/12%vol water; 1%wt/2%vol organic matrix) [17]. For this reason, enamel is considered to be a relatively static and predictable substrate regarding adhesive procedures. Conversely, dentin is a vital organic substrate, inherently hydrated and heterogeneous (70%wt/50%vol HAp; 12%wt/20%vol water; 18%wt/30%vol organic matrix—type I collagen) [18, 19], which makes adhesive procedures more complex when compared to enamel, depending mostly on its chemical composition, its level of moisture and regional morphological variation [20]. On the other hand, dentin suffers morphological and biomechanical changes related to physiological and/or pathological processes that involve alterations within its crystal and tubular structure, which cause changes to its permeability and its interfacial adhesion profile [21].

Adhesion reflects the capacity of a substance to adhere to another and may be the result of different processes, which can be mechanical, physical and/or chemical. Adhesive systems enhance interfacial strengths that aim at keeping organic solid substrates closely attached to the restorative materials such as composite resins [22]. The contact established between the substrate—enamel or dentin—and the adhesive depends on the level of the superficial wettability of the substrate by the adhesive, which is characterized by the contact angle formed when a drop of adhesive disperses on the substrate surface. The smaller the angle, the greater is the adhesive wettability. Thus, for the contact angle to come as close to zero degree as possible (which determines its spontaneous spreading), the surface tension of the liquid must be lower than the surface energy of the substrate, in this way ensuring a greater level of wettability between both components [23]. Compounds with high surface levels include very hard crystal structures with strong intermolecular strengths, as is the case with enamel. Conversely, the presence of an organic material, such as collagen, saliva components or organic resins provides less surface energy to their structure, similar to dentin [23, 24]. Hence, according to these concepts, it can be understood that adhesive wettability is potentially easier with enamel compared to dentin.

Regardless of how the different adhesive systems work, they all have similar compounds. However, their relative proportion varies according to the different adhesive system classes. Overall, their basic composition contains acids, several monomers, organic solvents, polymerization initiators and inhibitors and, sometimes, some inorganic content, each having a specific function. Knowing the chemical properties of these components is fundamental in understanding the adhesive procedure as well as being able to predict its behaviour [9, 25]. Monomers are the most important components of an adhesive system, which is why they are considered as the key constituents of the system. Basically, two distinct monomer types can be distinguished: functional monomers, which generally exhibit hydrophilic properties and, according to the functional specific group, may workout distinct actions, such as increase the wettability of the substrate (e.g. hydroxyl ethyl methacrylate; HEMA); initiate tissue demineralization (e.g. phenyl-P); determine anti-bacterial properties (e.g. MDPB) [9, 25] and the cross-linking monomers with a hydrophobic profile which form the polymeric matrix of the adhesive responsible for its biomechanical behaviour and for its interaction and co-polymerization with the restorative material [9].

In 1955, Buonocore [26] pioneered and boosted the research in the adhesion field when he reported the successful use of 85% phosphoric acid to improve the mechanical retention of an acrylic resin on enamel, being considered the father of modern adhesive restorative dentistry. Gwinnett [27] was the first to analyse the adhesive interface on an electron microscope and reported the ability of the resin to penetrate into the prismatic irregularities of the enamel after acid etching, describing a new structure resultant from the hybridization of enamel and resin. In 1982, Nakabayashi et al. [28] showed that resins could also infiltrate phosphoric acid-etched dentine, forming a superficial structural layer composed of a demineralized dentine organic matrix infiltrated with resinous monomers, which they called a hybrid layer, providing a new insight into dentin bonding [28, 29]. Later, Van Meerbeek [30] described this structure in detail using scanning electron microscopy (SEM) and transmission electron microscopy (TEM) and named it, the resin-dentin inter-diffusion zone.

The evolution of adhesive materials powered a change in the concept of cavity preparation based, nowadays, on a minimally invasive intervention coupled with maximal preservation of healthy dental structure [31]. Rotatory or manual instrumentation of the hard tissues produces organic and inorganic debris at their surface forming the so-called smear layer either in enamel (**Figures 1** and **2**) and dentin (**Figures 3** and **4**) [32–35]. On dentin, smear layer is a 1–2 μm adherent layer of debris composed of hydroxyapatite crushed aggregates, fragmented and denatured collagen, bacteria and their by-products [32, 34, 36]. Contiguously, the debris also occluded the orifices of the dentin tubules, identified as smear plugs, which may extend into the tubule to a depth of 1–10 μm [34]. Pashley [32] attributed two conflicting features to the smear layer. On one hand, the smear layer could act as a natural cavity protector by obliterating dentin tubules and reducing dentin permeability; on the other hand, it could present an obstacle to in-depth diffusion of monomers and hinder effective bonding of resins to substrates. Consequently, choosing between adhesive systems that remove or dissolve the smear layer is still one of the major discussion topics in adhesion.

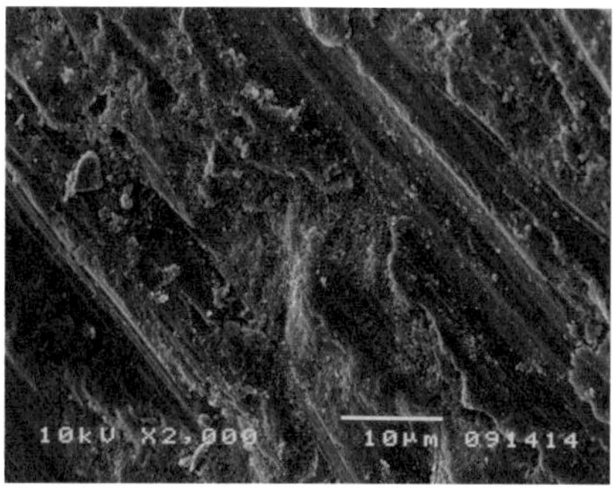

Figure 1. SEM image of ground enamel where the surface appears very rough, with grooves and covered with smear layer.

Figure 2. SEM image of fractured enamel showing the surface with a thick smear layer (sagittal view).

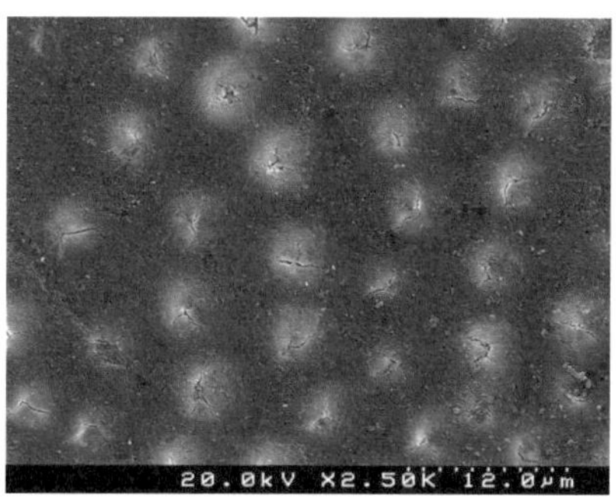

Figure 3. SEM image of prepared dentin where the surface is covered with smear layer and tubules are obliterated.

Figure 4. SEM image of fractured dentin showing the surface with a thick smear layer and smear plugs penetrating dentin tubules (sagittal view).

2.1. Adhesion strategies

Classification of adhesives is mandatory to provide an overview of the current field. The classification that best characterize adhesive systems was proposed by Van Meerbeek et al. [1] in 2003, as it considers the interaction mode with the substrate and the number of clinical application steps. Nowadays, two main adhesion strategies are identified: etch-and-rinse adhesive systems, which remove the smear layer and self-etching adhesive systems, which dissolve and incorporate the smear layer. A new group of materials called universal or multi-mode adhesive systems has recently been marketed. The main feature of these systems is that they can be applied according to different adhesion strategies (**Table 1**).

Category	Technique (generation)	Etch	Priming	Adhesive
Etch-and-rinse (ER)	Three steps (4th generation)	Phosphoric acid 32–40%	Primer	Adhesive resin
	Two steps (5th generation)	Phosphoric acid 32–40%	Primer + adhesive resin	
Self-etch (SE)	Two steps (6th generation)	Acidic monomers + primer		Adhesive resin
	One step (7th generation)	Acidic monomers + primer + adhesive resin		
Universal (U)	ER 2 steps	Phosphoric acid 32–40% (E+D)	Acidic monomers + primer + adhesive resin (E+D)	
	SE 1 step	Acidic monomers + primer + adhesive resin(E+D)		
	Selective etching enamel + SE dentin	Phosphoric acid 32–40% (E)	Acidic monomers + primer + adhesive resin(E+D)	

E: enamel; D: dentin.

Table 1. Classification of adhesive systems.

2.1.1. *Etch-and-rinse adhesive systems*

Etch-and-rinse adhesive systems can be applied in three or two steps, always using an initial acid-etching phase on both enamel and dentin. In the three-step systems, the acid etching and rinse step is followed by the intermediate application of a primer and ends with the application of a hydrophobic resin. For the simplified two-step etch-and-rinse approach, primer and resin are combined in a single bottle, which is why they are also known as one-bottle systems. Dental substrate etching is traditionally carried out with a gel of phosphoric acid at concentrations that vary between 30% and 40% (pH = 0.1–0.4), which is first applied to enamel and then to dentin, for a period of no longer than 15 seconds on dentin and no less than 15 seconds on enamel [33, 37].

Figure 5. SEM image of an enamel surface conditioned with 36% phosphoric acid. Different etching patterns can be observed either with dissolution of prism boundaries or dissolution of prism cores.

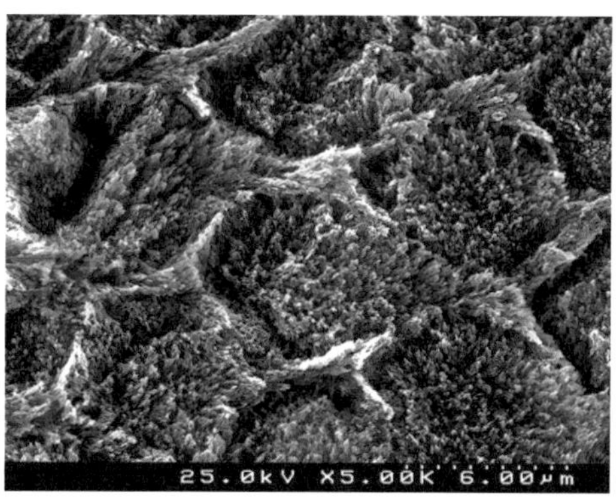

Figure 6. SEM image of an enamel surface conditioned with 36% phosphoric acid, showing hydroxyapatite crystallites forming a highly porous surface.

Phosphoric acid etching removes both the smear layer and approximately 10–50 μm of the enamel surface, creating a porous area with increased surface energy and wettability

(**Figures 5** and **6**) [35, 37–41]. This enables fluid resin dispersion into enamel microporosities by capillary attraction, forming macro and micro resin tags after polymerization [35, 38–40]. The macro-tags of resin fill the spaces around the enamel prisms, while the countless micro resin extensions cover each hydroxyapatite crystal individually. These micro-tags are probably the main contributors for enamel adhesion considering that they are numerous and provide a wide surface for adhesion (**Figures 7** and **8**) [1, 35, 38, 39]. The enamel structure and composition are relatively homogenous regardless of the depth and location, the reason why it is considered a static and predictable substrate regarding adhesive procedures [20].

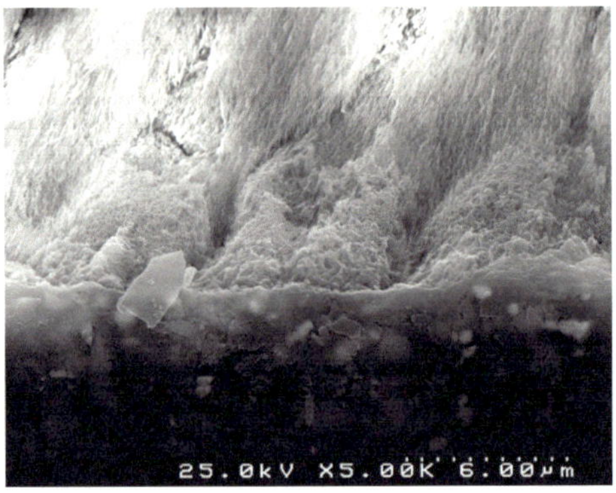

Figure 7. SEM image of an enamel interface formed with a three-step etch-and-rinse adhesive, showing intra-prismatic resin-penetration and intra-crystallite impregnation.

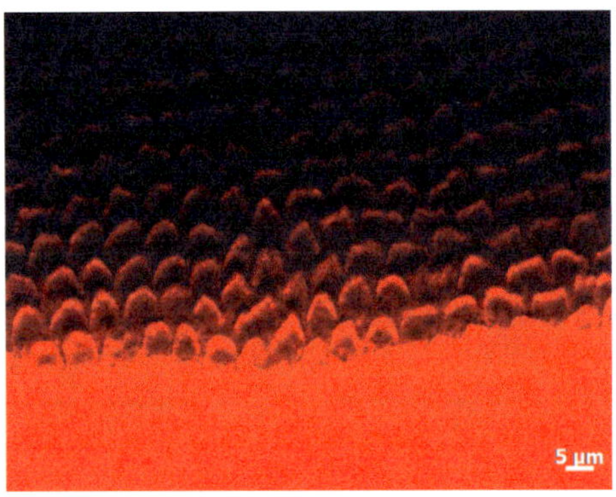

Figure 8. Confocal laser scanning microscope image of an enamel interface formed with a two-step etch-and-rinse adhesive, showing intra-prismatic resin penetration.

Etching dentin surfaces with phosphoric acid leads to a complete removal of the smear layer and demineralizes the inter-tubular dentin from 3 to 5 μm deep, leaving a tridimensional microporous mesh of collagen fibrils exposed, practically devoided of hydroxyapatite and randomly arranged. It also removes the smear plugs, dissolves the peritubular dentin and

http://dx.doi.org/10.5772/65605

opens the dentin tubules, which acquire a funnel shape, showing the collagen fibrils arranged in circles on the tubule walls. Intra-tubular demineralization can exceed the depth of the inter-tubular one reaching 6–7 μm in depth (**Figures 9** and **10**). Accordingly, as the inter-tubular distance decreases, deeper demineralization of dentin is expected [2, 24, 30, 33, 42, 43].

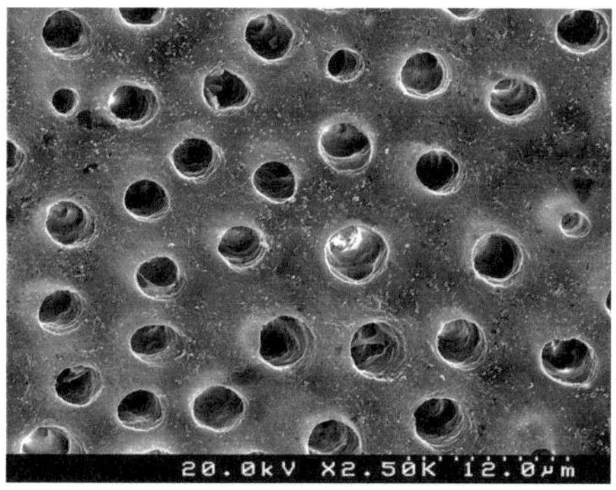

Figure 9. SEM image of a dentin surface conditioned with 36% phosphoric acid, showing complete smear layer removal and open dentin tubules.

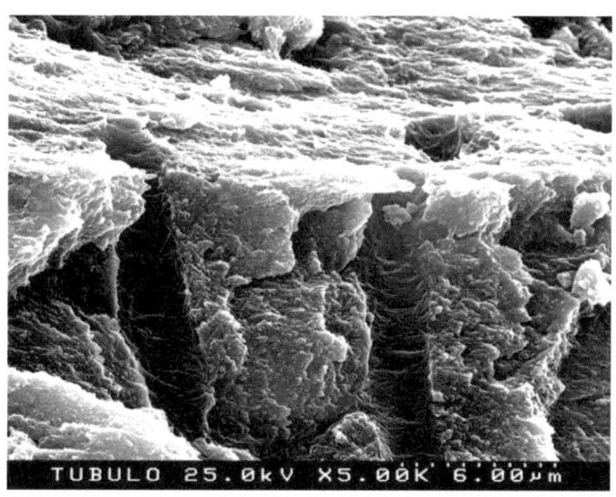

Figure 10. SEM image of a dentin surface conditioned with 36% phosphoric acid, showing a detailed view of demineralized open tubules.

After dentin etching, practically the whole mineral content (~50 vol%) of the surface and subsurface is solubilized, removed and replaced by water. The addition of this water to the volume of water that is originally a part of this dental tissue (~20 vol%) results in a total of ~70 vol% of water that is dispersed around the 30 vol% of the collagen fibrils that remain anchored to the subjacent mineralized dentin. Literally, this leaves a collagen scaffold suspended in water [2]. The exposed collagen acts like a mesh with nanometric-size porosities of about 10–40 nm, that should be able to mechanically retain the fluid resin upon *in situ* polymerization [2, 44–46]. In fact, demineralized dentin becomes a highly hydrophilic substrate of difficult clinical manipulation regarding adhesive procedures, as these materials are extremely sensitive to the presence of water [12, 33, 47–49].

For etch-and-rinse adhesive systems, one of the more delicate steps relates precisely to the acid etching, rinsing and the drying phase [45]. After the acid is rinsed off, the procedures to dry the tooth should allow a low level of humidity at enamel while keeping the dentin hydrated to optimize the adhesion process. However, this is a difficult clinical task to achieve. Dehydration of the demineralized dentin leads to the collapse of the exposed collagen mesh, reducing or nullifying the spaces between the fibrils and preventing them from acting as channels through which monomers may penetrate and spread within [24, 44, 46].

The main aim of the priming step is to alter the hydrophilic dentin surface into a hydrophobic phase, which will ensure an increase on the dentin surface energy making this substrate an adequate receptor of the subsequent hydrophobic bonding resin, allowing a better wettability and efficient penetration in the exposed collagen network [42]. The primers, adhesion-promoting agents, contain amphipathic monomers dissolved in organic solvents such as acetone, alcohol and/or water. These monomers have hydrophilic properties with affinity for the exposed collagen fibrils and hydrophobic properties to enable co-polymerization with the adhesive resin [9, 23, 25]. The hydroxyl ethyl methacrylate (HEMA) monomer is the most frequently incorporated in primers. Its low molecular weight together with its hydrophilic nature not only enhances dentin wettability, but also potentiates re-expansion of the collagen network, improving bonding capacity of adhesive systems [46]. Furthermore, it enables dispersion of hydrophobic monomers as it keeps them in solution [9, 25, 46, 50].

For etch-and-rinse adhesives, the wet bonding technique is recommended as a routine procedure essential for dentin bonding [51], particularly important when primers containing acetone as a solvent are used [45, 51–53], as acetone molecules are incapable of establishing hydrogen bonds and encourage collagen re-expansion [46, 51, 54]. This technique consists of leaving the dentin moist in the post-conditioning phase, preventing the collapse of the collagen network, which is crucial in ensuring a suitable infiltration of the adhesive monomers throughout the nano-sized voids in the organic mesh [44, 46]. For this, excess water in the cavity must be removed with a dry sponge or a short air blast [55]. However, perception of how moist the dentin must be kept to ensure complete removal of water afterwards by solvents is quite ambiguous [45, 56]. If dentin presents an over-wet condition, excess water determines phase separation of the hydrophobic and hydrophilic components, resulting in the formation of blister and resin globules at the resin-dentin interface. This potentially weakens the resin-dentin bonds and is particularly more concerning when simplified systems are applied [47, 52, 55, 57]. On the other hand, when the dentin surface is excessively desiccated, the collapsed and altered collagen structure determines an incomplete inter-tubular resin infiltration. Interfaces are characterized by infiltrated areas along the dentinal tubules and superficially, forming a very fine hybrid layer delimiting an incompletely infiltrated sub-superficial hybridoid region, rich in collagen and susceptible to premature breakdown [47, 52, 53, 55].

The solvents should be able to displace all the water trapped within the collagen network on dentin, allowing monomers to inter-diffuse completely along the surface demineralization depth [9, 25, 55]. In order to optimize this process, the primer must be in contact with dentin for at least 15 seconds before drying [58, 59]. In addition, the primer should be actively rubbed on the dentin surface with disposable brushes to improve monomer inter-diffusion and solvent evaporation, as an increase in mixture entropy and moieties kinetics is expected to occur [56, 60]. Afterwards, to complete primer application, a gentle air blast must be applied to promote evaporation of the remaining solvent. Nevertheless, the diffusion of monomers through the hydrated collagen decreases water vapour pressure, which may hinder its effective volatilization [9, 46]. Actually, the clinical time spent in this step seems insufficient to establish an effective solvent volatilization, even when a longer time period is allowed [61, 62]. Consequently, the residual presence of solvents associated with the centrifugal movement of dentinal fluid within the monomer hypertonic solutions may reduce the amount of monomers infiltrated in the collagen network. This may increase permeability of the adhesive layer and impair the conversion process of the monomers leading, ultimately, to the formation of more fragile and permeable polymers [2, 63, 64].

In three-step etch-and-rinse systems, an independent coat of a hydrophobic adhesive resin is applied to complete bonding procedure. In the simplified two-step systems, at least two layers of primer/bonding agent should be applied, whereby the first acts as a primer and the latter as bonding resin [2]. The diffusion of the adhesive resin throughout the inter-fibrillar collagen channels enables it to be micromechanically retained by hybridizing the inter-tubular dentin and tubule walls (**Figures 11** and **12**). This interpenetration enables a mixed structure to be formed, the hybrid layer, which is 4–6 m thick and where collagen fibres and hydroxyapatite crystals become surrounded by a resin matrix, which is the main bonding mechanism to dentin [1, 24, 30, 40, 65, 66]. Funnel-shaped resin tags result from tubule wall hybridization, but their importance in the bonding process is a matter of speculation [66]. In fact, there is a drop in bond strength values in deep dentin where the number and size of tubules is greater and inter-tubular dentin occupies a minor fraction of the total bonding area [65, 67].

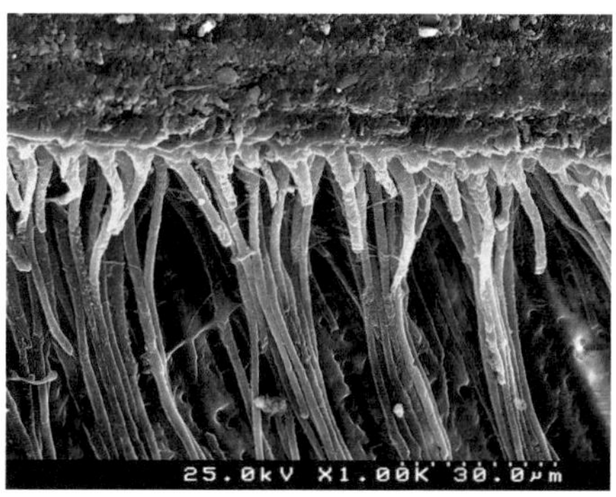

Figure 11. SEM image of a dentin interface formed with a two-step etch-and-rinse system, showing the hybrid layer along with a wide number of resin tags.

In theory, the hybrid layer and resin tags should be extended to the same depth to which dentin demineralization occurred as a result of the acid etching step. However, even considering clinical procedures that prevent acid over-etching or collapse of the collagen network, an incomplete in-depth infiltration of dentin is often observed [2, 12, 47–49, 68]. This may be related to a combined interaction of factors, such as the presence of a superficial amorphous phase at the demineralized dentin surface; the parallel arrangement of the collagen fibrils in relation to the surface and the presence of highly hydrated proteoglycans forming hydrogels in the spaces between fibrils [69]. These hydrogels act as filters for the molecules with a high molecular weight, such as dimethacrylates (Bis-GMA), allowing only smaller molecules such as HEMA to impregnate to a deeper level. In these conditions, the presence of linking monomers can be more restricted to the topmost part of the hybrid layer and, contrarily, its lower portion forms a more hydrophilic layer, susceptible to deterioration [2]. Indeed, the presence of a thin layer of water or hydrogel between the infiltrated bonding resin and the collagen fibrils, which are responsible for the formation of a hydrated network, was identified upon the use of the moist or water wet bonding technique (WWB). This hydrated network connects the resin tags to the base of the hybrid layer through the lateral ramifications of the dentin tubules and can spread throughout the whole thickness of the hybrid layer [2, 48]. Thus, the presence of hydrogels around the collagen fibrils may be responsible both for an infiltration differential and for phase separation of bonding components. This can lead to the presence of nano-sized porosities on bonding interfaces, impairing monomer conversion and mechanical properties development of the adhesive layer [63, 64] and reducing, consequently, its stability [2, 12, 70].

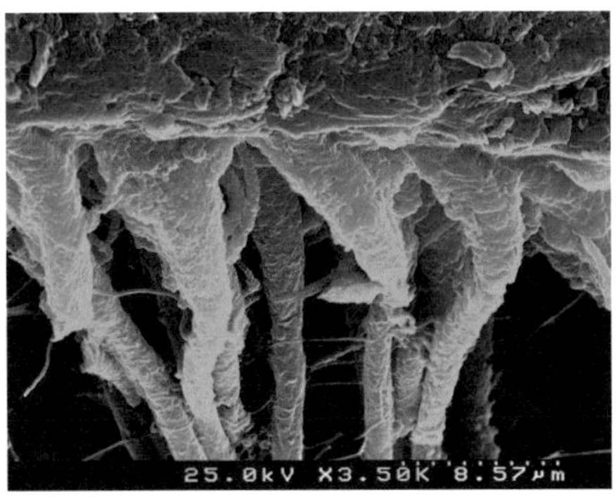

Figure 12. SEM image of a dentin interface formed with a two-step etch-and-rinse system showing the characteristic funnel-shaped resin tags with lateral anastomosis.

Deficient infiltration of the adhesive monomers leaves a strip of the collagen network in the hybrid layer incompletely impregnated, especially in the deeper areas, forming a microporous area characterized by the presence of sub-micrometric voids. This leads to the creation of hybrid layers that behave as semi-permeable membranes allowing water movements throughout the bonded interface even after adhesive polymerization. Those water pathways can be identified using the ammoniacal silver nitrate tracer under electron microscopy. The staining geometry observed determines different possible nanoleakage patterns running from a spotted to a reticular array that follow the discrepancy between dentin demineralization and adhesive impregnation along the resin-dentin interface [68]. Increased nanoleakage expression is more frequently linked to the use of two-step etch-and-rinse systems comparing to the three-step ones as they contain higher percentages of hydrophilic monomers and act as semi-permeable membranes after polymerization [2, 12, 47–49, 68]. Invariably, the progressive degradation of

http://dx.doi.org/10.5772/65605

the polymers and collagen increases the level of nanoleakage, inducing the presence of more fragile interfaces, which, in the long term, can compromise the durability of restorations [11, 12, 14, 15, 70–74].

The transactional effects of the different nanoleakage patterns result in the chemical degradation of the adhesive interfaces which follow a sequence of different phenomena, including water sorption of the polymers and the hydrolytic and enzymatic degradation of the unprotected collagen by matrix metalloproteinases (MMPs) and by cysteine cathepsins, particularly cathepsin K [3, 12, 70, 75]. Initially, the water sorption process leads to a swelling of the polymer and to the plasticization of the resin component that breaks the polymeric chains by splitting the ester linkages forming, in consequence, oligomers and monomers [76]. The ongoing deterioration of the polymer structure over time determines a change in its nuclear microstructure due to the formation of pores whereby degradation products are released, leading to a reduction of the polymer mass, which determines its solubility [72]. This inherent increase in porosity facilitates the movement of molecules, fluids, bacterial by-products and MMPs into and out of its structure, leading to a cumulative cycle of the water sorption rate and monomeric elution, which compromise polymer properties and the structural stability of the adhesive interface. Also, monomeric elution within the hybrid layer progressively exposes the collagen fibrils, which also become vulnerable to mechanical and chemical fatigue. Collagen depletion further debilitates the hybrid layer as porosities enlarge from a nanometric to a micrometric sized scale [12, 72, 75, 77–80]. Under occlusal function, resin composites may bend and allow compression of the water-filled voids in the degenerating hybrid layer, accelerating adhesive interface degeneration [81].

Researchers have focused attention on different molecules capable to inactivate or inhibit the action of the MMPs, which are responsible for collagen solubilization. Of these molecules, chlorhexidine has currently been the most studied [3, 75], with *in vivo* reports showing its ability in reducing the effects of hybrid layer degradation when a 2% chlorhexidine solution is applied for 60 seconds on the phosphoric acid etched dentin, before using an etch-and-rinse type bonding system [78, 79, 82]. It is also noteworthy that this procedure does not interfere with the immediate bond strengths of etch-and-rinse systems to the dentin substrate [71, 79, 82]. Consequently, along with its important antimicrobial activity, the application of chlorhexidine on etched dentin slows down the degradation of the exposed collagen fibrils, conditioning a more stable bonding interface in the long term.

Recently, the ethanol-wet-bonding (EWB) technique has been described as an alternative method to the water-wet-bonding technique, previously referenced. EWB relies on moistening demineralized dentin with ethanol and promote immediate hydrophobic monomer diffusion into dentin. The key question that nurtured the development of the EWB technique was whether it was possible to keep the demineralized dentin matrix saturated in ethanol without the occurrence of a simultaneous collapse of the collagen fibril while enabling hydrophobic monomer infiltration. This possibility was undertaken due to the convergence of the solubility parameters found between the substrate (ethanol and collagen) and the monomers solutions, determining their miscibility. For that, no phase separation is prone to occur [46, 83]. It was also verified that the EWB technique would cause a contraction of the collagen fibrils' diameter greater than the contraction of the volume of the collagen matrix resulting in a concomitant increase of the inter-fibrillar spaces and, consequently, an increased efficiency of monomeric infiltration [46, 84]. Indeed, some studies showed that the EWB technique applied to dentin optimizes the monomeric infiltration rate, improves the collagen matrix sealing, reduces permeability of the polymerized bonding layer, and minimizes water sorption by polymers forming more stable and less permeable bonding interfaces over time [2, 48, 84–86]. Despite this, the EWB technique cannot be optimized in a useful clinical time, as increasing concentrations of ethanol applied for long periods are required to ensure adequate dehydration of dentin [46, 83, 85, 86]. Because of this, a few simplified protocols were presented by using a two-concentration sequence or only one single solution of absolute ethanol to accomplish the process [83–85]. Nonetheless, no clinical study has confirmed the superiority of the EWB technique compared to the WWB technique [87, 88].

2.1.2. Self-etching adhesive systems

In early 1990s, development of self-etching adhesives containing acidic primers as promoting monomers for direct bonding to the smear layer covered enamel and dentin introduced a new dental bonding technology [3]. These systems do not require a separate etching step as acidic monomers simultaneously etch and prime the dental substrate. Thus, all by-products arising from the demineralization process and from the dissolved smear layer become incorporated in the bonding interface [3, 45]. Self-etch adhesives simplified the bonding protocol as they require less operative steps and exhibit reduced technique-sensitivity, as the level of substrate moistness is no longer a critical issue of the bonding procedure [47, 89]. Also, as they do not fully remove the smear layer and the smear plugs, immediate surface rehydration secondary to dentinal fluid movement is prevented and reduced post-operative sensitivity is likely to occur [90, 91]. Another significant advantage pointed out to these systems relies in the fact that demineralization and infiltration phenomena take place simultaneously and to the same depth, which would ensure, theoretically, complete penetration of the adhesive leaving no sub-infiltrated demineralized areas [3]. However, interfacial nanoleakage identifying micro-porous areas within the hybrid layer have also been verified, rejecting the supposition of adhesive resin complete infiltration [10, 49, 50, 57, 89, 90, 92].

The composition of the self-etch adhesive systems is unique, as they contain high concentrations of water. They are basically formed by aqueous mixtures of acidic functional monomers (the so-called self-etching primers), generally phosphoric acid- or carboxylic acid-esters with a higher pH than that of phosphoric acid used in the etch-and-rinse bonding systems. Water acts as an ionizing medium for dissociation of acidic protons and dissolution of calcium ions, promoting the etching process, while preventing, simultaneously, the collapse of demineralized collagen fibrils [90, 93]. Other hydrophilic functional monomers and hydrophobic cross-linking monomers are also present in their formulation [3, 9, 25]. Co-solvents, such as ethanol or acetone are frequently added to form an azeotropic mixture with water in order to promote solvation of hydrophobic moieties and to ensure proper thinning, spreading and dentin wettability [25, 93, 94].

http://dx.doi.org/10.5772/65605

HEMA is a frequently added monomer as it is a low molecular weight water-soluble monomer that improves the miscibility and solubility of the polar and non-polar adhesive components. Besides, HEMA can also stabilize the collagen fibril network, improving the dentinal permeability, wettability and monomer diffusion [25, 50]. However, the presence of HEMA further increases the hydrophilic nature of self-etch adhesives, which encourage water uptake. Some studies have concluded that HEMA-rich systems attract water from the underlying dentin thanks to an osmotic time-dependent process, inducing the presence of water droplets on the surface of the adhesive layer that behaves as a semi-permeable membrane [47, 57, 92]. Those droplets can reduce the monomer conversion rate, weakening the mechanical strength of the adhesive and potentially jeopardizing bond durability [63, 64]. Thus, for HEMA-rich adhesives immediate light curing of the lining composite should be performed as soon as possible before water droplets arise at the adhesive surface. It was shown that delayed curing influenced bond strength adversely [57, 92]. HEMA-free self-etching adhesives were also developed. However, another problem has been detected related to a phase separation process. Once ethanol/acetone starts to evaporate, the solvent-monomer balance is broken, with water separating from the other adhesive components. This is particularly more critical in acetone-based adhesives as acetone volatilizes faster and severe changes in the water/acetone ratio are likely to occur [25, 50, 90, 95]. This phenomenon was easily identified using light microscopy by dispensing a drop of adhesive onto a glass plate. As the co-solvent starts to evaporate, many small water droplets start gradually to emerge in the adhesive layer. Spontaneous disappearance can happen, but takes several minutes and entrapment of the droplets upon light curing becomes more likely. This can compromise bond strength and increase interfacial degradation susceptibility [45, 50, 89, 96]. Thus, a very strong air-drying step is highly recommended for this kind of adhesives in order to blow the droplets out, leaving only a thin layer of bonding resin [45, 50], although this prerequisite is a less ambiguous instruction than 'gentle air-drying'. This procedure seems effective on flat surfaces, but eliminating these droplets in the more complex cavity geometries may prove more difficult [3, 45, 50].

Self-etch bonding systems are classified as two-step and one-step adhesives. The two-step self-etching systems are based on the separate application of a hydrophilic self-etching primer followed by a hydrophobic bonding resin. One-step self-etching systems can be further subdivided into one- or two-component adhesives. In two-component systems, functional monomers are separate from water, increasing adhesive shelf life and hydrolytic stability, but additional and adequate mixing of both components is mandatory before clinical application. Single-component systems, also known as all-in-one adhesives combine all ingredients related to the acidic, priming and bonding functions into a single bottle [3, 45, 96]. This last group of one-step adhesives exhibits a complex mixture of hydrophilic and hydrophobic components that accounts for major shortcomings of the adhesion protocol [45, 89, 96].

The great variability reported to the performance of different self-etching adhesive systems can be attributed to several characteristics related to their composition and, particularly, to some properties of the functional monomers with respect to its acidity, hydrolytic stability and chemical interaction ability [5, 7, 9]. The composition and concentration of acidic resin monomers establish differences in the acidity and aggressiveness of self-etching systems and determine their effect on smear layer dissolution and demineralization ability on unground and ground enamel and dentin, determining distinct bonding interfacial ultra-morphology [3, 40, 41, 45, 47, 97–99]. Aggressiveness depends on the pH of the solution and self-etch systems can be categorized according to their pH into ultra-mild (pH > 2.5), mild (pH ≈ 2), intermediately strong (pH between 1 and 2) and strong (pH ≤ 1) (**Figures 13** and **14**) [3]. As acidic monomers diffuse in depth they are gradually buffered by the mineral content of dental substrates, loosing progressively their etching ability [10, 94, 100].

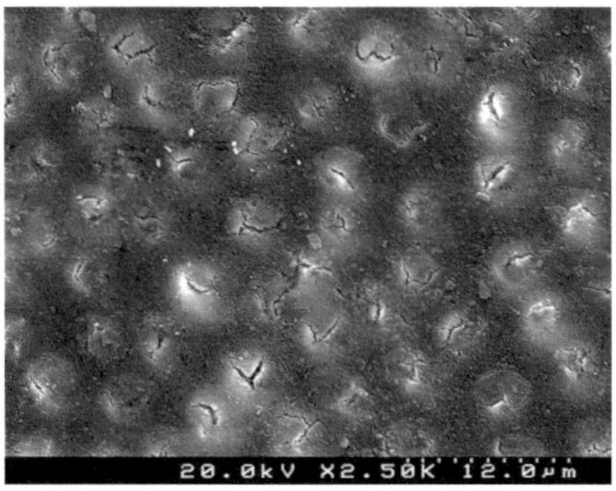

Figure 13. SEM image of a dentin surface conditioned with an ultra-mild two-step self-etch adhesive, showing a lightly dissolved smear layer and obliterated tubules.

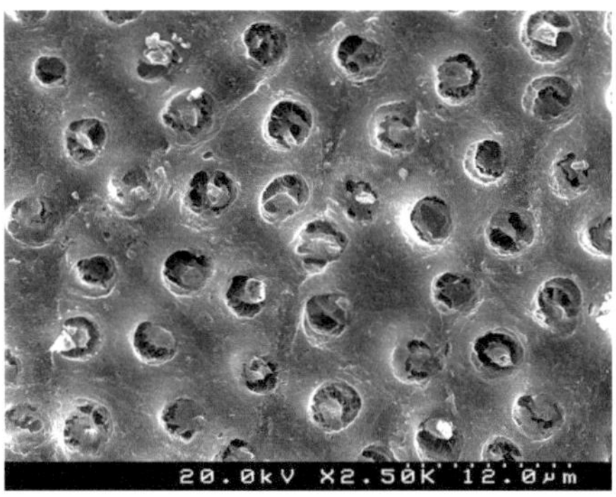

Figure 14. SEM image of a dentin surface conditioned with an intermediately strong one-step self-etch adhesive, showing incomplete smear layer dissolution and partially open dentin tubules.

Like etch-and-rinse systems, all self-etch adhesives warrants the three main bonding stages: etching, priming and bonding, ensuring micromechanical interlocking of resin monomers upon polymerization and consequent tissue hybridization [1, 3, 45]. Furthermore, some functional monomers evidence an additional chemical interaction potentiality with calcium hydroxyapatite, which seems to improve bond durability [3, 101–103]. In fact, functional acidic monomers as 10-MDP, phenyl-P or 4-MET are able to establish ionic bonds between their functional groups (phosphate, phosphonic or carboxylic group) and calcium of hydroxyapatite, but with different behaviours in accordance to the adhesion-decalcification concept [101].

According to this model, all acids are able to establish a chemical bond with hydroxyapatite calcium, but whether the produced molecule will remain bonded or will debond depends on the stability of the formed calcium salt, that is inversely related to the monomer acidity. Although phenyl-P, 4-META and phosphoric acid molecules strongly interact with calcium, the instability of the calcium phosphate/carboxylate compound formed implies ready de-bonding, resulting in a moderate to severe decalcification effect. On the contrary, 10-MDP forms a stable calcium-phosphate salt along with a shallow demineralization of the surface [101, 103].

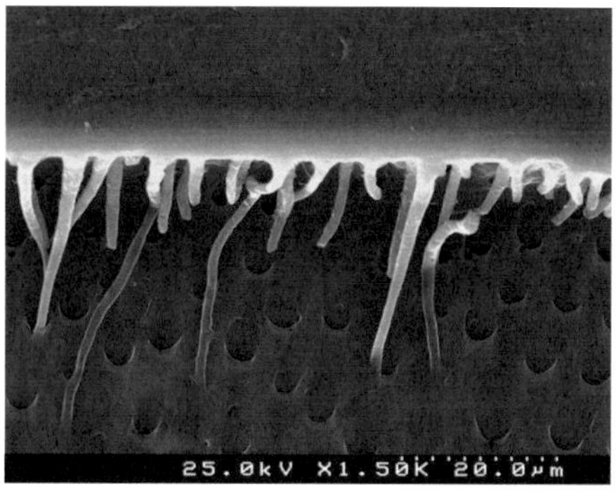

Figure 15. SEM image of a dentin interface formed with an ultra-mild two-step self-etch adhesive, showing the hybrid layer along with short cylindrical resin tags.

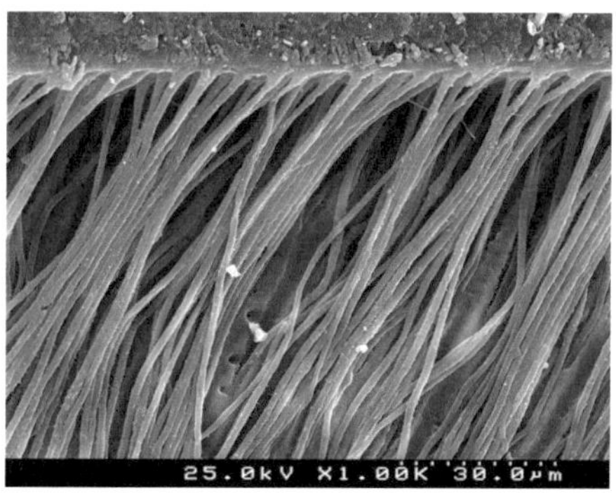

Figure 16. SEM image of a dentin interface formed with a strong one-step self-etch adhesive, showing a thick hybrid layer along with enlarged resin tags.

Ultra-morphological interfacial characteristics produced by self-etching systems depend both on their acidic aggressiveness and on their formulation. In general, strong self-etching systems with a more acidic pH, equal to or lower than 1, are associated with a high demineralization depth, determining an etching pattern typically approaching that of phosphoric acid. At the dentine level, a collagen mesh practically devoided of hydroxyapatite can be found. After micromechanical adhesive interlocking, a 2–5 μm thick hybrid layer is produced along with resin tags, whose ultra-morphological features are very similar to those provided by etch-and-rinse systems. The demineralization action of intermediately strong self-etching systems, with a pH ranging from 1 to 2, allows the identification of two zones; a superficial one with complete dentine demineralization and a partially demineralized basal zone. The latter retains some mineral or hydroxyapatite content with the ability to form chemical bonds with functional monomers. Mild and ultra-mild self-etching adhesive systems, with a pH higher than 2, were shown to demineralize dentin only partially and superficially, leaving part of the smear layer and smear plugs intact and a substantial amount of hydroxyapatite crystals around collagen fibrils, keeping them encapsulated and protected. Still, their 0.2–1 μm demineralization depth creates a microporous dentin surface capable of micromechanically retaining adhesives through a hybridization process. However, the resulting hybrid zones show much lower thicknesses than those obtained by strong and intermediately strong self-etching systems or etch-and-rinse systems (**Figures 15** and **16**) [1, 3, 40, 45, 47, 65, 97–99]. On the other hand, hydroxyapatite within the hybrid layer act as a receptor for the additional chemical adhesion process [3, 101, 103].

Figure 17. SEM image of an enamel surface conditioned with a mild two-step self-etch adhesive, showing a shallow inter-prismatic etching effect.

One of the greatest challenges of using self-etching systems relates to their effectiveness on simultaneous adhesion to enamel and dentin [3]. Generally, more acidic self-etching systems have the ability to completely dissolve the smear layer and their depth of demineralization into the enamel is similar to that of phosphoric acid. Instead, as the acidic degree of the systems decreases, morphological features on enamel surfaces become shallower providing enamel etching patterns less homogenous, especially with mild and ultra-mild systems (**Figures 17** and **18**) [38–41, 104–106]. Additionally, enamel-resin interfaces of less acidic systems have also proven to form thinner and relatively non-uniform hybrid layers with scarce and poorly defined inter- or intra-prismatic resin extensions, when compared to more acidic systems (**Figures 19** and **20**) [35, 38, 39, 41, 106]. Since enamel contains a high mineral fraction and larger hydroxyapatite crystals than dentin, and because the demineralization potential of self-etching systems is more limited than that provided by phosphoric acid, it is possible that more hydrogen ions released from the adhesive are neutralized by the enamel smear, limiting their interaction with the subsurface enamel [35, 94, 100]. Different studies reported the unpredict-

http://dx.doi.org/10.5772/65605

ability of the performance of self-etching adhesives in intact enamel, presenting evidence that previous mechanical preparation by grounding enamel can potentiate the adhesion capacity of these systems [38, 98, 107–110]. Also, the use of a selective enamel etching approach with phosphoric acid in intact or ground enamel seem to increase the enamel adhesion potential of these systems [111–114].

Figure 18. SEM image of an enamel surface conditioned with an intermediately strong one-step self-etch adhesive, showing a moderate inter- and intra-prismatic etching effect.

Figure 19. SEM image of an enamel interface formed with an ultra-mild two-step self-etch adhesive, showing a very superficial resin-impregnation within demineralized enamel along with a thick layer of adhesive.

In order to increase the acidity of adhesive solutions, the contents of acidic monomers and water is increased, which make stronger self-etch systems with increased hydrophilicity, leading to higher water sorption, and consequently, decreased hydrolytic stability [96]. Several studies have shown that despite more acidic self-etching systems exhibit a good bond ability to enamel, they invariably behave much worse at the dentine level, in spite of forming thick hybrid layers. Notwithstanding, a similar performance was reported for many less acidic systems, especially single-step systems. Although it has been claimed that discrepancies between the depths of demineralization and resin infiltration should not occur with self-etching systems, silver nitrate staining has been reported along resin/dentin interfaces produced by these systems, thereby challenging the original concept that no pathways for water movement would be present. More relevant in all-in-one systems, the resin/dentin interface function as a permeable membrane that allow water movement across the adhesive layer even after polymerization. Nanoleakage expression occurs within voids left in un-infiltrated areas of the hybrid layer, but also within the adhesive layers as water trees and water droplets, which have been linked with incomplete water and/or solvent removal and/or water sorption from the underlying hydrated dentin. Over time, bond strength and stability is invariably compromised [10–12, 14, 15, 47, 49, 50, 57, 72–74, 89, 90, 92, 98]. For many of these systems, nanoleakage expression have also been identified at the base of the hybrid layer, suggesting that mineralized dentine underlying the hybrid layer may undergo structural changes secondary to the adhesive procedure, becoming porous enough to retain water. This may be due to the progressive buffering of acidic monomers by the substrate, which gradually loses their etching ability. This may result in an increasingly irregular and sparse demineralization in depth, creating smaller open inter-fibrillar spaces at this level. These spaces may contain dissolution by-products, such as hydrolytically unstable calcium salts [100–102] that act as molecular sieves, preventing infiltration of higher molecular weight resin monomers and making deep areas more hydrophilic [10, 89]. Furthermore, accumulation of unpolymerized acidic monomers in this area due to the presence of water (either from the dentin substrate or from the adhesive system itself) may result in their hydrolysis, forming acidic reaction by-products, which may play an important role in subsequent dentin demineralization once the hybrid layer is formed [10, 92, 115].

Figure 20. SEM image of an enamel interface formed with an intermediately strong one-step self-etch adhesive, showing a moderate inter- and intra- prismatic resin-impregnation within demineralized enamel.

Similar to etch-and-rinse adhesive systems, for self-etching systems, the literature also favours systems applying a hydrophobic adhesive resin as an individualized step, that is, two-step self-etching systems. However, unlike three-step etch-and-rinse systems, which invariably report good clinical and laboratory outcomes, making them the gold-standard in adhesion, two-step etch systems report varying degrees of efficacy, mainly due to their different level of acidity and hydrophilicity. Still, applying a hydrophobic adhesive resin in an independent step is usually associated with higher bond strengths and hydrolytically more stable interfaces than simpler systems containing a mixture of both hydrophobic and hydrophilic components [9, 11, 49]. In fact, one-step self-etching adhesive systems create thinner adhesive films, which is why their application in two or more layers or employing an additional single-step application of a hydrophobic resin was shown to improve their performance regarding bond strength and interface stability [80, 116, 117]. The application mode of self-etching systems on dental substrates can also influence its performance. Some studies have shown that active application of the acidic self-etching system primers can improve adhesion to dentin and enamel by improving depth of demineralization, infiltration and chemical interaction as it promotes a more effective contact of monomers at the surface [74, 106, 118].

Although adhesive classification by categories regarding their clinical application is convenient and internationally accepted, adhesive behaviour depends much more on its composition and functional monomer specificity than on the category to which they belong [3, 5, 7, 9]. In fact, dentin bonding optimization with self-etching adhesive systems is associated with the presence of the 10-MDP monomer. This molecule has become the most promising acidic functional monomer for bonding, as it establishes readily and intensively an ionic bond with hydroxyapatite calcium and form a regular nano-layered structure on the dentin surface, within which calcium salts are deposited. The potentiality to chemically interact with interfacial hydroxyapatite is preponderant to achieve bond durability. The 10-MDP monomer was originally synthesized and patented by Kuraray (Osaka, Japan). The good *in vitro* and clinical outcome of the two-step self-etching systems containing 10-MDP, namely the Clearfil SE Bond and Clearfil Protect Bond, both developed by Kuraray may be partly attributed to the intense chemical adhesion established with tooth tissue [7, 15, 40, 89, 119–124]. Also, an acid-base resistant zone (ABRZ) formed beneath the hybrid layer after acid-base challenge was reported. Densely packed crystallites suggest that a hypothetical chemical reaction of hydroxyapatite and 10-MDP may take place in this area, enhancing resistance against demineralization from acid attack, which might play an important role in the prevention of secondary caries [125, 126].

Kuraray no longer holds the patent for 10-MDP, and many other manufacturers started to incorporate it into new adhesive formulations, including in many of the currently so-called universal or multi-mode systems [127–132].

2.1.3. Universal adhesive systems

It is clear that the trend in adhesive technology is to simplify bonding procedures by reducing application steps, shortening clinical application time and improving standardization [129]. Aiming at providing a single product for all situations, universal or multi-mode adhesive systems have been recently introduced and can be applied either with the etch-and-rinse or the self-etch mode. This multi-approach capability enables the clinician to apply an adhesive with the so-called selective enamel etching technique, which combines the advantages of an etch-and-rinse technique on enamel with a simplified self-etch approach on dentine [4, 132, 133].

The majority of universal adhesives are designed under the 'all-in-one' concept of the existing one-step self-etch adhesives, thus requiring water in their formulation to ionize hydrophilic acidic monomers [96]. Concerning etching capability, most universal adhesives present a 'mild' or 'extra-mild' profile [134]. It has been shown that adhesives within this pH range can be effective in bonding dentin. However, there is high concern about their behaviour when bonding to enamel [4]. Most often, phosphate esters are their primary functional monomer. Due to the high reported performance of the 10-MDP monomer, this specific molecule has been incorporated in a wide range of adhesives in this class. The 10-MDP molecule is an amphiphilic functional monomer with a long carbon chain backbone, which renders it fairly hydrophobic [9]. Additionally, stable MDP-calcium salts deposited in self-assembled nanolayers at dentin surfaces help to explain its high bond stability [103].

At present, there is only sparse literature reporting on the clinical efficacy of universal adhesives with short-term follow-ups. *In vitro* studies indicate that these adhesives show similar adhesive performance in dentin, regardless of their application mode: etch-and-rinse or self-etch. However, when the self-etch mode is performed a more superficial interaction might reduce the risk of post-operative sensitivity and delay the inevitable degradation phenomenon, as chemical interaction is likely to occur [4, 129, 135]. Concerning the enamel substrate, the literature suggests that selective etching prior to the application of a mild universal adhesive is an advisable strategy to optimize bonding [4, 132, 135–137]. Nevertheless, the clinical use of phosphoric acid only on enamel margins without accidentally reaching dentin is challenging, especially if low-viscosity etchants are used.

After all, the increased versatility of universal adhesives has not followed the technological advances, which could help to overcome the flaws associated with previous generations of adhesives. Currently, adhesives with effective bioactive behaviours are still lacking [138]. In fact, a larger quantity of solvents and hydrophilic monomers in those adhesive formulations leads to a greater number of residual solvents to be entrapped in the adhesive layer, which may reduce dentin bond strengths, and increase the permeability of the adhesive layer after polymerization, reducing stability over time [129, 130]. Therefore, likewise one-step self-etch adhesives, several modified application techniques to overcome these problems are being evaluated, such as extending solvent evaporation time, applying adhesives actively with extended times on enamel, and placing an additional hydrophobic resin coating to optimize bonding [137, 139–142].

2.2. Research in adhesion

The limited durability of restorations mainly results from exposure of adhesive interfaces to the oral environment. These interfaces are permanently subjected to mechanical, chemical and/or thermal stimuli that can compromise its longevity in the short, medium or long term. The action of occlusal forces on adhesive interfaces as a result of a cumulative effect of masticatory cycles may induce tensions exceeding interfacial fracture toughness. This will determine a site for failure initiation, which can in turn, act as a local point of increased stress concentration, contributing to the progression and spread of the initial failure [13]. Temperature oscillations in the oral cavity may also induce repetitive contraction and expansion stresses at the tooth/resin interface, caused by differences between their thermal expansion coefficients. Such stresses may exacerbate the occurrence of interfacial cracks [143]. Additionally, biochemical hydrolytic and enzymatic breakdown can also potentiate interface degradation mechanisms, in which both polymeric components and of collagen are enrolled. Finally, stresses developed during polymerization shrinkage of composite resins may also influence in adhesive interface failure [12, 13, 70].

The popularity of laboratory studies on adhesion is mainly associated with the speed with which new materials are emerging on the market, without a proper clinical evaluation of precursor systems, particularly in medium and long-term studies [6, 144]. Although *in vitro* studies cannot anticipate or replace results from clinical studies, they are an important step to compare and rank the adhesion potentiality. On the other hand, they are also useful to analyse distinct application variables that could hardly be tested in clinical studies due to both methodological details and ethics [7, 13, 145–148].

Currently, bond strength determination, micro-leakage assessment, quantitative marginal sealing evaluation and adhesive interfacial qualitative analysis are the most relevant experimental means to assess the performance of adhesive systems, particularly when associated with sample ageing methods [7, 145]. *In vitro* adhesion durability can be tested using different artificial ageing modes, namely water storage, thermal fatigue, mechanical fatigue, thermomechanical fatigue, enzymatic/chemical degradation among others [7]. However, most of these methods only mimics one of the factors involved in *in vivo* interfacial degradation, where all usually operate simultaneously [70]. Bond strength evaluation is a relatively simple method that allows testing different materials and/or techniques in a process where adhesive interfaces respond to an acute load. Dynamic experimental tests can potentially provide more relevant and realistic information to predict an adhesive behaviour, particularly when combined with complementary ageing procedures. Nevertheless, these methods demand considerably more technological investment and are time-consuming [7, 13, 16, 148–151].

Currently, one of the major issues discussed among researchers relates to the validation of *in vitro* tests results in order to determine whether they can correlate positively with clinical performance of adhesive restorations, i.e. to what extent is it possible to make clinical inference from the results of *in vitro* studies? In fact, several studies have evidenced that clinical results could be, in a very limited way, predicted by laboratory results [7, 145, 147, 148].

2.3. Adhesive protocols for clinical application: key steps

2.3.1. Etch-and-rinse systems

The three-step etch-and-rinse adhesive systems are considered as the gold standard for bonding to enamel and dentine because of their excellent performance in both *in vitro* and *in vivo* studies [1, 5–8].

Enamel etching by means of phosphoric acid results in a pattern of regular, deep demineralization with the overall production of interfaces with intensive inter- and intra-prismatic adhesive penetration along high and durable enamel bond strengths, which accounts for long-lasting clinical enamel marginal adaptation. Therefore, enamel conditioning with phosphoric acid is still considered as the gold standard procedure to optimize adhesion to this substrate [1, 3, 105, 120, 149, 150, 152].

However, resin-dentin interfaces proved to be much more unstable when etch-and-rinse adhesives are used. Dentin hybrid layer undergo degradation with time, more important for the two-step systems, even when ensuring an adequate enamel seal [11, 15, 71, 73, 78–80].

KEY STEPS:

Do not etch dentin for more than 15 seconds [33, 37, 44].

Why? Longer exposure of phosphoric acid on dentin can cause collagen denaturation and increase the demineralization depth, which can negatively affect the subsequent dentin hybridization process.

Leave the dentin surface moist before applying primer (three-step) or adhesive (two-step), particularly when acetone-based systems are used [46, 52–54, 153].

Why? Desiccation of the demineralized dentin leads to the collapse of the exposed collagen mesh, reducing or nullifying the spaces between the fibrils due to spontaneous establishment of inter-peptide hydrogen bonds. This reduces dentin permeability and monomer infiltration is hindered.

In case of excessive dentine desiccation, surface rehydration is mandatory and should be performed with water [46, 54].

Why? Water is a strong polar protic solvent and it has been shown to be the most efficient solution in the re-expansion of collapsed collagen.

Actively apply the primer/adhesive solution for at least 15 seconds to the demineralized dentin [56, 58–60].

Why? This is the time needed to allow solvents to act as water-chasers. Active application improves monomer diffusion and solvent evaporation before the light-curing step.

Promote extended solvent evaporation time [56, 58–60, 154–156].

Why? The residual presence of solvents reduces the amount of infiltrated monomers in the collagen network and impairs monomer conversion.

For two-step etch-and-rinse adhesives, place two or more coats of the primer/bonding solution and polymerize each separately [2, 61, 62, 64].

Why? The second layer act as a bonding resin and increases the thickness of the adhesive layer. This produces an area of greater flexibility and elasticity that can absorb some of the shrinkage stresses raised during composite resin polymerization.

2.3.2. *Self-etch systems*

Two-step self-etch adhesive systems, particularly those containing 10-MDP in their composition, are considered as the gold standard when it comes to bonding dentin, because of their excellent performance in both *in vitro* and *in vivo* studies [7, 15, 40, 89, 119–124]

However, the behaviour of self-etching systems largely depends on their composition, aggressiveness and hydrophilicity. Hence, even within the same category there is a wide

http://dx.doi.org/10.5772/65605

variability of the reported results regarding bond strength and adhesion durability [1, 3, 5, 7–9].

KEY STEPS:

Etch enamel separately with phosphoric acid, particularly for mild or extra-mild self-etch systems [1, 5–8].

Why? As the acidic degree of self-etch systems decreases, morphological features on enamel surfaces become shallower providing enamel etching patterns less homogenous and producing non-uniform hybrid layers with scarce and poorly defined inter- or intra-prismatic resin extensions. Phosphoric acid increases the efficacy of self-etching systems in bonding to the enamel, which is particularly more important in unground or intact enamel.

Apply self-etch adhesives by actively scrubbing them vigorously onto the surface [38–41, 104, 105].

Why? Active rubbing improves bonding performance to enamel and dentin as it promotes solvent evaporation, depth of demineralization, infiltration and chemical interaction due to a more effective contact of monomers with the surface.

For one-step self-etch adhesives, apply more than one adhesive coat and/or place a separate hydrophobic coating resin [74, 106, 118].

Why? Applying a greater number of layers and/or the extra placement of a hydrophobic resin in a single step has been found to improve the laboratory performance of these systems regarding bond strength and interface durability.

For HEMA-rich systems, promote accelerated composite resin build-up and polymerization [80, 116, 117].

Why? HEMA-rich systems exhibit enhanced capacity for water absorption from the underlying dentine. This is due to a time-dependent osmotic process that induces the presence of water droplets on the surface of the adhesive layer. Clinically, polymerization of the first composite resin increments must take place in the shortest time possible.

For HEMA-free systems, promote strong air-drying of the adhesive surface [47, 57, 92].

Why? Systems without HEMA undergo phase separation after solvent evaporation (ethanol/acetone) and droplets of water gradually rise to the surface and can be retained if adhesive polymerization starts immediately. Clinically, a strong air-drying step is highly recommended in order to blow the droplets out, leaving only a thin layer of bonding resin.

2.3.3. Universal systems

In essence, universal adhesives are single bottled, no mix all-in-one systems with a 'mild' or 'extra-mild' etching profile provided by acidic functional monomers. They present broader applications and can be used under different bonding strategies, adapting the most appropriate to each specific clinical situation. However, increased versatility of universal adhesives has not followed the technological advances, which could help to overcome the flaws associated with previous generations of adhesives [25, 45, 50, 89, 90, 95, 96].

KEY STEPS:

Etch enamel selectively with phosphoric acid [129, 130, 138].

Why? The low acidity of most of these systems leads to a very superficial enamel demineralization pattern and phosphoric acid increases the efficacy in bonding to this substrate.

Use the self-etch mode to optimize dentin bonding [4, 132, 135–137].

Why? Dentin etching with phosphoric acid demineralizes it extensively and may restrict the chemical bonding potential of hydroxyapatite calcium and the phosphate-based monomers.

Increase solvent evaporation time [4, 129, 135].

Why? Extended evaporation time reduces residual water and/or solvent and may improve the performance of universal adhesives.

Apply the adhesives by actively scrubbing them vigorously onto the surface [139].

Why? Active rubbing improves bonding performance to enamel and dentin as it promotes solvent evaporation, depth of demineralization, infiltration and chemical interaction due to a more effective contact of monomers with the surface.

Apply a separate hydrophobic coating resin [137, 142].

Why? The additional placement of a hydrophobic resin in a single step has been found to improve the laboratory performance of these systems regarding bond strength and interface durability.

3. Conclusions

Nowadays, dental adhesives are used for a wide range of clinical applications. Undoubtedly, the current concept of minimal-invasive dentistry was determined by a significant improvement in the dental adhesive technology. Nevertheless, the increasing development of different types of dentin/enamel adhesives creates uncertainty among clinicians about which type to use in daily practice for a specific clinical situation.

Not only it is mandatory to rigorously follow the recommended protocols for adhesive procedures due to their inherent technical sensitivity, but also it is critical to select the correct adhesive strategy as a function of the restorative materials to be used and the substrates taking part in the process. Etch-and-rinse adhesive systems perform better on enamel and two-step self-etching systems containing 10-MDP seem to be more suitable for bonding to dentin. Therefore, an etch-and-rinse approach may be more appropriate to establish adhesion in cavities of anterior teeth, particularly in situations where enamel prevails over dentin. In posterior cavities, a selective enamel etching with phosphoric acid followed by the active

http://dx.doi.org/10.5772/65605

application of a 10-MDP-based two-step self-etch adhesive to both enamel and dentin is probably the better strategy to achieve, currently, optimal restoration durability.

Future perspectives in the adhesion field include the development of molecules or strategies to reduce the flaws of bonding systems that yield resin-dentin bonds prone to degradation. The use of therapeutic bioactive ion-releasing restorative materials that initiate nucleation and growth of the newly formed mineral within micro- and nano-sized voids is currently one of the main issues of research. Nevertheless, multicentric clinical studies should be encouraged to support incoming laboratory data, in order to point out if these new or modified approaches may effectively provide increased restoration longevity.

Author details

Alexandra Vinagre* and João Ramos

*Address all correspondence to: avinagre@fmed.uc.pt

Operative Dentistry, Dentistry Area, Faculty of Medicine, University of Coimbra, Coimbra, Portugal

References

[1] Van Meerbeek B, De Munck J, Yoshida Y, Inoue S, Vargas M, Vijay P, et al. Buonocore memorial lecture. Adhesion to enamel and dentin: current status and future challenges. Oper Dent. 2003;28(3):215–35.

[2] Pashley DH, Tay FR, Breschi L, Tjaderhane L, Carvalho RM, Carrilho M, et al. State of the art etch-and-rinse adhesives. Dent Mater. 2011;27(1):1–16. DOI: 10.1016/j.dental.2010.10.016.

[3] Van Meerbeek B, Yoshihara K, Yoshida Y, Mine A, De Munck J, Van Landuyt KL. State of the art of self-etch adhesives. Dent Mater. 2011;27(1):17–28. DOI: 10.1016/j.dental.2010.10.023.

[4] Rosa WL, Piva E, Silva AF. Bond strength of universal adhesives: a systematic review and meta-analysis. J Dent. 2015;43(7):765–76. DOI: 10.1016/j.jdent.2015.04.003.

[5] Peumans M, Kanumilli P, De Munck J, Van Landuyt K, Lambrechts P, Van Meerbeek B. Clinical effectiveness of contemporary adhesives: a systematic review of current clinical trials. Dent Mater. 2005;21(9):864–81. DOI: 10.1016/j.dental.2005.02.003.

[6] Demarco FF, Correa MB, Cenci MS, Moraes RR, Opdam NJ. Longevity of posterior composite restorations: not only a matter of materials. Dent Mater. 2012;28(1):87–101. DOI: 10.1016/j.dental.2011.09.003.

[7] Van Meerbeek B, Peumans M, Poitevin A, Mine A, Van Ende A, Neves A, et al. Relationship between bond-strength tests and clinical outcomes. Dent Mater. 2010;26(2):e100–21. DOI: 10.1016/j.dental.2009.11.148.

[8] Heintze SD, Rousson V. Clinical effectiveness of direct class II restorations—a meta-analysis. J Adhes Dent. 2012;14(5):407–31. DOI: 10.3290/j.jad.a28390.

[9] Van Landuyt KL, Snauwaert J, De Munck J, Peumans M, Yoshida Y, Poitevin A, et al. Systematic review of the chemical composition of contemporary dental adhesives. Biomaterials. 2007;28(26):3757–85. DOI: 10.1016/j.biomaterials.2007.04.044.

[10] Carvalho RM, Chersoni S, Frankenberger R, Pashley DH, Prati C, Tay FR. A challenge to the conventional wisdom that simultaneous etching and resin infiltration always occurs in self-etch adhesives. Biomaterials. 2005;26(9):1035–42. DOI: 10.1016/j.biomaterials.2004.04.003.

[11] Reis AF, Giannini M, Pereira PN. Long-term TEM analysis of the nanoleakage patterns in resin-dentin interfaces produced by different bonding strategies. Dent Mater. 2007;23(9):1164–72. DOI: 10.1016/j.dental.2006.10.006.

[12] Breschi L, Mazzoni A, Ruggeri A, Cadenaro M, Di Lenarda R, De Stefano Dorigo E. Dental adhesion review: aging and stability of the bonded interface. Dent Mater. 2008;24(1):90–101. DOI:10.1016/j.dental.2007.02.009.

[13] De Munck J, Van Landuyt K, Peumans M, Poitevin A, Lambrechts P, Braem M, et al. A critical review of the durability of adhesion to tooth tissue: methods and results. J Dent Res. 2005;84(2):118–32. DOI: 84/2/118.

[14] Donmez N, Belli S, Pashley DH, Tay FR. Ultrastructural correlates of in vivo/in vitro bond degradation in self-etch adhesives. J Dent Res. 2005;84(4):355–9. DOI: 84/4/355.

[15] Van Landuyt KL, De Munck J, Mine A, Cardoso MV, Peumans M, Van Meerbeek B. Filler debonding & subhybrid-layer failures in self-etch adhesives. J Dent Res. 2010;89(10):1045–50. DOI: 10.1177/0022034510375285.

[16] De Munck J, Mine A, Poitevin A, Van Ende A, Cardoso MV, Van Landuyt KL, et al. Meta-analytical review of parameters involved in dentin bonding. J Dent Res. 2012;91(4):351–7. DOI: 10.1177/0022034511431251.

[17] Nanci A. Enamel: composition, formation and structure (Chapter 7) In: Ten Cate´s oral histology: development, formation and structure. 8th ed, Ten Cate AR, editor St Louis, MO: Mosby Inc (Elsevier Health Science). 2013. pp. 122–64.

[18] Tjaderhane L, Carrilho M, Breschi L, Tay F, Pashley D. Dentin basic structure and composition—an overview. Endod Topics. 2009;20(1):3–29.

[19] Nanci A. Dentin-pulp complex (Chapter 8). In: Ten Cate´s oral histology: development, formation and structure. 8th ed, Ten Cate AR, editor. St Louis, MO: Mosby Inc (Elsevier Health Science). 2013. pp. 165–204.

[20] Van Meerbeek B, Van Landuyt K, De Munck J, Inoue S, Yoshida Y, Perdigao J, et al. Bonding to enamel and dentin (Chapter 8). In: Fundamentals of operative dentistry. 3rd Revised ed, Summit J, Robbins J, Hilton T, Scwartz R, editors. USA: Quintessence Publishing Co Inc. 2006. pp. 183–260.

[21] Pashley DH, Pashley EL, Carvalho RM, Tay FR. The effects of dentin permeability on restorative dentistry. Dent Clin North Am. 2002;46(2):211–45, v–vi.

[22] Marshall SJ, Bayne SC, Baier R, Tomsia AP, Marshall GW. A review of adhesion science. Dent Mater. 2010;26(2):e11–6. DOI: 10.1016/j.dental.2009.11.157.

[23] Erickson RL. Surface interactions of dentin adhesive materials. Oper Dent. 1992;Suppl 5:81–94.

[24] Perdigao J. An ultra-morphological study of human dentine exposed to adhesive systems. Doctoral Thesis Leuven, Belgium: Catholic University of Leuven. 1995.

[25] Moszner N, Salz U, Zimmermann J. Chemical aspects of self-etching enamel-dentin adhesives: a systematic review. Dent Mater. 2005;21(10):895–910. DOI: 10.1016/j.dental. 2005.05.001.

[26] Buonocore MG. A simple method of increasing the adhesion of acrylic filling materials to enamel surfaces. J Dent Res. 1955;34(6):849–53.

[27] Gwinnett AJ, Matsui A. A study of enamel adhesives. The physical relationship between enamel and adhesive. Arch Oral Biol. 1967;12(12):1615–20.

[28] Nakabayashi N, Kojima K, Masuhara E. The promotion of adhesion by the infiltration of monomers into tooth substrates. J Biomed Mater Res. 1982;16(3):265–73. DOI: 10.1002/jbm.820160307.

[29] Nakabayashi N, Nakamura M, Yasuda N. Hybrid layer as a dentin-bonding mechanism. J Esthet Dent. 1991;3(4):133–8.

[30] Van Meerbeek B, Dhem A, Goret-Nicaise M, Braem M, Lambrechts P, VanHerle G. Comparative SEM and TEM examination of the ultrastructure of the resin-dentin interdiffusion zone. J Dent Res. 1993;72(2):495–501.

[31] Dalli M, Colak H, Mustafa Hamidi M. Minimal intervention concept: a new paradigm for operative dentistry. J Investig Clin Dent. 2012;3(3):167–75. DOI: 10.1111/j. 2041-1626.2012.00117.x.

[32] Pashley DH. Smear layer: physiological considerations. Oper Dent Suppl. 1984;3:13–29.

[33] Pashley DH, Horner JA, Brewer PD. Interactions of conditioners on the dentin surface. Oper Dent. 1992;Suppl 5:137–50.

[34] Pashley DH, Tao L, Boyd L, King GE, Horner JA. Scanning electron microscopy of the substructure of smear layers in human dentine. Arch Oral Biol. 1988;33(4):265–70. DOI: 0003-9969(88)90188-4.

[35] Mine A, De Munck J, Vivan Cardoso M, Van Landuyt KL, Poitevin A, Kuboki T, et al. Enamel-smear compromises bonding by mild self-etch adhesives. J Dent Res. 2010;89(12):1505–9. DOI: 10.1177/0022034510384871.

[36] Pashley DH. Smear layer: overview of structure and function. Proc Finn Dent Soc. 1992;88(Suppl 1):215–24.

[37] Lopes GC, Thys DG, Klaus P, Oliveira GM, Widmer N. Enamel acid etching: a review. Compend Contin Educ Dent. 2007;28(1):18–24; quiz 5, 42.

[38] Hashimoto M, Ohno H, Yoshida E, Hori M, Sano H, Kaga M, et al. Resin-enamel bonds made with self-etching primers on ground enamel. Eur J Oral Sci. 2003;111(5):447–53. DOI: 060.

[39] Hannig M, Bock H, Bott B, Hoth-Hannig W. Inter-crystallite nanoretention of self-etching adhesives at enamel imaged by transmission electron microscopy. Eur J Oral Sci. 2002;110(6):464–70.

[40] Perdigao J, Lopes MM, Gomes G. In vitro bonding performance of self-etch adhesives: II—ultramorphological evaluation. Oper Dent. 2008;33(5):534–49. DOI: 10.2341/07-133.

[41] Pashley DH, Tay FR. Aggressiveness of contemporary self-etching adhesives. Part II: etching effects on unground enamel. Dent Mater. 2001;17(5):430–44. DOI: S0109-5641(00)00104-4.

[42] Aguilar-Mendoza JA, Rosales-Leal JI, Rodriguez-Valverde MA, Cabrerizo-Vilchez MA. Effect of acid etching on dentin wettability and roughness: self-etching primers versus phosphoric acid. J Biomed Mater Res B Appl Biomater. 2008;84(1):277–85. DOI: 10.1002/jbm.b.30871.

[43] Ma S, Cai J, Zhan X, Wu Y. Effects of etchant on the nanostructure of dentin: an atomic force microscope study. Scanning. 2009;31(1):28–34. DOI: 10.1002/sca.20135.

[44] El Feninat F, Ellis TH, Sacher E, Stangel I. A tapping mode AFM study of collapse and denaturation in dentinal collagen. Dent Mater. 2001;17(4):284–8. DOI: S0109-5641(00)00083-X.

[45] Van Meerbeek B, Van Landuyt K, De Munck J, Hashimoto M, Peumans M, Lambrechts P, et al. Technique-sensitivity of contemporary adhesives. Dent Mater J. 2005;24(1):1–13.

[46] Pashley DH, Tay FR, Carvalho RM, Rueggeberg FA, Agee KA, Carrilho M, et al. From dry bonding to water-wet bonding to ethanol-wet bonding. A review of the interactions between dentin matrix and solvated resins using a macromodel of the hybrid layer. Am J Dent. 2007;20(1):7–20.

[47] Gregoire G, Guignes P, Nasr K. Effects of dentine moisture on the permeability of total-etch and one-step self-etch adhesives. J Dent. 2009;37(9):691–9. DOI: 10.1016/j.jdent.2009.05.010.

[48] Sauro S, Watson TF, Mannocci F, Miyake K, Huffman BP, Tay FR, et al. Two-photon laser confocal microscopy of micropermeability of resin-dentin bonds made with water or ethanol wet bonding. J Biomed Mater Res B Appl Biomater. 2009;90(1):327–37. DOI: 10.1002/jbm.b.31290.

[49] Malacarne-Zanon J, de Andrade ESSM, Wang L, de Goes MF, Martins AL, Narvaes-Romani EO, et al. Permeability of dental adhesives—A SEM assessment. Eur J Dent. 2010;4(4):429–39.

[50] Van Landuyt KL, De Munck J, Snauwaert J, Coutinho E, Poitevin A, Yoshida Y, et al. Monomer-solvent phase separation in one-step self-etch adhesives. J Dent Res. 2005;84(2):183–8. DOI: 84/2/183.

[51] Kanca J, 3rd. Effect of resin primer solvents and surface wetness on resin composite bond strength to dentin. Am J Dent. 1992;5(4):213–5.

[52] Tay FR, Gwinnett AJ, Pang KM, Wei SH. Resin permeation into acid-conditioned, moist, and dry dentin: a paradigm using water-free adhesive primers. J Dent Res. 1996;75(4): 1034–44.

[53] Hashimoto M, Ohno H, Kaga M, Sano H, Endo K, Oguchi H. The extent to which resin can infiltrate dentin by acetone-based adhesives. J Dent Res. 2002;81(1):74–8.

[54] Pashley DH, Carvalho RM, Tay FR, Agee KA, Lee KW. Solvation of dried dentin matrix by water and other polar solvents. Am J Dent. 2002;15(2):97–102.

[55] Tay FR, Gwinnett JA, Wei SH. Micromorphological spectrum from overdrying to overwetting acid-conditioned dentin in water-free acetone-based, single-bottle primer/adhesives. Dent Mater. 1996;12(4):236–44. DOI: S0109-5641(96)80029-7.

[56] Reis A, Pellizzaro A, Dal-Bianco K, Gones OM, Patzlaff R, Loguercio AD. Impact of adhesive application to wet and dry dentin on long-term resin-dentin bond strengths. Oper Dent. 2007;32(4):380–7. DOI: 10.2341/06-107.

[57] Tay FR, Pashley DH, Suh BI, Carvalho RM, Itthagarun A. Single-step adhesives are permeable membranes. J Dent. 2002;30(7–8):371–82. DOI: S0300571202000647.

[58] el-Din AK, Abd el-Mohsen MM. Effect of changing application times on adhesive systems bond strengths. Am J Dent. 2002;15(5):321–4.

[59] Cardoso Pde C, Loguercio AD, Vieira LC, Baratieri LN, Reis A. Effect of prolonged application times on resin-dentin bond strengths. J Adhes Dent. 2005;7(2):143–9.

[60] Dal-Bianco K, Pellizzaro A, Patzlaft R, de Oliveira Bauer JR, Loguercio AD, Reis A. Effects of moisture degree and rubbing action on the immediate resin-dentin bond strength. Dent Mater. 2006;22(12):1150–6. DOI: 10.1016/j.dental.2005.10.010.

[61] Ikeda T, De Munck J, Shirai K, Hikita K, Inoue S, Sano H, et al. Effect of evaporation of primer components on ultimate tensile strengths of primer-adhesive mixture. Dent Mater. 2005;21(11):1051–8. DOI: 10.1016/j.dental.2005.03.010.

[62] Garcia G, Fernandes KB, Garcia FC, D'Alpino PH, da Rocha Svizero N, Wang L. Solvent retention of contemporary commercial dentin bonding agents in a demineralized dentin matrix. Eur J Dent. 2010;4(3):293–7.

[63] Sadek FT, Calheiros FC, Cardoso PE, Kawano Y, Tay F, Ferrari M. Early and 24-hour bond strength and degree of conversion of etch-and-rinse and self-etch adhesives. Am J Dent. 2008;21(1):30–4.

[64] Cadenaro M, Antoniolli F, Sauro S, Tay FR, Di Lenarda R, Prati C, et al. Degree of conversion and permeability of dental adhesives. Eur J Oral Sci. 2005;113(6):525–30. DOI: 10.1111/j.1600-0722.2005.00251.x.

[65] Langer A, Ilie N. Dentin infiltration ability of different classes of adhesive systems. Clin Oral Investig. 2013;17(1):205–16. DOI: 10.1007/s00784-012-0694-4.

[66] Giachetti L, Bertini F, Scaminaci Russo D. Investigation into the nature of dentin resin tags: a scanning electron microscopic morphological analysis of demineralized bonded dentin. J Prosthet Dent. 2004;92(3):233–8. DOI: 10.1016/S0022391304004184.

[67] Pegado RE, do Amaral FL, Florio FM, Basting RT. Effect of different bonding strategies on adhesion to deep and superficial permanent dentin. Eur J Dent. 2010;4(2):110–7.

[68] Sano H, Shono T, Takatsu T, Hosoda H. Microporous dentin zone beneath resin-impregnated layer. Oper Dent. 1994;19(2):59–64.

[69] Pereira PN, Bedran-de-Castro AK, Duarte WR, Yamauchi M. Removal of noncollagenous components affects dentin bonding. J Biomed Mater Res B Appl Biomater. 2007;80(1):86–91. DOI: 10.1002/jbm.b.30572.

[70] Carvalho RM, Manso AP, Geraldeli S, Tay FR, Pashley DH. Durability of bonds and clinical success of adhesive restorations. Dent Mater. 2012;28(1):72–86. DOI: 10.1016/j.dental.2011.09.011.

[71] Carrilho MR, Carvalho RM, de Goes MF, di Hipolito V, Geraldeli S, Tay FR, et al. Chlorhexidine preserves dentin bond in vitro. J Dent Res. 2007;86(1):90–4. DOI: 86/1/90.

[72] Reis AF, Giannini M, Pereira PN. Influence of water-storage time on the sorption and solubility behavior of current adhesives and primer/adhesive mixtures. Oper Dent. 2007;32(1):53–9. DOI: 10.2341/06-13.

[73] De Munck J, Shirai K, Yoshida Y, Inoue S, Van Landuyt K, Lambrechts P, et al. Effect of water storage on the bonding effectiveness of 6 adhesives to Class I cavity dentin. Oper Dent. 2006;31(4):456–65. DOI: 10.2341/05-57.

[74] Loguercio AD, Stanislawczuk R, Mena-Serrano A, Reis A. Effect of 3-year water storage on the performance of one-step self-etch adhesives applied actively on dentine. J Dent. 2011;39(8):578–87. DOI: 10.1016/j.jdent.2011.06.005.

[75] Tjaderhane L, Nascimento FD, Breschi L, Mazzoni A, Tersariol IL, Geraldeli S, et al. Optimizing dentin bond durability: control of collagen degradation by matrix metal-

loproteinases and cysteine cathepsins. Dent Mater. 2013;29(1):116–35. DOI: 10.1016/j.dental.2012.08.004.

[76] Anusavice KJ, Shen C, Rawls HR. Resin-based composites (Chapter 13). In: Phillips' Science of Dental Materials. 12th ed. Elsevier Saunders; 3251 Riverport Lane, St Louis, Missouri 63043 Pages 275–306. 2013.

[77] Carrilho MR, Tay FR, Pashley DH, Tjaderhane L, Carvalho RM. Mechanical stability of resin-dentin bond components. Dent Mater. 2005;21(3):232–41. DOI: 10.1016/j.dental.2004.06.001.

[78] Hebling J, Pashley DH, Tjaderhane L, Tay FR. Chlorhexidine arrests subclinical degradation of dentin hybrid layers in vivo. J Dent Res. 2005;84(8):741–6. DOI: 84/8/741.

[79] Carrilho MR, Geraldeli S, Tay F, de Goes MF, Carvalho RM, Tjaderhane L, et al. In vivo preservation of the hybrid layer by chlorhexidine. J Dent Res. 2007;86(6):529–33. DOI: 86/6/529.

[80] Brackett WW, Ito S, Tay FR, Haisch LD, Pashley DH. Microtensile dentin bond strength of self-etching resins: effect of a hydrophobic layer. Oper Dent. 2005;30(6):733–8.

[81] Sano H. Microtensile testing, nanoleakage, and biodegradation of resin-dentin bonds. J Dent Res. 2006;85(1):11–4. DOI: 85/1/11.

[82] Brackett MG, Tay FR, Brackett WW, Dib A, Dipp FA, Mai S, et al. In vivo chlorhexidine stabilization of hybrid layers of an acetone-based dentin adhesive. Oper Dent. 2009;34(4):379–83. DOI: 10.2341/08-103.

[83] Tay FR, Pashley DH, Kapur RR, Carrilho MR, Hur YB, Garrett LV, et al. Bonding BisGMA to dentin--a proof of concept for hydrophobic dentin bonding. J Dent Res. 2007;86(11):1034–9. DOI: 86/11/1034.

[84] Hosaka K, Nishitani Y, Tagami J, Yoshiyama M, Brackett WW, Agee KA, et al. Durability of resin-dentin bonds to water- vs. ethanol-saturated dentin. J Dent Res. 2009;88(2):146–51. DOI: 10.1177/0022034508328910.

[85] Sadek FT, Castellan CS, Braga RR, Mai S, Tjaderhane L, Pashley DH, et al. One-year stability of resin-dentin bonds created with a hydrophobic ethanol-wet bonding technique. Dent Mater. 2010;26(4):380–6. DOI: 10.1016/j.dental.2009.12.009.

[86] Liu Y, Tjaderhane L, Breschi L, Mazzoni A, Li N, Mao J, et al. Limitations in bonding to dentin and experimental strategies to prevent bond degradation. J Dent Res. 2011;90(8):953–68. DOI: 10.1177/0022034510391799.

[87] Araujo JF, Barros TA, Braga EM, Loretto SC, Silva e Souza Pde A, Silva e Souza MH. One-year evaluation of a simplified ethanol-wet bonding technique: a randomized clinical trial. Braz Dent J. 2013;24(3):267–72. DOI: 10.1590/0103-6440201302128.

[88] Kuhn E, Farhat P, Teitelbaum AP, Mena-Serrano A, Loguercio AD, Reis A, et al. Ethanol-wet bonding technique: clinical versus laboratory findings. Dent Mater. 2015;31(9): 1030–7. DOI: 10.1016/j.dental.2015.05.010.

[89] Van Landuyt KL, Mine A, De Munck J, Jaecques S, Peumans M, Lambrechts P, et al. Are one-step adhesives easier to use and better performing? Multifactorial assessment of contemporary one-step self-etching adhesives. J Adhes Dent. 2009;11(3):175–90.

[90] Tay FR, King NM, Chan KM, Pashley DH. How can nanoleakage occur in self-etching adhesive systems that demineralize and infiltrate simultaneously? J Adhes Dent. 2002;4(4):255–69.

[91] Unemori M, Matsuya Y, Akashi A, Goto Y, Akamine A. Self-etching adhesives and postoperative sensitivity. Am J Dent. 2004;17(3):191–5.

[92] Van Landuyt KL, Snauwaert J, De Munck J, Coutinho E, Poitevin A, Yoshida Y, et al. Origin of interfacial droplets with one-step adhesives. J Dent Res. 2007;86(8):739–44. DOI: 86/8/739.

[93] Gregoire G, Dabsie F, Dieng-Sarr F, Akon B, Sharrock P. Solvent composition of one-step self-etch adhesives and dentine wettability. J Dent. 2011;39(1):30–9. DOI: 10.1016/j.jdent.2010.09.008.

[94] Dieng-Sarr F, Sharrock P, Dabsie F, Gregoire G. Modifications of the organic and mineral fractions of dental tissues following conditioning by self-etching adhesives. J Dent. 2011;39(2):141–7. DOI: 10.1016/j.jdent.2010.11.004.

[95] Hashimoto M, Fujita S, Endo K. Bonding of self-etching adhesives on dehydrated dentin. J Adhes Dent. 2011;13(1):49–54. DOI: 10.3290/j.jad.a18238.

[96] Salz U, Zimmermann J, Zeuner F, Moszner N. Hydrolytic stability of self-etching adhesive systems. J Adhes Dent. 2005;7(2):107–16.

[97] Tay FR, Pashley DH. Aggressiveness of contemporary self-etching systems. I: Depth of penetration beyond dentin smear layers. Dent Mater. 2001;17(4):296–308. DOI: S0109-5641(00)00087-7.

[98] De Munck J, Vargas M, Iracki J, Van Landuyt K, Poitevin A, Lambrechts P, et al. One-day bonding effectiveness of new self-etch adhesives to bur-cut enamel and dentin. Oper Dent. 2005;30(1):39–49.

[99] Gregoire G, Millas A. Microscopic evaluation of dentin interface obtained with 10 contemporary self-etching systems: correlation with their pH. Oper Dent. 2005;30(4): 481–91.

[100] Salz U, Mucke A, Zimmermann J, Tay FR, Pashley DH. pKa value and buffering capacity of acidic monomers commonly used in self-etching primers. J Adhes Dent. 2006;8(3):143–50.

[101] Yoshida Y, Nagakane K, Fukuda R, Nakayama Y, Okazaki M, Shintani H, et al. Comparative study on adhesive performance of functional monomers. J Dent Res. 2004;83(6):454–8.

[102] Van Landuyt KL, Yoshida Y, Hirata I, Snauwaert J, De Munck J, Okazaki M, et al. Influence of the chemical structure of functional monomers on their adhesive performance. J Dent Res. 2008;87(8):757–61. doi: 10.1177/154405910808700804.

[103] Yoshihara K, Yoshida Y, Nagaoka N, Fukegawa D, Hayakawa S, Mine A, et al. Nano-controlled molecular interaction at adhesive interfaces for hard tissue reconstruction. Acta Biomater. 2010;6(9):3573–82. DOI: 10.1016/j.actbio.2010.03.024.

[104] Tay FR, Pashley DH, King NM, Carvalho RM, Tsai J, Lai SC, et al. Aggressiveness of self-etch adhesives on unground enamel. Oper Dent. 2004;29(3):309–16.

[105] Frankenberger R, Tay FR. Self-etch vs etch-and-rinse adhesives: effect of thermo-mechanical fatigue loading on marginal quality of bonded resin composite restorations. Dent Mater. 2005;21(5):397–412. DOI: 10.1016/j.dental.2004.07.005.

[106] Vinagre A, Ramos J, Messias A, Marques F, Caramelo F, Mata A. Microtensile bond strength and micromorphology of bur-cut enamel using five adhesive systems. J Adhes Dent. 2015;17(2):107–16. DOI: 10.3290/j.jad.a34060.

[107] Ermis RB, De Munck J, Cardoso MV, Coutinho E, Van Landuyt KL, Poitevin A, et al. Bonding to ground versus unground enamel in fluorosed teeth. Dent Mater. 2007;23(10):1250–5. DOI:10.1016/j.dental.2006.11.005.

[108] Osorio R, Monticelli F, Moreira MA, Osorio E, Toledano M. Enamel-resin bond durability of self-etch and etch & rinse adhesives. Am J Dent. 2009;22(6):371–5.

[109] Perdigao J, Gomes G, Duarte S, Jr., Lopes MM. Enamel bond strengths of pairs of adhesives from the same manufacturer. Oper Dent. 2005;30(4):492–9.

[110] Reis A, Moura K, Pellizzaro A, Dal-Bianco K, de Andrade AM, Loguercio AD. Durability of enamel bonding using one-step self-etch systems on ground and unground enamel. Oper Dent. 2009;34(2):181–91.

[111] Van Landuyt KL, Kanumilli P, De Munck J, Peumans M, Lambrechts P, Van Meerbeek B. Bond strength of a mild self-etch adhesive with and without prior acid-etching. J Dent. 2006;34(1):77–85. DOI: 10.1016/j.jdent.2005.04.001.

[112] Frankenberger R, Lohbauer U, Roggendorf MJ, Naumann M, Taschner M. Selective enamel etching reconsidered: better than etch-and-rinse and self-etch? J Adhes Dent. 2008;10(5):339–44.

[113] Li N, Nikaido T, Alireza S, Takagaki T, Chen JH, Tagami J. Phosphoric acid-etching promotes bond strength and formation of acid-base resistant zone on enamel. Oper Dent. 2013;38(1):82–90. DOI: 10.2341/11-422-L.

[114] Luhrs AK, Guhr S, Schilke R, Borchers L, Geurtsen W, Gunay H. Shear bond strength of self-etch adhesives to enamel with additional phosphoric acid etching. Oper Dent. 2008;33(2):155–62. DOI: 10.2341/07-63.

[115] Wang Y, Spencer P. Continuing etching of an all-in-one adhesive in wet dentin tubules. J Dent Res. 2005;84(4):350–4. DOI: 84/4/350.

[116] Van Landuyt KL, Peumans M, De Munck J, Lambrechts P, Van Meerbeek B. Extension of a one-step self-etch adhesive into a multi-step adhesive. Dent Mater. 2006;22(6):533–44. DOI: 10.1016/j.dental.2005.05.010.

[117] Reis A, Albuquerque M, Pegoraro M, Mattei G, Bauer JR, Grande RH, et al. Can the durability of one-step self-etch adhesives be improved by double application or by an extra layer of hydrophobic resin? J Dent. 2008;36(5):309–15. DOI: 10.1016/j.jdent.2008.01.018.

[118] Torres CR, Barcellos DC, Pucci CR, Lima Gde M, Rodrigues CM, Siviero M. Influence of methods of application of self-etching adhesive systems on adhesive bond strength to enamel. J Adhes Dent. 2009;11(4):279–86.

[119] Perdigao J, Gomes G, Gondo R, Fundingsland JW. In vitro bonding performance of all-in-one adhesives. Part I —microtensile bond strengths. J Adhes Dent. 2006;8(6):367–73.

[120] Peumans M, De Munck J, Van Landuyt KL, Poitevin A, Lambrechts P, Van Meerbeek B. Eight-year clinical evaluation of a 2-step self-etch adhesive with and without selective enamel etching. Dent Mater. 2010;26(12):1176–84. DOI: 10.1016/j.dental.2010.08.190.

[121] Perdigao J, Carmo AR, Anauate-Netto C, Amore R, Lewgoy HR, Cordeiro HJ, et al. Clinical performance of a self-etching adhesive at 18 months. Am J Dent. 2005;18(2):135–40.

[122] Peumans M, De Munck J, Van Landuyt K, Lambrechts P, Van Meerbeek B. Five-year clinical effectiveness of a two-step self-etching adhesive. J Adhes Dent. 2007;9(1):7–10.

[123] Peumans M, Munck J, Van Landuyt K, Lambrechts P, Van Meerbeek B. Three-year clinical effectiveness of a two-step self-etch adhesive in cervical lesions. Eur J Oral Sci. 2005;113(6):512–8. DOI: 10.1111/j.1600-0722.2005.00256.x.

[124] van Dijken JW. A prospective 8-year evaluation of a mild two-step self-etching adhesive and a heavily filled two-step etch-and-rinse system in non-carious cervical lesions. Dent Mater. 2010;26(9):940–6. DOI: 10.1016/j.dental.2010.05.009.

[125] Inoue G, Nikaido T, Foxton RM, Tagami J. The acid-base resistant zone in three dentin bonding systems. Dent Mater J. 2009;28(6):717–21.

[126] Nurrohman H, Nikaido T, Takagaki T, Sadr A, Ichinose S, Tagami J. Apatite crystal protection against acid-attack beneath resin-dentin interface with four adhesives: TEM and crystallography evidence. Dent Mater. 2012;28(7):e89–98. DOI: 10.1016/j.dental.2012.04.025.

[127] Perdigao J, Sezinando A, Monteiro PC. Laboratory bonding ability of a multi-purpose dentin adhesive. Am J Dent. 2012;25(3):153–8.

[128] Hanabusa M, Mine A, Kuboki T, Momoi Y, Van Ende A, Van Meerbeek B, et al. Bonding effectiveness of a new 'multi-mode' adhesive to enamel and dentine. J Dent. 2012;40(6): 475–84. DOI: 10.1016/j.jdent.2012.02.012.

[129] Marchesi G, Frassetto A, Mazzoni A, Apolonio F, Diolosa M, Cadenaro M, et al. Adhesive performance of a multi-mode adhesive system: 1-year in vitro study. J Dent. 2013. DOI: 10.1016/j.jdent.2013.12.008.

[130] Munoz MA, Luque I, Hass V, Reis A, Loguercio AD, Bombarda NH. Immediate bonding properties of universal adhesives to dentine. J Dent. 2013;41(5):404–11. DOI: 10.1016/j.jdent.2013.03.001.

[131] Mena-Serrano A, Kose C, De Paula EA, Tay LY, Reis A, Loguercio AD, et al. A new universal simplified adhesive: 6-month clinical evaluation. J Esthet Restor Dent. 2013;25(1):55–69. DOI: 10.1111/jerd.12005.

[132] Perdigao J, Kose C, Mena-Serrano A, De Paula E, Tay L, Reis A, et al. A new universal simplified adhesive: 18-month clinical evaluation. Oper Dent. 2013. DOI: 10.2341/13-045-C.

[133] Wagner A, Wendler M, Petschelt A, Belli R, Lohbauer U. Bonding performance of universal adhesives in different etching modes. J Dent. 2014;42(7):800–7. DOI: 10.1016/j.jdent.2014.04.012.

[134] Alex G. Universal adhesives: the next evolution in adhesive dentistry? Compend Contin Educ Dent. 2015;36(1):15–26; quiz 8, 40.

[135] Goracci C, Rengo C, Eusepi L, Juloski J, Vichi A, Ferrari M. Influence of selective enamel etching on the bonding effectiveness of a new "all-in-one" adhesive. Am J Dent. 2013;26(2):99–104.

[136] Loguercio AD, de Paula EA, Hass V, Luque-Martinez I, Reis A, Perdigao J. A new universal simplified adhesive: 36-month randomized double-blind clinical trial. J Dent. 2015;43(9):1083–92. DOI: 10.1016/j.jdent.2015.07.005.

[137] Cardenas AM, Siqueira F, Rocha J, Szesz AL, Anwar M, El-Askary F, et al. Influence of conditioning time of universal adhesives on adhesive properties and enamel-etching pattern. Oper Dent. 2016. DOI: 10.2341/15-213-L.

[138] Chen C, Niu LN, Xie H, Zhang ZY, Zhou LQ, Jiao K, et al. Bonding of universal adhesives to dentine—old wine in new bottles? J Dent. 2015;43(5):525–36. DOI: 10.1016/j.jdent.2015.03.004.

[139] Luque-Martinez IV, Perdigao J, Munoz MA, Sezinando A, Reis A, Loguercio AD. Effects of solvent evaporation time on immediate adhesive properties of universal adhesives to dentin. Dent Mater. 2014;30(10):1126–35. DOI: 10.1016/j.dental.2014.07.002.

[140] Munoz MA, Sezinando A, Luque-Martinez I, Szesz AL, Reis A, Loguercio AD, et al. Influence of a hydrophobic resin coating on the bonding efficacy of three universal adhesives. J Dent. 2014;42(5):595–602. DOI: 10.1016/j.jdent.2014.01.013.

[141] Sezinando A, Luque-Martinez I, Munoz MA, Reis A, Loguercio AD, Perdigao J. Influence of a hydrophobic resin coating on the immediate and 6-month dentin bonding of three universal adhesives. Dent Mater. 2015;31(10):e236–46. DOI: 10.1016/j.dental.2015.07.002.

[142] Loguercio AD, Munoz MA, Luque-Martinez I, Hass V, Reis A, Perdigao J. Does active application of universal adhesives to enamel in self-etch mode improve their performance? J Dent. 2015;43(9):1060–70. DOI: 10.1016/j.jdent.2015.04.005.

[143] Gale MS, Darvell BW. Thermal cycling procedures for laboratory testing of dental restorations. J Dent. 1999;27(2):89–99. DOI: S0300-5712(98)00037-2.

[144] Manhart J, Chen H, Hamm G, Hickel R. Buonocore Memorial Lecture. Review of the clinical survival of direct and indirect restorations in posterior teeth of the permanent dentition. Oper Dent. 2004;29(5):481–508.

[145] Heintze SD. Systematic reviews: I. The correlation between laboratory tests on marginal quality and bond strength. II. The correlation between marginal quality and clinical outcome. J Adhes Dent. 2007;9(Suppl 1):77–106.

[146] Heintze S, Forjanic M, Cavalleri A. Microleakage of Class II restorations with different tracers--comparison with SEM quantitative analysis. J Adhes Dent. 2008;10(4):259–67.

[147] Heintze SD, Blunck U, Gohring TN, Rousson V. Marginal adaptation in vitro and clinical outcome of Class V restorations. Dent Mater. 2009;25(5):605–20. DOI: 10.1016/j.dental.2008.11.004.

[148] Frankenberger R, Kramer N, Lohbauer U, Nikolaenko SA, Reich SM. Marginal integrity: is the clinical performance of bonded restorations predictable in vitro? J Adhes Dent. 2007;9(Suppl 1):107–16.

[149] Blunck U, Zaslansky P. Enamel margin integrity of Class I one-bottle all-in-one adhesives-based restorations. J Adhes Dent. 2011;13(1):23–9. DOI: 10.3290/j.jad.a18445.

[150] Roggendorf MJ, Kramer N, Appelt A, Naumann M, Frankenberger R. Marginal quality of flowable 4-mm base vs. conventionally layered resin composite. J Dent. 2011;39(10):643–7. DOI: 10.1016/j.jdent.2011.07.004.

[151] Bortolotto T, Doudou W, Kunzelmann KH, Krejci I. The competition between enamel and dentin adhesion within a cavity: an in vitro evaluation of class V restorations. Clin Oral Investig. 2012;16(4):1125–35. DOI: 10.1007/s00784-011-0623-y.

[152] Peumans M, De Munck J, Van Landuyt KL, Poitevin A, Lambrechts P, Van Meerbeek B. A 13-year clinical evaluation of two three-step etch-and-rinse adhesives in non-

carious class-V lesions. Clin Oral Investig. 2012;16(1):129–37. DOI: 10.1007/s00784-010-0481-z.

[153] Pashley DH, Carvalho RM. Dentine permeability and dentine adhesion. J Dent. 1997;25(5):355–72. DOI: S0300571296000577.

[154] Perdigao J, Carmo AR, Geraldeli S. Eighteen-month clinical evaluation of two dentin adhesives applied on dry vs moist dentin. J Adhes Dent. 2005;7(3):253–8.

[155] Loguercio AD, Loeblein F, Cherobin T, Ogliari F, Piva E, Reis A. Effect of solvent removal on adhesive properties of simplified etch-and-rinse systems and on bond strengths to dry and wet dentin. J Adhes Dent. 2009;11(3):213–9.

[156] Loguercio AD, Raffo J, Bassani F, Balestrini H, Santo D, do Amaral RC, et al. 24-month clinical evaluation in non-carious cervical lesions of a two-step etch-and-rinse adhesive applied using a rubbing motion. Clin Oral Investig. 2011;15(4):589–96. DOI: 10.1007/s00784-010-0408-8.

Chapter 4

Adhesive Materials for Biomedical Applications

Andrea J Vernengo

Additional information is available at the end of the chapter

http://dx.doi.org/10.5772/64958

Abstract

Recently, polymeric bioadhesives have become known as promising alternatives to sutures, staples, and wires. Traditional wound closure techniques are time-consuming to apply and cause additional tissue damage. In instances of large-scale hemorrhage or minimally invasive laparoscopic surgery, sutures are impractical to apply. Alternatively, newly developed bioadhesives are polymers that can be dripped or sprayed over superficial or internal injuries, solidifying in situ to form a seal that apposes tissue or arrests bleeding over large areas. This review will outline the main categories of polymers that have been investigated for these applications. The chemistry, mechanisms of adhesion, and advantages and limitations of each category will be described. In addition, needs for next-generation adhesives in tissue engineering will be discussed. For the repair of certain load-bearing areas of the body, such as cartilage and the intervertebral disc, scaffold adhesion is necessary for anchoring the scaffold in place and providing adequate transmission of forces. Researchers continue developing new formulations that exhibit improved biocompatibility, strength, elasticity, and degradability. These advances promise to improve clinical outcomes by enhancing bleeding control and wound healing. In the long term, bioadhesives will play an important role in making orthopedic and musculoskeletal tissue engineering clinically feasible.

Keywords: surgical sealants, glues, bioadhesives, tissue adhesion, in situ forming hydrogels, tissue engineering

1. Introduction

Sutures, wires, and staples are routine ways for achieving closure of superficial or internal tissues, yet there are several limitations to these traditional closure methods. For instance, the application of sutures is time-consuming and requires the penetration of surrounding tissues, causing additional damage [1] and increasing potential for infection [2]. The suture points also cause high stress concentration [3], which can result in more pain for the patient. Gaps in the injury site left behind due to incomplete suturing can cause leakage, and depending on the type of fluid (bowel content, bile, or cerebral spinal fluid), serious clinical complications could result [4, 5]. Finally, there are surgical situations where sutures, staples, and wires are difficult, if not impossible, to implement, such as in minimally invasive laparoscopic surgery [2, 6], or when trauma results in large-scale incompressible hemorrhage [7].

Over the past few decades, a variety of new flexible polymers with bioadhesive properties have emerged, which offer alternatives to traditional closure methods. These polymers have the ability to fill irregularly shaped cavities or be spread over large areas of tissue to quickly form a seal that mitigates bleeding. Commonly, the monomer or macromer components are applied as liquids and solidify on the tissue via in situ polymerization or cross-linking reaction. Polymeric bioadhesives have been studied for approximating tissues in surface wounds (glues), controlling bleeding (hemostats), and closing of fluid or air leaks (sealants) [8]. This review will outline the main classes of bioadhesives that have been studied for these applications, both in the clinic and in scientific literature, including cyanoacrylates, proteins, hydrogels, polysaccharides, and nature-inspired synthetic polymers. An emphasis is placed on polymer chemistry, mechanisms of adhesion, and the advantages and disadvantages of each material class. Because it is impossible to compare quantitative data generated by various researchers in separate experiments [1], the descriptions will emphasize qualitative findings. The final component of this review focuses on emerging applications for bioadhesives in regenerative medicine and the unmet needs that still exist.

2. Main classes of adhesives

2.1. Cyanoacrylates

Cyanoacrylate tissue adhesives are monomers that polymerize in the presence of aqueous milieu to create a solid layer that joins apposed wound edges [9]. The liquid monomers have a high reactivity; thus, polymerization happens in seconds without addition of a catalyst or elevated temperature [1]. When the monomers are applied on tissue, the liquid flows into the existing crevices, sealing by a mechanical interlock upon curing [1]. The NH_3 groups in tissue also participate in the polymerization, contributing toward the strong bonding that cyanoacrylates form [10].

Cyanoacrylates have been successfully applied in a clinical setting. Dermabond®, composed of. 2-octyl-cyanoacrylate (**Figure 1A**), was approved by the FDA in 1998 for the closure of topical skin incisions. The polymer degrades in 7–10 days [1]. Dermabond was compared to the use of metal clips in the closure of wounds from thyroidectomy in 70 patients [11]. In terms of quality of skin closure and cosmetic outcomes, both closure methods yielded similar results. However, according to patient input, postoperative management was easier in the Dermabond group. Poly(2-octyl-cyanoacrylate) has also been reported to have antimicrobial properties for 72 h after it is first applied. The cured material, applied on superficial wounds, was an effective barrier against infection when used with best approximation of wound lips after plastic surgery [12].

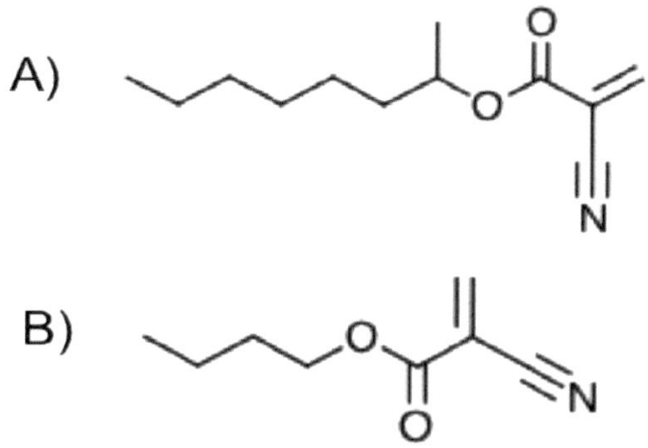

Figure 1. The structures of the cyanoacrylate tissue adhesives (A) 2-octyl-cyanoacrylate and (B) N-butyl-2-cyanoacrylate.

Another clinical study was conducted on 35 patients with inguinal hernias that were repaired with a prosthesis fixed with n-hexyl-α-cyanoacrylate. Results were compared with fixation with sutures. The use of cyanoacrylate adhesive instead of sutures did not alter the relapse rate, but it did significantly reduce surgical time and postoperative pain with no complications [13]. N-hexyl-α-cyanoacrylate was also used successfully in a rat model of microvascular anastomosis. In the experimental technique being evaluated, the thermoreversible polymer Polaxomer® 407 was injected into both ends of severed blood vessel. The ends were then approximated, and the glue was applied circumferentially. The method was reported as easy to apply, faster than sutures, and produced a lower foreign body response [14]. Idle et al. [15] reported successful use of N-butyl-2-cyanoacrylate (**Figure 1B**) as a hemostatic agent to prevent hemorrhage after tooth extraction. At 2 months, the glue was completely resorbed and the tooth socket healed.

There are some drawbacks to cyanoacrylate adhesives that limit the use to external, temporary applications. For instance, the mechanical properties of cyanoacrylate adhesives have been reported to be weaker than sutures [16]. In addition, the cured sealant is brittle, and thus, researchers have sought to modify the polymer structure to enhance flexibility. Poly(L-lactide-co-ε-caprolactone (PLCL) is a synthetic copolymer that was used as an additive in ethyl-2-cyanoacrylate (EC) and allyl-2-cyanoacrylate (AC) monomers. All formulations containing PLCL resisted breaking when subjected to 30 cycles of bending stress. In contrast, EC and AC broke after one and seven cycles, respectively [17].

There are also concerns over potential cytotoxicity and inflammatory responses associated with the use of cyanoacrylate adhesives. Degradation of cyanoacrylates occurs via hydrolysis. The by-products, cyanoacetate and formaldehyde, can potentially cause toxicity [18–20]. The first cyanoacrylate adhesives applied for medical purposes were ethyl-2-cyanoacrylate (**Figure 2A**) and methyl-2-cyanoacrylate (**Figure 2B**). These short-chain monomers exhibited fast degradation upon polymerization, allowing for the accumulation of higher amounts of formaldehyde and/or cyanoacetate in the tissue. These monomers have been abandoned [1] and replaced with those with longer alkyl chains, like n-hexyl-α-cyanoacrylate, N-butyl-2-cyanoacrylate, and 2-octyl-cyanoacrylate, although concerns over the use of these formulations exist as well. The biocompatibility of Glubran 2®, an n-butyl cyanoacrylate-based adhesive, was evaluated. Extracts of cured glue were applied over a monolayer of L929 fibroblast cells. Severe cytotoxicity was observed with the undiluted extracts, which disappeared when the extracts were diluted 1:10 [21]. In a clinical report, a foreign body response was observed 3 weeks after using of Dermabond to seal a superficial wound on the wrist of a 39-year-old [22]. The authors recommend informing patients of the potential risk of an inflammatory response to cyanoacrylate glues.

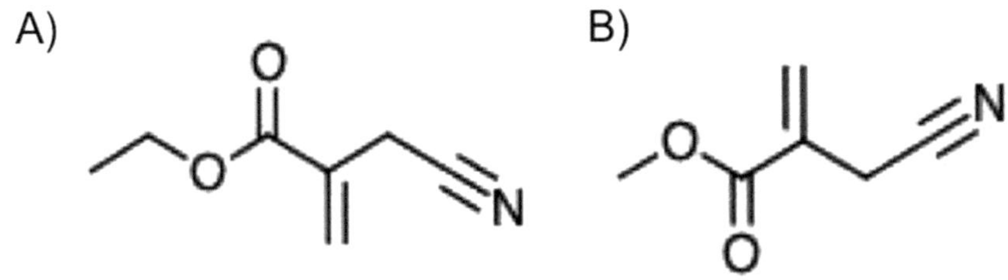

Figure 2. The structures of short-chain cyanoacrylate tissue adhesives, (A) ethyl-2-cyanoacrylate and (B) methyl-e-cyanoacrylate.

2.1.1. Notable advances in cyanoacrylate adhesives

Recently, cyanoacrylates were used in the development of self-healing materials. Self-healing materials are those that are engineered to prevent failure through an autonomous repair mechanism. In the first reported use of cyanoacrylate in this area, the polymer was combined with acrylate bone cement, poly(methyl methacrylate) (PMMA) [23]. In this work, polyurethane shells were used to encapsulate 2-octyl cyanoacrylate monomers, which were embedded in the PMMA to form a composite upon cure. The 2-octyl cyanoacrylate was released from the polyurethane shells and polymerized upon contact with moisture. In doing so, it filled cracks and defects that formed inside the PMMA, reinforcing the matrix. Incorporation of the 2-octyl cyanoacrylate slowed down progressive crack propagation under loading compared with unreinforced PMMA cement. In subsequent work [24], it was shown that reinforced specimens were able to withstand twice as many repetitive loading cycles before failure compared with unreinforced specimens. Cyanoacrylate films with a broad range of nanostructured architectures were also created [25]. In order to achieve this, N-octyl-2-cyanocrylate was electrospun and an air pump was used to precisely direct the nanosized polymer fibers to a wound bed. The deposited fibers formed a flexible, high strength membrane that was used to effectively arrest bleeding in in vivo models of liver and lung resection.

http://dx.doi.org/10.5772/64958

2.2. Protein adhesives

2.2.1. *Gelatin-based*

Gelatin is derived from the hydrolysis of collagen [6] and widely used in biomedical applications because of its biocompatibility and degradability [26]. Gelatin is capable of forming physically cross-linked hydrogels, but the mechanical strength is low and the hydrogels do not possess adhesive properties [27]. For this reason, gelatin-based tissue adhesives use chemical cross-linking and often also include additional polymer additives.

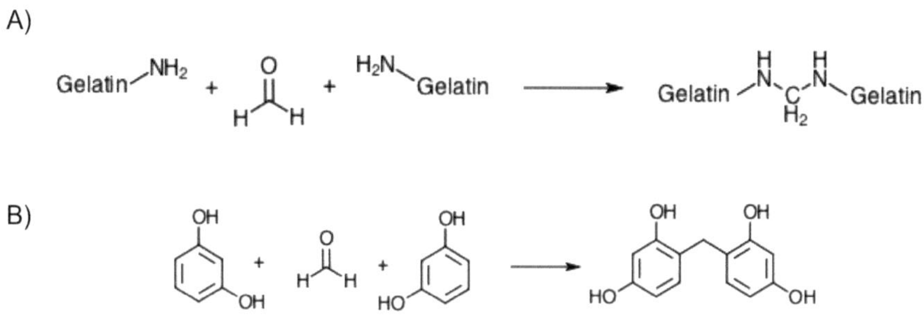

Figure 3. (A) The chemistry of cross-linking gelatin with formaldehyde and (B) the network that is formed by the reaction of resorcinol with formaldehyde.

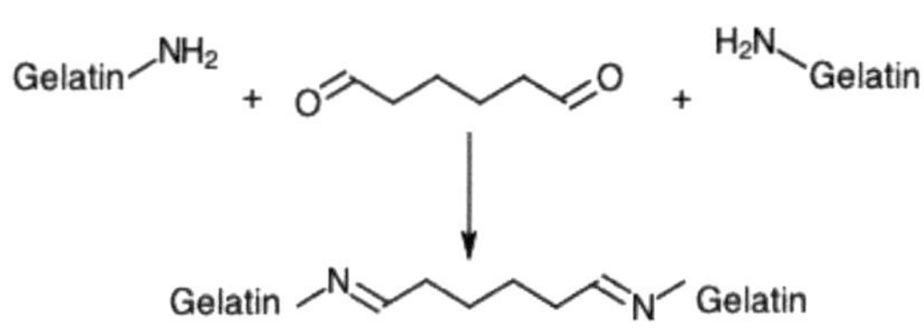

Figure 4. The cross-linking of gelatin with glutaraldehyde via a Schiff's base reaction.

The earliest gelatin-based adhesive is composed of gelatin, resorcinol, and formaldehyde. Referred to as GRF glue, it forms a seal in situ because the formaldehyde cross-links with the gelatin (**Figure 3A**) and the resorcinol (**Figure 3B**) to form a network. The resorcinol is added to augment the mechanical properties of the adhesive [10]. The composition of GRF glue comprises 18% by weight of formaldehyde, despite the well-known concerns over its toxicity [1, 6]. Braunwald et al. [28] evaluated GRF glue in an in vivo canine model, noting that the degree of irritation caused by the formaldehyde depended on the vascularity of the tissue. Excess formaldehyde may be diluted and washed away in areas where there is high enough blood flow. However, due to concerns over the safety of GRF glue, the formulation was later revised to replace a portion of the formaldehyde in the adhesive with glutaraldehyde, which is less histotoxic [29, 30]. Glutaraldehyde forms a Schiff's base with the free amines on gelatin, forming cross-links in the network just as formaldehyde does (**Figure 4**). This updated formulation was called gelatin-resorcinol-formaldehyde-glutaraldehyde (GRFG) glue. When applied in the liver of rats, no evidence of necrosis was found and it was shown to be an effective sealant and hemostat. Another group modified the GRFG formulation by adding 2.5% sodium carboxymethylcellulose (CMC). This created a "jelly" consistency for the glue, helping to prevent migration away from the target tissue site. The jelly was applied in humans to seal air leaks and found safe for shallow cuts in the lung [31]. Despite some successful outcomes, glutaraldehyde is technically classified as a toxic substance, so safety concerns prevent approval of GRF/GRFG glues by the FDA in the United States [32, 33].

2.2.2. *Advances in gelatin-based adhesives*

Researchers have looked into alternative cross-linkers for gelatin that are potentially less toxic than glutaraldehyde and formaldehyde. Rapidly polymerized elastic gelatin networks were also produced via photochemical cross-linking in the presence of ruthenium II trisbipyridyl chloride $[RuII(bpy)_3]^{2+}$ catalyst and ammonium persulfate oxidant [34]. Visible light causes the photolysis of the $[RuII(bpy)_3]^{2+}$. The Ru(II) and sulfate radicals oxidize tyrosine residues, present on gelatin, allowing formation of an intermolecular dityrosine cross-link. When gelatin was modified with additional phenolic (tyrosine-like) residues, the sealant stiffness increased by a factor of five and swelling was restricted compared with gels made from unmodified gelatin due to the increased cross-linking density [35]. Blends of modified and unmodified gelatin can give tunable elasticity and elastic modulus [33]. The reagents were found to be non-toxic at the concentrations used to make the gels [35].

In another study, cholesterol chloroformate was used to partially convert the amino groups on the gelatin to cholesterols. Cross-linking occurred by mixing a combination of cholesterol-modified gelatin with disuccinimidyl tartrate. Covalent bonding also occurred with collagen in the surrounding tissue extracellular matrix (ECM). Moreover, adhesion was enhanced due to the hydrophobic cholesterol groups, which promoted polymer chain penetration into the tissue and anchoring to cell membranes [36].

N-(3-dimethyl-aminopropyl)-N'-ethylcarbodiimide hydrochloride, a water-soluble carbodiimide (WSC), was shown to be an effective cross-linker of gelatin, but high concentrations (20 mg/mL) negatively impacted cell viability. The group was able to decrease EDC concentrations in the adhesive by functionalizing the gelatin with N-hydroxysuccinimide (NHS) groups [37]. N-hydroxysuccinimide (NHS) esters are known to react with primary or secondary amines (**Figure 5**). In this reported study, the addition of low concentrations of NHS (1 mg/mL) allowed the EDC concentrations to be decreased to 10 mg/mL without compromising adhesive strength and maintaining high cell viability (89–100%) [37].

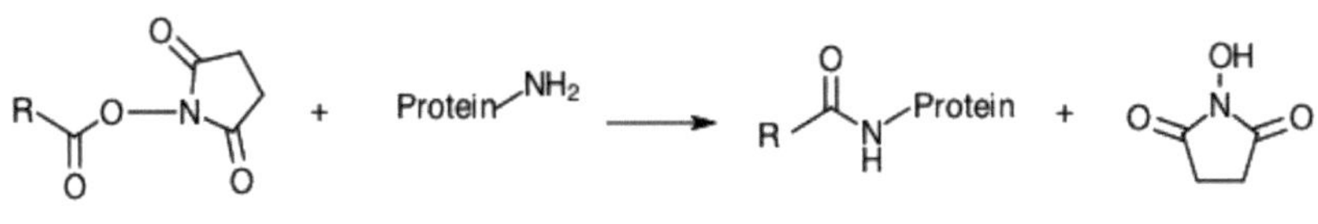

Figure 5. Reaction scheme for NHS ester conjugation to a primary amine.

Another approach involves the use of enzymatic cross-linking. The advantage of this route is that the gelation occurs under physiological conditions. Transglutaminase are ubiquitous in nature and catalyze the reaction between lysine and glutamine, resulting in cross-linking of proteins (**Figure 6**). Chen et al. [38] used this enzyme to introduce cross-links into gelatin under moist conditions. The adhesive did not cause cell death in rat retinal tissue when applied in vivo for 2 weeks. It is important to note that since the cross-linking reaction causes release of ammonia, local tissue damage or inflammation is possible.

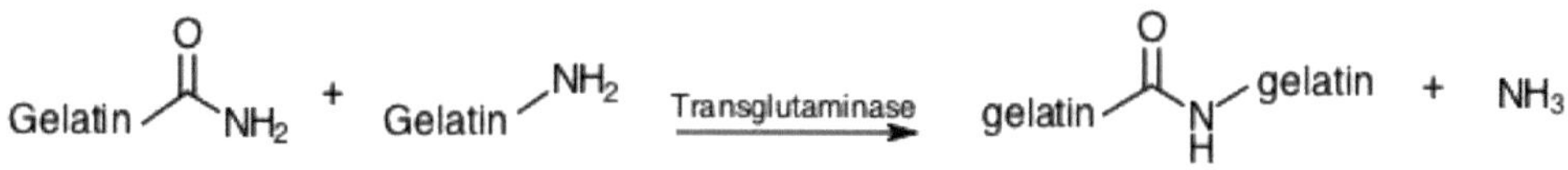

Figure 6. The cross-linking of gelatin via the enzyme transglutaminase [38].

2.2.3. Albumin-based adhesives

Another type of protein-based adhesive is based on albumin and glutaraldehyde. Albumin is an abundant protein found in the blood of mammals. Glutaraldehyde cross-links albumin to itself and to the proteins in the surrounding tissue at the repair site via a Schiff's base reaction. The albumin network forms a rigid, brittle solid [29]. Bioglue® is a bovine albumin-based glue available on the market that is approved for cardiac and vascular surgery [39]. Bioglue is less histotoxic than GRF glue and has superior bonding ability [40, 41]. Bioglue was successfully used to clinically treat bronchopleural fistula [42], partial renal nephrectomy [43], and vascular anastomoses [44]. Due to the presence of glutaraldehyde, there are still safety concerns with its application in certain areas of the body with high cellularity. For instance, adverse effects were reported on nerve function in a porcine model when the adhesive was applied directly onto the phrenic nerve [45].

2.2.3.1. Advances in albumin-based adhesives

De Somer et al. [46] developed an autologous albumin glue to avoid the immune response associated with bovine components. Fresh human plasma was ultrafiltered to concentrate the plasma proteins, which were then combined with glutaraldehyde. The autologous glue exhibited higher compliance than rigid Bioglue, likely due to the presence of fibrinogen in the concentrated blood plasma. The disadvantage of this method is that it requires 60 min of preparation time once the plasma is obtained. As an alternative to cross-linking with glutaraldehyde, another group prepared a modified tartaric acid with two active NHS ester groups and combined it with human albumin [47]. Subcutaneous implantation in mice indicated the adhesive was safe, eliciting a mild inflammatory response, but the results were not compared side by side with an adhesive cross-linked with glutaraldehyde.

2.2.4. Fibrin adhesives

Fibrin adhesives work by mimicking the biochemical reactions of the last stage of clotting [48, 49]. The resulting clot at the site of repair is a cross-linked network of proteins that also forms covalent linkages with the surrounding tissue, allowing adhesion to occur. Fibrin glues are applied as hemostats to stop bleeding, in addition to being used as sealants to achieve closure [6].

In general, fibrin sealants are composed of three major protein building blocks: fibrinogen, thrombin, and factor XIII, which are derived from human or bovine plasma [1]. Upon mixing the constituents, the thrombin converts the fibrinogen to fibrin monomers. The fibrin monomers will self-assemble into a fibrin polymer, forming a network weakly held together by hydrogen bonds. At the same time, thrombin activates factor XIII, which catalyzes the formation of cross-links in the fibrin polymer and between the polymer and the surrounding tissue, providing a stable the clot [49–51]. The sealants also contain calcium ions, since they are required for the reactions in the clotting cascade [52]. Commercial fibrin sealants, available in the United States, have components derived from blood banks and include TISSEEL®, Crosseal™, Beriplast®, and Evicel® [53]. Because fibrin glue is one of the oldest developed adhesives, it has a long list of reported applications, such as the repair of nerves [54], gastrointestinal tracts [55], topical wounds [56], and ophthalmology [57].

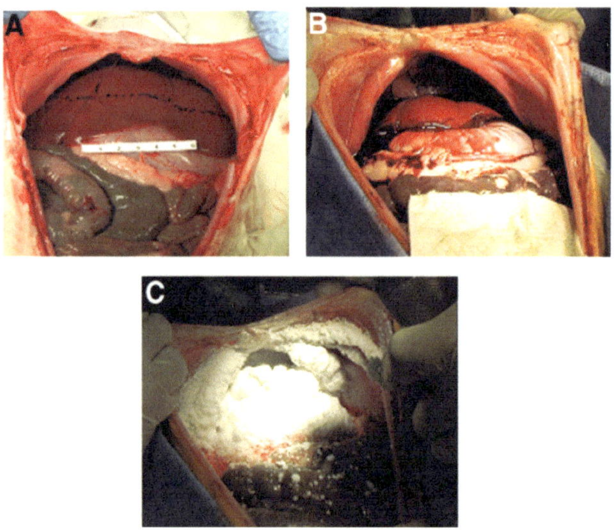

Figure 7. (A) Rabbit liver lobes shown before injury, (B) after the resection injury, and (C) after FSF application [62]. Copyright 2008. Reproduced with permission from Elsevier Inc.

The components are either lyophilized powders that require reconstitution prior to use, or liquid products that are stored frozen then thawed and mixed together in the surgical field [58]. The fibrinogen and thrombin components, supplied separately, can be applied successionally or simultaneously, with a dual-syringe or spray device. The material properties of the clot, and

http://dx.doi.org/10.5772/64958

thus its clinical performance, will be determined by its elasticity, tensile strength, and adhesion strength (Kjaerd 2000). For this reason, several groups have studied the polymerization and cross-linking characteristics of fibrin sealants. It was reported that key factors to ensuring glue performance are adequate fibrinogen content, effective mixing of the components, and maintenance of factor XIII activity [59]. Sierra et al. [60] found that the tensile strength, modulus of elasticity, burst strength, and failure strength, all increased with fibrinogen concentration.

Spray application of fibrin adhesive may have advantages over simultaneous and sequential drip application because spray application results in the most homogenous clots [61] and allows for better control over severe bleeding over large areas, which can occur after trauma. A self-expanding fibrin sealant foam (FSF) was characterized [62]. For preparation, the sealant was placed in a pressure-resistant bottle with liquefied gas propellant, which converted the fibrin components to foam at atmospheric pressure. It was used to arrest bleeding in a rabbit liver resection (**Figure** 7).

As another option, fibrin sealants have been made available as ready-to-use lyophilized powders that can be stored at room temperature. Fibrocaps (Raplixa; ProFibrix BV, Leiden, The Netherlands, a subsidiary of The Medicines Company) is dry powder thrombin and fibrinogen that can be applied straight to the wound site to form the clot [58]. For compressible injuries, lyophilized fibrinogen and thrombin have been combined with gauze (Larson 1995), as well as collagen sheets (Nistor 1997).

2.2.4.1. *Notable advances in fibrin sealants*

Fibrin glues are well known for being completely non-toxic and biodegradable [1], yet there are some disadvantages that limit their applicability in advanced applications. First, degradation in vivo will occur within days, a time frame that could allow for rebleeding. Aprotinin is a molecule consisting of 58 amino acids that is added to commercial fibrin glues to slow down its cell-mediated degradation [63]. When simply codissolved in a fibrin gel, aprotinin can freely diffuse out due to its small size. To prolong degradation time further, aprotinin has been covalently conjugated to fibrinogen [64]. Varying the levels of conjugated aprotinin allowed for control of the degradation time.

A second drawback to fibrin sealants is that they are mostly derived from human donors, or from the blood of animals with compatible clotting systems, limiting their availability, making them expensive, and introducing the possibility of transmission of blood borne pathogens. One option that has been investigated is deriving the protein components from animals where the evolutionary distance from humans is large, thus allowing for minimized risk of disease transmission. Fibrin components have thus been derived from salmon [65, 66] and crotalus durissus terrificus snake venom [67]. Another option is to use completely autologous fibrin glue. The Vivostat® System is an automated device that allows for the production of 5 mL of fibrin sealant from 120 mL of the patient's blood [68]. When Kjaergard et al. [69] compared its performance to commercially available Tissucol® and Beriplast®, autologous glue exhibited a higher elongation at break. However, the sealant takes about 30 min to prepare. Another report described a totally recombinant human sealant that exhibited comparable hemostatic efficacy to a commercial plasma-derived sealant in a porcine hepatic excision model [70].

A third important drawback to fibrin sealants is the poor adhesion properties, especially in wet environments, and low mechanical properties compared with most elastic tissues [2]. Researchers have been developing fibrin glues with alternative structures that lead to improved physical properties. A genipin cross-linked fibrin adhesive was described that was used to seal small defects in the annulus fibrosus of the intervertebral disc [71]. Genipin is a plant-based chemical cross-linker reactive with amines and reported to have low cytotoxicity [72]. Fibrin gels that were cross-linked with genipin were dimensionally stable over the 21-day in vitro study and exhibited higher shear stiffness than fibrin gels that were not cross-linked with genipin. In another study, elastic gels were produced via photochemical cross-linking of fibrinogen using $[RuII(bpy)_3]^{2+}$ and ammonium persulfate. Overall, fibrinogen gels cross-linked with this method are reported to have a higher tensile strength than Tisseel® [73].

2.2.5. Synthetic hydrogel sealants

2.2.5.1. Photocrosslinkable polyethylene glycol sealants

Although natural materials such as fibrin have the advantages of biodegradability and biocompatibility, the various shortcomings in fibrin properties have led to the development of synthetic polymer networks [74]. Engineers have more control over mechanical and degradation properties than with natural materials because polymer structure can be tailored. In these sealants, hemostasis is achieved using reagents that cross-link the polymer network, while binding with tissue, to close the injured site. Poly(ethylene glycol) (PEG) is an FDA-approved material that has been extensively used as the main component in synthetic adhesives [32].

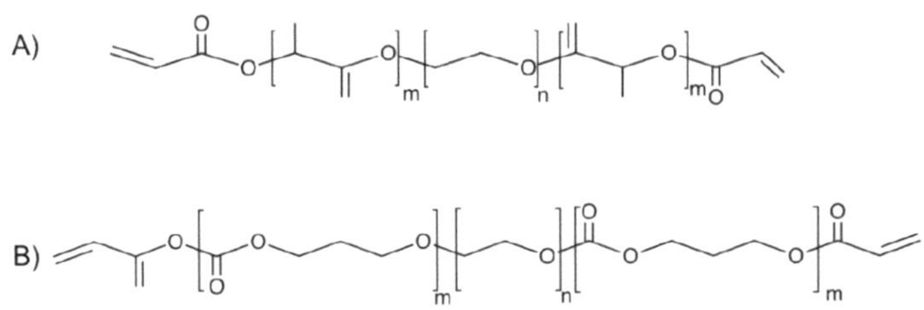

Figure 8. FocalSeal is formed from two different aqueous solutions of reactive macromers, (A) PLAm-PEGn-PLAm diacrylate and (B) PTMCm-PEGn-PTMCm diacrylate.

FocalSeal® was FDA-approved in 2000 for sealing air leaks after lung surgery. The sealant is provided to surgeons in two parts (primer and sealant). The primer is an aqueous solution of a triblock copolymer of poly(lactide) and polyethylene glycol (PLA-PEG-PLA) with acrylated end groups (**Figure 8A**). The sealant is an aqueous solution of poly(ethylene glycol)-co-trimethylene carbonate-co-lactide (PTMCm-PEGn-PTMCm) with acrylated end groups and a photoinitiator, eosin Y (**Figure 8B**). The primer and sealant solutions are combined immediately before application to the target area in the body. Upon exposure to light (450–550 nm), the macromers polymerize to form a cross-linked network due to the reaction between the acrylate groups [75]. The adhesion occurs due to mechanical interlocking, in other words, the liquid monomer solution infiltrates the pores and irregularities in the tissue surface, and, upon gel formation, adhesion occurs [32]. Over time, the lactide and trimethylene carbonate groups degrade by hydrolysis and the PEG chain are cleared by the kidneys. The sealant takes 3–4 weeks to resorb [1]. FocalSeal was evaluated as a dural substitute in a canine craniotomy model [75]. All treated sites remained free of leaks of cerebrospinal fluid (CSF) for the 56-day study. In a later clinical trial, 100% of the 46 patients receiving the dural sealant after cranial surgery remained free of CSF leaks [76]. FocalSeal was used with 100% success as a supplement to sutures or staples to repair air leaks after pulmonary resection in pigs [77]. It was also used successfully to repair retinal breaks [78].

A drawback to photopolymerization is that it may not be ideal for all parts of the body, as ultraviolet light cannot infiltrate tissue located greater than 5 mm of depth into the body [79]. Further, photoactivation makes application of the sealant difficult in the case of a hemorrhage [80]. There is a danger of releasing free radicals into the physiological environment [1]. Another disadvantage of FocalSeal specifically is that it can swell up to 300% of its original weight [81]. This could limit its application in areas of the body where there is potential for nerve compression [82].

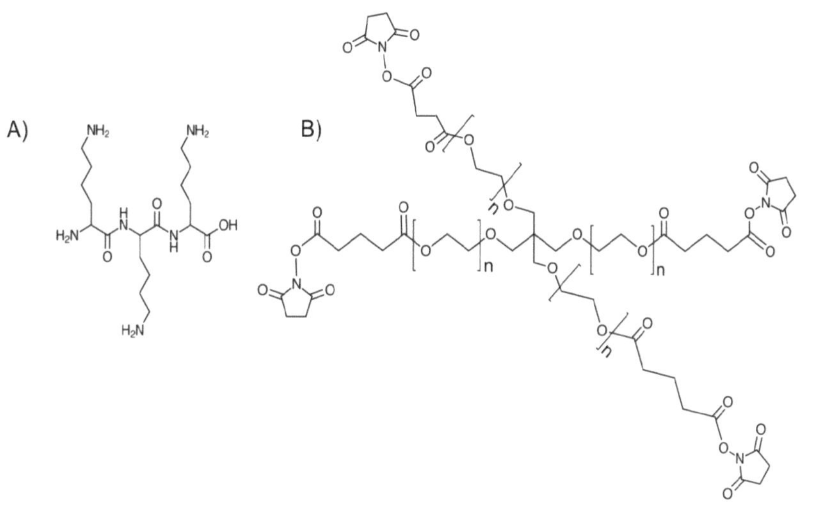

Figure 9. The reactive components of DuraSeal, (A) trilysine and (B) pentaerythritol poly(ethylene glycol) ether tetrasuccinimidyl glutarate.

DuraSeal® is a two-component system of trilysine (a tetramine cross-linker, **Figure 9A**), dissolved in a pH 10 borate buffer. The second is pentaerythritol poly(ethylene glycol) ether tetrasuccinimidyl glutarate (a 4-arm PEG encapped with N-hydoxysuccinimide (NHS) esters), dissolved in a sodium phosphate buffer (pH = 4) (**Figure 9B**) [1]. Upon mixing, the trilysine reacts with the NHS ester groups, generating the cross-linked hydrogel. In addition, upon contact with the amine groups in the extracellular matrix, the NHS-functionalized PEG chains will covalently bond with the surrounding tissue, providing adhesion combined with mechanical interlocking. Degradation takes place over 4–8 weeks [83]. One of the earliest reports of DuraSeal was in 2003 [84], when it was characterized to be an effective adjunct to sutures for dural closure in a canine model. A clinical trial with DuraSeal was reported in 2011 [85]. A total of 158 patients underwent spinal surgery requiring dural incision. 100% of the 102 patients that received DuraSeal combined with sutures had complete closure, compared with only 64% in the control group that received sutures alone.

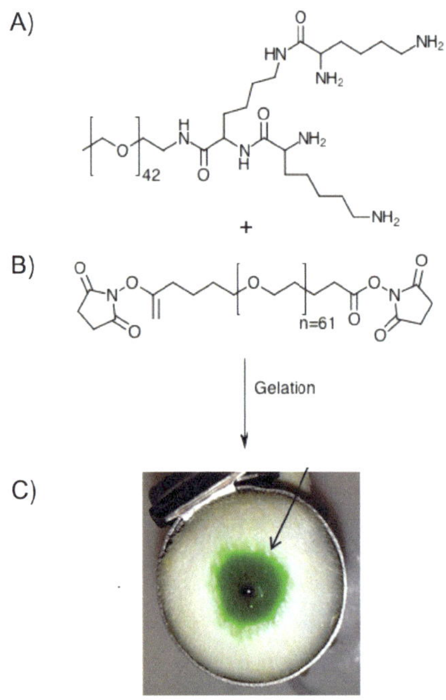

Figure 10. The combination of (A) a PEGylated lysine dendron and (B) PEG succinimidyl valerate results in the formation of a PEG-LysNH2 hydrogel, shown in (C) dyed with green food coloring [87]. Copyright 2015. Reproduced with permission from Biomed Central.

After gelation, DuraSeal can swell up to 50% after application [1]. To minimize swelling, another formulation was developed, DuraSeal® Xact Adhesion Barrier and Sealant System (DSX). Compared with DuraSeal, it has an increased cross-link density by modification of the ratio of PEG to trilysine [1, 86]. Villa-Camacho et al. [87] reported a novel sealant that uses similar chemistry to DuraSeal. Here, a PEGylated lysine dendron in dissolved in buffer at pH 9 is combined with solubilized PEG succinimidyl valerate at pH 6.5 (**Figure 10**). The hydrogel forms spontaneously upon mixing. The ability of the adhesive to withstand pressures analogous to human arterial pressures was demonstrated ex vivo.

CoSeal™ is a two-component tissue sealant. The first component 20% (w/v) buffer solution of a 4-arm PEG polymer of 10 kDa molecular weight end capped with thiol groups (pH 9.6) (**Figure 11**), and the second component is a 20% (w/v) buffer solution of pentaerythritol poly(ethylene glycol) ether tetrasuccinimidyl glutarate (pH 6.0) (**Figure 9B**). The components are reactive with each other and the proteins in the surrounding ECM [88], providing adhesion via chemical bonding and mechanical interlocking. The hydrogel degrades within several weeks due to hydrolysis. In one of the earliest studies with CoSeal, it lessened blood loss and time to hemostasis in bleeding rabbit arteries compared with using a tamponade [89]. CoSeal has been reported as cumbersome to use because it is supplied as a powder, which requires mixing back and forth 20 times to dissolve [90]. It has also been reported to swell to 400% of its original size [1, 90]. Likely related to the high swelling characteristics, the mechanical strength of CoSeal has been found to be weaker than DuraSeal and other commercially available protein-based and cyanoacrylate sealants [30].

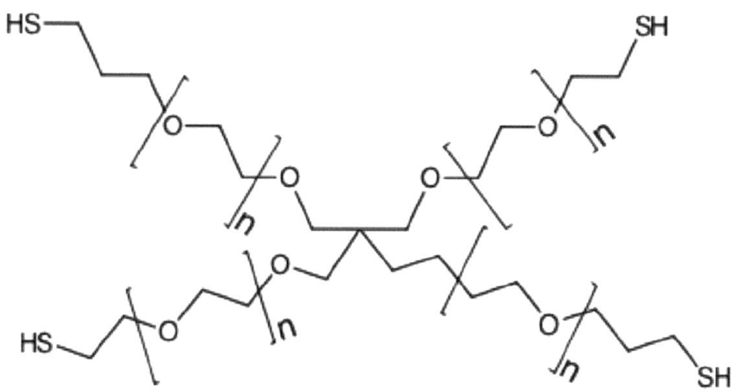

Figure 11. Structure of thiol end capped 4-arm PEG used in CoSeal®.

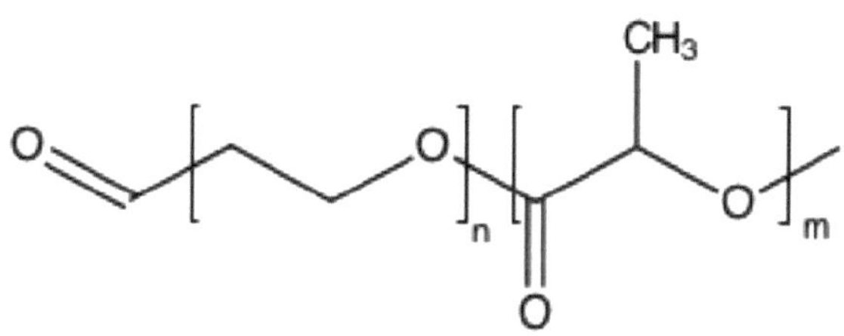

Figure 12. The structure of aldehyde-terminated poly(ethylene glycol)-poly(D,L-lactide) (PEG-PLA) block copolymers prepared by Murakami et al. [80].

Another group of PEG adhesives uses Schiff's base chemistry as the cross-linking reaction. For example, an 8-arm PEG molecule was functionalized with either aldehyde or amine groups. When combined in water, it spontaneously formed gels because of the reaction between the aldehyde and free amine groups [91]. Gels formed from PEG arms that were 10 kDa in size were stiffer than those formed from 20 kDa due to higher cross-link density. In vitro lap shear testing showed that the adhesive was stronger than fibrin glue (Tissucol). It swelled to about 200% of its original weight, which is less than CoSeal and FocalSeal. The results of a direct contact cell assay with L929 fibroblasts showed no cytotoxicity associated with the cured gels. Aldehyde-terminated poly(ethylene glycol)-poly(D,L-lactide) (PEG-PLA) micelles were also prepared (**Figure 12**) by dissolving the block copolymer in aqueous media. A hydrogel forms in ~2 s when a polyallylamine is added to the micelles. Adhesive strength of the network was found to be proportional with aldehyde concentration, with the strength of the high aldehyde content adhesives comparable to fibrin glue [80].

2.2.5.2. Notable advances in PEG adhesives

Researchers continue to innovate systems that allow them to achieve high adhesive strengths, ease of use, fast curing, biodegradability, and biocompatibility with alternate methods of cross-linking. Hu et al. [92] incorporated peptide substrates of transglutaminase into linear or branched polymers of PEG. The in situ formed gels exhibited adhesive strengths to porcine skin comparable to that of fibrin glue [92].

Systems have been developed that are combinations of synthetic and natural components. These have potential to exhibit improved biocompatibility compared to systems that are completely synthetic. For instance, cells do not adhere to PEG hydrogels. Hynes et al. [93] chose poly(L-lysine) (PLL) to be incorporated into PEG structure because the charged side chain promotes cell survival and adhesion. Macromers of PEG/PLL, shown in **Figure 13**, were formed hydrogels via photopolymerization and shown to support survival and differentiation of neural precursor cells. Another group [94] reported a gelation system consisting of the following components: calcium-loaded liposomes composed of 1,2-bis(pal-mitoyl)-sn-glycero-3-phosphocholine (DPPC), human recombinant factor XIII, thrombin, and 4-arm PEG, each arm terminating in a 20-mer sequence derived from fibrin. Heating the system to 37°C causes the liposome bilayer to melt, releasing the calcium and activating factor XIII, causing gelation to occur in 9 min.

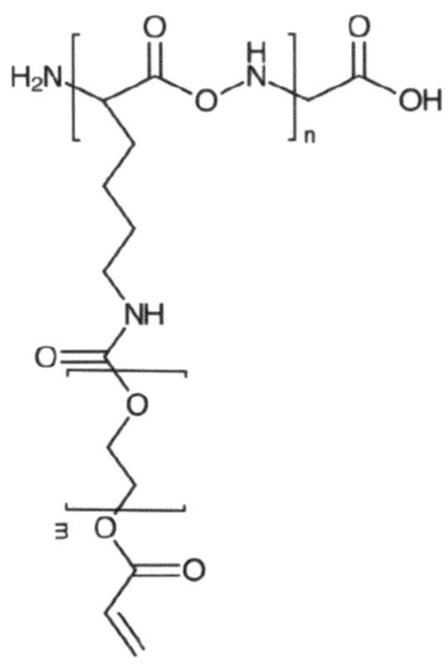

Figure 13. The molecular structure of PEG/PLL macromers for the formation of photopolymerized hydrogels developed by Hynes et al. [93].

Besides improving biocompatibility, researchers are working on developing stronger PEG adhesives. For example, by combining silk, which has a high mechanical strength, with PEG, a sealant was produced with fast curability, superior adhesion strength to CoSeal, and swelling ratios of only 60–70% due to the hydrophobicity of the silk [90].

2.3. Polysaccharides

In situ forming tissue sealants have also been prepared from natural polysaccharides, which are composed of repeat units of naturally occurring sugars, making them inherently biodegradable non-immunogenic [1]. Like synthetic polymers, the chemical structure of polysaccharides can be modified to induce spontaneous or light-triggered gelation in situ and enhance adhesion strength with surrounding tissue.

Photopolymerization of polysaccharides that are functionalized with acrylate or methacrylates along the backbone is one way to achieve in situ gel formation. In this case, the polysaccharide may be applied in liquid form, where it will penetrate the crevices in the tissue. Upon polymerization, reacted acrylates or methacrylates form covalent cross-links and the resulting gel seals the area [95]. An example is the methacrylation and photopolymerization of hyaluronic acid (HA), a linear polysaccharide consisting of alternating β-(1–4) linked 2-acetamine-2-deoxyglucose and β (1–3) linked d-glucuronic acid (**Figure 14A**). Methacrylate-functionalized hyaluronic acid (HA) (**Figure 14B**) was applied to corneal lacerations in rabbits to allow for network formation upon irradiation [96]. This produced a flexible clear patch that sealed 97% of the corneal lacerations and retained its shape and size for the duration of the 7-day study. The authors hypothesized that the high molecular weight, entangled 3D structure of the HA slowed down its enzymatic degradation in vivo.

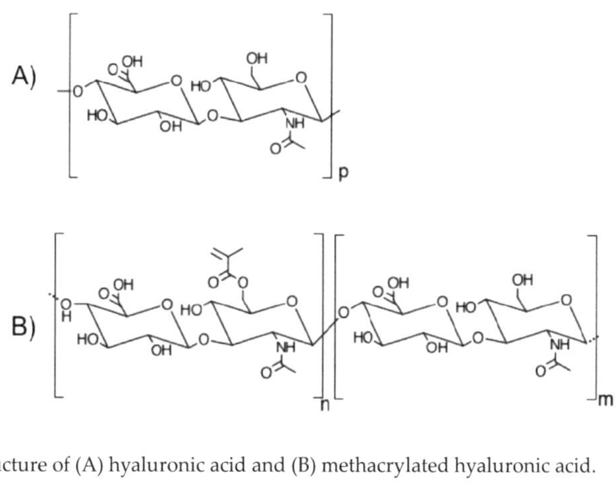

Figure 14. Chemical structure of (A) hyaluronic acid and (B) methacrylated hyaluronic acid.

As an alternative to photopolymerization, oxidation of polysaccharides to aldehyde groups using sodium periodate [97] is widely applied for forming polysaccharide networks. The aldehyde functionalities readily react amine-containing molecules via Schiff's base reaction to internally cross-link a polysaccharide to itself and surrounding tissue ECM [7]. Sodium alginate (**Figure 15**) derived from brown algae, functionalized with both methacrylate and aldehyde groups, was investigated as a tissue sealant. Cross-linking of the methacrylate groups was achieved by visible light. Simultaneously, the aldehydes reacted with the amines in the surrounding ECM, allowing for covalent attachment to surrounding tissue. Gels that were methacrylated, but not oxidized, maintained mechanical integrity during burst pressure tests, but delaminated from the substrates easily. On the other hand, materials that received both methacrylate and aldehyde groups did not exhibit delamination before failure [98].

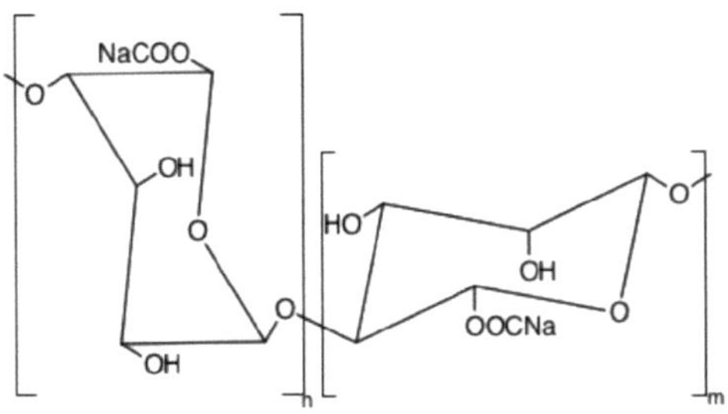

Figure 15. The chemical structure of sodium alginate.

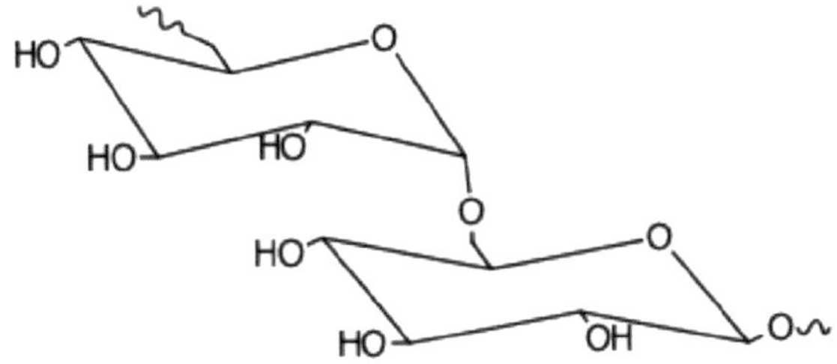

Figure 16. The chemical structure of dextran.

Dextran is another polysaccharide that has been extensively investigated as a sealant in various combinations with other polymers. Dextran is a microbial product consisting of about 95% α-1,6-linked d-glucopyranose and 5% α-(1,3) linkages (**Figure 16**). Araki et al. developed a new bioadhesive that forms from the reaction of dextran aldehyde and ε-poly(L-lysine). The components were applied in a pulmonary air leakage model in a canine model. The sealant had significantly higher burst pressure than fibrin glue and degraded more slowly [99]. Structure-property relationships in this adhesive were studied further [8]. It was found that as the degree of oxidation (DOO) of the dextran increased, gelation time decreased. Gelation times ranged between 3 s (40% oxidation) to 60 s (12% oxidation) [8]. Increased DOO of dextran was also associated with a slower degradation time [100].

In another study, dextran aldehyde was cross-linked with either 8-arm star PEG amine or linear PEG diamine [101]. A higher stability cohesion was observed for gels made from the 8-arm star PEG because the larger number of available amines allowed for a higher degree of cross-

http://dx.doi.org/10.5772/64958

linking. Another group, Bhatia et al. [102] studied adhesives based on dextran aldehyde and 8-arm PEG. They reported the system to cure in less than 1 min, adhere in a water environment, and degrade hydrolytically. The adhesive was utilized to seal a 5-mm corneal incision in eyes of New Zealand white rabbits [103]. They found that sealants produced from higher polymer contents led to better sealing and higher leak pressures. Degradation occurred within three days, appropriate time frame for corneal epithelial healing, and no cytotoxicity was observed from the cured adhesive. In a follow-up work, the degradation time was extended to 5 days by using an 8-arm PEG containing two primary amine groups at the end of each arm, instead of one [104].

Chondroitin sulfate (CS), a constituent of cartilage and corneas, is comprised of alternating glucuronic acid and N-acetyl-galactosamine groups (**Figure 17**) [1]. CS was functionalized with succinimidyl succinate (CS-NHS) and combined with 6-armed polyethylene glycol (PEG) amine as a corneal sealant [105]. This adhesive was developed as an alternative to using CS aldehyde, hypothesizing that using a CS with an intact backbone helps to preserve its biological activity. Upon mixing, the materials gelled within 1–2 min and cured polymers exhibited no cytotoxicity in vitro with rabbit primary epithelial, stromal, and endothelial cells. No adverse effects were found with implantation into in vivo swine cornea.

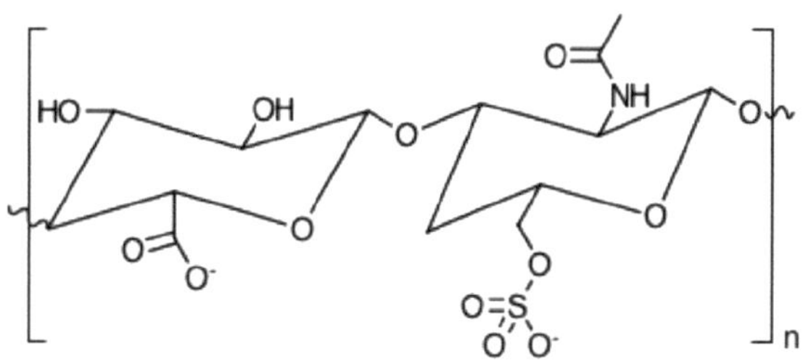

Figure 17. The chemical structure of chondroitin sulfate.

2.3.1. Notable advances in polysaccharide adhesives

Adhesives made out of chitin derivatives have the potential to be valuable additions to the presently developed spectrum of polysaccharide-based sealants. Chitin is comprised of repeat units of 1,4 b-linked N-acetyl-D-glucosamine (**Figure 18A**). Chitosan is formed from partially deacetylated chitin [106] (**Figure 18B**). Since chitin is highly hydrophobic and completely insoluble in water, chitosan is more often used in biomedical applications. Chitosan has been shown to have a number of biological advantages, including being beneficial for wound healing [107, 108], inherent hemostatic ability [109], and antimicrobial activity [110, 111]. Chitosan has adhesive properties due to the positively charged amino groups, which interact with negative charges in mucus [112] and cell membranes [113]. However, one downside to chitosan is that it has limited solubility in water at neutral pH. Thus, researchers modify chitosan with side groups that enhance solubility in water at physiological conditions. For example, chitosan was modified with lactose groups to enhance water solubility [106] and cross-linked in situ with the addition of photoreactive azide groups. The resulting rubber-like adhesive was an effective sealant in rabbit arteries and lungs. In another study, a new adhesive was developed based on 2-hydroxy-3-methacryloyloxypropylated carboxymethyl chitin derivatives (HMA-CM-chitin) and embedded chitosan nanofibers. It was shown in vitro to have comparable strength to cyanoacrylate since the nanofibers provided gel network reinforcement [108].

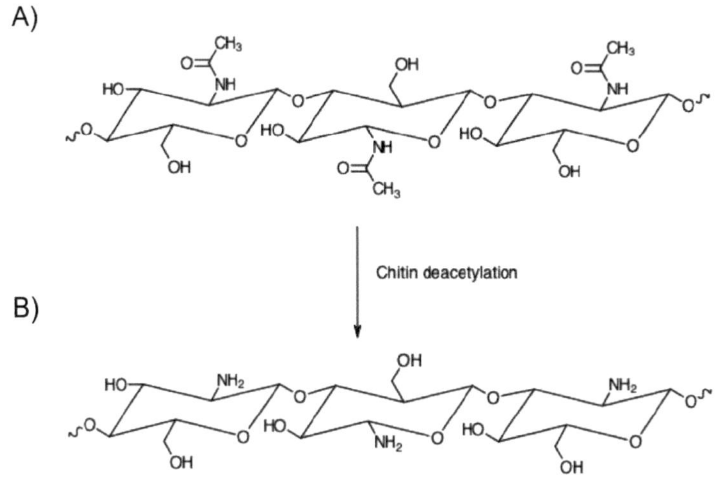

Figure 18. The structure of (A) chitin and (B) partially deacetylated chitin, also referred to as chitosan.

2.4. Nature-inspired materials

2.4.1. Mussel-inspired sealants

The most extensively investigated class nature-inspired materials uses mussel adhesive proteins (MAPs). Mussels can adhere strongly to underwater structures, even in environments with high mechanical stresses. Mussels attach to surfaces because they secrete a thread-like adhesive called byssus, which is composed of 25–30 different types of MAPs [114, 115]. An amino acid residue in these proteins, tyrosine, is posttranslationally modified to 3,4-dihydroxy-phenylalanine (DOPA) residues, which contains a catechol group (**Figure 19**). Catechols can hydrogen bond with other catechols [116], as well as amino, hydrogen and carboxyl resides [117]. Catechols can be oxidized by metal ions or enzymes to form quinones (**Figure 19**). Quinones form covalent bonds with amine or thiol groups [114, 116, 118, 119]. Finally, the catechol group complexes with multivalent ions, such as Fe3+ [120, 121] to form networks. Researchers have been developing materials that recapitulate these mechanisms in order to achieve strong bonding in wet environments.

Extracting MAPs from mussels or genetically engineering the proteins can result in low yields [122]. Thus, catechol pendant groups have been attached to the backbone of natural and synthetic polymers to yield biomimetic adhesives. Catechol auto-oxidizes at slightly alkaline pH >8, or in the presence of oxidizing reagents, such as periodates [123]. Researchers use these methods to produce quinones in situ, which leads to covalent bonding with tissue. In some of the early work with mussel-inspired synthetic polymers, linear and branched PEGs were functionalized with either 1,2, or 4 DOPA end groups [124]. Gels were formed with the use of three different oxidation reagents: sodium periodate, horseradish periodate, and mushroom tyrosinase. Rapid gelation occurred when 2 or more DOPA functionalities were present on the PEG molecule. However, PEG-based hydrogels can act brittle, fracturing at low elongation [125]. Improvements in the mechanical properties of PEG-catechol hydrogels were achieved by using block copolymers of PEG and poly(caprolactone) with DOPA groups. Because PCL is hydrophobic, it prevented swelling of the hydrogels, so they were not as fragile. Moreover, PCL made the gels less brittle, since it has a high fracture strain [126]. In another work, poly[(3,4-dihydroxystyrene)-co-styrene] copolymers (**Figure 20**) were prepared [119]. The adhesive characteristics in the presence of periodate on plastic and wood surfaces were investigated and there was found to be an optimal pendant catechol concentration along the polymer backbone. When the concentration of catechol was too low, there were insufficient cross-linking sites available at the surface. Too high a concentration would bias the system toward internal cross-linking, and surface attachment suffered.

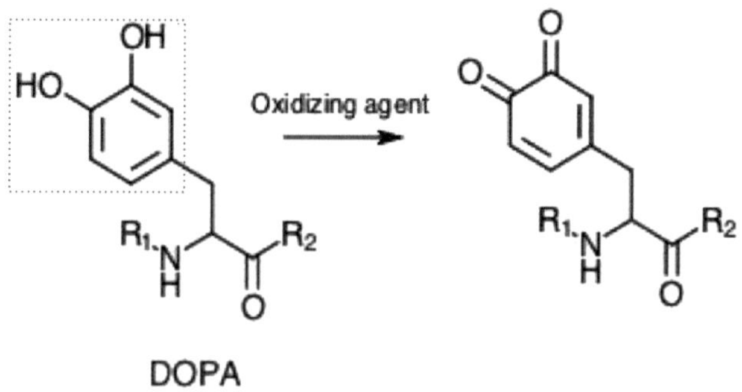

Figure 19. (A) 3,4-Dihydroxy-phenylalanine (DOPA) residues, with the catechol group indicated by the dotted line and (B) quinone formed from the oxidation of the catechol group.

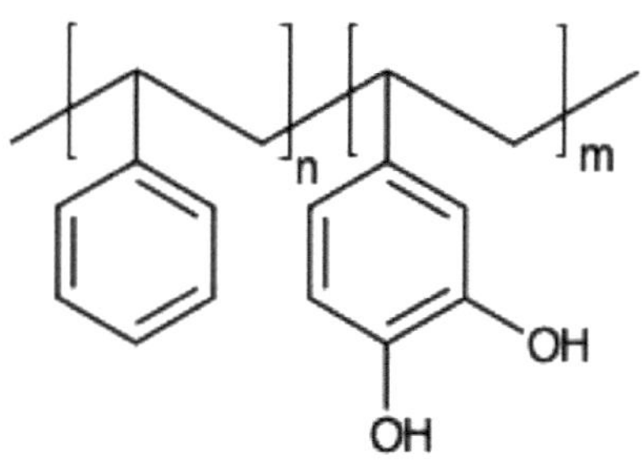

Figure 20. The chemical structure of poly[(3,4-dihydroxystyrene)-co-styrene] copolymers developed by Matos-Perez et al. [119].

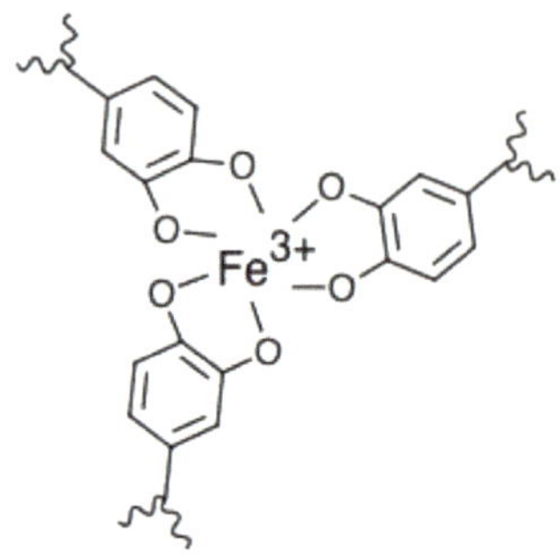

Figure 21. Fe^{3+} serves as a rapid cross-linker for DOPA-containing polymers by forming complexes with the catechol groups.

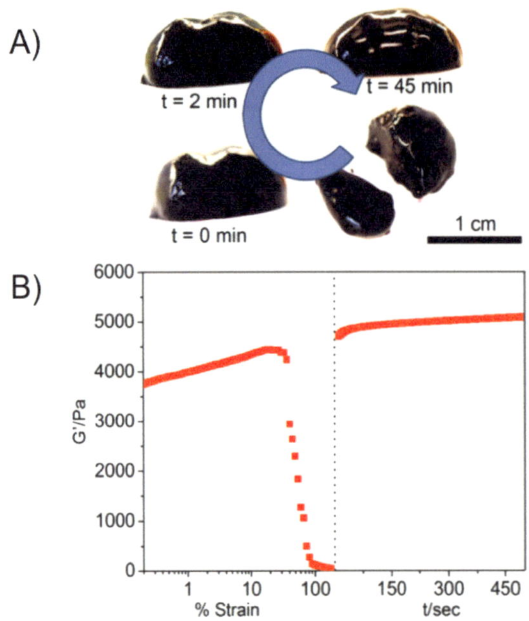

Figure 22. Qualitative (A) and quantitative (B) recovery tests for hydrogels composed of polyallylamine in solutions of $FeCl_3$. In the qualitative tests (A), a blob of hydrogel was fractured and the fragments healed upon contact with each other. This process was quantified with dynamic oscillatory rheology (B) by shearing a hydrogel at increasing strain at 1 s^{-1} from 0.01 until 200% strain. The recovery was recorded at 1% strain and 1 s^{-1} [120]. Copyright 2013. Reproduced with permission from American Chemical Society.

Oxidizing reagents like periodate are expected to exhibit toxicity in biological systems [118]. Interestingly, DOPA in its unoxidized state has bioadhesive properties [114]. For instance, functionalization of poly(ethylene oxide)-poly(propylene oxide)-poly(ethylene oxide) (PEO-PPO-PEO) block copolymers with DOPA end groups was performed. Gelation of these polymers in situ occurred because of the temperature-induced phase transition of the block copolymer. The results showed significantly improved adhesion over PEO-PPO-PEO alone

[118]. Ryu et al. functionalized PEO-PPO-PEO with terminal thiol groups [127] and blended it with catechol-functionalized chitosan. Again, immediate in situ gelation was achieved via the thermal transition of the PEO-PPO-PEO. Adhesive characteristics were imparted to the system via the catechol groups on the chitosan, which hydrogen bonded with other catechol groups on the chitosan, the thiols, and components in the tissues. A novel approach was used by Fan et al. [128]. The system they designed consisted of DOPA conjugated gelatin, a rapid cross-linker Fe^{3+}, and genipin. While the dopamine moieties impart interfacial adhesion with surrounding tissue, the Fe^{3+} served as a rapid cross-linker by forming complexes with the catechol groups (**Figure 21**). Because these complexes are unstable due to their reversibility, genipin was included in the formulation. It gradually reacted with the amino groups on the gelatin over hours, providing stable network cohesion over time. Metal-DOPA bonds can reform after breaking, so separated pieces of hydrogel can remerge when placed in contact with one another [129]. Krogsgard et al. [120] took advantage of this property to create "self-healing" DOPA-functionalized gels composed of polyallylamine in solutions of $FeCl_3$. Oscillatory rheology quantitatively shows the spontaneous reformation of separated hydrogel pieces (**Figure 22**).

2.4.2. *Gecko-inspired adhesives*

Another type of nature-inspired sealant comes from geckos. Geckos have an amazing ability to climb walls because their feet are covered with millions of tiny hairs that provide significant adhesion. Only in the last twenty years or so has the mechanism responsible for this adhesion

A)
B)

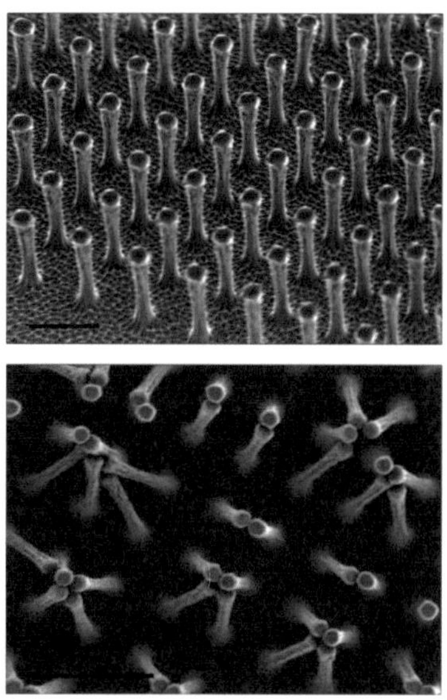

Figure 23. (A) Scanning electron micrographs of polyimide hairs in a gecko-inspired adhesive. (B) Example of bunching pillars, which could reduce adhesive strength of the adhesive (scale bar 2 μm) [134]. Copyright 2003. Reproduced with permission from Nature Publishing Group.

become understood. While any submicron object can stick to a surface, the strength of the adhesion depends on the shape and chemistry involved [130–133]. In the case of hydrophilic materials, capillary action with absorbed water plays a role [132, 133]. In certain sub-micron size ranges, van der Waals forces also promote adhesion to surfaces. The diameter of gecko hairs is such that both capillary action and van der Waals forces are maximized [131], so they can climb surfaces of varying hydrophilicity [134]. To create a gecko-inspired adhesive, one must manufacture a flat surface containing pillars that are sufficiently flexible so they can act in unison to adhere to a substrate (**Figure 23A**). The density of pillars must be maximized in order to generate sufficient adhesion force [134]. Bunching of the pillars (**Figure 23B**) is one mechanism that decreases adhesive strength [134].

Gecko adhesion disappears when submerged in water [135, 136]. An interesting approach to overcoming this limitation and developing a wet/dry adhesive was taken by a group who combined gecko- and mussel-inspired materials into one device. They fabricated pillars from poly(dimethyl siloxane) (PDMS). Separately, they synthesized a mussel mimetic polymer, poly(dopamine-methacrylamide-co-methoxy-acrylate) and used it to coat the PDMS. The addition of the mussel mimetic material to the PDMS increased its wet adhesion by 15 times. However, adhesion was only studied on inorganic surfaces [137]. Mahdavi et al. [138] developed a wet/dry gecko-inspired adhesive and examined adhesion to organic substrates. A micropatterned biodegradable (polyglycerol sebacate acrylate) elastomer was used. It was coated with oxidized dextran to introduce a mechanism for covalent cross-linking with tissue. Simultaneously, the aldehydes reacted with the free hydroxyl groups on the glycerol subunit of the elastomer. Coating the pillars of elastomer with the dextran significantly increased the interfacial adhesion strength in vitro and in vivo.

3. Next-generation applications in tissue engineering

Tissue engineering is a fast-growing area of research that seeks to rebuild damaged tissues and organs. Current strategies involve seeding biomaterials, called scaffolds, with cells and/or biologically active molecules that allow or induce tissue regrowth in a three-dimensional structure [139]. Tissue engineering of certain areas in the body, particularly cartilage [122] and the intervertebral disc [140, 141], is not thought to be clinically feasible without a scaffold that can form an interface with surrounding tissue, eliminating the risk of dislocation and restoring mechanical functionality by adequately transmitting forces to and from the surrounding tissue [122, 140–142]. A scaffold could potentially be sutured in place, but ideally, it would integrate with surrounding tissues via a bioadhesive mechanism while supporting encapsulated cellular function.

Fibrin was the first class of adhesives to be studied for tissue engineering because the gels are completely biocompatible, biodegradable, and capable of delivering growth factors [143, 144]. Additionally, fibrin has been shown to be permissible to in vitro mesenchymal stem cell differentiation toward a chondrogenic lineage [145]. Promising in vivo results with fibrin have been reported in adipose [146], bone [147, 148], cardiac [149], and neural tissue engineering

[150]. Despite these positive results, there are significant limitations with using fibrin adhesives for some tissue engineering applications. Fibrin glues are unstable over time, thus making them non-ideal for the long repair processes [151]. The low mechanical properties make the glues inappropriate for load bearing applications [152, 153], like cartilage and the intervertebral disc. Fibrin gels also have a low cohesive strength with tissue [154, 155].

Besides fibrin glue, the efficacy of other bioadhesive polymers as in situ forming cell carriers for tissue engineering has not been extensively studied. Rather, the biocompatibility of cured adhesives with surrounding cells is typically investigated in the development of new glue formulations. The presence of reactive molecules, such as glutaraldehyde, aldehydes, NHS esters, or the presence of toxic monomers, as in cyanoacrylates, is likely to adversely affect encapsulated cells prior to gel network formation. Only a few groups have sought to investigate this. Adhesive hydrogels composed of gelatin, coupled to oxidized hyaluronic acid (DOO 40%), and cross-linked with adipic acid dihydrazide were characterized as cell carriers. The adipic acid dihydrazide (**Figure 24**) bridged hyaluronic acid-gelatin molecules via an imide bond. Intervertebral disc cells were encapsulated within the gels by suspension in the macromer solution prior to cross-linking. Two days after gel formation, viable cells were shown to be attached to the material [156]. Chondrocytes were shown to survive for 3 weeks after being encapsulated within chondroitin sulfate-NHS networks and combined with bone marrow and recombinant bone morphogenetic protein-2 [157]. These are promising preliminary results, but it is important to note that most current bioadhesive polymers have functional groups that are not only reactive with components of ECM, but cell membranes, as well. More in-depth studies are warranted, as experimental outcomes are likely to be impacted by cell type and density, material reactivity, concentration of reactive groups, and the presence of biomolecular signals such as growth factors.

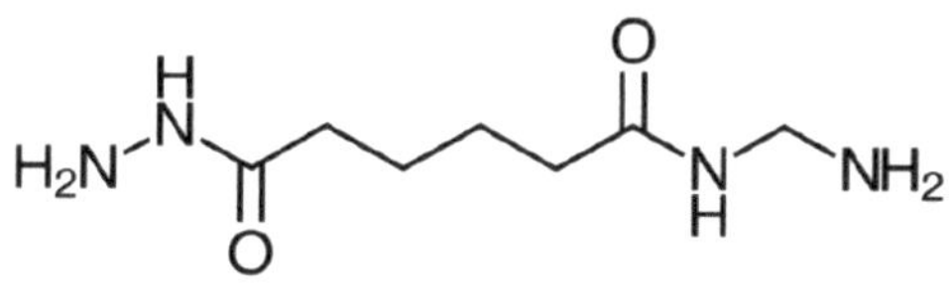

Figure 24. Chemical structure of adipic acid dihydrazide.

Wiltsey et al. [142] investigated the use of polysaccharides without reactive groups as bioadhesive tissue engineering scaffolds. Thermosensitive copolymers of poly(N-isopropylacrylamide) grafted with chondroitin sulfate (PNIPAAm-g-CS) containing mucoadhesive alginate microparticles as the adhesion mediators were developed. Below its lower critical solution temperature (LCST) around 30°C, PNIPAAm-g-CS forms a miscible solution with water. Above the LCST, the PNIPAAm becomes hydrophobic and an elastic hydrogel is formed without the use of monomers or cross-linkers [158]. Alginate microparticles and human embryonic kidney 293 cells were suspended in aqueous solutions of PNIPAAm-g-CS and encapsulated in the polymer network upon gelation at physiological temperature. Inclusion of the alginate microparticles significantly increased the tensile adhesive strength compared with PNIPAAm-g-CS alone and fibrin glue. Also, the formulation exhibited significantly less cytotoxicity than PNIPAAm-g-CS formulations containing CS-aldehyde as the adhesion mediator. The microparticle-hydrogel composites are now being investigated as scaffolds for intervertebral disc tissue engineering.

4. Conclusions

Due to differences in mechanical properties and cellularity in the various tissues of the body, it would be impractical to develop one sealant that can serve all applications [6]. However, there are some basic requirements that all sealants must have in order to meet present and future clinical needs:

- Easy to apply and rapid solidification to conform to irregular shaped defects or tissue areas.
- Able to provide sufficient mechanical support to the injured tissue as it heals.
- Strong adhesion in order to maintain approximation of tissue pieces, prevent dislocation, and/or ensure adequate transmission of forces across tissue/implant interface.
- Provoke a minimal degree of inflammation or toxicity from surrounding tissues with no risk of disease transmission.
- Biodegradation within a time frame that corresponds to the gradual formation of new tissue.
- Able to promote tissue growth and repair by acting as a suitable 3D culture system by promoting cell adhesion and extracellular matrix production.
- Possible to produce in large scale quantities and cost effective.

Presently, no current bioadhesives have been shown to fulfill all these requirements. Unanswered questions still exist on the suitability of the current formulations for tissue engineering applications. As the field of biomedical sealants evolves to encompass new applications, researchers are working to develop new formulations with increasing biocompatibility, adhesive strength, flexibility and degradability. These advances promise to improve clinical outcomes by enhancing bleeding control and wound healing. In the long term, the use of bioadhesives in tissue regeneration research will become more prominent. New innovations will pave the way towards making orthopedic and musculoskeletal tissue engineering clinically successful in the future.

Acknowledgements

The author is supported by a grant from the National Institute of Arthritis and Musculoskeletal and Skin Diseases and the National Institute of Biomedical Imaging and Bioengineering of the National Institutes of Health under Award Number 1R15 AR 063920–01. The content is solely the responsibility of the author and does not necessarily represent the official views of the

National Institutes of Health. The author also gratefully acknowledges Tobi Schwartz-Cassell for editing and proofreading assistance.

Author details

Andrea J Vernengo

Address all correspondence to: Vernengo@rowan.edu

Henry M. Rowan College of Engineering, Rowan University, Glassboro, NJ, USA

References

[1] Bouten PJM, Zonjee M, Bender J, Yauw STK, van Goor H, van Hest JCM, Hoogenboom R. The chemistry of tissue adhesive materials. Progress in Polymer Science. 2014;39:1375–405. DOI: 10.1016/j.progpolymsci.2014.02.001

[2] Annabi N, Tamayol A, Shin SR, Ghaemmaghami AM, Peppas NA, Khademhosseini A. Surgical materials: current challenges and nano-enabled solutions. Nano Today. 2014;9:574–89. DOI: 10.1016/j.nantod.2014.09.006

[3] Tajirian AL, Goldberg DJ. A review of sutures and other skin closure materials. Journal of Cosmetic and Laser Therapy. 2010;12:296–302. DOI: 10.3109/14764172.2010.538413

[4] Bruce J, Krukowski ZH, Al-Khairy G, Russell EM, Park KGM. Systematic review of the definition and measurement of anastomotic leak after gastrointestinal surgery. British Journal of Surgery. 2001;88:1157–68. DOI: 10.1046/j.0007-1323.2001.01829.x

[5] Pickleman J, Watson W, Cunningham J, Fisher SG, Gamelli R. The failed gastrointestinal anastomosis: an inevitable catastrophe? Journal of the American College of Surgeons. 1999;188:473–82. DOI: 10.1016/S1072-7515(99)00028-9

[6] Annabi N, Yue K, Tamayol A, Khademhosseini A. Elastic sealants for surgical applications. European Journal of Pharmaceutics and Biopharmaceutics. 2015;95:27–39. DOI: 10.1016/j.ejpb.2015.05.022

[7] Peng HT, Shek PN. Development of in situ-forming hydrogels for hemorrhage control. Journal of Materials Science: Materials in Medicine. 2009;20:1753–62. DOI: 10.1007/s10856-009-3721-5

[8] Hyon SH, Nakajima N, Sugai H, Matsumura K. Low cytotoxic tissue adhesive based on oxidized dextran and epsilon-poly-L-lysine. Journal of Biomedical Materials Research Part A. 2014;102:2511–20. DOI: 10.1002/jbm.a.34923

[9] Singer AJ, Thode HC, Jr. A review of the literature on octylcyanoacrylate tissue adhesive. American Journal of Surgery. 2004;187:238–48. DOI: 10.1016/j.amjsurg.2003.11.017

[10] Albes JM, Krettek C, Hausen B, Rohde R, Haverich A, Borst HG. Biophysical properties of gelatin-resorcin-formaldehyde/glutaraldehyde adhesive. Annals of Thoracic Surgery. 1993;56:910–5.

[11] Pronio A, Di Filippo A, Narilli P, Caporilli D, Vestri A, Ciamberlano B, Pelle F, Montesani C. Closure of cutaneous incision after thyroid surgery: a comparison between metal clips and cutaneous octyl-2-cyanoacrylate adhesive. A prospective randomized clinical trial. European Journal of Plastic Surgery. 2010;34:103–10. DOI: 10.1007/s00238-010-0477-6

[12] Silvestri A, Brandi C, Grimaldi L, Nisi G, Brafa A, Calabro M, D'Aniello C. Octyl-2-cyanoacrylate adhesive for skin closure and prevention of infection in plastic surgery. Aesthetic Plastic Surgery. 2006;30:695–9. DOI: 10.1007/s00266-006-0139-z

[13] Moreno-Egea A. Is sutureless hernia repair a safe option for treating abdominal wall hernias? A prospective study with a synthetic tissue adhesive (n-hexyl-alpha-cyanoacrylate). Cirugía Española (English Edition). 2013;91:243–9. DOI: 10.1016/j.cireng.2013.09.017

[14] Qassemyar Q, Michel G. A new method of sutureless microvascular anastomoses using a thermosensitive poloxamer and cyanoacrylate: an experimental study. Microsurgery. 2015;35:315–9. DOI: 10.1002/micr.22381

[15] Idle MR, Monaghan AM, Lamin SM, Grant SW. N-butyl-2-cyanoacrylate (NBCA) tissue adhesive as a haemostatic agent in a venous malformation of the mandible. British Journal of Oral Maxillofacial Surgergy. 2013;51:565–7. DOI: 10.1016/j.bjoms.2012.10.017

[16] Bresnahan KA, Howell JA, Wizorek J. Comparison of tensile strength of cyanoacrylate tissue adhesive closure of lacerations versus suture closure. Annals of Emergency Medicine. 1995;26:575–8.

[17] Lim JI, Lee Y-K, Shin J-S, Lim K-J. Cyanoacrylate adhesives curable to flexible polymeric materials by poly(l-lactide-co-ε-caprolactone). Materials Letters. 2010;64:2438–40. DOI: 10.1016/j.matlet.2010.07.064

[18] Quinn JV. Tissue Adhesives in Clinical Medicine. Lewiston, NY: PMPH-USA; 2005.

[19] Bhatia SK. Traumatic Injuries Chapter 10. Biomaterials for Clinical Applications. New York, NY: Springer New York; 2010. p. 213–58.

[20] Trott AT. Cyanoacrylate tissue adhesives: an advance in wound care. Journal of the American Medical Association. 1997;277:1559–60. DOI: 10.1001/jama.1997.03540430071037

[21] Montanaro L, Arciola CR, Cenni E, Ciapetti G, Savioli F, Filippini F, Barsanti LA. Cytoxicity, blood compatibility and antimicrobial activity of two cyanoacrylate glues for surgical use. Biomaterials. 2001;22:59–66.

[22] Dragu A, Unglaub F, Schwarz S, Beier JP, Kneser U, Bach AD, Horch RE. Foreign body reaction after usage of tissue adhesives for skin closure: a case report and review of the literature. Archives of Orthopaedic and Trauma Surgery. 2009;129:167–9. DOI: 10.1007/s00402-008-0643-5

[23] Brochu AB, Evans GA, Reichert WM. Mechanical and cytotoxicity testing of acrylic bone cement embedded with microencapsulated 2-octyl cyanoacrylate. Journal of Biomedical Materials Research Part B: Applied Biomaterials. 2014;102:181–9. DOI: 10.1002/jbm.b.32994

[24] Brochu AB, Matthys OB, Craig SL, Reichert WM. Extended fatigue life of a catalyst free self-healing acrylic bone cement using microencapsulated 2-octyl cyanoacrylate. Journal of Biomedical Materials Research Part B: Applied Biomaterials. 2015;103:305–12. DOI: 10.1002/jbm.b.33199

[25] Jiang K, Long YZ, Chen ZJ, Liu SL, Huang YY, Jiang X, Huang ZQ. Airflow-directed in situ electrospinning of a medical glue of cyanoacrylate for rapid hemostasis in liver resection. Nanoscale. 2014;6:7792–8. DOI: 10.1039/c4nr01412j

[26] Kuijpers AJ, Engbers GHM, Feijen J, De Smedt SC, Meyvis TKL, Demeester J, Krijgsveld J, Zaat SAJ, Dankert J. Characterization of the network structure of carbodiimide cross-linked gelatin gels. Macromolecules. 1999;32:3325–33. DOI: 10.1021/ma981929v

[27] Bae SK, Sung TH, Kim JD. A soft-tissue gelatin bioadhesive reinforced with a proteinoid. Journal of Adhesion Science and Technology. 2002;16:361–72.

[28] Braunwald NS, Gay W, Tatooles CJ. Evaluation of crosslinked gelatin as a tissue adhesive and hemostatic agent: an experimental study. Surgery. 1996;59:1024–30.

[29] Furst W, Banerjee A. Release of glutaraldehyde from an albumin-glutaraldehyde tissue adhesive causes significant in vitro and in vivo toxicity. Annals of Thoracic Surgery. 2005;79:1522–8; discussion 9. DOI: 10.1016/j.athoracsur.2004.11.054

[30] Saunders MM, Baxter ZC, Abou-Elella A, Kunselman AR, Trussell JC. BioGlue and Dermabond save time, leak less, and are not mechanically inferior to two-layer and modified one-layer vasovasostomy. Fertility and Sterility. 2009;91:560–5. DOI: 10.1016/j.fertnstert.2007.12.006

[31] Nomori K, Horio H, Morinaga S, Suemasu K. Gelatin-resorcinol-formaldehyde-glutaraldehyde glue for sealing pulmonary air leaks during thoracoscopic operation. Annals of Thoracic Surgery. 1999;67:212–6.

[32] Mehdizadeh M, Yang J. Design strategies and applications of tissue bioadhesives. Macromolecular Bioscience. 2013;13:271–88. DOI: 10.1002/mabi.201200332

[33] Vuocolo T, Haddad R, Edwards GA, Lyons RE, Liyou NE, Werkmeister JA, Ramshaw JA, Elvin CM. A highly elastic and adhesive gelatin tissue sealant for gastrointestinal surgery and colon anastomosis. Journal of Gastrointestinal Surgery. 2012;16:744–52. DOI: 10.1007/s11605-011-1771-8

[34] Fancy DA, Kodadek T. Chemistry for the analysis of protein-protein interactions: rapid and efficient cross-linking triggered by long wavelength light. Proceedings of the National Academy of Sciences of the United States of America. 1999;96:6020–4.

[35] Elvin CM, Vuocolo T, Brownlee AG, Sando L, Huson MG, Liyou NE, Stockwell PR, Lyons RE, Kim M, Edwards GA, Johnson G, McFarland GA, Ramshaw JA, Werkmeister JA. A highly elastic tissue sealant based on photopolymerised gelatin. Biomaterials. 2010;31:8323–31. DOI: 10.1016/j.biomaterials.2010.07.032

[36] Matsuda M, Inoue M, Taguchi T. Enhanced bonding strength of a novel tissue adhesive consisting of cholesteryl group-modified gelatin and disuccinimidyl tartarate. Journal of Bioactive and Compatible Polymers. 2012;27:31–44. DOI: 10.1177/0883911511434426

[37] Foox M, Raz-Pasteur A, Berdicevsky I, Krivoy N, Zilberman M. In vitromicrobial inhibition, bonding strength, and cellular response to novel gelatin-alginate antibiotic-releasing soft tissue adhesives. Polymers for Advanced Technologies. 2014;25:516–24. DOI: 10.1002/pat.3278

[38] Chen T, Janjua R, McDermott MK, Bernstein SL, Steidl SM, Payne GF. Gelatin-based biomimetic tissue adhesive. Potential for retinal reattachment. Journal of Biomedical Materials Research Part B: Applied Biomaterials. 2006;77:416–22. DOI: 10.1002/jbm.b.30439

[39] U.S. Food and Drug Administration: CryoLife BioGlue Surgical Adhesive – P010003 [Internet]. 2013. Available from: http://www.fda.gov/MedicalDevices/ProductsandMedicalProcedures/DeviceApprovalsandClearances/Recently-ApprovedDevices/ucm084009.htm [Accessed: July 18, 2016].

[40] Raanani E, Latter DA, Errett LE, Bonneau DB, Leclerc Y, Salasidis GC. Use of "BioGlue" in aortic surgical repair. The Annals of Thoracic Surgery. 2001;72:638–40.

[41] Herget GW, Kassa M, Riede UN, Lu Y, Brethner L, Hasse J. Experimental use of an albumin-glutaraldehyde tissue adhesive for sealing pulmonary parenchyma and bronchial anastomoses. European Journal of Cardio-Thoracic Surgery. 2001;19:4–9.

[42] Lin J, Iannettoni MD. Closure of bronchopleural fistulas using albumin-glutaraldehyde tissue adhesive. The Annals of Thoracic Surgery. 2004;77:326–8. DOI: 10.1016/s0003-4975(03)00750-1

[43] Hidas G, Kastin A, Mullerad M, Shental J, Moskovitz B, Nativ O. Sutureless nephron-sparing surgery: use of albumin glutaraldehyde tissue adhesive (BioGlue). Urology. 2006;67:697–700; discussion. DOI: 10.1016/j.urology.2005.10.064

[44] Coselli JS, Bavaria JE, Fehrenbacher J, Stowe CL, Macheers SK, Gundry SR. Prospective randomized study of a protein-based tissue adhesive used as a hemostatic and

structural adjunct in cardiac and vascular anastomotic repair procedures. Journal of the American College of Surgeons. 2003;197:243–52. DOI: 10.1016/S1072-7515(03)00376-4

[45] LeMaire S, Schmittling Z, Undar A. A new surgical adhesive (BioGlue) causes acute phrenic nerve injury and diaphragmatic paralysis. Journal of Surgical Research. 2000;93:354.

[46] De Somer F, Delanghe J, Somers P, Debrouwere M, Van Nooten G. Mechanical and chemical characteristics of an autologous glue. Journal of Biomedical Materials Research Part A. 2008;86:1106–12. DOI: 10.1002/jbm.a.31705

[47] Iwata H, Matsuda S, Mitsuhashi K, Itoh E, Yoshito I. A novel surgical glue composed of gelatin and N-hydroxysuccinimide activated poly(L-glutamic acid): part 1. Synthesis of activated poly(L-glutamic acid) and its gelation with gelatin. Biomaterials. 1998;19:1869–76.

[48] Alving B, Weinstein M, Finlayson J, Menitove J, Fratantoni J. Fibrin sealant: summary of a conference on characteristics and clinical uses. Transfusion. 1995;35:783–90.

[49] Martinowitz U, Saltz R. Fibrin sealant. Current Opinion in Hematology. 1996;3:395–402.

[50] Radosevich M, Goubran H, Burnouf T. Fibrin sealant: scientific rationale, production methods, properties, and current clinical use. Vox Sanguinis. 1997;72:133–43.

[51] Petersen B, Barkun A, Carpenter S, Chotiprasidhi P, Chuttani R, Silverman W, Hussain N, Liu J, Taitelbaum G, Ginsberg GG. Tissue adhesives and fibrin glues. Gastrointestinal Endoscopy. 2004;60:327–33. DOI: 10.1016/S0016-5107(04)01564-0

[52] Busuttil RW. A comparison of antifibrinolytic agents used in hemostatic fibrin sealants. Journal of the American College of Surgeons. 2003;197:1021–8.

[53] Duarte AP, Coelho JF, Bordado JC, Cidade MT, Gil MH. Surgical adhesives: systematic review of the main types and development forecast. Progress in Polymer Science. 2012;37:1031–50. DOI: 10.1016/j.progpolymsci.2011.12.003

[54] Albala DM. Fibrin sealants in clinical practice. Cardiovascular Surgery. 2003;11:5–11. DOI: 10.1016/S0967-2109(03)00065-6

[55] Truong S, Böhm G, Klinge U, Stumpf M, Schumpelick V. Results after endoscopic treatment of postoperative upper gastrointestinal fistulas and leaks using combined Vicryl plug and fibrin glue. Surgical Endoscopy and Other Interventional Techniques. 2004;18:1105–8. DOI: 10.1007/s00464-003-8286-7

[56] Ellis D, Shaikh A. The ideal tissue adhesive in facial plastic and reconstructive surgery. The Journal of Otolaryngology. 1990;19:68–72.

[57] Uy HS, Reyes JMG, Flores JD, Lim-Bon-Siong R. Comparison of fibrin glue and sutures for attaching conjunctival autografts after pterygium excision. Ophthalmology. 2005;112:667–71. DOI: 10.1016/j.ophtha.2004.08.028

[58] Gupta N, Chetter I, Hayes P, AH OY, Moneta GL, Shenoy S, Pribble JP, Zuckerman LA. Randomized trial of a dry-powder, fibrin sealant in vascular procedures. Journal of Vascular Surgery. 2015;62:1288–95. DOI: 10.1016/j.jvs.2015.05.038

[59] Dickneite G, Metzner H, Pfeifer T, Kroez M, Witzke G. A comparison of fibrin sealants in relation to their in vitro and in vivo properties. Thrombosis Research. 2003;112:73–82. DOI: 10.1016/j.thromres.2003.10.010

[60] Sierra DH, Eberhardt AW, Lemons JE. Failure characteristics of multiple-component fibrin-based adhesives. Journal of Biomaterials Applications. 2002;59:1–11.

[61] Kaetsu H, Uchida T, Shinya N. Increased effectiveness of fbrin sealant with a higher fibrin concentration. International Journal of Adhesion & Adhesives. 2000;20:27–31.

[62] Kheirabadi BS, Sieber J, Bukhari T, Rudnicka K, Murcin LA, Tuthill D. High-pressure fibrin sealant foam: an effective hemostatic agent for treating severe parenchymal hemorrhage. Journal of Surgical Research. 2008;144:145–50. DOI: 10.1016/j.jss.2007.02.012

[63] Sierra DH. Fibrin sealant adhesive systems: a review of their chemistry, material properties and clinical applications. Journal of Biomaterials Applications. 1993;7:309–52.

[64] Smith JD, Chen A, Ernst LA, Waggoner AS, Campbell PG. Immobilization of aprotinin to fibrinogen as a novel method for controlling degradation of fibrin gels. Bioconjugate Chemistry. 2007;18:695–701. DOI: 10.1021/bc060265o

[65] Wang LZ, Gorlin J, Michaud SE, Janmey PA, Goddeau RP, Kuuse R, Uibo R, Adams D, Sawyer ES. Purification of salmon clotting factors and their use as tissue sealants. Thrombosis Research. 2000;100:537–48.

[66] Michaud SE, Wang LZ, Korde N, Bucki R, Randhawa PK, Pastore JJ, Falet H, Hoffmeister K, Kuuse R, Uibo R, Herod J, Sawyer E. Purification of salmon thrombin and its potential as an alternative to mammalian thrombins in fibrin sealants. Thrombosis Research. 2002;107:245–54.

[67] Barros LC, Ferreira RS, Barraveira SRCS, Stolf HO, Thomazini-Santos IA, Mendes-Giannini MJS, Toscano E, Barraviera B. A new fibrin sealant from crotalus durissus terrificus venom: applications in medicine. Journal of Toxicology and Environmental Health, Part B. 2016;12:553–68. DOI: 10.1080/10937400903442514

[68] Schips L, Dalpiaz O, Cestari A, Lipsky K, Gidaro S, Zigeuner R, Petritsch P. Autologous fibrin glue using the vivostat system for hemostasis in laparoscopic partial nephrectomy. European Urology. 2006;50:801–5. DOI: 10.1016/j.eururo.2006.03.010

[69] Kjaergard HK, Velada JL, Pedersen H, Fleron H, Hollingbee DA. Comparative kinetics of polymerisation of three fibrin sealants and influence on timing of tissue adhesion. Thrombosis Research. 2000;98:221–8.

[70] Carlson MA, Calcaterra J, Johanning JM, Pipinos II, Cordes CM, Velander WH. A totally recombinant human fibrin sealant. Journal of Surgical Research. 2014;187:334–42. DOI: 10.1016/j.jss.2013.09.039

[71] Guterl CC, Torre OM, Purmessur D, Dave K, Likhitpanichkul M, Hecht AC, Nicoll SB, Iatridis JC. Characterization of mechanics and cytocompatibility of fibrin-genipin annulus fibrosus sealant with the addition of cell adhesion molecules. Tissue Engineering Part A. 2014;20:2536–45. DOI: 10.1089/ten.TEA.2012.0714

[72] Schek R, Michalek A, Iatridis J. Genipin-crosslinked fibrin hydrogels as a potential adhesive to augment intervertebral disc annulus repair. European Cells & Materials. 2011;21:373. DOI: 10.1155/2015/685690

[73] Elvin CM, Brownlee AG, Huson MG, Tebb TA, Kim M, Lyons RE, Vuocolo T, Liyou NE, Hughes TC, Ramshaw JA, Werkmeister JA. The development of photochemically crosslinked native fibrinogen as a rapidly formed and mechanically strong surgical tissue sealant. Biomaterials. 2009;30:2059–65. DOI: 10.1016/j.biomaterials.2008.12.059

[74] Cho E, Lee JS, Webb K. Formulation and characterization of poloxamine-based hydrogels as tissue sealants. Acta Biomaterialia. 2012;8:2223–32. DOI: 10.1016/j.actbio.2012.03.003

[75] Alleyene CH, Cawley MC, Barrow DL, Poff BC, Powell MD, Sawhney AS, Killehay DL. Efficacy and biocompatibility of a photopolymerized, synthetic, absorbable hydrogel as a dural sealant in a canine craniotomy model. Journal of Neurosurgy. 1998;88:308–13.

[76] Boogaarts JD, Grotenhuis JA, Bartels RHMA, Beems T. Use of a novel absorbable hydrogel for augmentation of dural repair: results of a preliminary clinical study. Operative Neurosurgery. 2005;57:146–51. DOI: 10.1227/01.neu.0000164384.05351.59

[77] Macchiarini P, Wain J, Almy S, Dartevelle P. Experimental and clinical evaluation of a new synthetic, absorbable sealant to reduce air leaks in thoracic operations. Cardiovascular Surgery. 1999;117:751–8.

[78] Hoshi S, Okamoto F, Arai M, Hirose T, Sugiura Y, Kaji Y, Oshika T. In vivo and in vitro feasibility studies of intraocular use of polyethylene glycol-based synthetic sealant to close retinal breaks in porcine and rabbit eyes. Investigational Ophthalmology and Visual Science. 2015;56:4705–11. DOI: 10.1167/iovs.14-15349

[79] Menon NG, Downing S, Goldberg NH, Silverman RP. Seroma prevention using an albumin-glutaraldehyde-based tissue adhesive in the rat mastectomy model. Annals of Plastic Surgery. 2003;50:639–43. DOI: 10.1097/01.SAP.0000054182.47311.17

[80] Murakami Y, Yokoyama M, Okano T, Nishida H, Tomizawa Y, Endo M, Kurosawa H. A novel synthetic tissue-adhesive hydrogel using a crosslinkable polymeric micelle. Journal of Biomedical Materials Research Part A. 2007;80:421–7. DOI: 10.1002/jbm.a.30911

[81] Torchiana DF. Polyethylene glycol based synthetic sealants. Journal of Cardiac Surgery. 2003;18:504–6.

[82] Lee G, Lee CK, Bynevelt M. DuraSeal-Hematoma: concealed hematoma causing spinal cord compression. Spine. 2010;35:E1522–E42. DOI: 10.1097/BRS.0b013e3181edfe2c

[83] Covidien: DuraSeal Dural Sealant System [Internet]. 2005. Available from: http://www.covidien.com/imageServer.aspx/doc178555.pdf?contentID=13915&contenttype=application/pdf [Accessed: July 18, 2016].

[84] Preul MC, Bichard WD, Muench TR, Spetzler RF. Toward optimal tissue sealants for neurosurgery: use of a novel hydrogel sealant in a canine durotomy repair model. Neurosurgery. 2003;53:1189–99. DOI: 10.3171/2009.10.SPINE09537

[85] Kim KD, Wright NM. Polyethylene glycol hydrogel spinal sealant (DuraSeal Spinal Sealant) as an adjunct to sutured dural repair in the spine: results of a prospective, multicenter, randomized controlled study. Spine (Phila Pa 1976). 2011;36:1906–12. DOI: 10.1097/BRS.0b013e3181fdb4db

[86] Pruel MC, Campbell PK, Garlick DS, Spetzler RF. Application of new hydrogel dural sealant that reduces epidural adhesion formation: evaluation in a large animal laminectomy model. Journal of Neurosurgery Spine. 2010;12:381–90.

[87] Villa-Camacho JC, Ghobril C, Anez-Bustillos L, Grinstaff MW, Rodriguez EK, Nazarian A. The efficacy of a lysine-based dendritic hydrogel does not differ from those of commercially available tissue sealants and adhesives: an ex vivo study. BMC Musculoskelet Disorders. 2015;16:116. DOI: 10.1186/s12891-015-0573-7

[88] Lequaglie C, Guidice G, Marasco R, Morte AD, Gallo M. Use of a sealant to prevent prolonged air leaks after lung resection: a prospective randomized study. Journal of Cardiothoracic Surgery. 2012;7:106. DOI: 10.1186/1749-8090-7-106

[89] Wallace DG, Cruise GM, Rhee WM, Schroeder JA, Prior JJ, Ju J, Maroney M, Duronio J, Ngo MH, Estridge TD, Coker GC. A tissue sealant based on reactive multifunctional polyethylene glycol. Journal of Biomedical Materials Research Part B: Applied Biomaterials. 2001;58:545–55.

[90] Serban MA, Panilaitis B, Kaplan DL. Silk fibroin and polyethylene glycol-based biocompatible tissue adhesives. Journal of Biomedical Materials Research Part A. 2011;98:567–75. DOI: 10.1002/jbm.a.33149

[91] Ghosh S, Cabral JD, Hanton LR, Moratti SC. Strong poly(ethylene oxide) based gel adhesives via oxime cross-linking. Acta Biomaterialia. 2016;29:206–14. DOI: 10.1016/j.actbio.2015.10.018

[92] Hu BH, Messersmith PB. Enzymatically cross-linked hydrogels and their adhesive strength to biosurfaces. Orthodontics Craniofacial Research. 2005;8:145–9. DOI: 10.1111/j.1601-6343.2005.00330.x

[93] Hynes SR, McGregor LM, Rauch MF, Lavik EB. Photopolymerized poly(ethylene glycol)/poly(L-lysine) hydrogels for the delivery of neural progenitor cells. Journal of Biomaterials Science Polymer Edition. 2007;18:1017–30. DOI: 10.1163/156856207781494368

[94] Sanborn TJ, Messersmith PB, Barron AE. In situ crosslinking of a biomimetic peptide-PEG hydrogel via thermally triggered activation of factor XIII. Biomaterials. 2002;23:2703–10. DOI: 10.1016/S0142-9612(02)00002-9

[95] Bitton R, Josef E, Shimshelashvili I, Shapira K, Seliktar D, Bianco-Peled H. Phloroglucinol-based biomimetic adhesives for medical applications. Acta Biomaterialia. 2009;5:1582–7. DOI: 10.1016/j.actbio.2008.10.004

[96] Daijiro M, Dastgheib K, Kim T, Pfiste-Serres A, Smeds KA, Inoue M, Hatchell DL, Grinstaff MW. A photopolymerized sealant for corneal lacerations. Cornea. 2002;21:393–9.

[97] Maia J, Ferreira L, Carvalho R, Ramos MA, Gil MH. Synthesis and characterization of new injectable and degradable dextran-based hydrogels. Polymer. 2005;46:9604–14. DOI: 10.1016/j.polymer.2005.07.089

[98] Charron PN, Fenn SL, Poniz A, Oldinski RA. Mechanical properties and failure analysis of visible light crosslinked alginate-based tissue sealants. Journal of the Mechanical Behavior of Biomedical Materials. 2016;59:314–21. DOI: 10.1016/j.jmbbm.2016.02.003

[99] Araki M, Tao H, Nakajima N, Sugai H, Sato T, Hyon SH, Nagayasu T, Nakamura T. Development of new biodegradable hydrogel glue for preventing alveolar air leakage. Journal of Thoracic Cardiovascular Surgery. 2007;134:1241–8. DOI: 10.1016/j.jtcvs.2007.07.020

[100] Matsumura K, Nakajima N, Sugai H, Hyon SH. Self-degradation of tissue adhesive based on oxidized dextran and poly-L-lysine. Carbohydrate Polymers. 2014;113:32–8. DOI: 10.1016/j.carbpol.2014.06.073

[101] Artzi N, Shazly T, Crespo C, Ramos AB, Chenault HK, Edelman ER. Characterization of star adhesive sealants based on PEG/dextran hydrogels. Macromolecular Bioscience. 2009;9:754–65. DOI: 10.1002/mabi.200800355

[102] Bhatia SK, Arthur SD, Chenault HK, Kodokian GK. Interactions of polysaccharide-based tissue adhesives with clinically relevant fibroblast and macrophage cell lines. Biotechnology Letters. 2007;29:1645–9. DOI: 10.1007/s10529-007-9465-8

[103] Bhatia SK, Arthur SD, Chenault HK, Figuly GD, Kodokian GK. Polysaccharide-based tissue adhesives for sealing corneal incisions. Current Eye Research. 2007;32:1045–50. DOI: 10.1080/02713680701767876

[104] Chenault HK, Bhatia SK, Dimaio WG, Vincent GL, Camacho W, Behrens A. Sealing and healing of clear corneal incisions with an improved dextran aldehyde-PEG amine tissue adhesive. Current Eye Research. 2011;36:997–1004. DOI: 10.3109/02713683.2011.606590

[105] Strehin I, Ambrose WM, Schein O, Salahuddin A, Elisseeff J. Synthesis and characterization of a chondroitin sulfate-polyethylene glycol corneal adhesive. Journal of Cataract and Refractive Surgery. 2009;35:567–76. DOI: 10.1016/j.jcrs.2008.11.035

[106] Ishihara M, Ono K, Saito Y, Yura H, Hattori H, Matsui T, Kurita A. Photocrosslinkable chitosan: an effective adhesive with surgical applications. International Congress Series. 2001;1223:251–7. DOI: 10.1016/S0531-5131(01)00430-7

[107] Prudden JF, Migel P, Hanson P, Friedrich L, Balassa L. The discovery of a potent pure chemical wound-healing accelerator. The American Journal of Surgery. 1970;119:560–4.

[108] Azuma K, Ifuku S, Osaki T, Okamoto Y, Minami S. Preparation and biomedical applications of chitin and chitosan nanofibers. Journal of Biomedical Nanotechnology. 2014;10:2891–920.

[109] Nie W, Yuan X, Zhao J, Zhou Y, Bao H. Rapidly in situ forming chitosan/epsilon-polylysine hydrogels for adhesive sealants and hemostatic materials. Carbohydrate Polymers. 2013;96:342–8. DOI: 10.1016/j.carbpol.2013.04.008

[110] Jayakumar R, Prabaharan M, Kumar PS, Nair S, Tamura H. Biomaterials based on chitin and chitosan in wound dressing applications. Biotechnology Advances. 2011;29:322–37. DOI: 10.1016/j.biotechadv.2011.01.005

[111] Nishimura K, Nishimura S, Nishi N, Tokura S, Azuma I. Immunological Activity of Chitin Derivatives. In: Muzzarelli R, Jeuniaux C, Gooday GW, editors. Chitin in Nature and Technology. Boston, MA: Springer US; 1986. p. 477–83.

[112] Fiebrig I, Harding S, Davis S. Sedimentation Analysis of Potential Interactions between Mucins and a Putative Bioadhesive Polymer. In: Lechner MD, editor. Ultracentrifugation. Springer; 1994. p. 66–73.

[113] Lih E, Lee JS, Park KM, Park KD. Rapidly curable chitosan-PEG hydrogels as tissue adhesives for hemostasis and wound healing. Acta Biomaterialia. 2012;8:3261–9. DOI: 10.1016/j.actbio.2012.05.001

[114] Yu M, Hwang J, Deming TJ. Role of L-3, 4-dihydroxyphenylalanine in mussel adhesive proteins. Journal of the American Chemical Society. 1999;121:5825–6. DOI: 10.1021/ja990469y

[115] Silverman HG, Roberto FF. Understanding marine mussel adhesion. Marine Biotechnology. 2007;9:661–81.

[116] Lee Y, Chung HJ, Yeo S, Ahn C-H, Lee H, Messersmith PB, Park TG. Thermo-sensitive, injectable, and tissue adhesive sol-gel transition hyaluronic acid/pluronic composite hydrogels prepared from bio-inspired catechol-thiol reaction. Soft Matter. 2010;6:977. DOI: 10.1039/b919944f

[117] Ye M, Jiang R, Zhao J, Zhang J, Yuan X, Yuan X. In situ formation of adhesive hydrogels based on PL with laterally grafted catechol groups and their bonding efficacy to wet

organic substrates. Journal of Materials Science: Materials in Medicine. 2015;26:273. DOI: 10.1007/s10856-015-5608-y

[118] Huang K, Lee BP, Ingram DR, Messersmith PB. Synthesis and characterization of self-assembling block copolymers containing bioadhesive end groups. Biomacromolecules. 2002;3:397–406. DOI: 10.1021/bm015650p

[119] Matos-Perez CR, White JD, Wilker JJ. Polymer composition and substrate influences on the adhesive bonding of a biomimetic, cross-linking polymer. Journal of the American Chemical Society. 2012;134:9498–505. DOI: 10.1021/ja303369p

[120] Krogsgaard M, Behrens MA, Pedersen JS, Birkedal H. Self-healing mussel-inspired multi-pH-responsive hydrogels. Biomacromolecules. 2013;14:297–301. DOI: 10.1021/bm301844u

[121] Holten-Andersen N, Harrington MJ, Birkedal H, Lee BP, Messersmith PB, Lee KYC, Waite JH, Iraelachvili JN. pH-induced metal-ligand cross-links inspired by mussel yield self-healing polymer networks with near-covalent elastic moduli. Proceedings of the National Academy of Sciences of the United States of America. 2011;108:2651–5. DOI: 10.1073/pnas.1015862108

[122] Wang J, Liu C, Lu X, Yin M. Co-polypeptides of 3,4-dihydroxyphenylalanine and L-lysine to mimic marine adhesive protein. Biomaterials. 2007;28:3456–68. DOI: 10.1016/j.biomaterials.2007.04.009

[123] Ryu JH, Lee Y, Do MJ, Jo SD, Kim JS, Kim BS, Im GI, Park TG, Lee H. Chitosan-g-hematin: enzyme-mimicking polymeric catalyst for adhesive hydrogels. Acta Biomaterialia. 2014;10:224–33. DOI: 10.1016/j.actbio.2013.09.014

[124] Lee BP, Dalsin JL, Messersmith PB. Synthesis and gelation of DOPA-modified poly(ethylene glycol) hydrogels. Biomacromolecules. 2002;3:1038–47. DOI: 10.1021/bm025546n

[125] Temenoff JS, Athanasiou KA, Lebaron RG, Mikos AG. Effect of poly (ethylene glycol) molecular weight on tensile and swelling properties of oligo (poly (ethylene glycol) fumarate) hydrogels for cartilage tissue engineering. Journal of Biomedical Materials Research. 2002;59:429–37.

[126] Murphy JL, Vollenwieder L, Xu F, Lee BP. Adhesive performance of biomimetic adhesive-coated biologic scaffolds. Biomacromolecules. 2010;11:2976–84. DOI: 10.1021/bm1007794

[127] Ryu JH, Lee Y, Kong WH, Kim TG, Park TG, Lee H. Catechol-functionalized chitosan/pluronic hydrogels for tissue adhesives and hemostatic materials. Biomacromolecules. 2011;12:2653–9. DOI: 10.1021/bm200464x

[128] Fan C, Fu J, Zhu W, Wang DA. A mussel-inspired double-crosslinked tissue adhesive intended for internal medical use. Acta Biomaterialia. 2016;33:51–63. DOI: 10.1016/j.actbio.2016.02.003

[129] Lee H, Scherer NF, Messersmith PB. Single-molecule mechanics of mussel adhesion. Proceedings of the National Academy of Sciences of the United States of America. 2006;103:12999–3003. DOI: 10.1073/pnas.0605552103

[130] Autumn K, Liang YA, Hsieh ST, Zesch W, Chan WP, Kenny TW, Fearing R, Full RJ. Adhesive force of a single gecko foot-hair. Nature. 2000;405:681–5. DOI: 10.1038/35015073

[131] Autumn K, Sitti M, Liang YA, Peattie AM, Hansen WR, Sponberg S, Kenny TW, Fearing R, Israelachvili JN, Full RJ. Evidence for van der Waals adhesion in gecko setae. Proceedings of the National Academy of Sciences of the United States of America. 2002;99:12252–6. DOI: 10.1073/pnas.192252799

[132] Cappella B, Dietler G. Force-distance curves by atomic force microscopy. Surface Science Reports. 1999;34:1–104. DOI: 10.1016/S0167-5729(99)00003-5

[133] Yamamoto K, Tanuma C, Gemma N. Competition between electrostatic and capillary forces acting on a single particle. Japanese Journal of Applied Physics. 1995;34:4176.

[134] Geim AK, Dubonos SV, Grigorieva IV, Novoselov KS, Zhukov AA, Shapoval SY. Microfabricated adhesive mimicking gecko foot-hair. Nature Materials. 2003;2:461–3. DOI: 10.1038/nmat917

[135] Huber G, Mantz H, Spolenak R, Mecke K, Jacobs K, Gorb SN, Arzt E. Evidence for capillarity contributions to gecko adhesion from single spatula nanomechanical measurements. Proceedings of the National Academy of Sciences of the United States of America. 2005;102:16293–6. DOI: 10.1073/pnas.0506328102

[136] Sun W, Neuzil P, Kustandi TS, Oh S, Samper VD. The nature of the gecko lizard adhesive force. Biophysical Journal. 2005;89:L14–L7. DOI: 10.1529/biophysj.105.065268

[137] Lee H, Lee BP, Messersmith PB. A reversible wet/dry adhesive inspired by mussels and geckos. Nature. 2007;448:338–41. DOI: 10.1038/nature05968

[138] Mahdavi A, Ferreira L, Sundback C, Nichol JW, Chan EP, Carter DJ, Bettinger CJ, Patanavanich S, Chignozha L, Ben-Joseph E, Galakatos A, Pryor H, Pomerantseva I, Masiakos PT, Faquin W, Zumbuehl A, Hong S, Borenstein J, Vacanti J, Langer R, Karp JM. A biodegradable and biocompatible gecko-inspired tissue adhesive. Proceedings of the National Academy of Sciences of the United States of America. 2008;105:2307–12. DOI: 10.1073/pnas.0712117105

[139] Ratner BD, Hoffman AS, Schoen FJ, Lemons JE. Biomaterials Science: An Introduction to Materials in Medicine. San Diego, CA: Academic Press; 2004.

[140] Klara PM, Ray CD. Artificial nucleus replacement: clinical experience. Spine. 2002;27:1374–7.

[141] Bertagnoli R, Schönmayr R, Vazquez RJ. Surgical and Clinical Results with the PDN® Prosthetic Disc-Nucleus Device. In: Gunzburg R, Mayer HM, Szpalski M, Aebi M, editors. Arthroplasty of the Spine. Berlin, Heidelberg: Springer; 2004. p. 87–92.

[142] Wiltsey C, Christiani T, Williams J, Scaramazza J, Van Sciver C, Toomer K, Sheehan J, Branda A, Nitzl A, England E. Thermogelling bioadhesive scaffolds for intervertebral disk tissue engineering: preliminary in vitro comparison of aldehyde-based versus alginate microparticle-mediated adhesion. Acta Biomaterialia. 2015;16:71–80. DOI: 10.1016/j.actbio.2015.01.025

[143] Briganti E, Spiller D, Mirtelli C, Kull S, Counoupas C, Losi P, Senesi S, Di Stefano R, Soldani G. A composite fibrin-based scaffold for controlled delivery of bioactive pro-angiogenetic growth factors. Journal of Controlled Release. 2010;142:14–21. DOI: 10.1016/j.jconrel.2009.09.029

[144] Zisch AH, Schenk U, Schense JC, Sakiyama-Elbert SE, Hubbell JA. Covalently conjugated VEGF-fibrin matrices for endothelialization. Journal of Controlled Release. 2001;72:101–13.

[145] Ahmed TA, Giulivi A, Griffith M, Hincke M. Fibrin glues in combination with mesenchymal stem cells to develop a tissue-engineered cartilage substitute. Tissue Engineering Part A. 2011;17:323–35. DOI: 10.1089/ten.TEA.2009.0773

[146] Cho S-W, Kim S-S, Rhie JW, Cho HM, Choi CY, Kim B-S. Engineering of volume-stable adipose tissues. Biomaterials. 2005;26:3577–85. DOI: 10.1016/j.biomaterials.2004.09.013

[147] Perka C, Schultz O, Spitzer R-S, Lindenhayn K, Burmester G-R, Sittinger M. Segmental bone repair by tissue-engineered periosteal cell transplants with bioresorbable fleece and fibrin scaffolds in rabbits. Biomaterials. 2000;21:1145–53.

[148] Yamada Y, Boo JS, Ozawa R, Nagasaka T, Okazaki Y, Hata K, Ueda M. Bone regeneration following injection of mesenchymal stem cells and fibrin glue with a biodegradable scaffold. Journal of Cranio-Maxillofacial Surgery. 2003;31:27–33. DOI: 10.1016/S1010-5182(02)00143-9

[149] Birla RK, Borschel GH, Dennis RG, Brown DL. Myocardial engineering in vivo: formation and characterization of contractile, vascularized three-dimensional cardiac tissue. Tissue Engineering. 2005;11:803–13. DOI: 10.1089/ten.2005.11.803

[150] di Summa PG, Kingham PJ, Raffoul W, Wiberg M, Terenghi G, Kalbermatten DF. Adipose-derived stem cells enhance peripheral nerve regeneration. Journal of Plastic, Reconstructive & Aesthetic Surgery. 2010;63:1544–52. DOI: 10.1016/j.bjps.2009.09.012

[151] Malafaya PB, Silva GA, Reis RL. Natural-origin polymers as carriers and scaffolds for biomolecules and cell delivery in tissue engineering applications. Advanced Drug Delivery Reviews. 2007;59:207–33. DOI: 10.1016/j.addr.2007.03.012

[152] Janmey PA, Winer JP, Weisel JW. Fibrin gels and their clinical and bioengineering applications. Journal of the Royal Society Interface. 2009;6:1–10. DOI: 10.1098/rsif.2008.0327

[153] Kjaergard H, Weis-Fogh US. Important factors influencing the strength of autologous fibrin glue; the fibrin concentration and reaction time-comparison of strength with commercial fibrin glue. European Surgical Research. 1994;26:273–6.

[154] Siedentop KH, Harris DM, Sanchez B. Autologous fibrin tissue adhesive: factors influencing bonding power. The Laryngoscope. 1988;98:731–3. DOI: 10.1288/00005537-198807000-00008

[155] Sierra DH, Feldman DS, Saltz R, Huang S. A method to determine shear adhesive strength of fibrin sealants. Journal of Applied Biomaterials. 1992;3:147–51. DOI: 10.1002/jab.770030210

[156] Chen YC, Su WY, Yang SH, Gefen A, Lin FH. In situ forming hydrogels composed of oxidized high molecular weight hyaluronic acid and gelatin for nucleus pulposus regeneration. Acta Biomaterialia. 2013;9:5181–93. DOI: 10.1016/j.actbio.2012.09.039

[157] Simson JA, Strehin IA, Lu Q, Uy MO, Elisseeff JH. An adhesive bone marrow scaffold and bone morphogenetic-2 protein carrier for cartilage tissue engineering. Biomacromolecules. 2013;14:637–43. DOI: 10.1021/bm301585e

[158] Wiltsey C, Kubinski P, Christiani T, Toomer K, Sheehan J, Branda A, Kadlowec J, Iftode C, Vernengo J. Characterization of injectable hydrogels based on poly (N-isopropylacrylamide)-g-chondroitin sulfate with adhesive properties for nucleus pulposus tissue engineering. Journal of Materials Science: Materials in Medicine. 2013;24:837–47. DOI: 10.1007/s10856-013-4857-x

Chapter 5

Adhesive Restorations and the Oral Environmental Behaviour

Egle Milia, Roberto Pinna, Enrica Filigheddu and Stefano Eramo

Additional information is available at the end of the chapter

http://dx.doi.org/10.5772/64973

Abstract

Adhesive restorations are based on the use of materials, which have the capacity to bond tooth effectively. This is possible due to a polymerizing hybrid layer interface created by the use of the Etch&Rinse (ERAs) and self-etching adhesives (SEAs). Bonding using ERAs include the acid-etching removal of the mineral phase from the substrates of enamel and dentine. A hybrid layer results by filling the voids left by minerals by means of adhesive monomers. However, etching dentine may result in too much demineralization and wetness with discrepancies in reinforcement at the bottom of hybrid layer. SEAs avoid the separate etching phase of ERAs using acidic functional monomers. In the two-step SEAs, hybridization is created by the application of a primer of different pH acidity, followed by an adhesive resin. In the 'One-Step SEAs', acidic and adhesive monomers are mixed in the same bottle thereby causing hybridization at the same time. 10-MDP mild SEAs represent the better bonding technology in dentistry due to the ability to form a strong chemical bond in tooth tissue. However, adhesive restorations have high vulnerability in the oral environment, which have been attributed to the esterase activity of *Streptococcus mutans* and hydrolysis by matrix metalloproteinase.

Keywords: 5 max adhesive restoration, Etch&Rinse adhesives, self-etching adhesives, oral biofilm, *Streptococcus mutans*, MMP

1. Introduction

Adhesion in dentistry is based on the use of materials which have the capacity to bond tooth tissues effectively.

It was the perception of Buonocore [1] who, in 1955, firstly thought to acidify the enamel tissue in order to 'make the tooth surfaces more receptive to an adhesion process'. During that period, adhesion was confined to the restoration of carious teeth, whereas today, most of the materials have become adhesive in dentistry completely changing the mind-set in the clinical approach.

In comparison with non-adhesive materials, which spread passively on the tooth surfaces, adhesive materials have the capacity to interlock chemically and micromechanically in tissues allowing a minimal invasion approach in therapy. To a large extent, adhesion concepts are now successfully applied in direct and indirect restorations to allow them to withstand masticatory stress [2], to add to the aesthetic and harmonious parameters in a smile [3–5], as well as to reinforce root canals after endodontic therapy, and, more recently, in the treatment of dental hypersensitivity [6] (**Figures 1** and **2**).

This huge evolution has been possible due to the presence of a polymerizing hybrid layer interface that is based on an exchange process by which inorganic tooth material is substituted by the synthetic monomers of adhesive systems [7]. On the surface of the tooth, the hybrid layer interlocks chemically and micromechanically to the tooth tissues, and on the surface of the filling, most frequently consisting of a polymerizing resin composite, it becomes integrated into the adhesive. On the composite, adhesion is achieved by a process of co-polymerization involving residual double bonds (-C=C-) in the oxygen inhibition layer of the hybrid layer in the tooth [7] (**Figure 3**).

However, adhesive resins, particularly dimethacrylates, have shown some adverse biological effects [8], which have directed attention to a possibly low biocompatibility with some mutagenic properties in addition to local cytotoxic effects. Some unreacted residual monomers could dissolve into saliva and be released into the oral environment causing mucosal damage. Moreover, a possible endocrine-disruptive effect of monomers has raised some concern [9, 10].

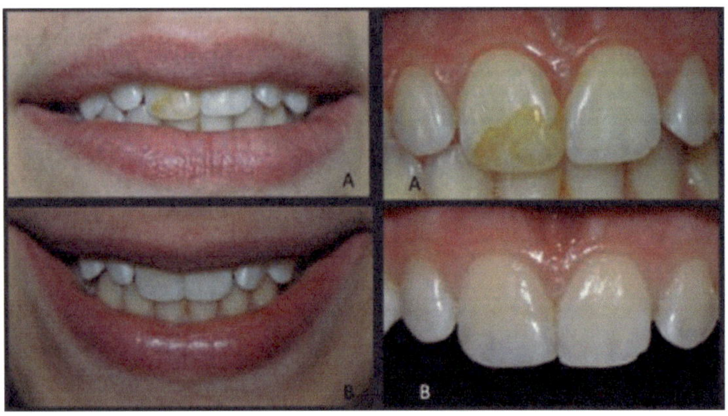

Figure 1. (A) The photograph shows a poorly shaped, IV class direct resin composite restoration in tooth 11 in a young girl. The restored tooth is not in accordance with the other upper maxillary teeth which are part of the smile. Also, the colours and shape of the composite in 11 did not match those of the surrounding unaltered tooth tissues. In addition, the margins of the restoration appear covered by a thick yellow layer, strongly contrasting to the surrounding sound tissues. (B) After replacing the new composite in 11, the incisal and proximal area of the restoration show quite natural opalescence and translucency, similar to the homologous tooth 21. In addition, the restored tooth is now in accordance with the other teeth which comprise the smile.

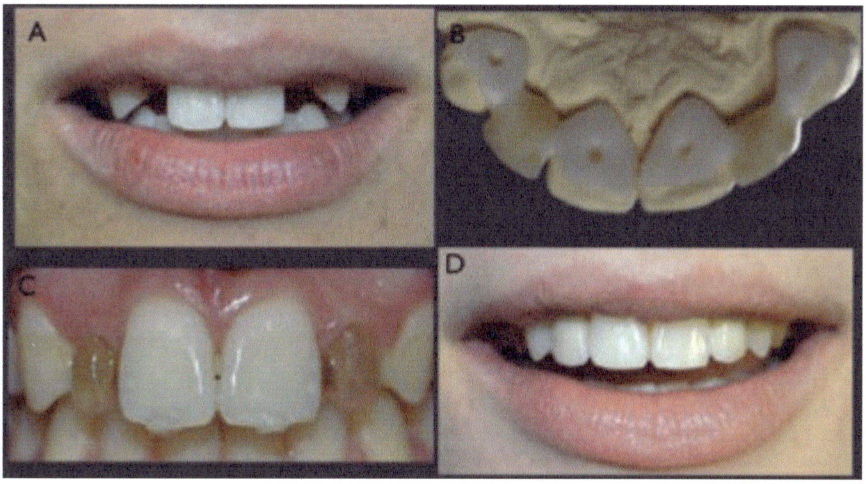

Figure 2. Adhesion techniques can be widely applied to recover the smile in complex situations, regardless of the age of the patient. (A) A 17-year-old girl, who has an agenesis of the upper maxillary lateral incisors (12 and 22). Until she reaches the age when implants are possible, a zirconia Maryland bridge has been employed (B), adhering to the no-prep palatal surfaces of the upper maxillary central incisors and canines, adjacent to the gaps (C), which supports the artificial porcelain teeth 12 and 22, recapturing the harmony of the smile (D).

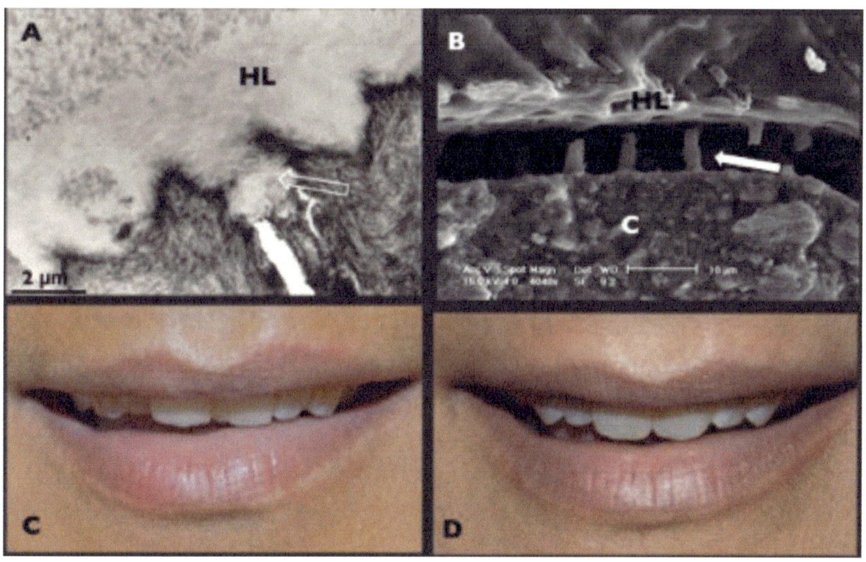

Figure 3. Adhesive restorations are based on a polymerizing hybrid layer interface between the tooth surface and the composite surface. (A) A non-decalcified transmission electron microscope (TEM) image showing the surface of a tooth in which the hybrid layer interlocks infiltrating dentin collagen and forming resin tags into the dentinal tubules (empty white arrow). (B) An scanning electron microscope (SEM) image of the bonding between the composite and the hybrid layer (white arrow), which is achieved by a process of co-polymerization involving residual double bonds in the oxygen inhibition layer of the hybrid layer in the tooth. (C and D) The application of composite in the recovery of smiling forces. HL = hybrid layer; RT = resin tag; C = composite.

Also, adhesive restorations have demonstrated a high vulnerability in the oral environment, which explains the high rate of failures [11] and over time the need for frequent replacement. Factors affecting durability could, to some extent, be ascribed to the chemistry of the materials, difficulties in the management of the tissues involved in bonding and on the choice of the most suitable adhesion procedure to be used [2]. In addition, breakdown of adhesive restorations has been attributed to the activity of the oral cariogenic biofilm [12, 13], particularly ascribed to the high susceptibility of resins to be colonized and degraded by the esterase activity of *S. mutans*. Also, hydrolysis of the bond has been connected to activation of the host-derived proteases matrix metalloproteinase (MMPs).

So, the intention of the present work is to examine adhesive materials and procedures in dental bonding whilst pointing out the main mechanisms of the breakdown of resins within the oral environment.

2. The adhesive systems

2.1. The chemistry of adhesive systems

Dental adhesives, as well as resin composites, both contain a similar mixture of resin monomers in order to ensure good covalent bonds between them. After photo-polymerization, resin monomers provide a structural, chemical-mechanical backbone to the adhesive restoration in teeth.

Like many other bioadhesive materials, adhesive monomers are composed of a mixture of hydrophobic-hydrophilic monomers, which should have (1) a high rate of polymerization; (2) sufficient stability in the monomeric form as well as in the subsequent form of the polymer; (3) minimum capacity of water sorption; and (4) absence of low toxicity in order to avoid damaging effects [14]. To ensure these qualities, monomers are supplemented by the addition of solvents and photo-initiators in adhesives. Some adhesives sometimes contain fluorine fillers and antibacterial agents with the purpose of promoting remineralization and the eradication of residual bacteria, particularly in caries affected dentine (CAD) bonds [15].

Moreover, the hydrophobic-hydrophilic monomers in bonding have the following functions: (1) in the monomer state, adhesive hydrophilicity allows infiltration of the wet, demineralized dentin matrix; and (2) in the polymerized state, adhesive hydrophobicity leads to lower fluid sorption. Also, hydrophobicity confers adequate stiffness to the polymer to minimize swelling and subsequent hydrolytic degradation [16].

Infiltration of the adhesive in tissues is caused by a physical combination of diffusion and convection phenomena, resulting from the external physical energy applied to disperse it in the tooth.

The spatial composition of the adhesive between the surface of the tooth and the resin composite is a conglomerate of many angstrom-scale structural units, which interact through various covalent bonds, forming linear, crosslinks and non-covalent interactions such as van

http://dx.doi.org/10.5772/64973

der Waals and hydrogen bonds, as well as physical interactions [16]. Despite this fact, compositional changes might be experienced by adhesive systems in the oral environment. Compositional changes, which have been related to the chemical composition and network structure of adhesives [17], may cause degradation of the hybrid layer over time [17, 18].

Adhesives are chemically formulated using two main types of monomers: (1) cross-linkers monomers, which have two or more vinyl groups or –C=C– polymerizable groups, after polymerization form cross-linked polymers with high mechanical strength; and (2) functional monomers, which have only one polymerizable group, mainly added to a functional group. They have the ability to enhance specific properties of the adhesive and after curing lead to linear polymer [19].

The chemical structure of the monomers can be described in the following way: two terminal groups separated by a spacer chain. One of the terminal groups consists of the polymerizable site of the monomer (generally methacrylate); the other terminal consists of a functional group [20].

Methacrylate monomers are widely used in dental adhesives and generally have hydrophobic behaviour. The hydrophobic bisphenol A glycerolate dimethacrylate (BisGMA) is widely utilized as a methacrylate. The hydrophobic aromatic rings in the backbone of BisGMA confer low-chain mobility and deformation under mechanical loading. In addition, the pendent hydroxyl groups along BisGMA chains form hydrogen bonds between molecules increasing the mechanical properties [21] (**Figure 4**). However, the molecular stiffness and H-bonding formation make BisGMA highly viscous, lowering the infiltration into the wet dentine. Additionally, the ester group (R1–CO–OR2), which is typical of all the methacrylate monomers, may render BisGMA susceptible to hydrolysis [22].

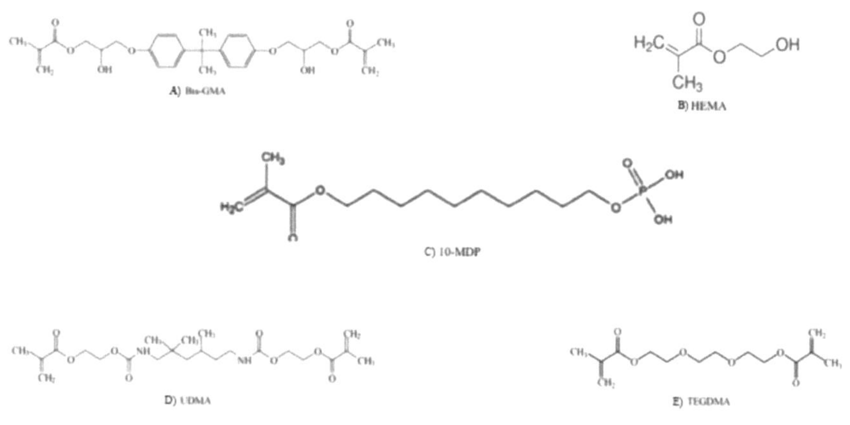

Figure 4. Chemical structures of (A) BisGMA (bisphenol A glycerolate dimethacrylate); (B) HEMA (ethylene glycol methacrylate); (C) 10 MDP (10-methacryloxy-decyl-dihydrogen-phosphate); (D) UDMA (urethane dimethacrylate); and (E) TEGDMA (triethylene glycol dimethacrylate).

These characteristics may lead to adverse reactions, such as phase separation of the hydrophobic/hydrophilic monomers in water, discrepancies in the reinforcement of the deepest region of demineralized collagen and poor polymerization [23]. Moreover, incomplete infiltration leads to water retention within the collagen/adhesive interface [24] and sub-optimal polymerization leads to loosely cross-linked domains that can be readily penetrated by water [25].

As regards the spacer, the length and composition of the chain between the polymerizable and the functional groups influence some physicochemical properties, such as the viscous solubility and wettability capacities. Sometimes the spacer chain may have effects on the degree of conversion and mechanical strength, as well as lowering water sorption. It usually consists of an alkyl chain, which contains esters, amides or aromatic groups that may be relatively hydrophobic, as in the case of the functional group 10-methacryloxy-decyl-dihydrogen-phosphate (10-MDP) (**Figure 4**). Moreover, the chain may consist of relatively hydrophilic polyethylene glycols, such as triethylene glycol dimethacrylate (TEGDMA).

As far as the functional group is concerned, it is generally a phosphate, carboxyl acid or alcohol group. All of these have abilities, such as wetting and demineralization, phase separation stabilization, improvement of the penetration of cross-linking monomers, fluoride release and antibacterial properties. Also, due to the hydrophilic properties, functional groups are considered adhesion-promoting agents in dentine, increasing infiltration and bond strength [26].

In particular, some acidic functional monomers of self-etching adhesives (SEAs) play an important role in bonding performance and the physicochemical properties of these adhesives.

HEMA (ethylene glycol methacrylate) (**Figure 4**) is a monomer very frequently used as a functional group. It is an excellent adhesion-promoting agent in dentine and is very soluble in water, ethanol and/or acetone solvents [19]. Also, HEMA has solvent-like properties, which causes stability in hydrophobic and hydrophilic solutions.

However, many adverse reactions have been ascribed to the use of HEMA, especially when added in high concentration. One of the most common is an increase in water sorption parameters in adhesives, particularly at basic pH [27]. Water sorption was reported in the uncured than in the cured state of HEMA [15]. Specifically, before polymerization, HEMA may cause water adsorption and dilution of the monomers affecting polymerization; after curing, the hydrophilic properties of HEMA still lead to water uptake into the polymer chain with consequent permeability of the hybrid layer, hypersensitivity and deterioration of the mechanical properties of the interface. A high concentration of HEMA may also prevent the complete evaporation of solvents from the adhesive due to a lowering of the vapour pressure of water and probably of the alcohol vehicle [7]. Moreover, adverse reactions are related to a possible toxicity of HEMA that, in the case of phase separation, may be trapped as a residual monomer as well as unreacted radicals [24]. The leaching of HEMA into the surrounding tissues may lead to apoptosis and production of reactive oxygen species (ROS) by gingival fibroblasts [28].

Solvents in adhesives are generally water, ethanol and acetone. They are necessary to bond dentine. Firstly, solvent molecules cause monomers to infiltrate the wetness of demineralized dentine [29, 30]. Secondly, solvents are useful in the re-expansion of the fibres in the case of shrunken after demineralization [31].

In particular, a water solvent is essential to the efficiency of both the Etch&Rinse adhesive systems (ERAs) and self-etching adhesive systems (SEAs). Water has a hydroxyl group that can form strong hydrogen bonds and re-expands the shrunken fibres by breaking the H-bonds and other forces, which keep the collagen in a collapsed state [31]. In SEAs, water provides ionization of the medium and leads to the activity of the self-etching monomers [32].

However, the low vapour pressure of water implies that this solvent may be difficult to remove from the adhesive after layering in teeth [25] resulting in sub-optimal polymerization. Although an air-drying procedure is recommended to increase evaporation of the solvent, it can be retained to some extent, hampering perfect conversion with dilution of the monomers, voids and permeability of the interface [15].

Photo-initiators are added to activate a radical polymerization reaction that allows conversion into a completely polymerized form [14]. Nevertheless, the conversion rates in adhesives and dental composites remain quite low and prevent a tight and impermeable resin interface.

Camphorquinone (CQ), 1,7,7-trimethyl-bicyclo[1,2,2]-heptan-2,3-dione, is a typical photoinitiator in dental adhesives and composites. It works in association with a co-initiator, usually a tertiary amine compound, modified by the addition of methacrylate to lower the toxic and mutagenic effects [19]. However, some noxious biological effects can be directly ascribed to CQ, which limit the addition in adhesives. Also, the yellow colour is another adverse effect, which may compromise the matching of the restoration and the tooth [33].

2.2. The chemistry of resin composites

Conventional composite contains a polymerizable organic matrix, inorganic reinforcing fillers and a silane coupling agent, which bond the organic matrix and inorganic fillers usually represented by glass, quartz or ceramic oxides [34] (**Figure 5**).

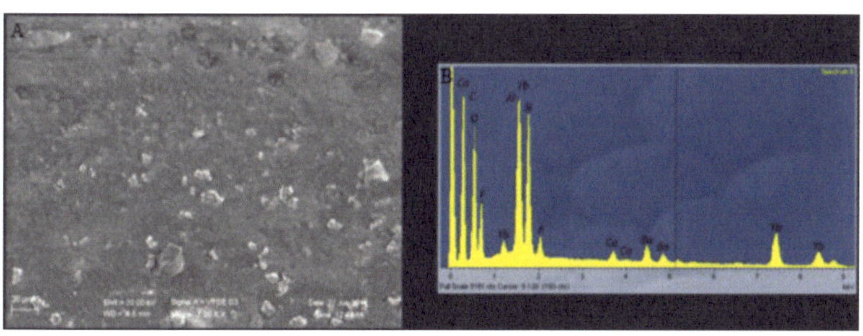

Figure 5. (A) An environmental scanning electron microscope (ESEM) micrograph of a composite (Vertis Flow, Kerr Corporation, Orange, CA) showing the resin matrix, which is very rich in nanoparticle fillers. Highly visible randomly distributed 5–40 μm particles can also be observed in the matrix. (B) An EDX (X-ray Energy Dispersive Spectrometer) image, which reveals the ionic nature of these filler particles. In this composite, Si, Yb and F are the most abundant elements in the matrix, in which Al, P, Ca and Ba are also present.

The organic matrix consists of several monomers, generally BisGMA and urethane dimethacrylate (UDMA) in addition to ethylene glycol di-methacrylate compounds, such as triethylene glycol dimethacrylate (TEGDMA) (**Figure 4**). Furthermore, the organic matrix contains an initiator/inhibitor polymerization system, which causes conversion through free radical chain polymerization of di-vinyl oligomers, usually initiated by photochemical and chemical means [35].

BisGMA monomer is the most common monomer used in the polymeric matrix. As in adhesive systems, BisGMA confers to the composite a high capacity to resist deformation and withstand the mechanical loading, due to its hydrophobic aromatic rings and H-bonding capacity [21]. However, the high viscosity of BisGMA rather limits the addition of a high concentration of fillers and the polymerization of the resin mass [21]. In fact, many studies have found that the rigidity of the material has a direct effect on the polymerization stress of composites.

In contrast, low viscous monomers, such as TEGDMA, are added as diluents. By reducing the overall viscosity, the low-molecular weight di-vinyl monomer of TEGDMA enhances the efficiency of polymerization, mixing of the different constituents, as well as handling of the composite [36]. However, restrictions to the addition of TEGDMA are a result of the presence of triethylene oxide spacers that increase water sorption in the composite [35]. Also, TEGMA increases volumetric shrinkage in the composite with the possibility of marginal gaps in restorations [36].

Gaps and/or tooth distortion still represent quite common negative effects of composite polymerization and are related to nanoleakage and breakdown of restoration till today. This is because conversion is accomplished by replacing the van der Waals spaces in monomers by covalent bonds in polymer and consequently, by a reduction of the free volume of the composite [37]. In addition, the so-called C-factor and compliance of the dental substrate may be considered in the amount of the final polymerization stress [38, 39]. One other cause of degradation connected to the conversion process is the presence of condensation type bonds within the resin. Condensation bonds, which include esters, urethanes and amides, are predominantly found in the di-vinyl monomers and are all prone to chemical hydrolysis catalysed by acids, bases or enzymes [21].

Different methods have been proposed to reduce shrinkage. Among them, a variety of techniques of cavity filling, such as incremental placement of the composite, the soft-start curing technique and the use of stress absorbing cavity liners (**Figure 6**). Moreover, modification in composite composition has been studied with the intention of reducing stress [40]. As polymerization shrinkage can be considered indirectly related to the concentration of loaded filler, the mass of filler particles has reached a range of 60–80% of the weight, becoming the major agent of the resin.

Fillers are usually inorganic particles and confer to the composite mechanical properties such as compressive strength and modulus of elasticity. In addition, fillers lower water absorption and the thermal expansion coefficient of composite [41].

Moreover, a variety of nanomaterials have attracted attention as fillers. Nanomaterials have advantages in higher reduction of the rate of wear, polymerization shrinkage in comparison to traditional fillers, at the same time improving mechanical properties of the composite restoration [42]. The increase in efficiency by the use of nanoparticles has been related to the high specific surface area and richness of functional groups [42].

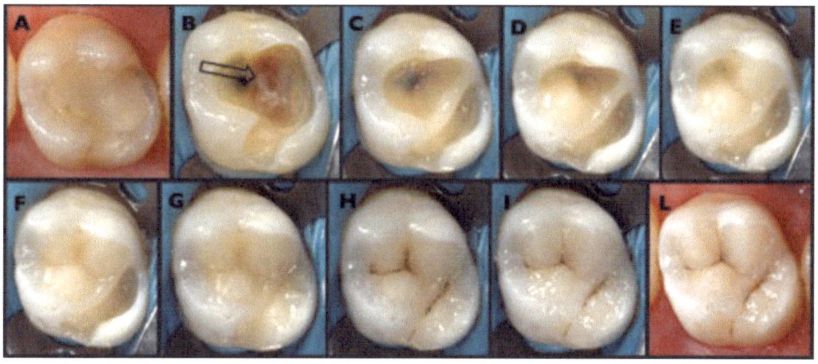

Figure 6. (A) A direct resin composite restoration in which mechanical wear clearly suggests the need for replacement. (B) The bottom cavity of the tooth after remover of the old composite. The dark, yellow affected dentine (arrow) appears surrounded by sound dentine and enamel margins. (C–I) The incremental placement technique of a new composite aiming to contrast volumetric contraction of the resin mass happening under polymerization. (L) The final result of the new restoration after removal of the dental gum (Courtesy of Dr. Stefano Serra).

Recently, there have been studies of nanocomposites using calcium-phosphate and calcium-fluoride nanoparticles [43]. Also, amorphous calcium phosphate particles of 116 μm were incorporated into composites. An advantage of the addition of calcium-phosphate particles is the release of supersaturating levels of calcium and phosphate ions at acidic pH with remineralization of initial tooth lesions. More recently, a quaternary ammonium dimethacrylate was incorporated into the calcium-phosphate nanocomposite combining remineralizing with antibacterial properties in the same material [44].

Moreover, nanofibers have become popular as fillers because of the increase in stiffness of composite due to both the high surface area and high aspect ratio of nanofibres. Studies [45] reported that zirconia-silica or zirconia-yttria-silica ceramic nanofibres, as well as hydroxyapatite (HAP) nanofibres [46], are able to improve mechanical properties and fracture toughness of composites. In addition, the associated overlapping of the fibres lowers polymerization shrinkage [47].

More recently, the incorporation of pre-polymerized resin fillers (organic fillers) has been proposed with the advantage of decreasing the volume fraction of the polymerizable resin, at the same time allowing the addition of increased filler volume fraction [48].

2.3. Adhesion techniques

2.3.1. Smear layer

Cutting teeth creates a layer of smear debris, which completely covers the surfaces of the enamel and dentine (49) and plugs the orifices of the dentinal tubules.

Using transmission electron microscope (TEM) and micro-RAMAN spectroscopy, the smear layer was described as a fibrous layer, composed of well-arranged and undisrupted collagen fibrils [49] that because of the inherent weakness can interfere with good adhesion [50].

So, adhesive systems have to interact with the smear to establish high bond strength in teeth (**Figure 7**). In fact, when the smear was removed by etching there was better adhesion performance [51]. However, the manner of smear creation might be directly related to interferences in bonding [52].

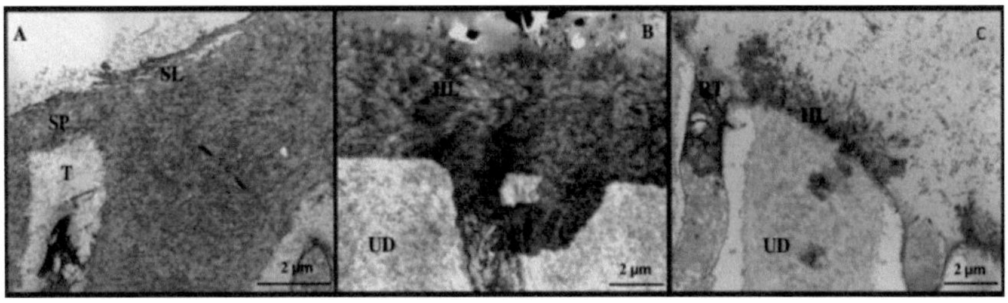

Figure 7. (A) A TEM photomicrograph of a dentine surface covered by a porous, amorphous smear layer. The tubule at the front of the dentine is smear plugged. (B) A demineralized TEM image of the hybrid layer formed by an Etch&Rinse adhesive (OptiBond Solo, Kerr Corporation, Orange, CA, USA). In this case, the hybrid layer is completely deprived of the smear layer because of the etching procedure of the Etch&Rinse. The hybrid layer appears as a dense resin micro-infiltration into the totally demineralized collagen, also showing micro-mechanical interlocking in tubules by long resin tags. (C) A demineralized TEM image of the hybrid layer created using a mild self-etching adhesive (Clearfil Protect Bond, Kurary, Osaka, Japan). In this case, the smear layer is included in the uppermost part of this hybrid layer, which is formed by resin infiltration of the water rich channels of the smear reaching the partially demineralized superficial dentine. Short resin tags are present into the tubules by percolation of the smear plug at the orifice. The self-etching hybrid layer in (C) is a dentine resin interaction of approximately 1.5 µm of thickness (between white arrows) and is less deeper than that formed by the Etch&Rinse adhesive in (B), which is about 4.5 µm depth (between black arrows). HL = hybrid layer; T = tubule; RT= resin tag; UD = unaltered dentine.

It was observed that the way smear layer was created has an influence on the thickness, density and attachment of the smear layer to the underlying tissue [53], while composition strongly indicates the characteristics of the tissue and type of cut through which the smear derived [54].

The composition of the smear layer influences the capacity of the etching to dissolve it. This has been justified by the presence of the highly disorganized collagen component of the smear that traps minerals causing a permanence of smear layer on the cavity floor even when acid-etching was used by ERAs [55]. Moreover, a greater quantity of residue, compared with the etched sound dentine, may persist on the surface of carious affected dentine (CAD) in a form of 'collagen smear layer' because acids only solubilize the mineral component of the layer [56]. Leaving this collagen smear layer at the bottom cavity, homogeneous infiltration of the monomers in the underlying dentine may be impeded, affecting the quality of bonding, which finally derives from the homogeneity of strengthening in the demineralized dentine [54, 57]. Poor infiltration of demineralized collagen can lead to hydrolysis due to activation of the MMPs over time [58].

Furthermore, the smear layer might adversely affect homogeneicity of hybridization when SEAs are used. SEAs hybrid layer is created by infiltration of the water-rich channels of the smear layer reaching the underline partially demineralized dentine. This means that the smear is included in the hybrid interdiffusion of SEAs [59]. However, in the case of thick smear layers, superficial demineralization and reinforcement of dentinal collagen might be compromised through early neutralization of the acidic primers of SEAs by the dentine buffering components of the smear [53].

Still, as far as we are currently aware, smear layer interferences remain a controversial issue irrespective of the adhesive techniques [60]. Some studies reported low dentine bond strengths over thick dentin smear layers, [52, 61], while others reported no influence on strength [33, 62], even if using mild SEAs [63], particularly in the early bond strength values [64].

2.3.1.1. ERA bonding

The fundamental processes involved in bonding enamel and dentine using ERAs include the removal of the mineral phase from the substrates without altering the collagen matrix in dentine. The hybrid layer results by filling the voids left by minerals using adhesive monomers that undergo complete *in situ* polymerization resulting in a tight adhesive interface.

Therefore, bonding by ERAs consists of a first phase of acid etching followed by the rinsing of the acid and then the infiltration of adhesive monomers in the demineralized tissue. The resulting hybrid layer is defined as a sort of dental tissue engineering formed by a mixture of inorganic resin monomers and organic demineralized dental tissue with resin tags in the dentinal tubules [56, 65, 66] (**Figure 8**). This chemical and micromechanical bonding has the role to seal the surface from leakages of fluids and bacteria [67].

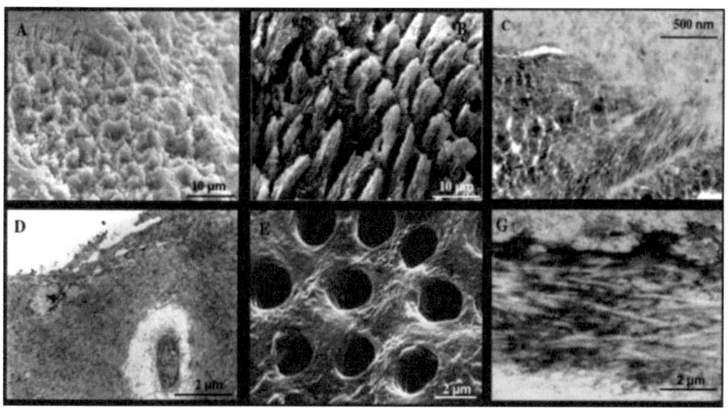

Figure 8. (A) An SEM image of an enamel surface, in which prisms are covered with a smear layer. (B) After the acid-etching procedure, the SEM shows dissolution of the smear layer and of the prism core with residual micro-craters and also of the interprismatic enamel border in a geometrical hexagonal ring (honeycomb appearance). (C) A non-demineralized TEM ultrathin section of the hybrid layer in enamel formed by the penetration of a low-viscosity BisGMA resin into the etched prisms. (D) A demineralized TEM section of a dentine surface covered by a smear. (E) An SEM image of acid-etched dentine, which appears completely deprived of the smear layer exposing the intertubular collagen matrix with nanometre-sized porosity and opened tubular apertures. (G) A demineralized TEM image of the fully diffusion of adhesive monomers into the completely demineralized collagen, which leads to the formation of a hybrid layer in which the composite will be able to adhere.

Phosphoric acid treatment completely exposes the enamel morphology of the prisms and interprismatic substance typically creating a *honeycomb* porosity that can be penetrated by low-viscosity resins by capillary attraction [68]. After polymerization, a durable attachment of the monomers in the enamel is achieved with the evidence of resin enveloping the prisms rendering them acid-resistant (**Figure 8**).

In dentine, phosphoric treatment exposes the 5–8 μm of the dentinal floor constituted by tubules and intertubular collagen matrix creating a nanometre-sized porosity within the matrix and opens the tubular orifices. In this way, the acid permits infiltration of monomers into and around collagen fibrils and tubular resin tags, gaining the following retention for the composite filling (**Figure 8**).

However, many problems have been claimed as a result of acid-etching dentine. Firstly, etching dentine may result in too much demineralization [54] compared with the concentration gradient of monomer infiltration with poor resin infiltration and discrepancies in reinforcement at the bottom of hybrid layer [54, 69].

Secondly, removing smear layer and smear plug from the dentine surface, etching increases tubular permeability causing outwards of pulpal fluids into the cavity [70].

Additional factors may interfere with the tight bond using etching when the cavity floor includes the very wet tissue of CAD. The soft, already demineralized collagen [71], the high degree of porosity and wetness [72], a lack of minerals around and within the fibrils [73], as well as spectral changes in the secondary structure of the collagen [49], may cause much aggressive etching in CAD [74]. At the same time, the use of strong acid cannot dissolve intratubular minerals, thus affecting infiltration of monomers and tubular tag formations [72]. Moreover, the low buffer of the tubular minerals may allow high demineralization and increase wetness in peritubular dentine with porosities in interdiffusion [15]. These facts result in a susceptibility of the affected interface to acid and base treatments, with higher degradation of CAD hybrid layers in comparison with sound dentine [75, 76].

Other adverse effects of etching dentine are related to the water-rinsing phase. In fact, the solubilized mineral is replaced by water rinsing, which results in an increase of the global amount of wetness in dentine. This wet dentine has difficulty to be tightly reinforced by the hydrophobic adhesive monomers even if resin tags can be formed in the opened tubules.

It has been shown that water is helpful in keeping the demineralized interfibrillar channels physically expanded allowing monomer percolation [56]. However, at the same time, water may produce (1) a lower degree of resin monomer conversion [49, 77]; (2) interference with the reinforcement of the hydrophobic BisGMA adhesives [15]; (3) phase separations between the hydrophobic and hydrophilic components of adhesives [78]. All of these may result in non-homogeneity and porosities at the interface as an expression of sub-optimal sealing [79–81].

To reduce interference by water, evaporation of the water rinsing in dentine is suggested using air-drying [56]. However, air-drying technique may shrink the demineralized collagen, and narrowing the interfibrillar channels [82] may impede infiltration by the monomers. In this event, bond strength would be limited to the strength of surface adhesion [56], leaving behind exposed and non-infiltrated fibrils.

This incomplete infiltration appears as voids or microporosity underneath and within the hybrid layer, which have been described through percolation of silver nitrate as nanoleakages [83] and pathway for hydrolytic and enzymatic degradation of the hybrid layer [84–86].

To increase the capacity of the monomers to re-expand the collapsed collagen network, studies evaluated the solubility parameters of chemicals with the intention of breaking collagen inter-peptide hydrogen bonds and then, the stiffened and collapsed state collagen [87]. Interactions were studied using the Hoy's solubility parameters [88], which provide estimation of the relative contribution of dispersion forces, polar forces and hydrogen-bonding forces to the overall cohesive forces that hold the polymers together. As a result, water and ethanol solvents were suggested to break the inter-peptide hydrogen bonds of the collagen, allowing the softening and expanding of the network in a rapid way [87]. These solvents have been generally blended with HEMA [89].

2.3.1.2. SEA bonding

Avoiding the separate etching phase of ERAs, SEAs seemed to be a good alternative to the problems related with.

SEAs are differentiated in view of the step needed to the hybrid layer formation. In the 'Two Step SEAs', hybridization is formed in two different procedures, the first of which is the layering of a primer of different acidity, followed by the application of an adhesive resin, usually BisGMA based. In the 'One-Step SEAs', acidic and adhesive monomers are mixed in the same solution, thereby causing hybridization at the same time.

Characteristics of SEAs are the acidic functional monomers in their formulation, which are demineralizing and infiltrating agents at the same time. Specifically, functional monomers have the capacity to interact with HAP and collagen by a series of chemical atomic-level interactions with an advantage of tissue strength. In dentine, these nano-interactions differ from a few hundreds of nanometres in the case of an ultramild self-etching ($pH \geq 2$) to one micrometer in the case of a mild ($pH \geq 2$) (**Figure 9**) and between one and two nanometres depth in the case of a strong-self-etching (pH around 1) [26, 90]. However, the resulting bond strength has been inversely related to the pH acidity. The interaction of 10-MDP, mild functional monomer (pH ≥ 2), has shown higher bonding and longevity compared to the strong 4-methacryloxyethyl trimellitic acid (4-MET) and phenyl-P [91, 92]. This is true despite the strong acidity producing a more retentive enamel etching (**Figure 10**) in comparison to the milder and the resulting hybrid layer is deeper than that of the mild monomer. Also, the retention of unstable calcium-phosphate salts makes the strong SEAs hybrid layer susceptible to hydrolysis [92, 93]. Furthermore, due to the great dissolution of calcium phosphate in intertubular dentine [56, 90, 91], a greater increase in wetness and porosity has to be considered in dentinal collagen, which has difficulty being infiltrated by hydrophobic BigGMA resins [63, 94].

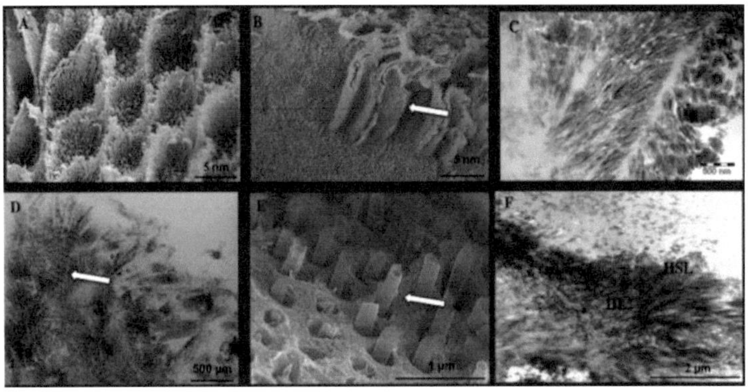

Figure 9. (A) A SEM micrograph of an enamel transversal section obtained by fracture after application of the primer of a two-step, mild-etching system adhesive system (Clearfil SE Bond, Kuraray, Osaka, Japan). This acidic priming is 10 MDP based and produces partial dissolution with infiltration of the core and boundaries of the prisms. (B) SEM section shows the infiltration of the interprismatic spaces in a longitudinal section with initial formation of tags between the prisms (arrow). (C) A non-decalcified TEM image, the application of the adhesive resin BisGMA based completes the hybrid layer in the enamel, which shows a nano-retentive appearance with hybridization of the inter- and intra-crystallite spaces of the prisms. (D) A demineralized TEM image of dentine priming using a 10 MDP primer. (D) A partial demineralization with infiltration of the collagen fibrils, which maintain dense transversal hydroxyapatite banding (arrow). (E) An SEM image, a preliminary tubular tag formation can be observed after priming (arrow). The application of the hydrophobic BisGMA adhesive resin in (F), a demineralized TEM image, completes the formation of the hybrid layer in dentine, which in the uppermost part includes hybridized smear layer residues. The dense appearance of the interdiffusion demonstrates the richness in chemical interactions with hydroxyapatite, which increases bonding of highly insoluble calcium salts. HL = hybrid layer.

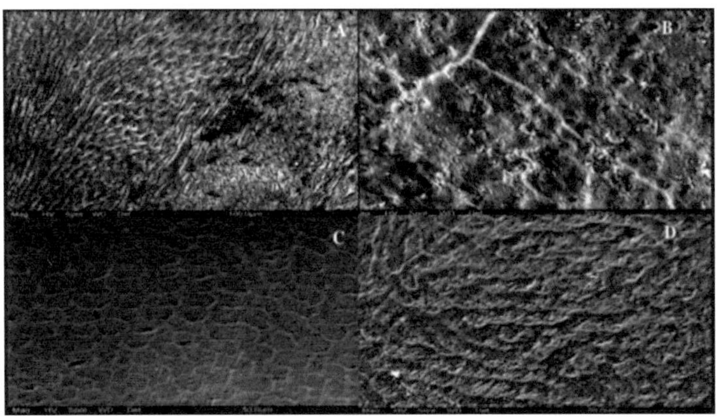

Figure 10. SEM images of the enamel-etching pattern obtained using four self-etchings of different pH acidity. (A) A phenyl-P-based primer, pH 0.5, (Resulcin AquaPrime, Merz Dental, Lütjenburg, Germany), which strongly etches the enamel prisms and interprismatic spaces in a similar way to phosphoric acid. (B) Modified methacrylates, phosphoric acid ester primer, pH 1.3 (Artegal One, Merz Dental GmbH, Eetzweg, Luütjenburg), the etching of which produces a geometrical pattern of enamel with shallow, crater-like depressions of the prisms and a less visible boundary wall appearance. (C) The enamel etching of a 10 MDP-based primer, pH 2.8 (Clearfil Protect Bond, Kuraray, Osaka, Japan); and (D) The etch produced by a 10-MDP-based one-step self-etching system, pH 2.7 (Clearfil S3 Bond, Kuraray, Osaka, Japan) both leading to a regular pattern of etching with a microporous appearance of the inter- and intraprismatic surfaces.

However, the mild acidity of 10 MDP causes a regularly layered structure on enamel and dentine, which is rich in chemical interactions with HAP and in highly insoluble calcium salts [63, 94]. Also, the stability of the 10 MDP hybrid layer in the oral environment is increased by the presence of the hydrophobic spacer carbon chain that prevents water sorption and degradation of the interface [20].

For this reason, 10 MDP mild SEAs have been suggested as the better bonding technology in dentistry particularly in the case of CAD.

In fact, using mild SEAs the residual HAP crystals in CAD will remain attached around the already demineralized collagen, thus allowing remineralization and at the same time, preventing discrepancies in the reinforcement (**Figure 11**) [58, 71, 81, 94–98]. In addition, the mineral deposits in CAD tubules cannot be dissolved without a high pH level of SEAs [56, 99] and thus resin tags cannot be an advantage to the CAD hybrid layer, neither in the case of mild nor using strong SEAs [2].

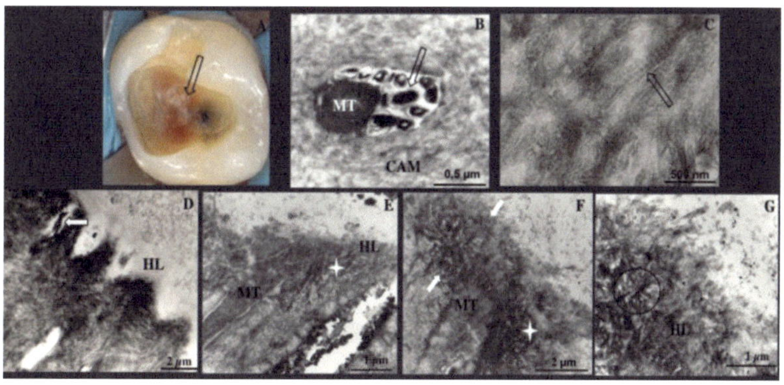

Figure 11. (A) The clinical appearance of a bottom cavity including the dark, yellow tissue of caries affected dentine (arrow). (B) A demineralized TEM image of a mineral occluded tubule also contains mineralized, residual bacteria (arrow) and (C) The TEM aspect of the partially demineralized fibrils of caries affected dentine with lose transversal banding (arrow). (D) A non-decalcified TEM image shows the hybrid layer in caries affected dentine formed by the use of an Etch&Rinse adhesive (OptiBond Solo, Kerr Corporation, Orange, CA, USA). Despite the use of a strong etching, tubular mineral occlusions cannot be dissolved (arrow) and complete resin tags cannot reinforce the hybrid layer of the Ech&Rinse adhesives. In (E, F and G), there are different decalcified TEM magnifications of the hybrid interface created using a mild self-etching adhesive (Clearfil Protect Bond, Kurary, Osaka, Japan) in affected dentine. Leaving crystals in and around the already demineralized collagen of caries affected dentine, 10 MDP functional monomer has the ability to interact with the residual hydroxyapatite with a series of chemical atomic-level interactions forming a strong chemical bond. (F) The magnified hybrid layer (between arrows) shows an irregular top of infiltrated smear layer plugging the tubular orifices, which still retain mineral occlusions. Nevertheless, hybridization of peritubular dentine is clearly discernible (asterisk). (G) A higher magnification of (F) shows attached, resin infiltrated hydroxyapatite crystals around the collagen of affected dentine (circle). This fact can prevent the affected fibrils from being exposed and hydrolysed in environmental fluids. MT = mineralized tubule; CAM = caries affected matrix; HL = hybrid layer.

Moreover, the use of mild SEAs in CAD leads to the evidence of an inhibitory effect on secondary caries [95]. Morphologically, an electron dense zone was described underlining the hybrid layer created by 10-MDP containing SEAs after exposure to an artificial demineralizing solution (pH 4.5) for 90 mins and then 5% sodium hypochlorite for 20 mins. This 'acid-base resistant layer' [95] was attributed to densely packed crystallites, which were probably formed by resin-infiltrated dentine. The observation accredited the chemical reactions that might take place between HAP and 10 MDP in dentine, which have the capacity of increasing the resistance to acid attacks of oral bacteria and, consequently, secondary caries.

As regards the use of one-step SEAs, they are complex mixtures of hydrophilic and hydrophobic components, which lead to acidification, priming and bonding in a same time.

These systems result in very thin hybrid layers, which are prone to less polymerization [100] and high permeation by fluids [28, 101, 102] (**Figure 12**). The incorporation of high concentrations of hydrophilic HEMA monomers [101, 102] explains absorption of water from the dentin fluids towards the interface [103–105]. Particularly, water permeability is enhanced in the case of which the composite interface would not have been immediately polymerized [106].

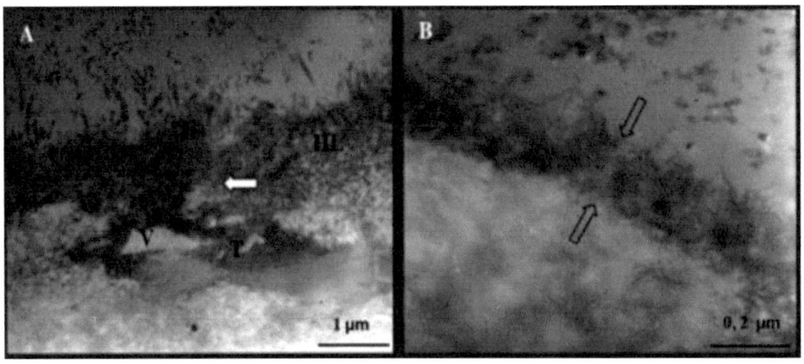

Figure 12. (A and B) Demineralized TEM sections of the very thin hybrid layer (between arrows) formed by a one-step self-etching system (Clearfil S3 Bond, Kurary, Osaka, Japan). In this case, samples were prepared in *in vivo* conditions of pulpal pressure and were extracted 10 min after completion. The hybrid layer shows a high hydrophilic HEMA appearance and exhibits water channels (white arrow) and porosities within the interface and voids in the tubule, which are a result of the poorly sealed tubules using these HEMA-rich, one-step self-etching adhesive systems. HL = hybrid layer; T = tubule; V = void.

In vitro studies [106] reported that the tubular occlusion in CAD might block the permeation of water fluids in the one-step hybrid layer. However, in clinical conditions of pulpal pressure, OSA's hybrid interdiffusion in CAD has been shown to be permeable [15].

Thus, it is not surprise that these hybrid layers are likely to be the cause of high hydrolysis [103, 107] and changes in their formulation have been proposed to overcome the problem. Hydrolytically resistant acrylamide-based monomers have been added in the complex mixture of One-steps promising better performance [67]. However, clinical evaluations reported inferior performance using acrylamide-based adhesives in comparison with those obtained in laboratory tests [108].

2.3.1.3. Degradation of adhesive restorations

Nowadays, the reasons for adhesive failure are of two kinds: (1) failures which originate from the surface of the hybrid layer and the upper polymerizing composite restoration; and (2) failures that take place in the depths of the hybrid layer, in which it interlocks chemically and micromechanically to the tooth tissues.

Failures related to the upper surface of the hybrid layer and to the surface of composite have been mainly correlated to microleakage of salivary fluids and proteins [109, 110]. Also, these failures have been connected to the high susceptibility of BisGMA-containing resins to accumulate dental plaque, which degrades resins as a result of the biofilm of cariogenic bacteria (Finer et al 2004).

Adhesive resins permitted infiltration of salivary fluids and proteins, mainly in the case of poor polymerization. Also, water fluids may be trapped in resins during the curing process. Biological breakdown is due to the addition of water to condensation type bonds within the resin [111]. These bonds, which include esters, urethanes and amides, are all prone to chemical hydrolysis, catalysed by acids, bases or enzymes.

Particularly, hydrolysis of the hybrid layer is linked to the addition of water to the ester bonds of the methacrylate polymer matrix [12]. Water breaks the covalent bonds in ester bonds with production of carboxylic acids, the same functional group that causes demineralization in the presence of lactic acid. Moreover, percolation of water fluids into the hybrid layer promotes the leaching of adhesive components, which have infiltrated the collagen matrix. This reaction is accelerated in pH excursions, which cause an increase of transient acid or base catalysis. Over time, hydrolytic degradation exposes susceptible bonds located deeper within the matrix, as well as allowing the infiltration of salivary enzymes, among which esterase is the most important class which are able to alter the stability of resins in the oral environment [21].

In composites, the main cause of degradation is the gap, which is a result of polymerization shrinkage. When salivary fluids and proteins diffuse into the polymer matrix, this causes a reduction of the glass-transition temperature, polymer plasticization and a decrease in thermal stability [112]. Resin undergoes hydrothermal degradation, hydrolysis and leaching of ions, which ultimately leads to surface softening and mechanical wear during mastication. Additionally, the ion leaching may directly cause permeation of water molecules into the spaces previously occupied by these ions [112–114].

A progressive deterioration of the resin mass permits the leakage of biofilm bacteria, which grow on the surface of composite. In particular, *S. mutans* has an accelerated growth on composites, which may be due to the leaching out of unreacted monomers or due to the physical property of surface roughness [6, 115]. Esterase activities of *S. mutans* are able to catalyse enzymatic hydrolysis of the polymer. Moreover, when sucrose from foodstuff is present, adhesion of *S. mutans* causes the formation of insoluble glucans [116] leading to intense local acidification and chemical hydrolysis [18].

Modification of the resin composition has been examined in an effort to prevent attachment of bacteria to the polymer surface. Quaternary ammonium compounds have demonstrated antibacterial properties and have been incorporated into methacrylate polymers for this purpose [117].

2.4. Activation of MMPs

Failures taking place at the bottom surface of the hybrid layer are related to hydrolytic activity by host-derived proteases. More specifically, MMPs are endopeptidases normally present in human saliva and capable of degrading extracellular matrix proteins [79].

MMPs are a family group of multi-domain calcium- and zinc-dependent endopeptidases that participate in physiological tissue development, remodelling of dentine matrices and remain entrapped in the dentin matrix during tooth development [118–120]. MMPs require metal ions such as calcium or zinc to bind to the active site for their catalytic activation through a so-called cysteine switch [13]. Moreover, MMPs need an acid micro-environment to become active [121, 122]. Activity has been documented during the carious process, in which MMPs digest carious collagen after having been activated by lactate released by the cariogenic bacteria. Furthermore, host-derived hydrolases can degrade sub-optimally infiltrated collagen fibres once they have been activated during bonding procedures.

In fact, it has been reported that acidic priming may induce the necessary low pH micro-environment capable of altering the conformation of the pro-peptide in the active form [119, 120]. More recently, it has been suggested that MMPs may become activated at the tooth and restoration interface by bacterial acid production [80]. In these situations, exposed and not protected collagen of chemical polymers from both the ERA and SEA systems might be degraded by MMPs.

Water is essential for the hydrolytic function of the enzymes because of the hydrolysis of peptide bonds in collagen, resulting in degradation of the polymer. [123]. The importance of water was accredited by experimentations showing no loss of dentine bonding over time when mineral oil was used as a storage medium instead of water [107].

MMP inhibitors have been suggested to treat dentine after acid priming. Among these (1) calcium and zinc chelators from acid-etched dentine, as the presence of calcium and zinc ions is necessary for MMPs to became activated [124, 125]; (2) protein cross-linking agents to cross-link their peptide chains immediately after acid etching [126]; (3) specific and nonspecific inhibitors of proteases added directly to primers [119, 126–129].

Furthermore, the use of the ethanol-wet bonding technique can be useful for preventing hydrolysis [129, 130]. Ethanol acts as a solvated primer to dehydrate the demineralize collagen allowing fibrils to be densely infiltrated by resin, keeping them free of water uptake.

Also, the application of chlorhexidine to acid-etched human dentin can save the collagen fibrils in the hybrid layer when etching and rinsing are used [131]. The addition of chlorhexidine as a primer in ERAs increases the capacity to maintain the structural integrity of the collagen probably by its zinc cation-chelating property [131]. However, it is necessary to define whether its effect is adhesive system specific. In fact, this is dependent upon the composition of the applied adhesive resin and generally results in a statistically significant decrease in the degree of conversion.

http://dx.doi.org/10.5772/64973

3. Conclusions

Despite the great evolution, dental adhesive restorations need to be improved in an effort to obtain stable restorations within the oral environment.

Even if, enamel etching by ERAs can represent the gold standard in enamel bond strength, at the same time, etching dentine may leave exposed and non-reinforced fibrils with all the consequences related to.

The 10 MDP technology of mild SEAs has been suggested as the better bonding technology due to the capacity of increasing the resistance to biological breakdown of adhesion.

However, modification of the resin materials is necessary to antagonize infiltration of salivary fluids and proteins, which are the causes of hydrolysis, particularly by the esterase activities of *S. mutans* and by MMPs.

Author details

Egle Milia[1*], Roberto Pinna[1], Enrica Filigheddu[1] and Stefano Eramo[2]

*Address all correspondence to: emilia@uniss.it

1 Department of Biomedical Science, University of Sassari, Sassari, Italy

2 Department of Surgery and Biomedical Science, University of Perugia, Perugia, Italy

References

[1] Buonocore, M. G. A simple method of increasing the adhesion of acrylic filling materials to enamel surfaces. J Dent Res. 1955;34:849–853.

[2] Pinna R, Maioli M, Eramo S, Mura I, Milia E. Carious affected dentine: its behaviour in adhesive bonding. Aust Dent J. 2015;60:276–293. doi: 10.1111/adj.12309

[3] Vanini L, Mangani FM. Determination and communication of color using the five color dimensions of teeth. Pract Proced Aesthet Dent. 2001;13:19–26.

[4] Magne P, Belser UC. Bonded porcelain restorations in the anterior dentition. Chicago: Quintessence; 2002.

[5] Pinna R, Colonna A, Milia E. Recovery of the fullness of the upper lip after laminate veneer positioning: A case report. In: Conseuro Congress; 9-11 May 2013; Paris. Clin Oral Investig; 2013. p. 1050

[6] Pinna R, Bortone A, Sotgiu G, Dore S, Usai P, Milia E. Clinical evaluation of the efficacy of one self-adhesive composite in dental hypersensitivity. Clin Oral Investig. 2015;19:1663–1672. doi: 10.1007/s00784-014-1390-3

[7] Milia E, Cumbo E, Cardoso RJ, Gallina G. Current dental adhesives systems. A narrative review. Curr Pharm Des. 2012;18:5542–5552. DOI: 10.2174/138161212803307491

[8] Schmalz G, Concepts in biocompatibility testing of dental restorative materials. Clin Oral Invest. 1997;1:154–162.

[9] Lewis JB, Rueggeberg FA, Lapp CA, Ergle JW, Schuster GS. Identification and characterization of estrogen-like components in commercial resin-based dental restorative materials. Clin Oral Invest. 1999;3:107–113.

[10] Demirci M, Hiller KA, Bosl C, Galler K, Schmalz G, Schweikl H. The induction of oxidative stress, cytotoxicity, and genotoxicity by dental adhesives. Dent Mater. 2008;24:362–371. doi:10.1016/j.dental.2007.06.009

[11] Hickel R, Manhart J, Garcia-Godoy F. Clinical results and new developments of direct posterior restorations. Am J Dent. 2000;13:41D–54D.

[12] Bourbia M, Ma D, Cvitkovitch DG, Santerre JP, Finer Y. Cariogenic bacteria degrade dental resin composites and adhesives. J Dent Res. 2013;92:989–994. doi: 10.1177/0022034513504436.

[13] Spencer P, Ye Q, Misra A, Goncalves SE, Laurence JS. Proteins, pathogens, and failure at the composite-tooth interface. J Dent Res. 2014;93:1243–1249. doi: 10.1177/0022034514550039.

[14] Fabrizio Cordeschi Sistemi adesivi con applicazioni in campo biomedicale. [PhD Thesis]. Bologna: University of Bologna; 2011.

[15] Milia E, Pinna R, Castelli G, Bortone A, Marceddu S, Garcia-Godoy F, Gallina G. TEM morphological characterization of a one-step self-etching system applied clinically to human caries-affected dentin and deep sound dentin. Am J Dent. 2012;25:321–326.

[16] Misra A, Parthasarathy R, Ye Q, Singh V, Spencer P. Swelling equilibrium of dentin adhesive polymers formed on the water–adhesive phase boundary: Experiments and micromechanical model. Acta Biomater. 2014;10:330–342. doi: 10.1016/j.actbio.2013.09.017.

[17] Van Krevelen DW, Te Nijenhuis K. Properties of polymers: their correlation with chemical structure; their numerical estimation and prediction from additive group contributions. Amsterdam: Elsevier; 2009.

[18] Donmez N, Belli S, Pashley DH, Tay FR. Ultrastructural correlates of in vivo/in vitro bond degradation in self-etch adhesives. J Dent Res. 2005;84:355–9. doi: 10.1177/154405910508400412

[19] Van Landuyt KL, Snauwaert J, De Munck J, Peumans M, Yoshida Y, Poitevin A, Coutinho E, Suzuki K, Lambrechts P, Van Meerbeek B. Systematic review of the

chemical composition of contemporary dental adhesives. Biomat. 2007;28:3757–3785. doi:10.1016/j.biomaterials.2007.04.044

[20] Feitosa VP, Sauro S, Ogliari FA, Stansbury JW, Carpenter GH, Watson TF, Sinhoreti MA, Correr AB. The role of spacer carbon chain in acidic functional monomers on the physicochemical properties of self-etch dental adhesives. J Dent. 2014;42:565–74. doi: 10.1016/j.jdent.2014.02.009.

[21] Delaviz Y, Finer Y, Santerre J. Biodegradation of resin composites and adhesives by oral bacteria and saliva: A rationale for new material designs that consider the clinical environment and treatment challenges. Dent Mater. 2014;30:16–32. doi: 10.1016/j.dental.2013.08.201.

[22] Salz U, Zimmermann J, Zeuner F, Moszner N. Hydrolytic stability of self-etching adhesive systems. J Adhes Dent. 2005;7:107–116.

[23] Milia E, Campus G, Santini A. Resin adhesion to sclerotic dentin: an *in vivo* TEM study of a new antibacterial system. J Dent Res. 2003;82:255. Abstract #1954.

[24] Abedin F, Ye Q, Good H, Parthasarathy R, Spencer P. Polymerization- and solvent-induced phase separation in hydrophilic-rich dentin adhesive mimic. Acta Biomater. 2014;10:3038–3047 doi: 10.1016/j.actbio.2014.03.001.

[25] Yiu CKY, King NM, Carrilho MRO, Sauro S, Rueggeberg FA, Prati C, et al. Effect of resin hydrophilicity and temperature on water sorption of dental adhesive resins. Biomaterials. 2006;27:1695–703. doi:10.1016/j.biomaterials.2005.09.037

[26] Watanabe I, Nakabayashi N, Pashley DH. Bonding to ground dentin by a phenyl-P self-etching primer. J Dent Res. 1994;73:1212–1220. doi: 10.1177/00220345940730061301

[27] Van Landuyt K, De Munck J, Snauwaert J, Coutinho E, Poitevin A, Yoshida Y, Inoue S, Peumans M, Suzuki K, Lambrechts P, Van Meerbeek B. Monomer-solvent phase separation in one-step self-etch adhesives. J Dent Res. 2005;84:183–188. doi: 10.1177/154405910508400214

[28] Ito S, Hashimoto M, Wadgaonkar B, Svizero N, Carvalho RM, Yiu C, Rueggeberg FA, Foulger S, Saito T, Nishitani Y, Yoshiyama M, Tay FR, Pashley DH. Effects of resin hydrophilicity on water sorption and changes in modulus of elasticity. Biomaterials. 2005;26:6449–6459. doi:10.1016/j.biomaterials.2005.04.052

[29] Spagnuolo G, D'Antò V, Cosentino C, Schmalz G, Schweikl H, Rengo S. Effect of N-acetyl-l-cysteine on ROS production and cell death caused by HEMA in human primary gingival fibroblasts. Biomaterials. 2006;27:1803–1809. doi:10.1016/j.biomaterials.2005.10.022

[30] Maciel KT, Carvalho RM, Ringle RD, Preston CD, Russell CM, Pashley DH. The effects of acetone, ethanol, HEMA, and air on the stiffness of human decalcified dentin matrix. J Dent Res. 1996;75:1851–8. doi: 10.1177/00220345960750110601

[31] Pioch T, Staehle HJ, Duschner H, Garcıa-Godoy F. Nanoleakage at the composite-dentin interface: A review. Am J Dent. 2001;14:252–58.

[32] Nakaoki Y, Nikaido T, Pereira PN, Inokoshi S, Tagami J. Dimensional changes of demineralized dentin treated with HEMA primers. Dent Mater. 2000;16:441–446. doi: 10.1016/S0109-5641(00)00042-7

[33] Tay FR, Pashley DH. Aggressiveness of contemporary self- etching systems. I: Depth of penetration beyond dentin smear layers. Dent Mater. 2001;17:296–308. doi:10.1016/S0109-5641(00)00087-7

[34] Nie J, Bowman CN. Synthesis and photopolymerization of N,N'- dimethyl,-N,N' -di(methacryloxy ethyl)-1,6-hexanediamine as a polymerizable amine coinitiator for dental restorations. Biomaterials. 2002;23:1221–1226.

[35] Ferracane IL. Current trends in dental composites. Crit Rev Oral Biol Med 1995;6:302–318. doi: 10.1177/10454411950060040301

[36] Kleverlaan CJ, Feilzer AJ. Polymerization shrinkage and contraction stress of dental resin composites. Dent Mater. 2005;21:1150–1157. doi:10.1016/j.dental.2005.02.004

[37] Gajewski VES, Pfeifer CS, Froes-salgado NRG, Boaro LCC, Braga RR. Monomers used in resin composites: degree of conversion, mechanical properties and water sorption/solubility. Braz Dent J. 2012;23:508–514. doi.org/10.1590/S0103-64402012000500007

[38] Braga RR, Ballester RY, Ferracane JL. Factors involved in the development of polymerization shrinkage stress in resin-composites: a systematic review. Dent Mater. 2005;21:962–970. doi:10.1016/j.dental.2005.04.018

[39] Davidson CL, De Gee AJ. Relaxation of polymerization contraction stresses by flow in dental composites. J Dent Res. 1984;63:146–148. doi: 10.1177/00220345840630021001

[40] Feilzer AJ, De Gee AJ, Davidson CL. Quantitative determination of stress reduction by flow in composite restorations. Dent Mater. 1990;6:167–171. doi: 10.1016/0109-5641(90)90023-8

[41] Watts DC, Schneider LFJ, Marghalani HY. Bond-disruptive stresses generated by resin composite polymerization in dental cavities. J Adhes Sci Technol. 2009;23:1023–1042.

[42] Sancerre JP, Shajii L, Leung BW. Relation of dental composite formulations to their degradation and the release of hydrolyzed polymeric-resin-derived products. Crit Rev Oral Biol Med. 2001;12:136–151. doi: 10.1177/10454411010120020401

[43] Moszner K, Klapdohr K. Nanotechnology for dental composites. Int J Nanotechnol. 2004;1:130–156.

[44] Langhorst SE, O'Donnell JN, Skrtic D. *In vitro* remineralization of enamel by polymeric amorphous calcium phosphate composite: quantitative microradiographic study. Dent Mater. 2009;25:884–891. doi: 10.1016/j.dental.2009.01.094.

[45] Cheng L, Weir MD, Xu HH, Antonucci JM, Kraigsley AM, Lin NJ, Lin-Gibson S, Zhou X. Antibacterial amorphous calcium phosphate nanocomposite with quaternary ammonium salt and silver nanoparticles. Dent Mater. 2012;28:561–572. doi: 10.1016/j.dental.2012.01.005.

[46] Guo G, Fan Y, Zhang JF, Hagan JL, Xu X, Novel dental composites reinforced with zirconia–silica ceramic nanofibers. Dent Mater. 2012;28:360–368. doi: 10.1016/j.dental.2011.11.006.

[47] Chen L, Yu Q, Wang Y, Li H. BisGMA/TEGDMA dental composite containing high aspect-ratio hydroxyapatite nanofibers. Dent Mater. 2011;27:1187–1195. doi: 10.1016/j.dental.2011.08.403.

[48] Tezvergil, Lassila LV, Vallittu PK. The effect of fiber orientation on the polymerization shrinkage strain of fiber-reinforced composites. Dent Mater. 2006;22:610–616. doi: 10.1016/j.dental.2005.05.017.

[49] Kinney JH, Balooch M, Marshall GW, Marshall SJ. A micromechanics model of the elastic properties of human dentine. Arch Oral Biol. 1999;44:813–822. doi:10.1016/S0003-9969(99)00080-1

[50] Spencer P, Wang Y, Walker MP, Swafford JR. Molecular structure of acidetched dentin smear layers—in situ study. J Dent Res. 2001;80:1802–1807. doi: 10.1177/00220345010800090601

[51] Watanabe I, Saimi Y, Nakabayashi N. Effect of smear layer on bonding to ground dentin. Relationship between grinding conditions and tensile bond strength. J Jpn Dent Mater. 1994;13:101–108.

[52] Koibuchi H, Yasuda N, Nakabayashi N. Bonding to dentin with a self-etching primer: the effect of smear layers. Dent Mater. 2001;17:122–126. doi:10.1016/S0109-5641(00)00049-X

[53] Pinna R, Usai P, Arrica M, Milia E. Effectiveness of two self-etchings bonded clinically in caries affected dentin with omogeneous smear layer. J Nanomat. 2015;1:7. doi.org/10.1155/2015/489183

[54] Oliveira SS, Pugach MK, Hilton JF, Watanabe LG, Marshall SJ, Marshall GW Jr. The influence of the dentin smear layer on adhesion: a self-etching primer vs. a total-etch system. Dent Mater. 2003;19:758–767. doi:10.1016/S0109-5641(03)00023-X

[55] Pashley DH, Carvalho RM. Dentine permeability and dentine adhesion. J Dent. 1997;5:355–372. doi:10.1016/S0300-5712(96)00057-7.

[56] Al Sunbul H, Silikas N, Watts DC. Polymerization shrinkage kinetics and shrinkage-stress in dental resin-composites. Dent Mater. 2016;32:998–1006 doi: 10.1016/j.dental.2016.05.006.

[57] Pashley DH, Ciucchi B, Sano H. Permeability of dentin to adhesive agents. Quintessence Int. 1993;24:618–631.

[58] Wang Y, Spencer P. Analysis of acid-treated dentin smear debris and smear layers using confocal Raman microspectroscopy. J Biomed Mater Res. 2002;60:300–308. doi: 10.1002/jbm.10108

[59] Hashimoto M, Ohno H, Kaga M, Endo K, Sano H, Oguchi H. In vivo degradation of resin-dentin bonds in humans over 1 to 3 years. J Dent Res. 2000;79:1385–1391. doi: 10.1177/00220345000790060601

[60] Milia E, Santini A. Ultrastructural transmission electron microscopy (TEM) study of hybrid layers formed beneath a one-bottle adhesive system using the total-etch technique and a self-etching system. Quintessence Int. 2003;34:447–452.

[61] Scholtanus JD, Purwanta K, Dogan N, Kleverlaan CJ, Feilzer A. Microtensile bond strength of three simplified adhesive systems to caries-affected dentin. J Adhe Dent. 2010;12:273–278. doi: 10.3290/j.jad.a17545.

[62] Ogata M, Harada N, Yamaguchi S, Nakajima M, Pereira PN, Tagami J. Effects of different burs on dentin bond strengths of self-etching primer bonding systems. Oper Dent. 2001;26:375–382.

[63] Tani C, Finger WJ. Effect of smear layer thickness on bond strength mediated by three all-in-one self-etching priming adhesives. J Adhes Dent. 2002;4:283–289.

[64] Tay FR, Carvalho R, Sano H, Pashley DH. Effect of smear layers on the bonding of a self-etching primer to dentin. J Adhes Dent. 2000;2:99–116.

[65] Reis A, Grandi V, Carlotto L, et al. Effect of smear layer thickness and acidity of self-etching solutions on early and long-term bond strength to dentin. J Dent. 2005;33:549–559. doi:10.1016/j.jdent.2004.12.003

[66] Nakabayashi N, Takarada K. Effect of HEMA on bonding to dentin. Dent Mater. 1992;8:125–130. doi:10.1016/0109-5641(92)90067-M

[67] Nakabayashi, N, Watanabe A, Ikeda W. Intra-oral bonding of 4-META-MMA-TBBO resin to vital human dentin. Am J Dent. 1995;8:37–42.

[68] Santini A, Milia E, Miletic V. A review of SEM and TEM studies on the hybridisation of dentine. In: Mendez-Vilas A, Diaz J, editors. Microscopy: Science, Technology, Applications and Education. Badajoz: Formatex Research Center; 2010. p. 256–268.

[69] Kanca JA. Improving bond strength through acid etching of dentin and bonding to wet dentin surfaces. J Am Dent Assoc. 1992;123:35-43. doi:10.14219/jada.archive.1992.0248

[70] Attal, JP, Asmussen, G, Degrange, M. Effects of surface treatment on the free surface of dentine. Dent Mater. 1994;10:259–264. doi:10.1016/0109-5641(94)90071-X

[71] Perdigão J. Dentin bonding—variables related to the clinical situation and the substrate treatment. Dent Mater. 2010;26:e24–e37. DOI: 10.1016/j.dental.2009.11.149.

[72] Ogawa K, Yamashita Y, Ichijo T, Fusayama T. The ultrastructure and hardness of the transparent layer of human caries dentin. J Dent Res. 1983;62:7–10 doi: 10.1177/00220345830620011701

[73] Yoshiyama M, Urayama A, Kimochi T, Matsuo T, Pashley DH. Comparison of conventional vs self-etching adhesive bonds to caries-affected dentin. Oper Dent. 2000;25:163–169.

[74] Spencer P, Wang Y, Katz JL, Misra A. Physicochemical interactions at the dentin/adhesive interface using FTIR chemical imaging. J Biomed Opt. 2005;10:031104. doi: 10.1117/1.1914844

[75] Marshall GW, Balooch M, Kinney JH, Marshall SJ. Atomic force microscopy of conditioning agents on dentin. J Biomed Mater Res. 1995;29:1381–1387. DOI: 10.1002/jbm.820291109

[76] Nakajima M, Hosaka K, Iwamoto N, Ikeda M, Foxton RM, Tagami J. Improving the effect of NaOCl pretreatment on bonding to caries-affected dentin using self-etch adhesives. J Dent. 2009;37:769–775. doi: 10.1016/j.jdent.2009.06.005.

[77] Kunawarote S, Nakajima M, Foxton RM, Tagami J. Pretreatment effect of mild acidic HOCl solution on adhesive to caries-affected dentin using self-etch adhesive. Eur J Oral Sci. 2011;119:86–92. DOI: 10.1111/j.1600-0722.2010.00788.x

[78] Jacsobsen T, Soderholm KJ. Some effects of water on dentin bonding. Dent Mater. 1995;11:132–136. doi:10.1016/0109-5641(95)80048-4

[79] Tay FR, Gwinnett JA, Wei SH. Micromorphological spectrum from overdrying to overwetting acid-conditioned dentin in water-free acetone-based, single-bottle primer/adhesives. Dent Mater. 1996;12:236–244. doi:10.1016/S0109-5641(96)80029-7

[80] Erhardt MC, Toledano M, Osorio R, Pimenta LA. Histomorphologic characterization and bond strength evaluation of caries-affected dentin/resin interfaces: Effects of long-term water exposure. Dent Mater. 2008;24:786–798. doi:10.1016/j.dental.2007.09.007

[81] Mohsen NM, Craig RG, Filisko FE. The effects of moisture on the dielectric relaxation of urethane dimethacrylate polymer and composites. J Oral Rehabil. 2001;28:376–392. DOI: 10.1046/j.1365-2842.2001.00669.x

[82] Sano H, Shono T, Takatsu T, Hosada H. Microporous dentin zone beneath resin-impregnated layer. Oper Dent. 1994;19:59–64

[83] Milia E, Lallai MR, Garcia-Godoy F. In vivo effect of a self-etching primer on dentin. Am J Dent. 1999;4:167–171.

[84] Doi J, Itota T, Torii Y, Nakabo S, Yoshiyama M. Micro-tensile bond strength of self-etching primer adhesive systems to human coronal carious dentin. J Oral Rehabil. 2004;31:1023–1028. DOI: 10.1111/j.1365-2842.2004.01339.x

[85] Yazici AR, Akca T, Ozgünaltay G, Dayangaç B. Bond strength of a self-etching adhesive system to caries-affected dentin. Oper Dent. 2004;29:176–181.

[86] Yoshiyama M, Doi J, Nishitani Y, et al. Bonding ability of adhesive resins to caries-affected and caries-infected dentin. J Appl Oral Sci. 2004;12:171–176. doi.org/10.1590/S1678-77572004000300002

[87] Nakajima M, Kitasako Y, Okuda M, Foxton RM, Tagami J. Elemental distributions and microtensile bond strength of the adhesive interface to normal and caries-affected dentin. J Biomed Mater Res B Appl Biomater. 2005;72:268–275. DOI: 10.1002/jbm.b.30149

[88] Sonoda H, Banerjee A, Sherriff M, Tagami J, Watson TF. An in vitro investigation of microtensile bond strengths of two dentine adhesives to caries-affected dentine. J Dent. 2005;33:335–342. doi:10.1016/j.jdent.2004.09.009

[89] Pereira PN, Nunes MF, Miguez PA, Swift EJ Jr. Bond strengths of a 1-step self-etching system to caries-affected and normal dentin. Oper Dent. 2006;31:677–681. doi.org/10.2341/05-131

[90] Moher D, Liberati A, Tetzlaff J, Altman DG; PRISMA Group. Preferred reporting items for systematic reviews and meta-analyses: the PRISMA statement. J Clin Epidemiol. 2009:62:1006–1012. doi: 10.1016/j.jclinepi.2009.06.005.

[91] Yoshida Y, Nagakane K, Fukuda R, Nakayama Y, Okazaki M, Shintani H, Inoue S, Tagawa Y, Suzuki K, De Munck J, Van Meerbeek B. Comparative study on adhesive performance of adhesive monomers. J Dent Res. 2004;83:454–458. doi: 10.1177/154405910408300604

[92] Yoshida Y, Van Meerbeek B, Nakayama Y, Yoshioka M, Snauwaert J, Abe Y, Lambrechts P, Vanherle G. Okazaki M. Adhesion to and decalcification of hydroxyapatite by carboxylic acids. J Dent Res. 2001;80:1565–1569. doi: 10.1177/00220345010800061701

[93] Waidyasekera K, Nikaido T, Weerasinghe DS, Ichinose S, Tagami J. Reinforcement of dentin in self-etch adhesive technology: a new concept. J Dent. 2009;37:604–609. doi: 10.1016/j.jdent.2009.03.021.

[94] Sano H, Takatsu T, Ciucchi B, Horner JA, Matthews WG, Pashley DH. Nanoleakage: leakage within the hybrid layer. Oper Dent. 1995;20:18–25.

[95] Kwong SM, Tay FR, Yip HK, Kei IH, Pashley DH. An ultrastructural study of the application of dentine adhesives to acid-conditioned sclerotic dentine. J Dent. 2000;28:515–528. doi:10.1016/S0300-5712(00)00032-4

[96] Ceballos L, Camejo DG, Victoria Fuentes M, et al. Microtensile bond strength of total-etch and self-etching adhesives to caries-affected dentine. J Dent. 2003;31:469–477. doi: 10.1016/S0300-5712(03)00088-5

[97] Sano H, Yoshikawa T, Pereira PN, et al. Long-term durability of dentin bonds made with a self-etching primer, in vivo. J Dent Res. 1999;78:906–911. doi: 10.1177/00220345990780041101

[98] Hashimoto M1, Ohno H, Sano H, Tay FR, Kaga M, Kudou Y, Oguchi H, Araki Y, Kubota M. Micromorphological changes in resin-dentin bonds after 1 year of water storage. J Biomed Mater Res. 2002;63: 306–11.

[99] Van Meerbeek B, De Munck J, Yoshida Y, et al. Buonocore memorial lecture. Adhesion to enamel and dentin: current status and future challenges. Oper Dent. 2003;28:215–235.

[100] Inoue G, Nikaido T, Foxton RM, Tagami J. The acid-base resistant zone in three dentin bonding systems. Dent Mater J. 2009;28:717–721. doi.org/10.4012/dmj.28.717

[101] Van Landuyt K, Snauwaert J, Peumans M, De Munck J, Lambrechts P, Van Meerbeek B. The role of HEMA in one-step self-etch adhesives. Dent Mater. 2008;24:1412–1419. doi: 10.1016/j.dental.2008.02.018

[102] Tay FR, Pashley DH. Water treeing – A potential mechanism for degradation of dentin adhesives. Am J Dent. 2003;16:6–12.

[103] Itthagarun A, Tay FR, Pashley DH, Wefel JS, Garcia-Godoy F, Wei S. Single-step, self-etch adhesives behave as permeable membranes after polymerization. Part III. Evidence from fluid conductance and artificial caries inhibition. Am J Dent. 2004;17:394–400.

[104] Sauro S, Mannocci F, Toledano M, Osorio R, Thompson I, Watson TF. Influence of the hydrostatic pulpal pressure on droplets formation in current etch-and-rinse and self-etch adhesives: a video rate/TSM microscopy and fluid filtration study. Dent Mater. 2009;25:1392–1402. doi: 10.1016/j.dental.2009.06.010.

[105] Hashimoto M, Ito S, Tay FR, et al. Fluid movement across the resin-dentin interface during and after bonding. J Dent Res. 2004;83:843–848. doi: 10.1177/154405910408301104

[106] Grégoire G, Joniot S, Guignes P, Millas A. Dentin permeability: Self-etching and one-bottle dentin bonding systems. J Prosthet Dent. 2003;90:42–49. doi:10.1016/S0022-3913(03)00258-0

[107] Tay FR, Pashley DH, Hiraishi N, et al. Tubular occlusion prevents watertreeing and through-and-through fluid movement in a single-bottle, one-step self-etch adhesive model. J Dent. 2005;84:891–896. doi: 10.1177/154405910508401004

[108] Carrilho MR, Carvalho RM, Tay FR, Yiu C, Pashley DH. Durability of resin-dentin bonds related to water and oil storage. Am J Dent. 2005;18:315–319.

[109] Van Meerbeek B, Yoshihara K, Yoshida Y, Mine A, De Munck J, Van Landuyt KL. State of the art of self-etch adhesives. Dent Mater. 2011;27:17–28. doi:10.1016/j.dental.2010.10.023.

[110] Nakajima M, Sano H, Zheng L, Tagami J, Pashley DH. Effect of moist vs. dry bonding to normal vs. caries-affected dentin with Scotchbond Multi-Purpose Plus. J Dent Res. 1999;78:1298–1303. doi: 10.1177/00220345990780070301

[111] Finer Y, Santerre JP. The influence of resin chemistry on a dental composite's biodegradation. J Biomed Mater Res A. 2004;69:233–46. DOI: 10.1002/jbm.a.30000

[112] Santerre JP, Shajii L, Leung BW. Relation of dental composite formulations to their degradation and the release of hydrolyzed polymeric-resin-derived products. Crit Rev Oral Biol Med. 2001;12:136–151.

[113] Ferracane JL. Hygroscopic and hydrolytic effects in dental polymer networks. Dent Mater. 2006;22:211–222. doi:10.1016/j.dental.2005.05.005

[114] Milia E, Castelli G, Bortone A, Sotgiu G, Manunta A, Pinna R, Gallina G. Short-term response of three resin-based materials as desensitizing agents under oral environmental exposure. Acta Odontol Scand. 2013;71:599–609 doi: 10.3109/00016357.2012.700063

[115] Kawai K, Tsuchitani Y. Effects of resin composite components on glucosyltransferase of cariogenic bacterium. J Biomed Mater Res. 2000;51:123–127.

[116] Beyth N, Bahir R, Matalon S, Domb AJ, Weiss EI. *Streptococcus mutans* biofilm changes surface-topography of resin composites. Dent Mater. 2008;24:732–736. doi: 10.1016/j.dental.2007.08.003

[117] Cagetti MG, Mastroberardino S, Milia E, Cocco F, Lingström P, Campus G. The use of probiotic strains in caries prevention: a systematic review. Nutrients. 2013;5:2530–2550. doi: 10.3390/nu5072530.

[118] Li Y, Carrera C, Chen R, Li J, Lenton P, Rudney JD, Jones RS, Aparicio C, Fok A. Degradation in the dentin–composite interface subjected to multi-species biofilm challenges. Acta Biomaterialia 2014;10:375–383 doi: 10.1016/j.actbio.2013.08.034.

[119] Linde A. Dentin matrix proteins: composition and possible functions in calcification. Anat Rec. 1989;224:154–166.

[120] Hebling J, Pashley DH, Tjäderhane L, Tay FR. Chlorhexidine arrests subclinical degradation of dentin hybrid layers in vivo. J Dent Res. 2005;84:741–746. doi: 10.1177/154405910508400811

[121] Tersariol IL, Geraldeli S, Minciotti CL, et al. Cysteine cathepsins in human dentin-pulp complex. J Endod. 2010;36:475–481. doi: 10.1016/j.joen.2009.12.034.

[122] Chaussain-Miller C, Fioretti F, Goldberg M, Menashi S. The role of matrix metalloproteinases (MMPs) in human caries. J Dent Res. 2006;85:22–32.

[123] Fusayama T, Terachima S. Differentiation of two layers of carious dentin by staining. J Dent Res. 1972;51:866.

[124] Carrilho MR, Geraldeli S, Tay F, et al. In vivo preservation of the hybrid layer by chlorhexidine. J Dent Res. 2007;86:529–533. doi: 10.1177/154405910708600608

[125] Visse R, Nagase H. Matrix metalloproteases and tissue inhibitors of metalloproteinases. Cir Res. 2003;92:827–839. doi: 10.1161/01.RES.0000070112.80711.3D

[126] Osorio R, Erhardt MCG, Pimenta LAF, Osorio F, Toledano M. EDTA treatment improves resin–dentin bonds resistance to degradation. J Dent Res. 2005;85:736–40. doi: 10.1177/154405910508400810

[127] Pashley DH, Swift EJ Jr. Dentin bonding. J Esthet Restor Dent. 2008;20:153–154. doi: 10.1111/j.1708-8240.2008.00169.x.

[128] Brackett MG, Tay FR, Brackett WW, Dib A, Dipp FA, Mai S, Pashley DH. In vivo chlorhexidine stabilization of an acetone-based dentin adhesives. Oper Dent. 2009;34:381–385. doi: 10.2341/08-103.

[129] Bedran-Russo AKB, Vidal CMP, Santos PHD, Castellan CS. Long-term effect of carbodiimide on dentin matri and resin–dentin bonds. J Biomed Mater Res B Appl Biomater. 2010;94:250–255. doi: 10.1002/jbm.b.31649.

[130] Almahdy A, Koller G, Sauro S, et al. Effects of MMP inhibitors incorporated within dental adhesives. J Dent Res. 2012;91:605–611. doi: 10.1177/0022034512446339.

[131] Nishitani Y, Yoshiyama M, Donnelly AM, Agee KA, Sword J, Tay FR, Pashley DH. Effects of resin hydrophilicity on dentin bond strength. J Dent Res. 2006;85:1016–1021. doi: 10.1177/154405910608501108

Section 2

Properties of Adhesives

Chapter 6

Experimental Investigation on the Self-Healing Efficiency of Araldite 2011 Adhesive Reinforced with Thermoplastic Microparticles

Halil Özer and Engin Erbayrak

Additional information is available at the end of the chapter

http://dx.doi.org/10.5772/65167

Abstract

Newly developed self-healing technologies allow self-repair of adhesively bonded joints without the need for replacing the damaged joint with a new one. This study addresses to define experimentally the self-healing ability and efficiency of the Araldite 2011 epoxy adhesive reinforced with the thermoplastic co-polyester (TPC). Heating the joint results in melting the co-polyester in adhesive, and then it is expected to repair the damaged region by the melted co-polyester. Firstly, before applying the self-healing process, a preliminary study was applied to define whether selected adhesive is compatible with the thermoplastic particles in terms of self-healing. From the initial results, it is seen that Araldite 2011 adhesive is suitable for use in the self-healing mechanism. In the healing cycle, initial crack in the reinforced adhesive was propagated until 30 mm during the double cantilever beam (DCB) testing. The fractured specimens were repeatedly healed in terms of the close-then-heal (CTH) scheme until no healing has taken place. After the healing process was completed, the healing efficiency was defined using the fracture energy values. In this study, the healing process was repeated two times with the acceptable healing efficiencies. It is concluded that the damaged reinforced adhesive can repair itself with a considerable healing efficiency.

Keywords: Self-healing, Araldite 2011, Thermoplastic additive, Healing efficiency, DCB test

1. Introduction

Structural adhesives have been used extensively in the space, aviation, automotive, and naval industries. Many techniques have been proposed to reduce high stress concentrations at the lap joints and to improve the joint strength. One of them is adding the additives into adhesive, such as graphene, rubber, nanoparticles, etc. Khan et al. [1] studied the effect of incorporation of different weight fractions of graphene on adhesive mechanical properties. They stated that addition of 0.7 % graphene directly increases both adhesive strength and toughness. In addition, they claimed that the graphene was the most appropriate additive for polyvinyl acetate (PVAc) adhesive. Sadigh and Marami [2] offered the new method of increasing the strength of adhesively bonded joints by reinforcing a small part of reduced graphene oxide (RGO). Finally, they stated that inclusion of 0.5 wt.% RGO increased both ultimate tensile and compressive strength of bulk specimens by rate of 30 and 26 %, respectively.

In addition, an alternative technique has been recently proposed to reduce peel and shear stress concentrations in the lap joints. Different names for this type of joint are used in literature such as mixed-adhesive, modulus-graded, bi-adhesive, and hybrid-adhesive joints. In this technique, stiff and flexible adhesives are used together along the overlap region, without mixing the adhesives. In application of that method, the stiff adhesive is located in the middle and flexible adhesive at the ends of the overlap. Ozer and Oz [3] investigated the effect of the bi-adhesive bondline on the shear and peeling stresses of a double lap joint using a three-dimensional finite element model based on solid and contact elements. Their results show that the stress components can be optimized using appropriate bond-length ratios.

Recently, new technologies proposed to increase the strength and repair the damaged zone of the adhesively bonded joints. Newly developed self-healing technology allows self-heal and self-repair of adhesively bonded joints without the need for replacing the damaged joint with a new one. Banea et al. [4] presented an overview of the recent developments in the use of smart adhesive technology and summarized the different strategies and approaches to obtain smart adhesive joints. The use of self-healing materials in adhesives is also discussed. Yin et al. [5] studied the preparation of epoxy microcapsules by amino resins. In their work, mechanical properties of the epoxy filled with the healing system were evaluated. From the preliminary result of double cantilever beam (DCB) testing, they showed that the plain weave glass fabric laminates using the self-healing epoxy as the matrix received a healing efficiency of 68 %. Long et al. [6] prepared microcapsules with low formaldehyde and evaluated the morphology, mean particle diameter, particle size, wall thickness, mechanical properties, and encapsulation efficiency as compared to microcapsules. They concluded that significant reduction in the levels of formaldehyde content is possible, while only marginally reducing the mechanical properties and still maintaining the encapsulation efficiency of ~75 %. (For more detailed literature on self-healing with microcapsule, see Refs. [5, 7–9].) D'Elia et al. [10] confined their study to a supramolecular polymer in a graphene ultralight network to form robust, electrically conductive composites able to self-repair. They stated that these composites can sense pressure and flexion and fully restore their properties after damage multiple times and without an external stimulus. Chen et al. [11] employed plasma bombardment to introduce structural defects in monolayer graphene films. They performed thermal annealing to study healing effects on these defects. Luo et al. [12] presented the development of a unique self-adhesive material that maintains a high degree of rigidity at the "adhesive" state while possessing the ability to easily de-bond upon heating. In their study, this adhesive layer can be melted again by heating to easily de-bond, and subsequent rebonding capacity was demonstrated, indicating repeated availability of PCL melt adhesive to the surface by the differential expansive bleeding (DEB) mechanism.

Another method to make self-healing joints for adhesive is to use thermoplastic particles in adhesive. Li et al. [13] modified a thermosetting adhesive by incorporating thermoplastic particles. The prepared joints were fractured per the peel testing of double cantilever beam (DCB) configurations. The fractured specimens were healed in terms of the close-then-heal (CTH) scheme. It is found that the CTH scheme can repeatedly heal the joint with a considerable healing efficiency.

In this study, self-healing efficiency of the Araldite 2011 epoxy adhesive reinforced with the thermoplastic co-polyester (TPC) was experimentally investigated. The weight of the thermoplastic co-polyester was chosen about 10 wt.% of the total composition. Adhesives were cured with two different temperatures including the room temperature (RT) and the temperature above the melting point of the co-polyester. Firstly, for defining the effects of the reinforcement on mechanical properties of adhesives, tensile tests were performed on reinforced and unreinforced dogbone-shaped bulk specimens. Secondly, fracture energy of the unreinforced adhesive was firstly defined from the results of double cantilever beam (DCB) test for Mode I. However, for the adhesive reinforced with TPC, after defining the fracture energy, a 30 mm crack occurred in the reinforced adhesive during the DCB test was closed both by hand and the self-weight of the upper mold, and then the DCB specimen was heated at 120 °C for half an hour in an oven. This fracture healing cycle was repeated until no healing has taken place, i.e., the fractured specimens were repeatedly healed in terms of the close-then-heal (CTH) scheme. After the healing process was completed, the healing efficiency was defined using the fracture energy values of the reinforced adhesive and the self-healed adhesives. In addition, using the Zeiss Evo Ls10 scanning electron microscope, microstructural analyses were also implemented in order to verify whether the co-polyester particles were melted during the healing process. Comparison between our results and the results reported in literature was given.

2. Materials

Material properties of Araldite 2011 adhesive (Huntsman) and co-polyester thermoplastic (Schaetti Fix 376, Switzerland) were tabulated in **Tables 1** and **2**, respectively.

It must be stated that the reason for choosing the co-polyester as additive material is that the lap joint including the co-polyester in adhesive is suitable for repairing the damaged lap joint by heating the joint, as reported in Li et al. [13]. Heating the joint results in melting the co-polyester in adhesive, and then damaged region in adhesive will be repaired by the melted co-polyester. In this study, the weight of the thermoplastic co-polyester is about 10 wt.% of the total composition, i.e., about 10 wt.% adhesive and co-polyester materials were mixed in a vacuum medium to prevent air bubbles and micro voids using a vacuum device designed and manufactured by authors (**Figure 1**).

		Araldite 2011
Elasticity modulus	MPa	1600
Poisson's ratio		0.41
Tensile strength	MPa	33
Tensile yield strength	MPa	36

Table 1. Mechanical properties of Araldite 2011 adhesive [14–17].

Particle size (μm)	Density (g/cm^3)	Melting point (°C)
80	1.29	109–119

Table 2. Material properties of thermoplastic co-polyester [18].

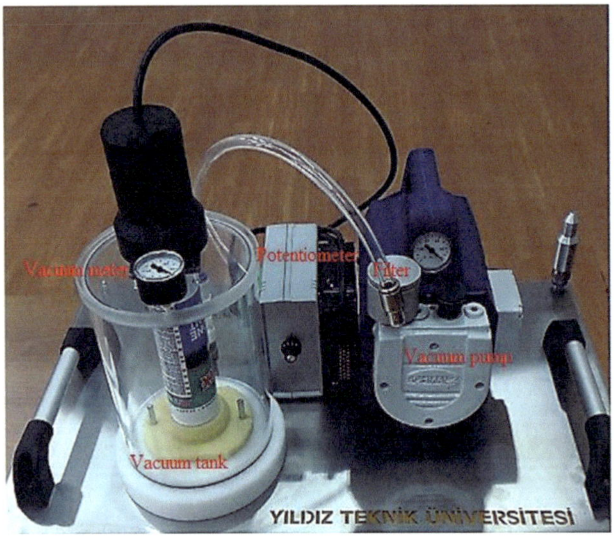

Figure 1. Vacuum device.

Considering the glass transition temperature (Tg) of Araldite 2011 and the melting point of co-polyester material, curing temperature and times were specified as RT (23 °C)/48 h and 120 °C/ 7 min from the material data sheets [17, 18]. (Here, RT corresponds to room temperature.) It is seen that the first temperature is below and the second one is above the melting point of the co-polyester.

http://dx.doi.org/10.5772/65167

Secondly, four bulk specimens were prepared for each reinforced and unreinforced bulk specimen accompanied by two different curing temperatures. Therefore, a total of 16 dogbone samples were produced. Then, all four groups of dogbone specimens were cured with two different temperatures (RT and 120 °C). PID-controlled oven was used for curing and self-healing operations. Digital thermocouple was used to record and control the internal temperature of the oven.

Reinforced and unreinforced adhesives were manufactured in a special mold as a thin plate (**Figure 2a**). The plates were then cut into the required dimensions by using KMT waterjet machine (**Figure 2b**). During the cutting process, pressure water (3800 bar) enriched with the abrasive-waterjet-additive sand (80 μ) was used.

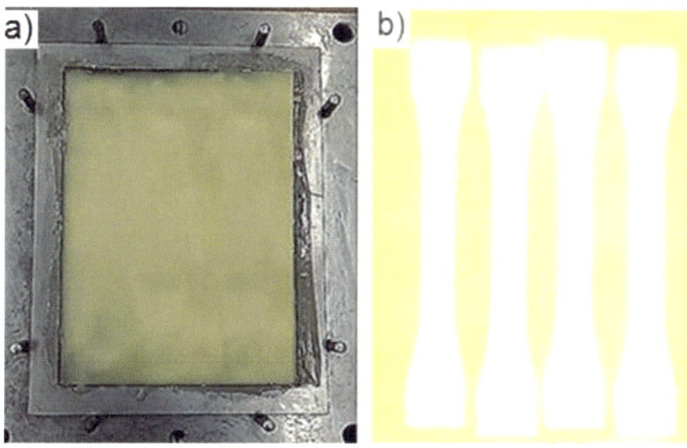

Figure 2. (a) Adhesive sheet within the mold and (b) adhesive sheet with dogbone cutouts.

3. Tensile tests

Reinforced and unreinforced dogbone-shaped bulk specimens were prepared in accordance with BS 2782 standard (**Figure 3**). For each curing temperature, four dogbone specimens were prepared.

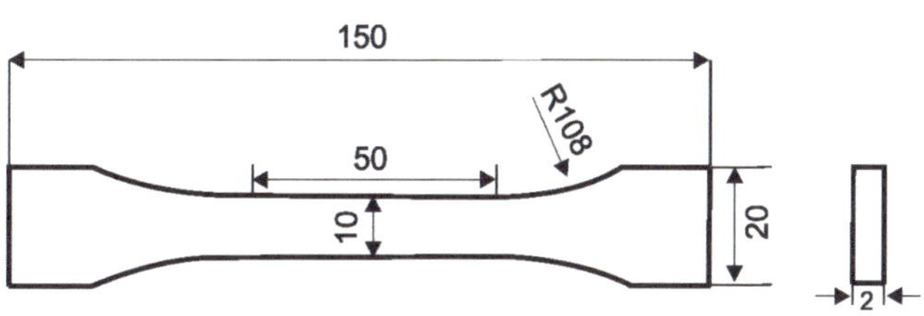

Figure 3. Dogbone specimen (all dimensions in mm).

Tensile tests were performed on test specimens with Instron 5982 at 1 mm/min tensile rate speed (**Figure 4**). The longitudinal strains were recorded during the tests using the video extensometer.

Figure 4. Tensile tester with video extensometer.

Stress-strain curves were obtained both for reinforced and unreinforced dogbone specimens (see **Figure 5**). (In the figure, "unreinforced & RT" corresponds to the tensile test results of the unreinforced adhesive cured at RT.) It is known that Araldite 2011 adhesive is ductile adhesive. As seen in **Figure 5**, under tensile load, the reinforced adhesive has higher deformation capacity than the unreinforced adhesive. Moreover, curing the adhesive at 120 °C causes the adhesive to become more ductile, with respect to curing it at RT. This can be explained by the fact that curing the adhesive at 120 °C results in melting the co-polyester inside the adhesive and leads to increasing the ductility of adhesive. However, as seen in **Figure 5**, the tensile strength is decreasing with increasing the curing temperature. It is concluded that curing temperature and reinforcement effect on the mechanical properties of adhesives.

http://dx.doi.org/10.5772/65167

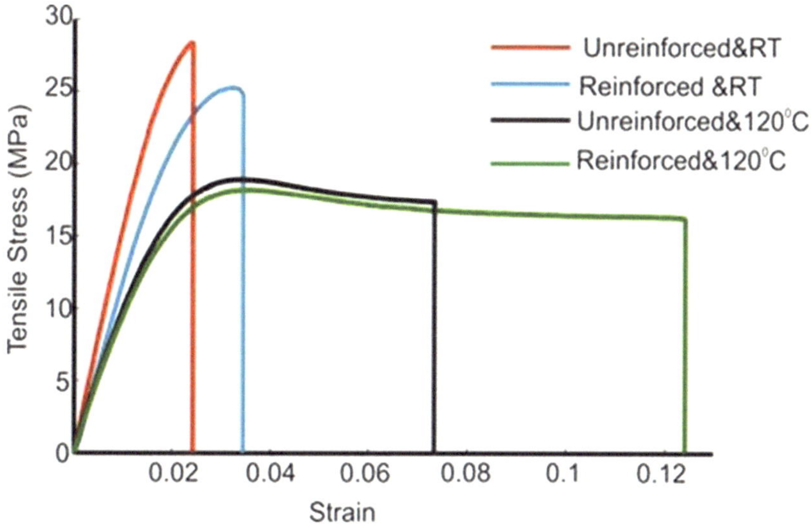

Figure 5. Stress-strain curves of the reinforced and unreinforced specimens cured at RT/48 h and 120 °C/7 min.

In addition, some mechanical properties were presented in **Tables 3** and **4** using the tensile test data of the unreinforced and reinforced dogbone specimens, cured at RT/48 h and at 120 °C/7 min. Comparisons between the mechanical properties were also carried out on the basis of the percent difference. The absolute value of the percent difference is defined as follows:

$$Difference\ \% = \left| \frac{Unreinforced - Reinforced}{Unreinforced} \right| x100 \tag{1}$$

	Unreinforced	Reinforced	Difference (%)
Elasticity modulus (MPa)	1667.7	1234.549	25.97
Tensile strength (MPa)	28.351	25.25	10.93

Table 3. Mechanical properties of dogbone samples cured at RT/48 h.

	Unreinforced	Reinforced	Difference (%)
Elasticity modulus (MPa)	1114.93	1054.331	5.43
Tensile strength (MPa)	18.92	18.20	3.805

Table 4. Mechanical properties of dogbone samples cured at 120 °C/7 min.

As seen in **Tables 3** and **4**, the tensile strengths of the reinforced adhesives are lower than those of the unreinforced ones at the related curing temperatures. In addition, the tensile strength is decreasing with increasing the curing temperature. When it comes to the elasticity moduli, the elasticity modulus of the unreinforced sample cured at 120 °C is higher than that of the reinforced one with the percent difference of 5.43 % (see **Table 3**). The difference between the moduli was maximum, with the percent difference of 25.97 %, for the dogbone sample cured at RT.

It is reported in literature that the curing temperature has effects on the strength and stiffness of adhesives in relation to their glass transition temperatures (Tg). The Tg value of the Araldite 2011 adhesive cured at RT was reported as 67 °C by DSC method in manufacturer's data sheet [17]. However, the Tg value of the adhesive cured at 120 °C is about in the range between 50 and 60 °C [14]. Therefore, if one compares the variation of the elasticity modulus and tensile strength with respect to the curing temperatures, as seen in **Tables 3** and **4**, curing above the Tg temperature of the unreinforced adhesive results is decreasing in the adhesive strengths and elasticity moduli [14]. However, Tg values of the reinforced adhesive are not available in the open literature and have not also been determined in this study. Despite the lack of the Tg data for the reinforced adhesive, similar decreasing trend was also seen in both the strength and stiffness values of the reinforced adhesive (see **Figure 5** and **Tables 3** and **4**).

As discussed above, a preliminary study was performed to define whether selected adhesive is compatible with the thermoplastic particles. From the initial results, it is seen that Araldite 2011 adhesive is suitable for use in the self-healing mechanism. The self-healing mechanism is discussed in the next section.

4. Self-healing

The fracture energy of the unreinforced adhesive was firstly defined from the results of the double cantilever beam (DCB) test. DCB tests were performed on test specimens with Mecmesin tensile tester at 1 mm/min tensile rate speed. DCB tests were carried out on the configuration depicted in **Figure 6**. An initial pre-crack of 40 mm was introduced to the specimens. The geometric properties of the DCB specimen are as follows: the adherend length $L = 230$ mm, adherend thickness $h = 12$ mm, adherend width $b = 25$ mm, and adhesive thickness $t = 0.2$ mm.

St37 steel was used as an adherend material. The elasticity modulus and Poisson's ratio of the adherend are 210 GPa and 0.3, respectively. It is known that surface treatment mainly affects the strength of adhesively bonded joints [19]. With the help of vacuum blasting machine, sandblasting was applied to the adherend surfaces by using silica carbide sand of size 400 μm. Sandblasted adherend surfaces were cleaned by using an ultrasonic bath in liquid pure acetone. The adhesive thickness between two adherends of DCB specimen was controlled by using the Mylar tape, and the tape of 0.2 mm thickness was fixed on one adherend surface. In **Figure 7**, the adherend surface is seen after performing the sandblasting treatment.

http://dx.doi.org/10.5772/65167

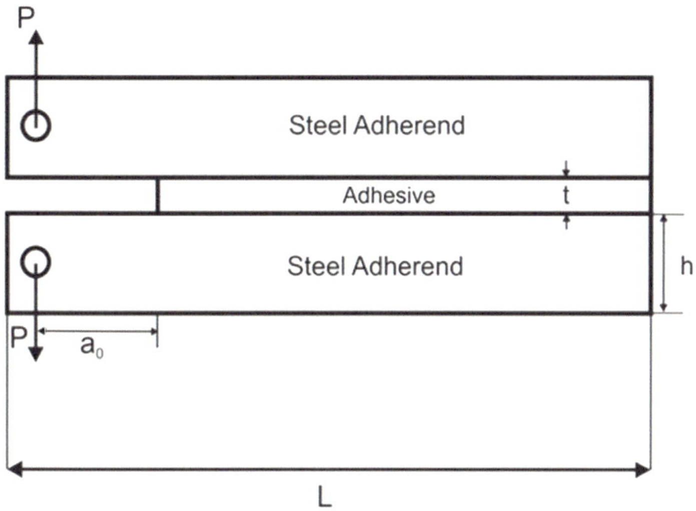

Figure 6. Configuration of double cantilever beam specimen.

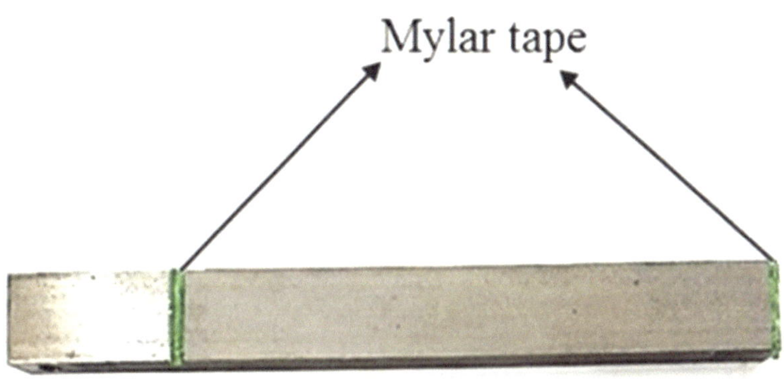

Figure 7. Adherend surface after performing the sandblasting treatment.

In the first stage, also can be named as pre-healing process, fracture energies of the reinforced and unreinforced adhesives were defined from the DCB tests, under the room temperature (23 °C). Tests were continued until the Mode I failure attained.

In the second stage, by following the steps reported in the study of Li et al. [13], when the crack of about 30 mm occurred in the reinforced adhesive during the DCB test, the test was stopped and specimen was removed from the Mecmesin tensile tester. For the four identical DCB specimens, these tests were separately repeated to reach the same crack length of about 30 mm. For measuring and following the crack growth, the adhesive layer over the DCB specimen was painted with the white correction fluid. The DCB specimen test setup is seen in **Figure 8**.

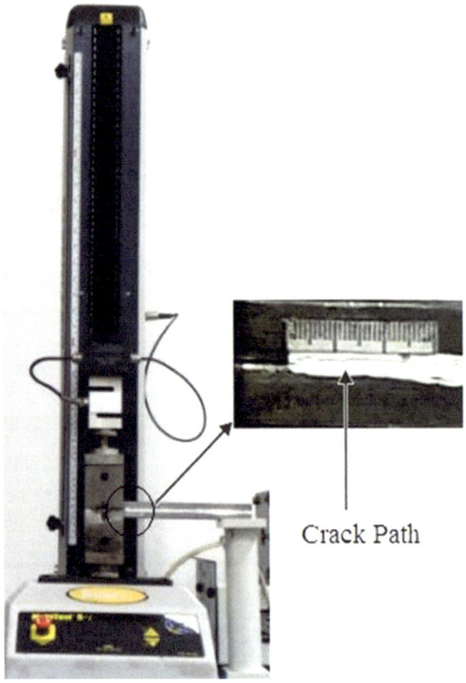

Figure 8. DCB specimen test setup.

After completing the four DCB tests, the first two cracks with the length of 30 mm in two DCB specimens were closed by hand, and the other two cracks having the same lengths were closed by compressive force using the upper part of the mold (self-weight of the upper mold, 45N). The four DCB specimens were then heated at 120 °C for half an hour in an oven and then removed from the oven. It has waited a day to cool the samples to room temperature (RT). By using these four specimens, the DCB tests were performed by continuing tests until the Mode I failure occurred. Loads and corresponding displacements were recorded throughout the tests, and the load-displacement curves were obtained. This cycle is defined the first healing cycle. After the first healing, some cohesive interfacial failures occurred at the adherend surfaces (see **Figure 9**).

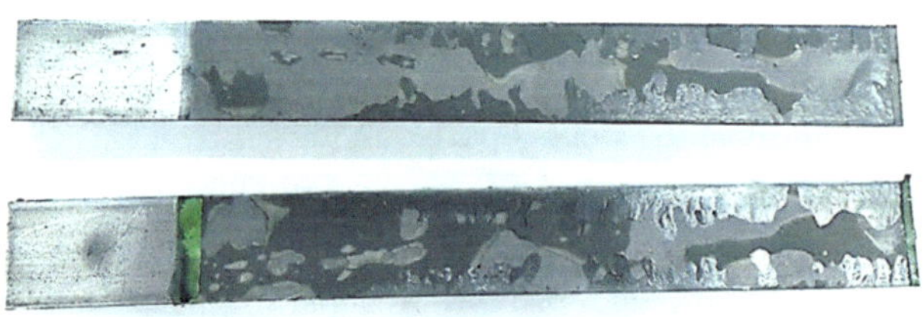

Figure 9. Failure surfaces for the DCB specimen.

http://dx.doi.org/10.5772/65167

For the second healing step, new four DCB specimens were prepared again using the reinforced adhesive. For four specimens each, the DCB tests were repeated up to crack reached about 30 mm. After completing the tests, the first two cracks inside the adhesive layer were closed by hand and the second two cracks by the self-weight of the upper part of the mold. The DCB specimens having closed cracks were heated at 120 °C for half an hour in an oven. The specimens were removed from the oven and cooled to room temperature by waiting a day. In this stage, the last process above should be repeated one more time in the next step. Therefore, by using the same specimens, the DCB tests were performed again up to reaching the cracks of about 30 mm; the four cracks were then closed; specimens having closed cracks were secondly cured at 120 °C for half an hour, in which the co-polyester inside the adhesive was melted again by heating; and the specimens were finally removed from the oven and cooled slowly back to RT at a day. Therefore, the last process was repeated twice, without reaching the final crack. The DCB tests were then performed by continuing tests until the Mode I failure attained and the load-displacement curves were obtained using the data of tests. This cycle is named as the second healing cycle.

The healing cycle was repeated until no healing has taken place. In this study, the healing process was repeated two times. The energy release rates (fracture energies) were calculated by using the compliance-based beam method [20] and calculating the area under the load-displacement curve. The healing efficiency was evaluated using the fracture energy values of the reinforced adhesive and the self-healed adhesives. The results were presented and discussed in the next section.

5. Results and discussion

A preliminary study was firstly performed to define whether selected adhesive is compatible with the thermoplastic particles. From the initial results, it is seen that Araldite 2011 adhesive is suitable for use in the self-healing mechanism. The healing process was then started by following the literature procedures [13]. The healing cycle was repeated until no healing has taken place. As stated above, the healing process was repeated two times with the acceptable healing efficiencies. In addition, it is concluded that, during the healing process, closing the crack within DCB specimen by hand gives much better results than closing it by the self-weight of the upper part of mold, in terms of healing efficiency. Aydın et al. [21] experimentally investigated the effect of curing pressure on the strength of adhesively bonded joints. They concluded that the residual thermal stresses occurring due to the curing pressure at elevated temperature need to be taken into account in order to simulate accurately the mechanical behaviors of adhesively bonded joints. Considering that closing the crack within DCB specimen by hand gives much better results, the results presented below are related to closing it by hand, but not closing it by self-weight.

Loads and corresponding displacements were recorded throughout the DCB tests, by continuing the tests until the Mode I failure occurred, and the load-displacement curves were obtained. **Figure 10** shows load-displacement curves. (In **Figure 10**, "unreinforced & DCB" and “reinforced & DCB” denote the DCB test results of the unreinforced and reinforced adhesives cured at RT. The “first healing” correspond to the DCB test results after performing the first healing cycle.)

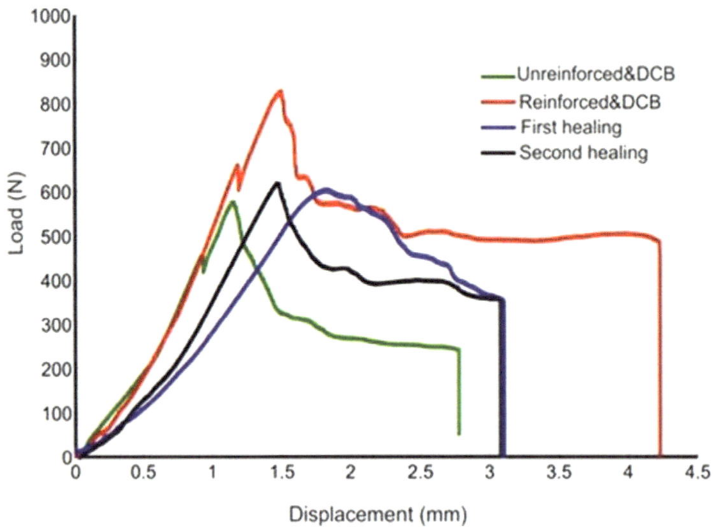

Figure 10. Load-displacement curve.

As seen in **Figure 10**, the reinforced adhesive reaches the highest peak load and has more deformation capacity among all the other. It can be also said that it has also the highest fracture energy, when considering the areas under the curves. Therefore, adding the co-polyester to Araldite 2011 adhesive caused the adhesive to be more ductile than that of the unreinforced adhesive. In addition, it is interesting that the peak load for the adhesive healed twice is higher than that of the unreinforced adhesive. The pick load after the second healing is higher than that of the unreinforced adhesive with the ratio of 7.9 %. From **Figure 10**, it is calculated that the peak displacement value of the reinforced adhesive is higher than that of the unreinforced adhesive by 52 %. Interestingly, the final deformation values remain approximately constant after the first and second healings.

Moreover, the fracture energies were evaluated from the compliance-based beam method, by using the data of **Figure 10** [20]. The main advantage of the method is that there is no need to control the crack lengths during tests. Compliance and flexural modulus values were calculated from the data of **Figure 10**. Using the compliance-based beam method, the fracture energy can be obtained by the relation below [20]:

$$J_{IC} = \frac{6P^2}{hB^2}\left(\frac{2a_e}{E_f h^2} + \frac{1}{5G}\right) \tag{2}$$

where a_e is the equivalent crack length, B the adherend width, h the adherend thickness, G the adherend shear modulus, and E_f the flexural modulus. **Figure 11** shows the fracture energies. (In addition, software built in Mecmesin tensile tester can also automatically calculate the areas under the curves. This gave also the opportunity to check our results.)

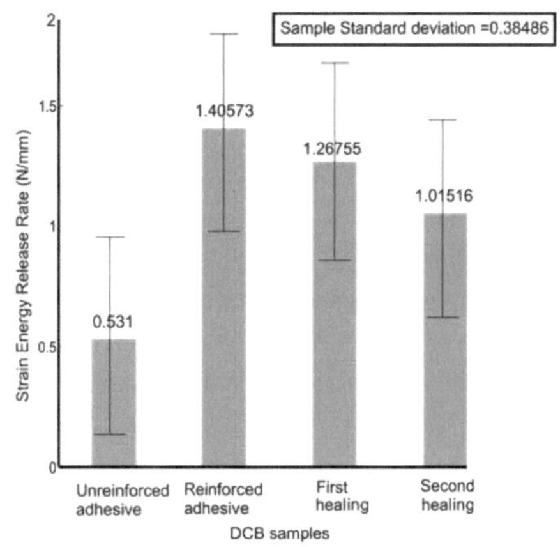

Figure 11. Strain energy release rate values for the DCB samples.

As seen in **Figure 11**, the energy becomes peak for the reinforced adhesive specimen (1.40573 N/mm). The energy is decreasing with increasing the number of healing. It is interesting that the energy at the second healing is approximately two times higher than that of the one at the unreinforced adhesive. In addition, the sample standard deviations of the energy release rates were calculated and given in **Figure 11** [22].

In the final step, the healing efficiency was evaluated using the fracture energy values of the reinforced adhesive and the self-healed adhesives by using the relation below [13]:

$$\eta = \frac{J_{IC}^{healed}}{J_{IC}^{reinforced}} x100 \tag{3}$$

where J_{IC}^{healed} and $J_{IC}^{\text{reinforced}}$ are the fracture energies of the healed and non-healed reinforced specimens, respectively. From the relation above, it is evaluated that the reinforced adhesive displays 90.166 % healing efficiency after the first healing and 74.812 % after the second healing.

In addition, using the Zeiss Evo Ls10 scanning electron microscope, microstructural analyses were also implemented in order to verify whether the co-polyester particles were melted during the healing cycle (**Figure 12**).

Before curing treatment, the additive TP particles in adhesive can be easily seen in **Figure 12a**. However, when the mixed adhesive was cured at 120 °C, as seen in **Figure 12b**, the co-polyester particles were melted in adhesive.

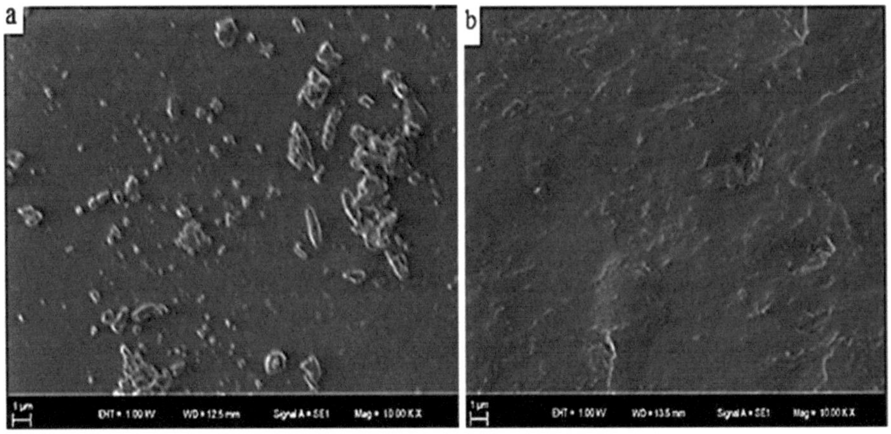

Figure 12. Microstructural view at x 10.000 magnification for the reinforced adhesive (a) cured at RT, (b) cured at 120 °C.

6. Conclusion

Firstly, a preliminary study was applied to define whether selected adhesive is compatible with the thermoplastic particles in terms of self-healing. From the initial results, it is seen that Araldite 2011 adhesive is suitable for use in the self-healing mechanism. Adding the co-polyester to Araldite 2011 adhesive caused the reinforced adhesive to be more ductile than that of the unreinforced adhesive. In this study, the healing process was repeated two times with the acceptable healing efficiencies. Heating the joint resulted in melting the co-polyester in adhesive and allowed the damaged adhesive to repair itself. The following conclusions can be made about the experimental results:

1. The fracture energy is decreasing with increasing the number of healing cycle. The energy at the second healing is approximately two times higher than that of the unreinforced virgin adhesive.

2. The peak load for the adhesive healed twice (i.e., after the second healing) is higher than that of the unreinforced virgin adhesive.

3. During the healing process, closing the crack by hand gives much better results than closing it by the self-weight of the upper part of mold, in terms of healing efficiency.

4. Interestingly, the final deformation values remain approximately constant after the first and second healings.

5. It is concluded that the reinforced adhesive is a reversible adhesive and can repeatedly heal itself with a considerable healing efficiency.

http://dx.doi.org/10.5772/65167

Acknowledgements

This work was funded by Yıldız Technical University under research project no. 2015-06-01-DOP01. The authors would like to thank Professor Hüseyin Üvet, the laboratory director of Yıldız Technical University, for giving the opportunity to use the laboratory facilities.

Author details

Halil Özer* and Engin Erbayrak

*Address all correspondence to: hozer@yildiz.edu.tr

Mechanical Engineering Department, Yıldız Technical University, Istanbul, Turkey

References

[1] Khan U, May P, Porwal H, Nawaz K, Coleman JN. Improved adhesive strength and toughness of polyvinyl acetate glue on addition of small quantities of graphene. Applied Materials & Interfaces. 2013;5:1423-1428. DOI: 10.1021/am302864f.

[2] Sadigh MAS, Marami G. Investigating the effects of reduced graphene oxide additive on the tensile strength of adhesively bonded joints at different extension rates. Materials and Design. 2016;92:36-43. DOI: 10.1016/j.matdes.2015.12.006 0264-1275.

[3] Ozer H, Oz O. Three dimensional finite element analysis of bi-adhesively bonded double lap joint. International Journal of Adhesion & Adhesive. 2012;37:50-55. DOI: 10.1016/j.ijadhadh.2012.01.016.

[4] Banea MD, Da Silva LMF, Campilho DSG, Sato C. Smart adhesive joints: An overview of recent developments. International Journal of Adhesion & Adhesive. 2014;90:16-40. DOI: 10.1080/00218464.2013.785916.

[5] Yin T, Rong MZ, Zhang MQ, Yang GC. Self-healing epoxy composites preparation and effect of healant consisting of microencapsulated epoxy and latent curing agent. Composites Science and Technology. 2007;67:201-212.DOI: 10.1016/j.compscitech.2006.07.028.

[6] Long Y, York D, Zhangc Z, Preece JA. Microcapsules with low content of formaldehyde: Preparation and characterization. Journal of Materials Chemistry. 2009;19:6882-6887. DOI: 10.1039/b902832c.

[7] Kirkby EL, Michaud VJ, Månsonn JAE, Sottos NR, White SR. Performance of self-healing epoxy with microencapsulated healing agent and shape memory alloy wires. Polymer. 2009;50:5533-5538. DOI: 10.1016/j.polymer.2009.05.014.

[8] Blaiszik BJ, Caruso MM, McIlroy DA, Moore JS, White SR, Sottos NR. Microcapsules filled with reactive solutions for self-healing materials. Polymer. 2009;50:990-997. DOI: 10.1016/j.polymer.2008.12.040.

[9] Jin H, Mangun CL, Griffi AS, Moore JS, Sottos NR, White SR. Thermally stable autonomic healing in epoxy using a dual-microcapsule system. Advanced Materials. 2014; 26:282-287.DOI: 0.1002/adma.201303179

[10] D'Elia E, Barg S, Ni N, Rocha V.G, Saiz E. Self-Healing graphene-based composites with sensing capabilities. Advanced Materials. 2015;27:4788-4794. DOI: 10.1002/adma. 201501653.

[11] Chen J, Shi T, Cai T, Sun TXL, Wu X, Yu D. Self-healing of defected graphene. Applied Physics Letters. 2013;102:103-107. DOI: 10.1063/1.4795292.

[12] Luo X, Lauber KE, Mather PT. A thermally responsive rigid and reversible adhesive. Polymer. 2010;51:1169-1175. DOI: 10.1016/j.polymer.2010.01.006.

[13] Li G, Ji G, Ouyang Z. Adhesively bonded healable composite joint. International Journal of Adhesion & Adhesives. 2012;35:59-67. DOI: 10.1016/j.ijadhadh.2012.02.004.

[14] Carbas RJC, Marques EAS, Da Silva LFM, Lopes AM. Effect of cure temperature on the glass transition temperature and mechanical properties of epoxy adhesives. The Journal of Adhesion. 2014;90:104-119. DOI: 10.1080/00218464.2013.779559.

[15] Zahurul Islam SM, Young B. FRP Strengthened aluminium tubular sections subjected to web crippling. Thin-Walled Structures. 2011;49:1392-1403. DOI: 10.1016/j.tws. 2011.06.007.

[16] Van Tooren MJL, Gleich DM, Beukers A.Experimental verification of a stress singularity model to predict the effect of bondline thickness. Journal of Adhesion Science and Technology. 2012;18:395-412. DOI: 10.1163/156856104323016315.

[17] Hunstman. Araldite 2011 Multi Purpose Epoxy Adhesive [Internet]. 2007. Available from: https://apps.huntsmanservice.com/WebFolder/ui/browse.do?pFileName=/opt/TDS/Huntsman%20Advanced%20Materials/English/Long/Araldite%202011_eur_e.pdf.

[18] Schaetti. High Performance Thermoplastic Copolyester [Internet]. 2015. Available from: http://www.schaetti.com/fileadmin/content/pdf/maerkte_produkte/4_Construction-Bausektor.pdf.

[19] Rudawska A. Selected aspects of the effect of mechanical treatment on surface roughness and adhesive joint strength of steel sheets. International Journal of Adhesion & Adhesives. 2014;50:235-243. DOI: 10.1016/j.ijadhadh.2014.01.032 0143-7496.

[20] De Moura MFSF, Campilho RDSG, Goncalves JPM. Crack equivalent concept applied to the fracture characterization of bonded joints under pure mode I loading. Composites Science and Technology. 2008;68:2224-2230. DOI: 10.1016/j.compscitech.2008.04.003.

[21] Aydın MD, Temiz S, Ozel A. Effect of curing pressure on the strength of adhesively bonded joints. Journal of Adhesion. 2007;83:553-571. DOI: 10.1080/00218460701453536.

[22] Newbold P, Carlson W, Thorne B. Statistics for Business and Economics. 4th ed. New Jersey, USA . Prentice Hall; 1994. 864 p.

Chapter 7

Research Progress on Formaldehyde-Free Wood Adhesive Derived from Soy Flour

Chengsheng Gui, Jin Zhu, Zhongtao Zhang and Xiaoqing Liu

Additional information is available at the end of the chapter

http://dx.doi.org/10.5772/65502

Abstract

Soy-based adhesives have been regarded as the most suitable candidates for wood industry. For a widespread use of soy-based adhesives, new technologies need to be developed to improve the water resistance. An overview on the methods to improve water resistance of soy-based adhesives is presented. Denaturants were once considered necessary to modify soy protein. However, water-resistant soy adhesives could be prepared by simply removing water-soluble carbohydrates and low molecular peptides from soy flour. In addition, proper grafting and cross-linking agents help to prepare water-resistant soy-based adhesives, which are used widely to bond interior wood composites. In particular, a new type of polyamidoamine (PADA) resin and an itaconic acid-based polyamidoamine-epichlorohydrin (IA-PAE) resin were synthesized to perform as cross-linking agents for soy-based adhesives. This review concludes that soy-based adhesives have great potential for use in numerous applications. However, future work is still needed to make soy-based adhesives more competitive with synthetic adhesives.

Keywords: soy adhesive, formaldehyde-free adhesive, denaturation, cross-linking agent

1. Introduction

Soy-based adhesives were once the dominant glue for manufacturing plywood in the 20th century [1]. The soy flour (SF) adhesives used at that time were made under highly alkaline conditions. They were also modified by casein, blood, borax, sodium silicates or carbon disulfide to give better water resistance. Such soy adhesive typically had a characteristic of high viscosity, low-solid content, and short usable pot life [1]. Plywood bonded by such soy flour-based adhesives (SAs) was adequate for interior applications. However, fossil fuel–based adhesives, such as urea-formaldehyde (UF), replaced soy-based adhesives for interior applications because UF adhesives were cheaper, easier to use, and more water resistant.

Formaldehyde-based polymers are currently dominated in the adhesive systems to bond wood composites, including plywood, blockboard, particleboard, fiberboard etc. The cured adhesives are easily decomposed to emit formaldehyde, especially under elevated temperature and high humidity conditions. Many events were reported that human health was threatened by the toxicity of exposure to formaldehyde. The harm of formaldehyde to humans was proved by a large quantity of research. In 2004, formaldehyde was classified by WHO as a carcinogen to both human and animals [2]. Many countries carried out strict regulations to limit the emission of formaldehyde from wood composites. For example, Japan implemented F★★★★ standards in 2003. California implemented CARB standards in 2007. And the "Formaldehyde Standards for Composite Wood Product Act" was carried out in America since 2011. Besides the hazardous issues associated with formaldehyde-based adhesives, finite fossil reserves generate a commercial interest in bio-based technology. Therefore, many bioresources were considered as the feedstock to prepare wood adhesives. Research focused on dextrin, starch, tannin, lignin, and soy [3–7]. Among them, soy protein-based adhesives have been regarded as the most suitable candidates for wood industry [7].

Older soy adhesive formulations, made under alkaline conditions and using calcium compounds [8], were good adhesives for wood under dry conditions, but their water resistance was typically poor. None of these formulations made products that could fully meet current performance standards, such as ANSI/HPVA HP1-2009 4.6 three-cycle soak test for decorative plywood. Thus, for a widespread use of soy adhesives, new technologies need to be developed to improve the water-resistance.

2. Modification of soy-based adhesives

Based on their protein contents, soybean products are mainly divided into soy flour (SF), soy protein concentrate (SPC), and soy protein isolate (SPI) [9]. SF, which is produced by extracting the oil from soybeans, contains about 50% protein and 35% carbohydrates (water-soluble carbohydrates [WSCs] and water-insoluble carbohydrates [WISCs]). SPC containing 70% protein is produced by simply removing soluble carbohydrates from SF using ethanol extraction. SPI containing 90% protein is further purified product where all carbohydrates have been removed. Apparently, SPI is much more expensive than SPC and SF due to the increased processing costs and lower yield as a result of removal of carbohydrates.

2.1. Technologies to modify soy proteins

Proteins are the main adhesion components in soy flour. With over 90% protein content, SPI is the commercial soy product used in many soy reaction studies. A large number of published research focused on the method to modify soy protein isolate to improve its water resistance. SPIs have been altered using a variety of denaturants, including surfactants, amino-containing agents, alkali, and enzymes, in efforts to improve adhesive properties.

http://dx.doi.org/10.5772/65502

Native soy proteins have a secondary structure of α-helices and β-sheets, and a three-dimensional tertiary structure where α-helices and β-sheets fold into compact globules that interact with the surfaces of other globules forming a quaternary structure [10]. The physical and chemical properties of proteins are influenced by this complex structure.

To use soy proteins as wood adhesives, denaturation was once considered to be necessary to expose more polar groups for solubilization and bonding via hydrogen bonds [11]. Denaturation is a process that changes the multilevel structure of the protein molecule without breaking covalent bonds (**Figure 1**). The hydrophilic groups of soy proteins are uncoiled and exposed after denaturation. Proteins can be denatured by exposure to heat, acid/alkali, organic solvents, surfactants, or urea. Dispersion of soy proteins to denaturants' solution can break apart the native quaternary state into individual polypeptides folded into their native tertiary structure. This tertiary structure can be further opened into short-range and then long-range expansion, leaving the secondary structure intact. Further disruption of α-helices and ß-sheets provides a normal polymer chain [12].

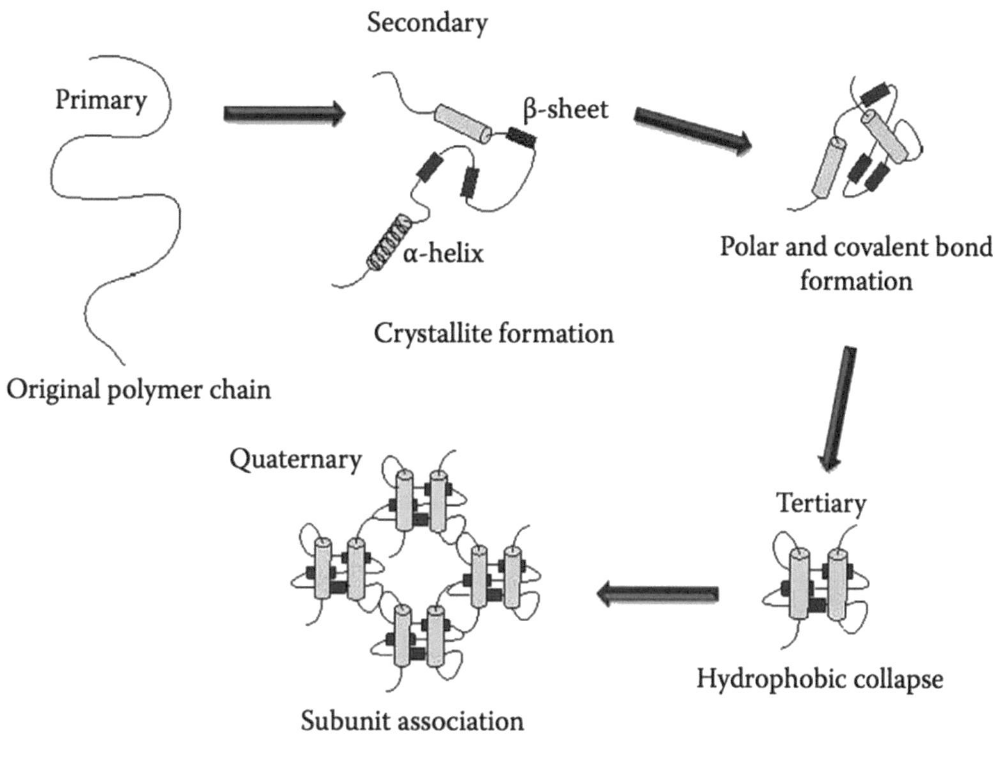

Figure 1. Multilevel structure of soy proteins [12].

Research on soy protein isolates had shown that both the pH value and the surfactant concentration have significant effects on the adhesive strength of modified soy protein isolate. Hettiarachchy et al. [13] used moderate alkali (pH=10) at 50°C to unfold soy protein isolate. Results showed that the adhesion strength and water resistance of modified SPI were dramatically improved. It is also possible to unfold protein complexes with urea, as its oxygen and hydrogen atoms interact with hydroxyl groups of the proteins and break down the hydrogen bonding in the protein body [14]. Too high a concentration of urea also breaks the secondary structure of the protein, and this can have a negative effect on the adhesive properties of the protein. The best adhesion has been found when the tertiary structure is disrupted, but disruption of the secondary structure is limited. Soy proteins can also be treated with sodium dodecyl sulfonate (SDS), sodium dodecyl benzene sulfonate (SDBS), and bioenzyme to break apart the quaternary protein structure while still retaining some secondary structure [15].

However, recent studies [16–18] showed that strong water-resistant soy protein-based adhesives could be prepared by adjusting the pH value to the protein's isoelectric point and they were much more stable than the traditional denatured soy adhesives when stored at room temperature.

Hunt et al.'s results also showed that SPI itself has a good water resistance even without any modification [19]. The results of our investigation also revealed that lab-prepared SPI could be used as adhesive directly and had a good wet strength without further modification. The wet strength of SPI was 1.09 MPa. The wet strength requirement for type II plywood (indoor application) was only >0.7 MPa according to Chinese National Standard [20]. We further found that the water-soluble carbohydrates and low molecular peptides (WSCs-LMPs) were the main causes of poor water resistance within soy flour, removing these fractions results in adhesives having less water attracting, less swelling (which can lead to the wet debonding of the adhesives). But the water-insoluble carbohydrates (WISCs) did not influence the water resistance of soy protein adhesive [21]. Based on these recent findings, we obtained water-resistant soy flour-based adhesives (SAs) by simply removing WSC-LMP from soy flour (**Table 1**, unpublished results) leaving the multilevel structure of soy protein retained.

	Protein content (%)	Wet strength (MPa)	Water uptake	Viscosities of adhesives (cP)	Water-insoluble content of cured adhesive (%)
SF	56.3	0.43 ± 0.07 (c)	70%	12,500	64.5 ± 0.9 (d)
SA	69.8	1.02 ± 0.08 (a)	80%	10,200	86.2 ± 1.2 (b)
SPI	93	1.09 ± 0.05 (a)	72%	13,500	90.3 ± 0.5 (a)

Wet strengths are means ± standard deviations of three replicates. Means within a column followed by different letters are significantly different at $p < 0.05$.

Table 1. Relationship between wet strength and adhesive properties.

2.2. Soy adhesives modified with grafting and cross-linking agents

Traditional methods of preparing soy adhesives have not provided as much water resistance as their competing synthetic adhesives. Uncross-linked soy proteins are good adhesives for wood under dry conditions, but their water resistance is typically poor. In recent years, more emphasis has been placed upon covalent cross-linking.

A number of grafting agents were also applied to improve water resistance of soy proteins. OSA, dopamine, undecylenic acid, and $POCl_3$ were used to graft soy proteins [22–25]. After grafting, the attached groups acted as cross-linking agents, either via covalent esterification with hydroxyl groups on wood chips or via ionic and hydrogen-bonding interactions with functional groups in wood chips.

Epoxy resins were studied as effective cross-linking agents to modify soy adhesives [26–28]. Gao et al. [26] used DETA and EGDE as the modifiers to cross-link soy protein. The results showed that DETA reacted with EGDE to form a long-chain structure with epoxy groups, which cross-linked the soy protein molecules to form a denser cured adhesive layer to improve the water resistance of the resultant adhesive (**Figure 2**). Incorporating EGDE/DETA, the wet shear strength of the plywood bonded by the resultant adhesive was improved by 30.7% to 1.15 MPa, which met the requirement of Chinese National Standard for interior use.

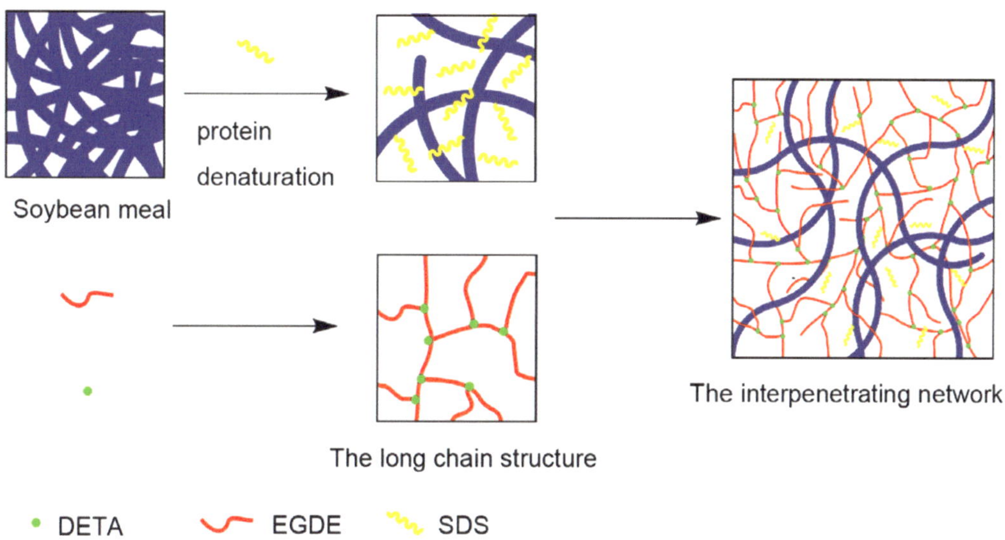

Figure 2. The diagram of DETA reacted with EGDE to form a long-chain structure [26].

However, epoxy-modified soy adhesives have the problem of low dry bond strength. Luo et al. [29] introduced polyisocyanate (pMDI) into epoxy-resin-modified soy-based adhesive to address the issue of low dry bond strength. By adding 2% pMDI, the dry and wet strength of the plywood was improved 29.5% and 39.7%, respectively. The pot life of resultant adhesive reached in 4 h which could meet the processing requirements of most manufacturers. The reactions between epoxy group and carbonyl as well as isocyano and amino were the main reasons for forming the cross-linking network of cured adhesive, thus, improved the dry and wet strength of the resultant plywood.

Soy flour adhesives using a polyamidoamine-epichlorohydrin (PAE) resin as a cross-linking agent are used increasingly as wood adhesives for interior products in North America [30]. The soy adhesive systems, SOYAD® [31], are easy to prepare in the panel mills and do not require harsh reaction conditions (**Figure 3**).

However, low-solid content and/or high viscosity are the main characteristics of most commonly used commercial PAE (C-PAE) [32]. For the purpose to improve the solid content of

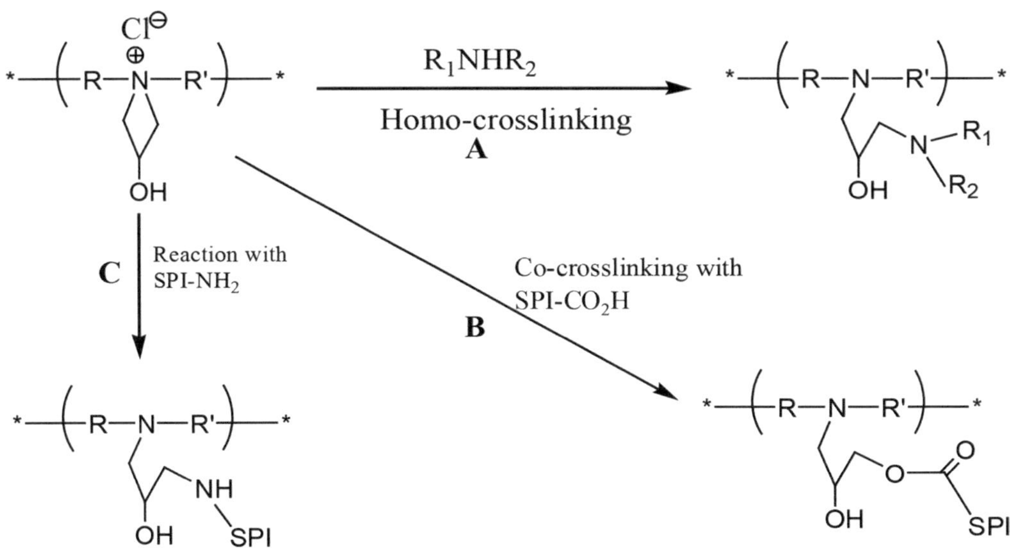

Figure 3. Proposed reactions of soy protein-PAE adhesives [30].

soy-based adhesives, curing agents which have a high-solid content and low viscosity are desired [32, 33]. Therefore, many studies had been done to prepare the curing agents for soy-based adhesives with a high-solid content and low viscosity. The results showed that a poly-amine solution which had a solid content of 40–60% and a viscosity of less than 600 cP was preferable to prepare soy-based adhesives [32]. To solve the problems of low-solid content and high viscosity of curing agents, our group synthesized a new type of polyamidoamine (PADA) resin [34] for producing soy flour-based adhesives (**Figure 4**). The obtained PADA solution had high-solid content of 50 wt% and low viscosity of 270 cP. This PADA combined with maleic anhydride (MA) was employed as curing agent in SF-PADA-MA adhesive systems. The wet strength of plywood prepared at the optimum weight ratio was 0.82 MPa, which meant that the plywood could be used as type II plywood according to the Chinese National Standard GB/T 9846.7-2004. The cross-linking network formed by the reactions of PADA and MA was discussed as the main reason for improved water resistance (**Figure 5**).

The polyamines talked above are currently derived from fossil resources. In view of sustain-ability and environmental protection it is desirable to develop bio-based polyamines to per-form as the cross-linking agents of soy flour-based adhesives. As epichlorohydrin can now be made from bio-based glycerol, Jang et al. [2] synthesized a bio-based polyamine by the reaction of ammonia with epichlorohydrin. PAE can also be obtained from renewable resources since adipic acid can now be made from bio-based materials [35]. Using other types of renewable raw materials, such as citric acid, to replace adipic acid is also an effective way to produce bio-based PAE resins [32, 33]. In our group, we used readily available renewable itaconic acid to synthesize bio-based PAE. Itaconic acid is a nontoxic compound which is considered as one of the top value-added building block chemicals that can be produced from sugars [36]. The obtained bio-based PAE will bring a sustainable development to wood product industry.

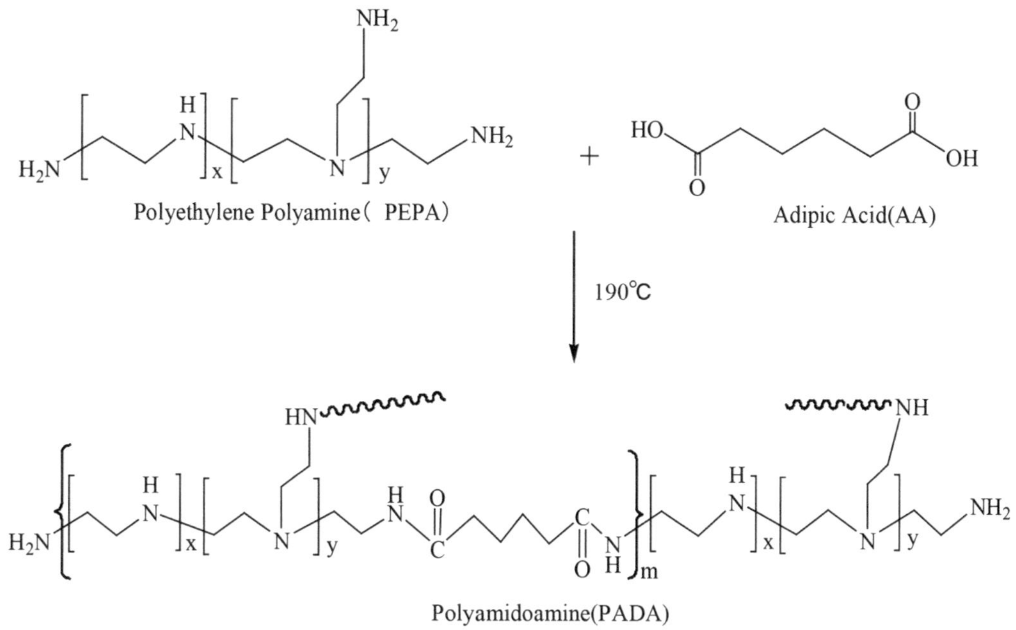

Figure 4. Synthesis of polyamidoamine resin [34].

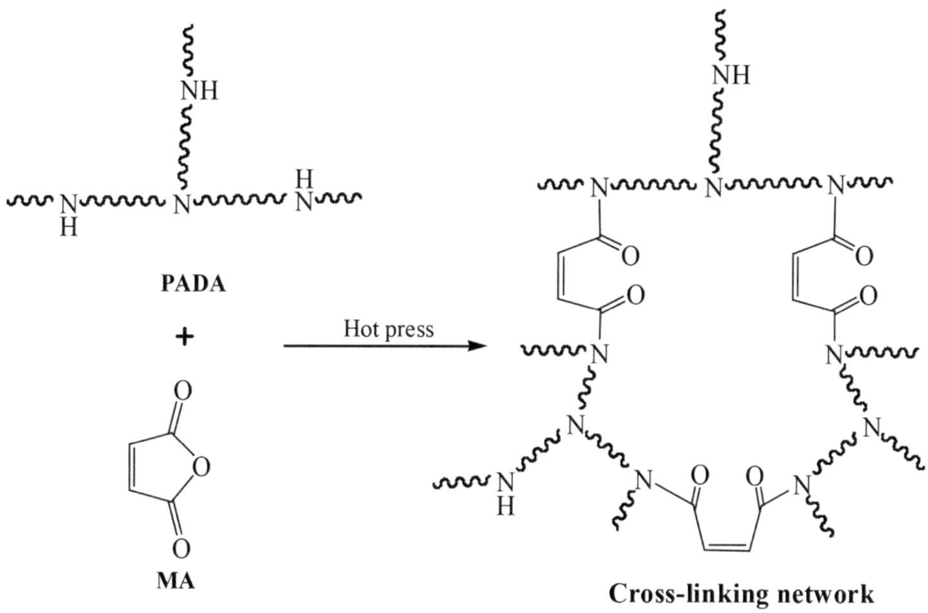

Figure 5. Proposed main curing reactions between polyamidoamine and maleic anhydride [34].

High molecular weight of PAE is not necessary for modifying soy-based adhesives, which is unlike the commercial PAE (C-PAE) used in papermaking process. Some PAE solutions with high-solid contents and low viscosities were synthesized [32, 33]. When preparing IA-PAE

[37], our main focus was to functionalize IA-PADA with ECH without causing much increase in molecular weight, which would limit the solid content of the resultant solution (**Figure 6**). Therefore, the obtained IA-PAE solution had a high-solid content of 50 wt % and low apparent viscosity of 144 cP. Wet strength of IA-PAE-SF on plywood was 0.95 MPa which was comparable to that of C-PAE-SF and met the requirements of Chinese National Standards for interior applications.

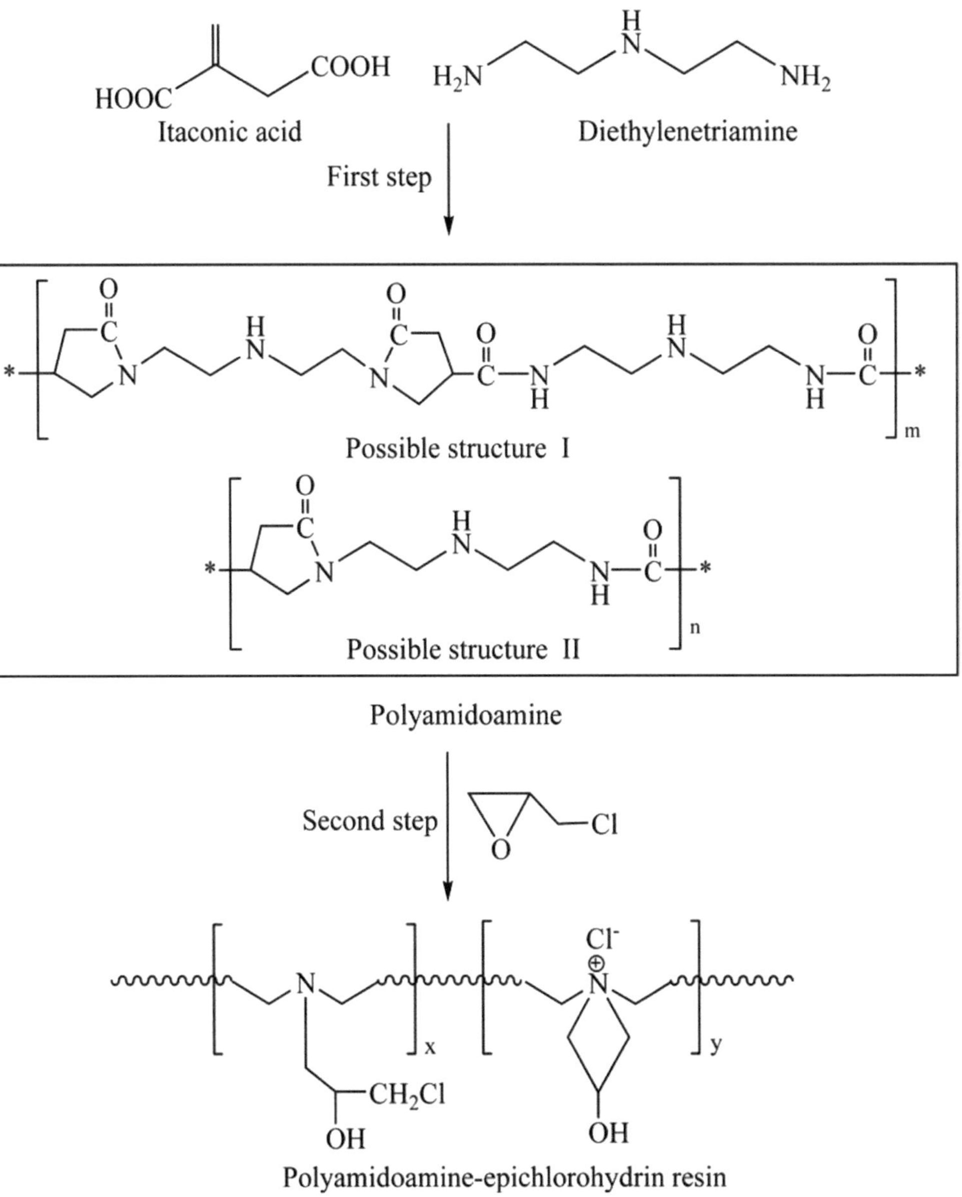

Figure 6. Synthesis process of itaconic acid-based polyamidoamine-epichlorohydrin resin.

3. Conclusions and perspective

Because of the low cost and wide availability of soy flour, soy adhesives have the greatest potential for widespread use in the wood product industry. At the end of 2006, Columbia Forest Products, the largest manufacturer of hardwood veneer and hardwood plywood in the United States, converted all of its standard hardwood plywood production to produce formaldehyde-free panels called PureBond using SOAYD adhesive. Since 2005, more than 70 million hardwood plywood panels have been produced in North America with SOYAD adhesive technology. In fact, the Hardwood Plywood Veneer Association of North America estimated that more than half of the hardwood plywood panels manufactured in North America during 2010 were made with SOYAD adhesives.

In China, OZERO adhesive is widely used by Nature Flooring, Der Flooring, and Tubaobao Flooring to produce engineered wood flooring. OZERO adhesive is also used by Paterson to produce blockboard and by Furen Group Co. Ltd. to produce particleboard.

Although performance of soy adhesives has vastly improved, there is still a need for better products to make them more competitive with synthetic adhesives.

3.1. Solid content/fluidity

Soy proteins in water have high viscosities and are very shear thinning. The SOYAD®, OZERO®, and SOYBABY® [38–40] used in the wood industry only have a solid content of 28–40%. Making soy adhesives with higher solid contents is an important goal in order to compete more effectively with fossil fuel–based adhesives. This involves understanding how proteins and carbohydrates contribute to the high viscosity and designing methods to reduce the viscosity.

3.2. Carbohydrates

There are about 35% carbohydrates in soy four, including water-soluble carbohydrates and water-insoluble carbohydrates. Understanding the role of these carbohydrates will provide further insight into the use of soy four as a protein-based adhesive. Studies have shown that water-soluble portion is the main cause of poor water resistance. Removing the water-soluble carbohydrates can result in soy adhesives having higher water resistance. On the other way, methods to modify the carbohydrates in the soy flour should be developed so that the carbohydrates also contribute to the strength of the adhesive network.

3.3. Mould

The bondlines of soy-based adhesives are easily affected by mould, including *Rhizopus oryzae* and *Penicillium citrinum*. Sodium diacetate, sodium borate, sodium nitrite etc. were investigated as effective mildew preventives by Zhai et al. [41]. Results from the application of OZERO® adhesive showed that Kathon has an effective mould-proof property for soy adhesives. Caution needs to be drawn that the situation may be different when the wood products are exposed to specific environment conditions.

3.4. Aging properties

As derived from natural resources, bondlines of soy adhesives may be easily decomposed, especially under elevated temperature and high humidity. Zeng et al. [42] employed eight wet-dry cycles of 25°C, 63°C, and 95°C to accelerate the aging of poplar plywood bonded by soy-based adhesives. After eight wet-dry cycles under all three conditions, the strength of soy adhesive decreased. But the loss of strength was not as much as UF adhesive. Thus the aging resistance property was not a big problem of soy adhesive when used for interior applications.

Acknowledgements

This work was funded by National Promotion Project of Forestry Science and Technology (Grant No. 2016–26) and the Technology Innovation Alliance of Wood & Bamboo Industry (Grant No. TIAWBI2015-01).

Author details

Chengsheng Gui[1,2], Jin Zhu[1], Zhongtao Zhang[3] and Xiaoqing Liu[1]*

*Address all correspondence to: liuxq@nimte.ac.cn

1 Ningbo Institute of Materials Technology and Engineering, CAS, Ningbo City, China

2 Ningbo Zhaolu & CAS New Materials Co., Ningbo City, China

3 China State Forestry Planning and Design Institute of Forest Products Industry, Beijing, China

References

[1] Lambeth A, Protein adhesives for wood. In: Handbook of Adhesive Technology. A. Pizzi and K.L. Mittal (Eds), 2nd ed., pp. 457–478, Marcel Dekker, New York (2003).

[2] Jang Y, Huang J, Li K. A new formaldehyde-free wood adhesive from renewable materials. Int J Adhes Adhes 2011; 31(7):754–759. DOI: 10.1016/j.ijadhadh.2011.07.003.

[3] Liu X, Yi W, Yang C, et al. Study of dextrin-derived curing agent for waterborne epoxy adhesive. Carbohyd Polym 2011; 83(3):1180–1184. DOI: 10.1016/j.carbpol.2010.09.019.

[4] Lei H, Pizzi A, Du G. Environmentally friendly mixed tannin/lignin wood resins. J Appl Polym Sci 2008; 107(1):203–209. DOI:10.1002/app.27011.

[5] Mansouri N, Pizzi A, Salvado J. Lignin-based polycondensation resins for wood adhesives. J Appl Polym Sci 2007; 103(3):1690–1699. DOI: 10.1002/app.25098.

[6] Wang Z, Li Z, Gu Z, et al. Preparation, characterization and properties of starch-based wood adhesive. Carbohyd Polym 2012; 88(2):699–706. DOI: 10.1016/j.carbpol.2012.01.023.

[7] Lépine E, Riedl B, Wang X, et al. Synthesis of bio-adhesives from soybean flour and furfural: relationship between furfural level and sodium hydroxide concentration. Int J Adhes Adhes 2015; 55(5):6270–6273. DOI: 10.1016/j.ijadhadh.2015.08.007.

[8] Johnson O. Adhesive. 1923. US Patent 1,406,757.

[9] Kim J, Netravali A. Mechanical, thermal, and interfacial properties of green composites with ramie fiber and soy resins. J Agric Food Chem 2010; 58:5400–5407. DOI: 10.1021/jf100317y.

[10] Wool PR, Sun XS. 10-Soy protein adhesives. Bio-based Polymers and Composites 2005: 327-368. Elsevier-Academic Press, Burlington, MA. Print Book ISBN : 9780127639529. DOI: 10.1016/B978-012763952-9/50011-3.

[11] Hemmilä V, Trischler J, Sandberg D. Bio-based adhesives for the wood industry—an opportunity for the future? Pro Ligno 2013; 9(4):118–125.

[12] Frihart CR. Soy protein adhesives, McGraw-Hill yearbook of science & technology, 2010. New York : McGraw-Hill, c2010. p. 354–356: ISBN: 9780071639286: 0071639284

[13] Hettiarachchy N, Kalapathy U, Myers D. Alkali-modified soy protein with improved adhesive and hydrophobic properties. J Am Oil Chem Soc 1995; 72(12):1461–1464. DOI: 10.1007/BF02577838.

[14] Huang W, Sun X. Adhesive properties of soy proteins modified by urea and guanidine hydrochloride. J Am Oil Chem Soc 2000; 77(1):101–104. DOI: 10.1007/s11746-000-0016-6.

[15] Huang W, Sun X. Adhesive properties of soy proteins modified by sodium dodecyl sulfate and sodium dodecylbenzene sulfonate. J Am Oil Chem Soc 2000; 77:705–708. DOI: 10.1007/s11746-000-0113-6.

[16] Sun X, Xhu L, Wang D. Latex adhesives derived from ionic strength induced soy protein complexes. 2008. US Patent 2008/0287635.

[17] Qi G, Sun X. Soy protein adhesive blends with synthetic latex on wood veneer. J Am Oil Chem Soc. 2011; 88(2):271–281. DOI: 10.1007/s11746-010-1666-y.

[18] Qi G, Li N, Wang D, et al. Physicochemical properties of soy protein adhesives obtained by in situ sodium bisulfite modification during acid precipitation. J Am Oil Chem Soc 2012; 9(2):301–312. DOI: 10.1007/s11746-011-1909-6.

[19] Hunt C, Wescott J, Lorenz L. Soy adhesive-moisture interactions. In: Wood Adhesives, 2009; Madison, WI: Forest Products Society.

[20] Chinese National Standard GB/T 17657-1999. Test methods of evaluating the properties of wood-based panels and surface decorated wood-based panels.

[21] Gui C, Wang G, Zhu J, et al. Effect of constituents of soy flour-based adhesive on its viscosity and wet strength. China Forest Products Industry, 2015; 42(3), 36-39

[22] Qi G, Li N, Wang D, et al. Physicochemical properties of soy protein adhesives modified by 2-octen-1-ylsuccinic anhydride. Ind Crop Prod 2013, 46:165–172. DOI: 10.1016/j.indcrop.2013.01.024.

[23] Liu Y, Li K. Modification of soy protein for wood adhesives using mussel protein as a model: the influence of a Mercapto Group. Macromol Rapid Comm 2004, 25(21):1835–1838. DOI: 10.1002/marc.200400363.

[24] Liu H, Li C, Sun XS. Improved water resistance in undecylenic acid (UA)-modified soy protein isolate (SPI)-based adhesives. Ind Crop Prod 2015, 74:577–584. DOI: 10.1016/j.indcrop.2015.05.043.

[25] Zhu D, Damodaran S. Chemical phosphorylation improves the moisture resistance of soy flour-based wood adhesive. J Appl Polym Sci 2014, 131(13):378–387. DOI: 10.1002/app.40451.

[26] Gao Q, Shi S, Li J, et al. Soybean meal-based wood adhesives enhanced by modified polyacrylic acid solution. Bioresources 2011, 7(1):946–956. DOI: 10.15376/biores.7.1.0946-0956.

[27] Lei H, Du G, Wu Z, et al. Cross-linked soy-based wood adhesives for plywood. Int J Adhes Adhes 2014, 50(4):199–203. DOI: 10.1016/j.ijadhadh.2014.01.026.

[28] Zhu H, Xiao L. Influence of moist heat treatment or cross linking on the water resistance properties and structure of chitosan/PVA blend films. J Zhejiang Ocean Univ, 2005, 6, 48–52

[29] Luo J, Luo J, Li X, et al. Effects of polyisocyanate on properties and pot life of epoxy resin cross-linked soybean meal-based bioadhesive. J Appl Polym Sci 2016. DOI: 10.1002/app.43362.

[30] Li K, Svetlana P, Geng X. Investigation of soy protein-Kymene® adhesive systems for wood composites. J Am Oil Chem Soc 2004; 81:487–491. DOI: 10.1007/s11746-004-0928-1.

[31] http://solenis.com/en/industries/specialties-wood-adhesives/innovations/soyad-adhesive-technology/ [accessed: 2016-08-20].

[32] Spraul B, Brady R, Allen A. Adhesive composition of low molecular weight polyaminoamide-epichlorohydrin (PAE) resin and protein. US Patent 2008/0050602.

[33] Varnell D, Spraul B, Evans M. Adhesive compositions. US Patent 2011/0190423.

[34] Gui C, Liu X, Wu D, et al. Preparation of a new type of polyamidoamine and its application for soy flour-based adhesives. J Am Oil Chem Soc 2013; 90(2):265–272. DOI: 10.1007/s11746-012-2160-5.

[35] Picataggio S, Beardslee T. Biological methods for preparing adipic acid. 2012. US Patent 8,241,879 B2.

[36] Werpy T, Petersen G. Top value added chemicals from biomass. Volume I: Results of Screening for Potential Candidates from Sugars and Synthesis Gas. Pacific Northwest National Laboratory (PNNL), National Renewable Energy Laboratory (NREL), Office of Biomass Program (EERE) for the Office of the Biomass Program, Nato Advanced Science Institutes, 2007(2):263-275

[37] Gui C, Wang G, Wu D, et al. Synthesis of a bio-based polyamidoamine-epichlorohydrin resin and its application for soy-based adhesives. Int J Adhes Adhes 2013, 44(10):237–242. DOI: 10.1016/j.ijadhadh.2013.03.011.

[38] Yang G, Yang B, Yuan C, Geng W, Li H. Effects of preparation parameters on properties of soy protein-based fiberboard. J Poly Environ 2011; 19:146–151. DOI: 10.1007/s10924-010-0246-4.

[39] http://www.soy-baby.com/ [accessed: 2016-08-20].

[40] http://ozero-inside.com/ [accessed: 2016-08-20].

[41] Zhai Y, Yang G, Yang B, et al. The optimization of mildew preventives of soy-based adhesive. Appl Chem Ind 2012; 41(1):61–65.

[42] Zeng X, Luo J, Hu J, et al. Aging resistance properties of poplar plywood bonded by soy protein-based adhesive. Bioresources 2016, 11(2):4332–4341.

Chapter 8

Evaluation of Addition of Reactive Resin for an Adhesive Formulation of Pressure-Sensitive Adhesive

Tzeitel Hernández-Martínez,
Beatriz Adriana Salazar Cruz,
José Luis Rivera-Armenta,
María Yolanda Chávez-Cinco,
María Leonor Méndez-Hernández and
Ulises Paramo-García

Additional information is available at the end of the chapter

http://dx.doi.org/10.5772/64941

Abstract

Nowadays, adhesive industry is growing, and its development will be important in a short future because it offers good returns, and in some cases it is a better option for packaging and sealing with advantages in prices, productivity and weight reduction. In terms of joining and/or sealing, adhesives are well positioned among joining systems; however, knowledge about adhesives is need for their efficient use and only through proper design of the union can be achieved satisfactory results. In this chapter, a development of a formulation of pressure-sensitive adhesive based on styrene-butadiene copolymers using a reactive resin is reported. Non-aromatic solvents were used in adhesive formulation with the aim of avoiding the emission of harmful solvents into the Atmosphere, and the adequate combination and amount of solvents were found. The effect of addition of a phenolic resin in the adhesive formulation as a crosslinking agent was evaluated. By means Fourier Transform Infrared spectroscopy (FTIR), the crosslinking reaction was also studied. The performance of adhesive formulation was evaluated by means of dynamic mechanical analysis (DMA).

Keywords: reactive resin, crosslinking, adhesive properties, radial SBS

1. Introduction

Adhesives have been used since ancient times, since the discovery of materials with adhesive properties from natural and animal resources. Then, through the study of the properties and appearance of new materials, several kinds of adhesives with synthetic components have been developed with the aim of improving their performance properties in specific application areas.

An adhesive is a material capable of joining two substrates through mechanisms of adhesion and cohesion through surface contact; it is applied over surfaces to keep them joined permanently. Also, the adhesives can be used as sealants.

Adhesion referred as the state between two bodies in intimal interfacial contact that transferred as interfacial forces in substrate area to carry out the joining or adhesion.

The adhesion phenomena can be understood by some theories: the adhesion chemical theory refers to intermolecular and Van Der Waals forces and chemical bonds that are generated between the adhesive and substrate. The diffusion theory explains the adhesion based on polymeric chain movements through the interphase, determining higher mobility according to molecular weight resulting in partial penetrations between materials. The weak limit layer theory proposes the existence of zones in the adhesion surface with cohesive breakages or weak joint forces originated by impurities in adhesive or adherent. The mechanical theory explains that adhesion can be achieved according to porosity and rugosity of the substrate surface when the adhesive penetrates forming anchor zones. The electrostatic theory proposes the formation of electrostatic forces at the adhesive-substrate interphase and is commonly applied in polymer-metal adhesion. The adsorption theory refers to the forces generated in the adhesive-substrate interface considering the superficial tension of adhesive to wetting the substrate and forming a contact angle [1].

The adhesive joint can be considered of two types: structural and non-structural. The structural adhesive joint is applied on substrates that experiment high deformations, over its yield point. This type of joints can be capable of transferring strain without loss in the integrity of the system in the design limits. The non-structural adhesive joints are those that do not require support substantial charges; its function is to keep materials of light weight in established place in the system of joint design. The sensitive adhesives to pressure and package adhesives are examples of non-structural adhesives.

1.1. Adhesive formulations

There are several types of adhesives that are developed in function of its final application, among contact adhesive base solvent with Polychloroprene for applications in wood, canvas and felt, etc. A disadvantage of these adhesives is when it comes to the environmental field, due to the kind of polymer, so it is necessary to find alternatives to replace Polychloroprene, styrene-butadiene-styrene copolymers (SBS) an option, using a reactive resin to obtain contact adhesive base solvent from equal or better performance. However, in attempts to establish a formulation with these components, there are some problems in stability and precipitation during storage; therefore, a formulation is required based on a proposal of resin and polymer that makes adhesives capable to cover all the performance parameters during storage and application.

On the other hand, development of adhesive formulations that are environmental friendly is a requirement according to several organizations, with the aim of covering regulations, because it is necessary for incorporation of substitution of harmful materials. Adhesives can be formulated by several paths, according to applications and compliance, which depends on cohesive and adhesive properties that can be tailored with a good polymer selection or amount of reactive resin or with the presence of another additive. Polychloroprene is one of the most used polymers in pressure-sensitive adhesive (PSA) formulations, because there are regulations that the material needs to be replaced. Among the materials that have been used in PSA formulations, styrene-butadiene copolymers represent a good option [2, 3]. In adhesive formulations, crosslinking reaction is the main step that controls the adhesive properties, and it can be carried out by means of addition of different agents; the most common crosslinking agent is sulphur [4], but organic peroxides and ester, ether and phenolic resins can also promote the crosslinking agent with appropriate conditions, for instance radiation, heat, change in pH, etc.

SP-154 is a heat-reactive phenolic resin that was developed for adhesive formulation; it is soluble in aromatic hydrocarbons and ketones and partially soluble in alcohols. Adhesive formulations with SP-154 resin have excellent heat resistance and high cohesive strength [5].

Styrene-butadiene copolymers (SBC) have a wide field of applications as shoe soles, impact modifiers, asphalt modifiers, adhesives and sealants. Several kinds of SBC have been used in adhesive formulations and as films in adhesives [2, 6, 7]; however, study of radial structure is not reported before.

There is a report that studied the adhesion properties of SBS copolymer and an epoxy resin after crosslinking with ultraviolet light, and that crosslinking grade of polybutadienic block was high and the adherence strength depends on the crosslinking level [6].

The amount and resin type in an adhesive formulation affect its performance, so it is necessary to evaluate the corresponding amounts to obtain the desired properties. Previous report [8] carried out the shear test for single overlap in cured probes using as substrate aluminium with adhesive formulations prepared with phenolic resin in different concentrations. According to the results, a higher resin concentration increases the shear strength; it was also found that the failure kind depends on resin concentration in formulation, so phenolic resin was shown to be an effective additive to improve adhesive properties.

An important characteristic to be considered in adhesive formulations is the polymer structure. Researchers prepared two copolymers, one of radial structure and the second one random from poly (methyl methacrylate) and poly (2-ethylhexylacrylate-co-methylacrylate), reporting high differences between block copolymer and random, due to the ability of the last one to form physical networks by phase separation [9].

1.2. Resin effect on adhesive

According to international standards ISO 472, ISO 4618/3, the resins are defined as "organic solids, semisolids or pseudo solids that have an indefinite and relative high molecular weight, and often melt over a wide temperature range".

The resins have the following characteristics: average molecular weight lower than 10,000 g/mol, amorphous and complex structures, and do not exhibit a well-defined melt point, but have a glass transition temperature and softening temperature.

The resins can be divided into natural and synthetic. The first are from animal and vegetal sources; the second result from controlled chemical reactions and are divided into:

- Hydrocarbon resins that are produced by means of polymerization, and they are added to polymer to provide adhesion, flow and toughness.

- Synthetic resins obtained by addition polymerization and polycondensation; they are intermediates in plastic synthesis of high molecular weight [10].

The selection of resin for use in thermoplastic rubbers was based on its tendency to associate with intermediate or final block, depending on the desired effect on adhesive properties. The resin with softening point over 85°C tends to associate with the final copolymer blocks. This category of resins tends to provide rigidity to formulation and hold the cohesive strength in high temperature, and is used to improve the specific adhesion of the final block phase and adjust the flow viscosity [11].

1.3. Phenolic resins

The phenolic resins are used widely in industrial level due to its high thermal stability, toughness and stress resistance, and provide wide application areas especially in coatings, adhesives, insulation and laminates. The classic phenolic resins are resol and novalac; its structure can be modified to obtain desired properties in several applications [12].

The resol and novalac resins are intermediate reactive that under some transformations variety (for instance, novalac in production of poly (aril ether amides) and resol by means of reaction with acetals to produce wire varnish) among some improvements and in some cases improved resin systems that produce reactive resins for value-add products [13].

The phenolic resins are used also in combination with reinforcers such as carbon or fiberglass. In these systems, the mechanical properties are lower than when epoxy resin is used, but with better fire behaviour. One of the most important applications is in aircraft interiors.

The novalac resins are stable, thermoplastic, solid, soluble and meltable. They are prepared under acidic conditions with less than equimolar proportion of formaldehyde with phenol, and these kinds of resins are preferred to moulding materials [14]. **Figure 1** shows the obtaining of a contact adhesive using a resol resin.

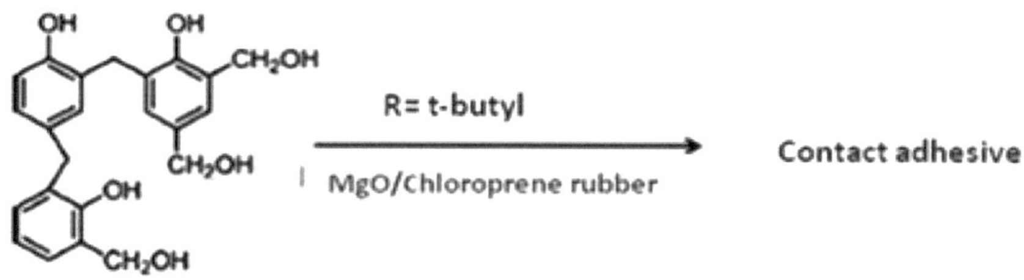

Figure 1. Obtaining of contact adhesive from a resol resin [14].

In this kind of contact adhesives based on chloroprene (**Figure 1**) are added magnesium oxide and zinc, with the aim of acting as HCl acceptors and crosslinkers. The magnesium oxide plays an important role to form and no meltable but soluble metallic complex with phenolic resins. This complex improves the adhesive cohesive strength at high and room temperatures. The resins family of Terc-butylphenolics is the most widely used in neoprene adhesives because it provides high thermal resistance when reacting with magnesium oxide present in these formulations [14]. The rubber properties change when crosslinked. These changes are reflected in thermoplasticity reduction, improving the abrasion and increasing tension strength.

1.4. Curing kinetic study

The chemical reactions involve formation and rupture of chemical bonding. The associate energy to a chemical bonding is a potential energy form, so that the chemical reactions involve potential energy changes. Often, for a reaction to occur, certain covalent bonds must be broken for others to emerge, which only occurs if the molecule orientation is appropriate to collide with enough kinetic energy to overcome the stabilization of potential energy of the bonds. The average kinetic energy of a molecule group is proportional to absolute temperature.

The kinetic contemplates the chemical reaction study and laws that govern that rate. The experimental fact that is based on the reaction rate of a system, at specific pressure and temperature, is only a state function. According to that, the basic equation of rate is:

$$\frac{da}{dt} = kf(a) \tag{1}$$

Eq. (1) relates reaction rate *dα/dt*, at pressure *P* and constant temperature *T*, with a function that shows the reactive species concentration *f (a)*, through the rate constant *k*. The *k* units depend on global reaction order; the calculated value corresponds to the reaction at a particular temperature and varies if temperature changes. Experimentally, it has been probed that for most of the homogeneous and heterogeneous reactions, the reaction constant depends on temperature. This dependence follows the Arrhenius equation [3]:

$$K = A^{(-E_a/\mathrm{RT})} \tag{2}$$

where K is the reaction constant, E_a is the activation energy, A is the preexponential factor or frequency factor, R is the gas constant (8.314 J/mol K), T is the absolute temperature and often E_a and A are considered temperature independents.

The activation energy E_a represents the energy limit to be overcome to carry out a reaction. The frequency factor is understood as the constant of rate to an infinite temperature. For a first-order reaction, the logarithmic form of Arrhenius equation is reordered for a unique temperature and is compared with straight line equation [3]:

$$\ln k = -\left(\frac{Ea}{R}\right)\left(\frac{1}{T}\right) + \ln A \tag{3}$$

The frequency factor value of the collisions of A is almost constant in moderated temperature changes; thus, ln A can be determined with the constant term of the equation (the interception). The line slope is obtained when a graph of ln k in function of $1/T$ is $-Ea/R$ is plotted; in this way, it is possible to determine the E_a value from the slope.

The reaction conversion is defined as a function $f(\alpha)$ that depends on the reaction mechanism. The model can be iso-conversional or named of n order and autocatalytic model [15, 16].

In the present work is reported the adhesive formulation based on a styrene-butadiene copolymer using a phenolic reactive resin. The effect of resin amount and combination of non-aromatic solvents was also studied. By means of infrared spectroscopy, the crosslinking reaction was monitored, and adhesive properties were evaluated by means of dynamic mechanical analysis and tensile test.

2. Methodology

2.1. Materials

A thermoplastic copolymer styrene-butadiene-styrene (SBS) of radial block structure with polystyrene content of 30% and density of 0.94 g/cm^3 provided by Dynasol Elastomers was used. As reactive resin, a commercial phenolic type resin named SP-154 from Ameripol Chemical S.A. de C.V. was used, with phenolic content >95%, formaldehyde content 200–400 ppm, melt point 79.4°C, flammability point 176.7°C and specific gravity 1.1 g/cm^3; it was non-water soluble. As solvents were used ethyl acetate ≥99.5%, cyclohexane HPLC grade, Aldrich, and acetone ACS grade, Analytyka. The solvent selection was according to solubility parameter.

After proper solvent selection, solubility parameter of mix (δ_{mix}) was determined, according to Eq. (4), and it is possible to determine the amount of each solvent.

http://dx.doi.org/10.5772/64941

$$\delta_{mix} = \Sigma_{i=1}^{n} \delta_i f_i \tag{4}$$

where δ_i is the solubility parameter of each component and *(f)* is the volumetric fraction. **Table 1** shows the individual solubility parameter, mix solubility parameter and volume fraction.

Solvent	Solubility parameter	Phr	Volume fraction
Ethyl acetate	9.1	140	0.367
Cyclohexane	8.2	203	0.533
Acetone	10.0	38	0.1
Σ		381	1.0
δ_{mix}			8.71

Table 1. δ_{mix} for SBS polymer and solubility parameter of used solvents.

Adhesive formulation-based solvent consists of SBS elastomer, phenolic resin with commercial name SP-154, zinc oxide powder as filler and a mix of solvents (**Table 2**).

Component	Phr
Polymer	100
Phenolic resin (SP-154)	0, 30, 50, 70
Zinc oxide	0, 1.0
Cyclohexane	203
Acetone	38
Ethyl acetate	140

Table 2. Experimental matrix for adhesive formulation.

2.2. Adhesive preparation

Adhesive preparation consists of mixing all components in a Kettler glass reactor using a mechanical stirrer (IKA Euro-STPCV 600051 model) at 3500 rpm as maximum speed, at room temperature and 50% relative humidity. The material addition sequence was established according to several experimentations, based on solubility, dispersion, among others; first, SBS was added to solvent mix, after the zinc oxide and finally the phenolic resin and was continuously stirred for 2 h.

The thermogravimetric analysis (TGA) was carried out on materials in SDT2960 TA instrument equipment under a N_2 atmosphere with 20 ml/min flow, in a temperature range from 40 to 600°C with 10°C/min rate, using 15 mg sample amount. DMA analysis was carried out in a DMA 2980 TA Instruments equipment, with a film tension clamp with multifrequency mode in a temperature range from −100 to 400°C with heat ramp of 5°C/min, 1 Hz of frequency and sample dimensions of 17 × 6 × 1 mm. Infrared (IR) analysis was carried out to evaluate crosslinking reaction and possible secondary reactions. IR spectra were recorded in Spectrum One Perkin Elmer equipment using the ATR technique with 12 scans and resolution of 4 cm^{-1} in wavenumber range of 4000–500 cm^{-1}. Adhesive films were cured at 193°C, and different times since 0 until 60 min, with the aim of evaluating crosslinking process.

3. Results and discussion

3.1. Thermogravimetric analysis

From thermogravimetric analysis, with initial decomposition temperature, can be predicted the thermal stability of materials, and from the derivative curve of loss weight can be determined the maximum decomposition temperature.

Figure 2 shows the TGA thermogram where derivative indicates that region decomposition begins at 193°C, while at 217, 356 and 457°C other resin composites are degraded. Previous reports [17, 18] found a decomposition temperature at 199°C when a phenolic resin was used in adhesive formulation, mainly due to metylol and hydroxyl groups from phenolic structure. With higher phenolic group content, the temperature decomposition is lower. Around 350°C, this resin type can present active sites that condensed with metylol groups showing a constant degradation and at 400°C can be produced the breaking of phenolic groups with methyl groups from phenol structure.

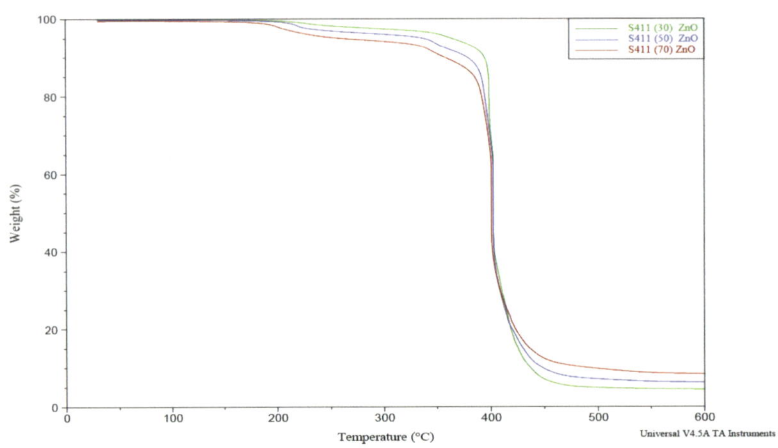

Figure 2. TGA thermogram of adhesive formulations 30, 50 and 70 phr of resin SP-154 and zinc oxide.

3.2. Dynamic mechanical analysis (DMA)

DMA results were compared to evaluate the effect of resin in adhesive mixtures and the effect of addition of ZnO, monitoring the storage modulus and Tan δ. **Figure 3** left shows the DMA

http://dx.doi.org/10.5772/64941

thermogram for SBS, SBS with 30 and 70 phr of reactive resin. At lower temperatures, the storage modulus (E') increases with resin content from 0 to 70 phr at low temperatures, and when temperature increases E' decreases. It can be observed that over 193°C, the E' values increase, which is more evident for adhesive with 30 phr of reactive resin which presents higher E' value, attributed to recovering behaviour from crosslinking reaction of elastomer with resin, indicating that higher resin content is not favourable for crosslinking. When zinc oxide was added in adhesive formulation (**Figure 3** right), the behaviour was similar to adhesive formulations without zinc oxide but with higher E' values due to addition of fillers, such as zinc oxide. Improvement of the elastic properties of a polymeric matrix at higher temperatures in rubber region is attributed to interactions between particles and polymer matrix due the presence of a high surface area. This behaviour can be corroborated with Tg displacement to higher temperatures [19].

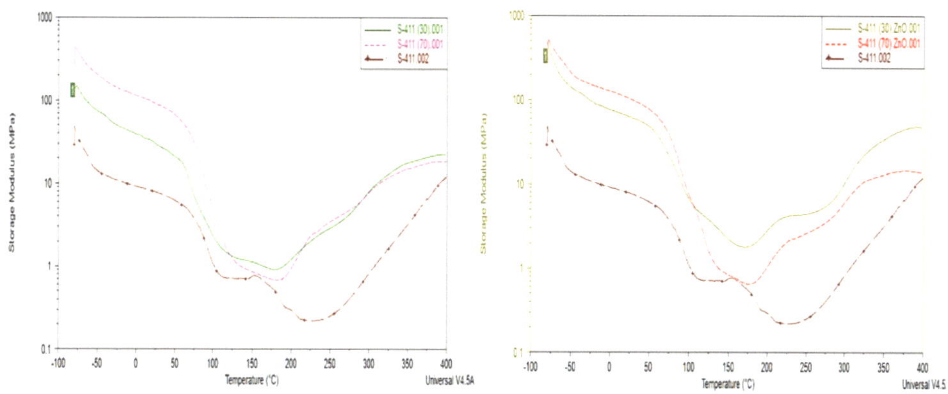

Figure 3. DMA thermogram of adhesive formulations with 30, 50 and 70 phr of resin SP-154 without (left) and with zinc oxide (right).

A non-uniform crosslinking has no considerable effect over the macroscopic properties of a material, such as E', but in fracture properties the affectation must be bigger due to the presence of weak localizable zones that act as failure initiation points [20].

The transitions of adhesive formulations were determined by means of Tan δ curves from DMA (**Figure 4**). It was possible to observe that adhesive formulation with 70 phr resin has higher Tan δ quotient compared with others, in the polystyrene Tg region. The deformations from this are attributed to non-homogeneous crosslinking between elastomer and resin. The increasing in Tg value is due to the crosslinking reaction that generates more rigidity [2]. In the crosslinking process, a three-dimensional network is generated, and the chain flexibility controls the physical and thermo-mechanical properties [20]. These changes are more notable when zinc oxide is added to adhesive formulation, because the particles cause an increase of interfacial areas in the system diminishing the crosslinking grade. The peak over 80°C is attributed to crosslinking process of elastomer, and it can be seen that it is affected by resin addition. **Figure 4** shows the Tan δ curve from DMA for adhesive formulations with zinc oxide.

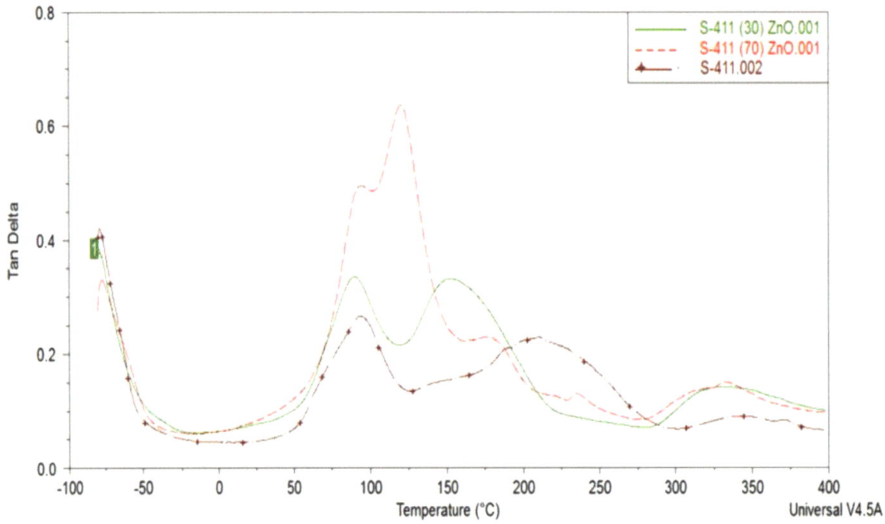

Figure 4. Tan δ curve from DMA for adhesive formulations with elastomer and 30 and 70 phr of resin SP-154 and zinc oxide.

3.3. Kinetic study of crosslinking process

By means of DSC analysis, the curing parameters of adhesive formulations were determined, and the measurements were carried out in isothermal conditions at 180, 190, 200 and 210°C. From general reaction rate equation is possible to realize kinetic studies by following the reactions that were carried out in solid state in function of time at constant temperature according to the next equation [3]:

$$\gamma = \frac{d\alpha}{dt} \tag{5}$$

where γ is the reaction rate, α is the conversion grade or reaction extension and dt is the differential time.

By means of DSC, it was assumed that in material crosslinking during an exothermic process, the conversion rate is directly proportional to heat flow with the next equation [3]:

$$\frac{d\alpha}{dt} = \frac{dH/dt}{H_T} = kf(\alpha) \tag{6}$$

where *dH/dt* represents heat flow, H_T is the total enthalpy when crosslinking reaction has been completed, *k* is the rate constant and *f(α)* is a function that depends on the reaction model. α can be expressed as with equation [21]:

http://dx.doi.org/10.5772/64941

$$\alpha = \frac{H_t}{H_T} \tag{7}$$

where H_t is enthalpy at time t. The total enthalpy can be estimated by means of integration of the main exothermic peak of DSC non-isothermic curves [21].

For the crosslinking process to adjust to a model of order, the reaction kinetic is described as a function of conversion grade $f(\alpha)$, which provides a dependence of the reaction rate by means of the next equation [21]:

$$f(\alpha) = k(1-\alpha)^n \tag{8}$$

where n is the reaction order. In the kinetic order n, the maximum conversion rate $d\alpha/dt$ corresponds to $t = 0$ and is proportional to material concentration that has been reacted [15]:

$$\frac{d\alpha}{dt} = k(1-\propto)^n \tag{9}$$

It is assumed that reaction rate depends on the temperature; thus, the constant rate can be determined by means of the Arrhenius equation [15]:

$$k = Ae^{-E_a/RT} \tag{10}$$

where A is the pre-exponential factor, E_a is the activation energy, T is the absolute temperature and R is the gas constant. Reordering Eq. (9) [15]:

$$\ln\frac{d\alpha}{dt} = \ln k + n\ln(1-\alpha) \tag{11}$$

By means of lineal regression, the $\ln(d\alpha/dt)$ vs $\ln(1-\alpha)$ is possible to determine the kinetic parameters k and n. Studies have been conducted for crosslinking reactions from dynamic and isothermal methods in which the very similar results are obtained in the values of the kinetic parameters [15].

The kinetic parameters n and H_T from model of order reaction n were calculated with Pyris software of Perkin Elmer using isothermal curves. The H_T values can be introduced in the software if reaction heat or melting heat of sample is known, however. All adhesive formulations were carried out by duplicate at corresponding temperatures 180, 190, 200 and 210°C for comparing the kinetic values and obtaining an average.

Using a multilinear regression, parameters A, E_a and n were obtained from the next equation [22]:

$$\ln\left(\beta\frac{d\alpha}{dt}\right)=ln(A)-\frac{E_a}{RT}+n\cdot\ln(1-\alpha) \tag{12}$$

where β is the heating rate.

The adjust quality of dates from kinetic model can be noticed in graph ln k vs $1/T$. In the ideal case, the lines passing through the points (the slope for calculation of activation energy) form parallel straight lines, as is required by the theory of kinetic model applied [22].

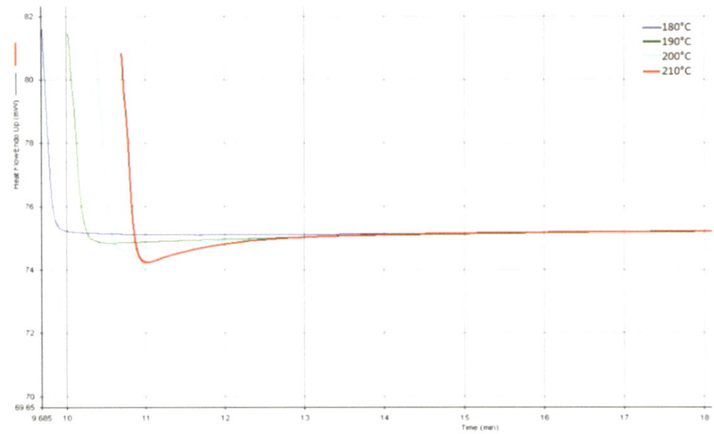

Figure 5. DSC analysis in isothermal mode for adhesive formulation with 70 phr of resin and zinc oxide at different temperatures.

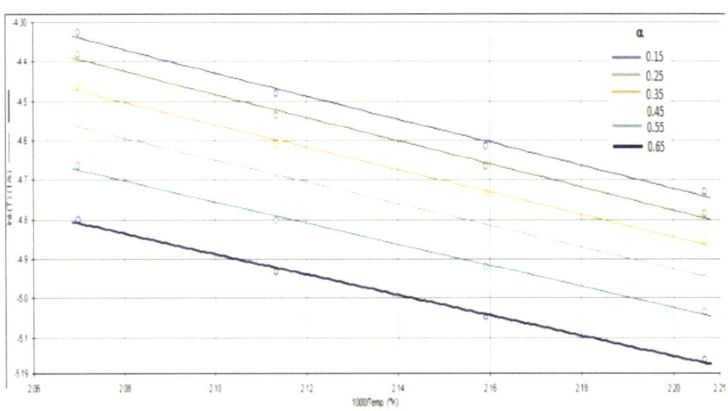

Figure 6. ln k vs 1000/Ta at different values for adhesive formulation with 70 phr of resin and zinc oxide.

http://dx.doi.org/10.5772/64941

In **Figure 5** are presented the isothermal curves obtained for adhesive formulation with 70 phr and zinc oxide at different temperatures, showing the influence of temperature in exothermic region, and **Figure 6** shows the corresponding graph of calculating activation energy (ln *k* vs *1/T*), obtaining a lineal range in the extension of reaction ensuring that the data are convincing [21].

The selected temperature to obtain the kinetic parameters was 190°C, because it was closer to crosslinking reaction temperature. **Table 3** shows the obtained kinetic values.

Adhesive formulation	***T*, °C**	***N***	**ln *A***	**ln *k* (s^{-1})**	**E_a, KJ/mol**
Adhesive 0 phr	190	1.72	19.4	−3.8	98.07
Adhesive 30 phr resin	190	1.3	0.52	−4.77	24.91
Adhesive 50 phr resin	190	1.2	−1.44	−5.07	17.86
Adhesive 70 phr resin	190	1.28	−1.87	−5.27	15.38
Adhesive 0 resin phr ZnO	190	1.32	0.98	−4.14	61.98
Adhesive 30 phr resin ZnO	190	1.34	2.27	−4.61	33.46
Adhesive 50 phr resin ZnO	190	1.35	0.92	−4.77	26.85
Adhesive 70 phr resin ZnO	190	1.27	−1.24	−5.08	21.46

Table 3. Kinetic parameters obtained from DSC isotherms.

From **Table 3** the order reaction is $1 \leq n \leq 2$. The reaction order value, *n*, affects the reaction rate of the formulation allowing to determine the proper period of crosslinking of materials. On the other hand, the reaction order is indicative of molecule number that can react in crosslinking reaction, so that the resin affects in predominant way the crosslinking kinetic process [2].

About the activation energy values from **Table 3**, it can be observed that resin content decreases the E_a values, and the zinc oxide causes an increase in E_a values compared with adhesives without zinc oxide.

The generous concentration of reactive sites and low viscosities in the initiation stage of crosslinking reaction contribute to increasing the degree of conversion after a time interval decelerates showing changes in the time reaction due the decreasing of availability of active sites and the increasing of viscosity, so the decreasing on reaction rate carry to the crosslinking reaction pass from chemical to diffusion control [21].

The physical properties of thermoplastics and elastomers used for crosslinking depend on the crosslinked structure, grade, time and crosslinking temperature; these parameters are related to network formation that provides substantial information about network structure and properties and processability [23].

3.4. FTIR characterization

FTIR technique was used to evaluate the crosslinking at different times (0, 20, 40 and 60 min, shorter times do not show significant effect), identifying the main functional groups that form or disappear when crosslinking reaction was carried out.

The crosslinking process is carried out by means of reaction of methylol and hydroxyl groups with unsaturations of polybutadiene [2]. Initially, the decomposition reaction of resin by thermal effect generates active sites (methylol and hydroxyl) that stimulate the reaction to the formation of benzocarbenium ions with the double bond of the polybutadiene units to form cromane ring (**Figure 7**). Previous works [24, 25] prepare thermoplastic vulcanizate from natural rubber and high density polyethylene with and without compatibilizer, using an octyphenol formaldehyde resin using a compatibilizer obtaining good tensile strength and high elongation values, besides lower deformation compared with thermoplastic vulcanized without compatibilizer, which can be attributed to the formation of cromane ring of reaction. SBS copolymer and phenolic resin are present in an adhesive formulation.

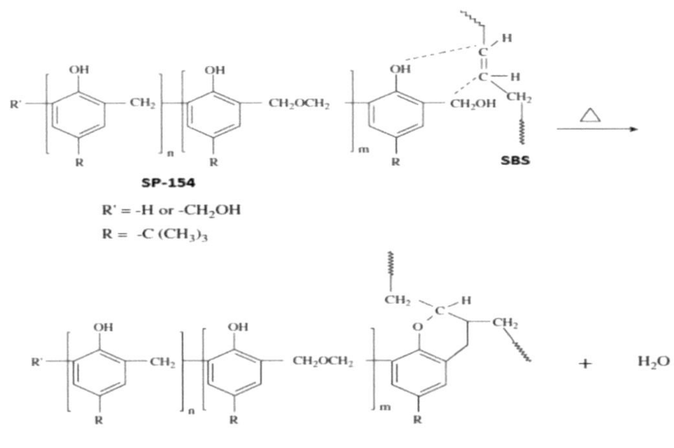

Figure 7. Crosslinking reaction for phenolic resin and SBS elastomer [25, 26].

Figure 8 shows the IR spectra for adhesive formulations with 30 phr resin without zinc oxide at different times. According to crosslinking reaction proposal, double bond of polybutadiene block has to decrease. The characteristic peak of C=C is at 1639 cm^{-1} and does not show a significant change with time; however, there is another peak attributed to =C—H trans and $=CH_2$ vinyl at 963 and 910 cm^{-1}, which present a decrease in intensity. The peaks from =C—H cis at 727 cm^{-1} also show a decrease in intensity with time, which is indicative that crosslinking reaction of phenolic resin and SBS copolymer is carried out, mainly in cis structures. The polystyrene peaks at 1602 and 697 cm^{-1} do not show significant changes, which implies that crosslinking is not taking place in that polymer block. On the other hand, peaks attributed to phenolic resin are present at 3319, 1450, 1210 and 1026 cm^{-1}, which present significative changes in intensity corroborating that crosslinking reaction is carried out with elastomer compared with elastomer without curing.

http://dx.doi.org/10.5772/64941

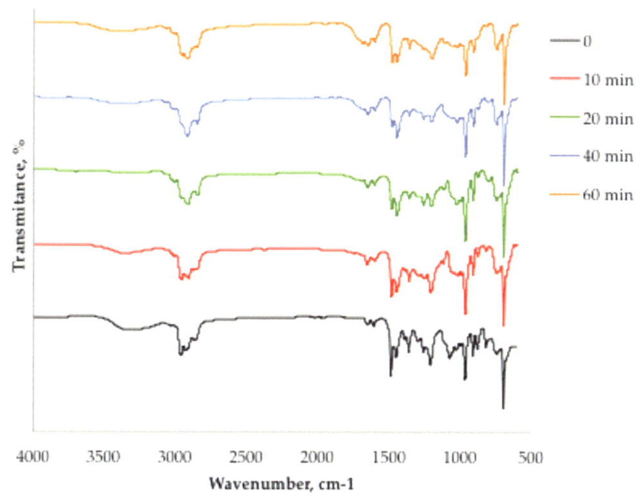

Figure 8. FTIR spectra of adhesive formulation with 30 phr resin at different times.

The peak at 1450 cm^{-1} corresponding to CH_2 group increases due to the formation of methylene bonding between methylol groups of phenolic resin and some double bonds from polybutadiene block. At 1210 cm^{-1} peak attributed to C—O of phenols presents a decrease in intensity due to the influence of phenol in the crosslinking reaction. **Figure 9** shows the IR spectra for adhesive formulations with zinc oxide. The changes in peaks are similar to those observed when zinc oxide is not added to adhesive formulation, which is indicative that it does not affect the crosslinking reaction. There are reports about the appearance of symmetric bands of OH groups that indicate breaking of hydrogen bonds and that over 200°C the OH group peak is indistinguishable [19, 26].

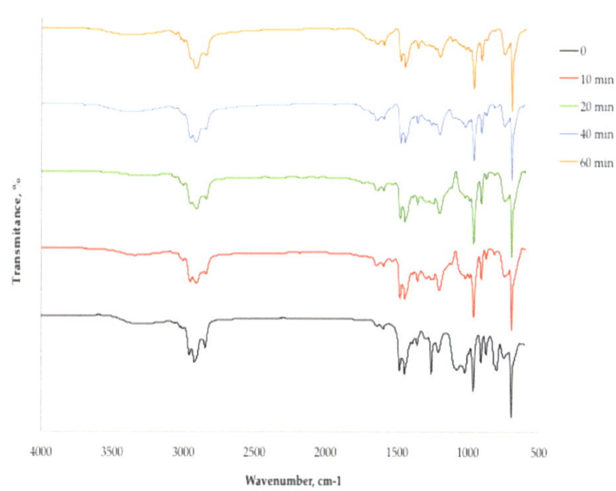

Figure 9. FTIR spectra for adhesive formulations with 30 phr resin and zinc oxide at different curing times.

4. Conclusions

According to results, it is possible to obtain an adhesive formulation using solvents as an alternative for less environmental impact based on the solubility parameter that depends on the polymer used; in this case, solubility parameter found favours the polystyrene and polybutadiene blocks. The use of a reactive resin is an option to prepare adhesive formulations based on SBS elastomers. The inclusion of reactive resin affects the viscoelastic properties, due to the crosslinking process that was carried out as DMA results demonstrate; the best viscoelastic properties were observed for adhesive formulation with 30 phr of resin, and addition of zinc oxide was not favourable for adhesives. The results of kinetic study of crosslinking process indicate that addition of reactive resin is favourable for decreasing the E_a values. Also, it is evident that polymeric structure with available insaturations at the end of chain favours the reduction in E_a values. FTIR demonstrates that crosslinking reaction carried out in metylol and hydroxyl groups presents in resin and double bonds of polybutadiene block of elastomer.

Acknowledgements

The present research work is registered in Tecnológico Nacional de México with code MAD-PYR-2015-0033. Two of the authors THM and MLMH thank CONACYT for the scholarship.

Author details

Tzeitel Hernández-Martínez, Beatriz Adriana Salazar Cruz, José Luis Rivera-Armenta*, María Yolanda Chávez-Cinco, María Leonor Méndez-Hernández and Ulises Paramo-García

*Address all correspondence to: jlriveraarmenta@itcm.edu.mx

Secondary Petrochemical Research Center, Instituto Tecnologico de Ciudad Madero, Ciudad Madero, Tamaulipas, Mexico

References

[1] Packham DE, editor. Handbook of Adhesion. 2nd ed. Addison-Wesley Longman Ltd; 1992. 17 pp. 10.1002/0470014229

[2] Salazar-Cruz BA, Mendoza-Martínez AM, Esquivel-de la Garza AC, Moctezuma Espíricueto SA, Rivera-Armenta JL. Evaluation of crosslinking reaction in adhesive based styrene-butadiene elastomers using infrared spectroscopy. In: Theophanides

Theophile, editor. Infrared spectroscopy-anharmonicity of biomolecules, crosslinking ofbiopolymers, food quality and medical applications. InTech; 2015. p. 117-135. DOI 10.5772/58911.ch7.

[3] Salazar-Cruz BA, Rivera-Armenta JL, García-Alamilla R, Mendoza-Martínez AM, Esquivel de la Garza A, Moctezuma Espiricueto S. Evaluación térmica del curado de adhesivos base SBR usando peróxido de dicumilo. Quimica Nova. 2015;38:651-656. DOI 10.539/0100-4042.20150067.

[4] Zielinska AJ. Cross-linking and Modification of Saturated Elastomers Using Functionalized Azides [Thesis]. Eindhoven, The Netherlands, University of Twente; 2011.

[5] SI group, technical data sheet [Internet]. 2009. Available from: https://www.pharosproject.net/uploads/files/sources/1828/SP-154-1.pdf [Accessed: 2016-03-31].

[6] Kim J, Kim WH, Lee DH. Adhesion properties of UV crosslinked polystyrene-block-polybutadiene block-polystyrene copolymer and tackifier mixture. Polymer. 2002;43:5005-5010. DOI: 10.1016/S0032-3861(02)00327-0.

[7] Navarro-Bañón MV, Pastor-Blas MM, Martín-Martínez JM. Elimination of the reactivation process in the adhesion of chlorinated SBS rubber with polychloroprene adhesives. eXPRESS Polymer Letters. 2007;1:236-244. DOI: 10.3144/expresspolymlett.2007.36.

[8] Sturiale A, Vazquez A, Cisilino A, Manfredi LB. Enhancement of the adhesive joint strength of the epoxy–amine system via the addition of a resole-type phenolic resin. International Journal of Adhesion and Adhesives. 2007;27:156-164. DOI: 10.1016/j.ijadhadh.2006.01.007.

[9] Simal F, Jeusette M, Leclère PH, Lazzaroni R, Roose P. Adhesive properties of a radial acrylic block co-polymer with a rosin ester resin. Journal of Adhesion Science and Technology. 2007; 21:559-574. DOI:10.1163//156856107781192355.

[10] Fink JK. Reactive Polymers Fundamentals and Applications: A Concise Guide to Industrial Polymers. 2nd ed. William Andrew. 2013. 29pp.

[11] Skeist I, editor. Handbook of Adhesives. 3rd ed. International Thomson Publishing; 2011. 556 p.

[12] Reghunadhan Nair CP, Bindu R L, Ninan KN. Thermal characteristics of addition-cure phenolic resins. Polymer Degradation and Stability. 2001;73:251-257.DOI:10.1016/S0141-3910(01)00076-3.

[13] Pilato L. Phenolic resins: 100 Years and still going strong. Reactive and Functional Polymers. 2013; 73:270-277. DOI:10.1016/j.reactfunctpolym.2012.07.008.

[14] Scheneberger, G.L. editor. Adhesives in Manufacturing. 1st ed. Marcel Dekker. 1983. 353pp.

[15] Tan D, Wang Y, Li Z, Xing H. Synthesis and cure kinetics of diphenyl (diphenylethynyl) silane monomer. Research on Chemical Intermediates. 2013;39:3427-3440. DOI: 10.1007/s11164-012-0855-8.

[16] Pérez JM, Oliet M, Alonso MV, Rodriguez F. Cure kinetics of lignin–novolac resins studied by isoconversional methods. Thermochimica Acta. 2009;487:39-42. DOI: 10.1016/j.tca.2009.01.005.

[17] Ginic-Markovic M, Choudhury NR, Dimopoulos M, Williams DR, Matisons J. Characterization of elastomer compounds by thermal analysis. Thermochimica Acta. 1998: 316:87-95. DOI: 10.1016/S0040-6031(98)00290-1.

[18] Feng S, Yuan Z, Leitch M, Xu CC. Thermal degradation performance of bark based phenol formaldehyde adhesives. Journal of Analytical and Applied Pyrolysis. 2015;115:184-193. DOI: 10.1016/j.jaap.2015.07.015.

[19] Mangal R, Srivastava S, Archer LA. Phase stability and dynamics of entangled polymer-nanoparticle composites. Nature Communications. 2015:6:1-9. DOI:10.1038/ncomms8198.

[20] Rahul R, Kitey R. Effect of cross-linking on dynamic mechanical and fracture behavior of epoxy variants. Composites Part B: Engineering. 2016;85,336-342. DOI: 10.1016/j.compositesb.2015.09.017.

[21] Ren R, Xiong X, Ma X, Liu S, Wang J, Che P, Zeng Y. Isothermal curing kinetics and mechanism of DGEBA epoxy reins with phtalide-containing aromatic diamine. Thermochimica Acta. 2016; 623:15-21. DOI: 10.1016/j.tca.2015.11.011.

[22] PerkinElmer Instruments LLC. Pyris Kinetics Software User Guide. 2002. Part Number 0996-6500, Release A.

[23] Cai H, Li P, Sui G, Yu Y, Li G, Yang X, Ryu S. Curing kinetics study of epoxy resin/flexible amine toughness systems by dynamic and isothermal DSC. Thermochimica Acta. 2008;473:101-105. DOI: 10.1016/j.tca.2008.04.012

[24] Nakason C, Nuansomsri K, Kaesaman A, Kiatkamjornwong S. Dynamic vulcanization of natural rubber/high-density polyethylene blends: effect of compatibilization, blend ratio and curing system. Polymer Testing. 2006;25:782-796. DOI: 10.1016/j.polymertesting.2006.05.001.

[25] Scagliusi S, Cardoso E, Luga, A. Radiation-induced degradation of butyl rubber vulcanized by three different crosslinking systems. Radiation Physics and Chemistry. 2012;81,991-994. DOI: 10.1016/j.radphyschem.2012.01.011.

[26] Yue S, Juanfeng H, Hong H, Hequing F, Hanwei Z, Huanqin C. Synthesis, properties and application of a novel epoxidized soybean oil-toughened phenolic resin. Chinese Journal of Chemical Engineering. 2007;15:418-423. DOI:10.1016/S1004-9541(07)60101-3.

Chapter 9

What are the Health Risks of Occupational Exposure to Adhesive in the Shoe Industry?

Imed Gargouri, Moncef Khadhraoui and
Boubaker Elleuch

Additional information is available at the end of the chapter

http://dx.doi.org/10.5772/64936

Abstract

The term "shoe" covers a wide range of products made from various materials. Organic solvents (OS) are components of various products such as the adhesives which are used in many industries. The shoe industry consumes large amounts of adhesives. This chemical risk assessment (CRA) is to validate the hypothesis of decreasing gradient of RA craft enterprises to industrial enterprises through the semi-industrial. The simplified methodology CRA of INRS was applied. For each chemical, a hazard class is assigned based on the sign, and then the potential exposure class is determined according to parameters "quantity and frequency of use". Our RA process is complete with ambient measurements for solvents to which employees are most frequently exposed. Adhesives being constituted as solvent mixture, we have conventionally used an exposure index which is compared with the limit values. The simplified method CRA of INRS was chosen because it is one of the reference methods in RA established from the Kinney model. If these atmospheric samples and toxicology tests were made for the first time in the shoe industry in Sfax, they concerned a sample of companies of the three shoe manufacturing processes preceded by a preliminary RA with post-study and inventory of products handled.

Keywords: risk assessment, chemical risk, shoes manufacturing, organic solvents, adhesive, occupational health

1. Introduction

The term "shoe" covers a wide range of products made from various materials. Boots, shoes, sandals, slippers, clogs, and others are made entirely or in part with leather, rubber, synthetic and plastic, cloth, rope, and wood [1]. Our study deals with the shoe industry, that is to say one based on traditional manufacturing methods. The rubber boots factory (or equivalent synthetic materials), which is one of the branches of the rubber industry, is excluded from this study [1, 2].

Shoes, boots and leather sandals, felt, or other materials have been for century's hand. Quality shoes are still made entirely or partly in this way by craftsmen, but in all industrial countries, today the mass production predominates [2, 3].

Organic solvents (OS) are components of various products such as paints, inks, pesticides, degreasers, solvents, and glues, which are used in many industries [1]. The shoe industry consumes large amounts of adhesives, diluents, and strippers OS bases. Consequently, the employees in this sector are likely to present particular nerve pathologies [2].

2. History of the shoe in the Sfax city

Sfax is an industrial and agricultural pillar of the Tunisian economy [4]. Among its industries, shoemaking is highly developed sector, which employ about 10000 employees or 10% the workforce of the nonagricultural professional environment which 6000 artisans' shoes located in the old city "the Medina."

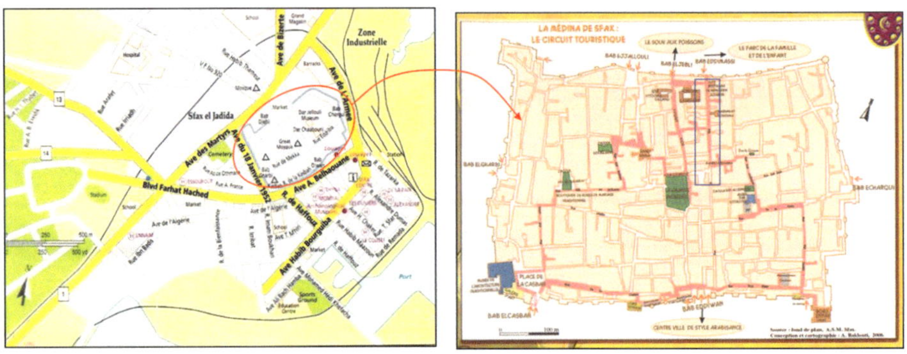

Figure 1. Map of (a) the center of Sfax city and (b) the old town of Sfax.

As in many cities, Sfax had a craft of the shoe. But in addition, Sfax was a shoe industry. The manufacture of the shoes to form Sfax "industrial" dates back to the nineteenth century. It was well structured in the Medina in the shoe area called "Souk Blaghgias." It consists of the main street shoes exhibition: The current street Mongi Slim known under its former name "Street of Bay." Until now, along this road for approximately 300 m shoe shops displaying their products. In the level one called "the Ally" in every store or in the neighboring streets were their own manufacturing facilities.

In 1960, after independence, observers were predicting the rapid decline and disappearance of the craft before the competition of the modern shoe industry. Now, after more than 30 years, this craft, far from having disappeared, has prospered and it feeds today, modern shoes, more than half of the Tunisian market. The artisan district of the boot has grown in the old town of Sfax; we hear all the noise of machines whose artisans are equipped to modernize the sensitive points of their manufacturing (**Figure 1**) [4].

3. Manufacturing processes of the shoe

The realization of a shoe includes several operations. Each step has been mechanized, but the manual production is used as reference. The introduction of new materials has modified manufacturing processes without changing the outline [1, 2].

The manufacture of rods requires sorting and preparation of leather or other materials. The stems are cut with cutters on stitching presses (or burning). The various parties, including the linings, are then "assembled," that is to say, stitched or glued together. Perforation, eyeletting, or making buttonholes are the procedures carried out further.

In the lower part, the outer and inner soles, heels and welts are cut using cutters to moving blades or molding presses. It manufactures heels compressing leather or wooden slats. The assembly is then cut, formed, cleaned, and labeled.

The upper and lower parts of the shoe are then assembled and then stitched, glued, nailed, or screwed together. After that the shoes are shaped and smoothed by means of rollers. The finishing of the shoe includes waxing, staining, spraying, polishing, and packaging.

Among the raw materials used in manufacturing, adhesives, including adhesive solids and liquids and natural adhesive solutions prepared from organic solvents, represent the largest occupational hazards [3].

4. The basic products used in the shoe industry

In the industry of the adhesive and its derivatives, products intended for the manufacture of the shoe are divided into three groups [2, 5–11]:

- Four groups of adhesive,
- Primers, hardeners, and additives,
- Solvents, thinners, and strippers.

4.1. Four groups of adhesive (Table 1)

- The polyurethane (PU) adhesive is used for assembling thermoplastic rubber soles (TR), rubber, polyvinyl chloride (PVC), leathers high in fat or synthetic fabrics requiring the use of high-performance adhesives.

- The polychloroprene adhesives (neoprene) are used for making coatings soles, fast repair, and bonding of natural materials (leather, rubber). They are made from synthetic resins dissolved in ketone solvents.
- The adhesives in aqueous emulsion (latex adhesives or white glues) may be natural or synthetic:
  - Natural latex adhesives are applied by machine gun or by hand. These are contact adhesives to attach soles, stitching, and lining.
  - The synthetic latex adhesives are used for laying the ends and bonding liners.

Products	Composition	Use
Polychloroprene adhesives (neoprene)	– Solids based on polychloroprene (15–30%) depending on use	– Welded shoe soles
Implemented by contact with the following:	– *Organic solvents* (mixture): the most used are, by decreasing importance order:	
– Coating the two sides to stick,	* MEK	
– Respect the drying time	* Gasoline C (up to 30% hexane)	
(in ways that the two adhesive	* Acetone	
films release their solvents)	* Ethyl acetate	
– Assembly of parts (pressure)	* Toluene	
after reactivation	* Cyclohexane	
	* The alcohols and chlorinated hydrocarbons (*little used*)	
The polyurethane adhesives (PU)		
1. based polyurethane elastomers	Polyvinyl chloride Elements (PCV)	– Contact adhesive which is taken after solvent evaporation
	– Solids of polyurethane elastomers	
	– *Organic solvents*: mixture	
	* Ketones and / or	
	* Overheads	
2. Two-component	*Part A:*	Very flexible assemblies
	– Polypol (polyester or polyester)	The mixture of A with B is carried out at the time of use
	– *Organic solvents* (mixture) :	
	* Ketones and / or	– The adhesion is after solvent evaporation
	* Overheads	

Products	Composition	Use
	Part B: – Isocyanate prepolymers (small amount of free isocyanate: Ex diphenylmethane diisocyanate ....)	
Adhesives in aqueous emulsion: latex adhesives or white glues	– Solids (40–60% may be: * Vinyl acetate * polyisoprene * Acrylic polymers * Copolymers: acrylic, isoprene, chlorobutadiene * polyurethane *– Solvents:* * In general: WATER * Optionally and low proportions: – Alcohol – Other solvents	– Installation of liners: manual application by simply coating and touch – Water evaporation Making contact
Hot melt adhesives or "hot melt" *Equipment cadence "very expensive"*	Copolymer Ethylene-vinyl acetate ± possibly: plasticizers or antioxidants	Applying on one of the faces to be bonded by a polymer melt temperature rise

Table 1. The basics products in the shoe manufacturing (part welded) [11].

The hot melt adhesive is a new generation of glue developed in order to reduce the use of solvents. It consists in applying on one of the faces to be bonded by a polymer melt temperature rise. It requires clock equipment which is expensive.

4.2. Primers, hardeners, and additives

- The halogenations operation of TR soles with new functions (rubber look and matt or transparent), must be strictly control any change of model. There are several formulas ready to use (one or two components) for receiving an additive to control the regular application on the soles.
- Catalysts, slow or fast, are used in polyurethane adhesives, polychloroprene or dispersion, as well as direct injection. They provide good resistance to the migration of fat leathers.

4.3. Solvents, thinners, and strippers

Beside adhesives preparations, there are other necessary preparations and complementary type of strippers, thinners, and solvents (**Table 2**).

Preparations	Applications and examples
The cleaning agent	– PU soles: remove traces of mold release agent ; – PCV soles: remove surface plasticizer contained in PCV compounds
The diluents	– Dilute polychloroprene adhesives ; – Clean the machine to glue and brushes, spray guns stained with paint, screens for screen printing ink
The solvents	– Clean the rubber soles, spray guns stained with paint, screens for screen printing ink; – Dilute paints and glues.

Table 2. Other preparations used in the manufacture of shoes [13].

5. Risks of the solvents and the adhesives on the health

According to their properties, the solvents can be used in the manufacture of the shoe, of:

- Degreasing (cleaning soles, textiles…)
- Additives and solvents (paints, varnishes, inks, adhesives…)
- Stripping (removal of paints, varnishes, glues…)

OS hydrocarbons are divided into eight main groups [12–14]. Six families are used in the manufacturing industry adhesives and footwear (**Table 3**).

The intensive use of flammable liquids causes a high risk of fire, and the widespread use of presses and assembling machines has increased also the risk of accidents. The main danger to the health of workers is using toxic solvents, high concentrations of dust in the air, the risks due to ergonomic shortcomings, and noise from machinery.

OS can cause acute and chronic effects on the central nervous system. Benzene, which was once used in adhesives and solvents, has been replaced by toluene, xylene, hexane, methylethylketone (MEK), and methylbutylketone. The n-hexane and MEK can cause peripheral neuropathy and should be replaced by heptane or other solvents [2].

In many factories, there were outbreaks of a disease known as the "paralysis of the shoemaker" and characterized by clinical signs of a form of more or less serious paralysis. This is a flange type of paralysis, located in the upper and/or lower limbs resulting in deep tendon atrophy with areflexia, without altering the superficial or deep sensitivity. From the clinical point of view, this syndrome is due to an inhibition or a functional lesion of the lower motor neurons of the voluntary motor system (pyramidal system). Most often, it leads to a neurological regression with extensive proximodistal functional recovery [1, 2].

Groups of solvents	Applications and examples
Aromatic hydrocarbons	Homologues of benzene (toluene, xylenes, ethylbenzene …) are used as lacquer thinners, adhesives and inks.
Petroleum solvents	Used as solvents in paints, adhesives and coatings: – **Alkanes** such as **hexane (C6H14)**; – **Cycloalkanes**: The most used is **cyclohexane (C_6H_{12})**
Alcohols	Very used as **diluents** inks, resins, varnishes, paints and **glues**. These are excellent dehydrating agents having good degreasing Action Example: **isopropanol** (or isopropyl alcohol).
Ketones	Mainly used as solvent paints, lacquers, varnishes, glues and adhesives. They are sometimes impregnated on wipes for small cleaning. They are good agents drying damp rooms. The most commonly encountered are: **acetone** or **MEK**.
Acetates (esters)	Often mixed with other solvents, for example, in the curing products and halogenation
Halogenated hydrocarbons	Used as diluents **glues**, **adhesives**, paints. Examples: **Trichloroethylene** (degreaser leather, thinner glue) or **dichloromethane** (thinner adhesives and cleaner)

Table 3. The families of solvents used in the manufacture of adhesives and shoes [13, 15].

In the manufacture of shoes and boots, handling and leather processing can cause diseases by exposure to certain chemical substances mentioned above and in tanning and finishing. In addition, various chemicals can produce other diseases. Exposure to toxic solvents contained in adhesives and cleaning products, and to leather dust in the air is of particular concern. The use of benzene can cause thrombocytopenia (decreased number of red blood cells, platelets, and white blood cells in the blood) or pancytopenia. It has been almost eliminated from the footwear industry [3]. It was also found in some shoe factories cases of peripheral neuropathy due to n-hexane content in adhesives. This substance has also been largely replaced by less toxic solvents. Cases of electroencephalographic changes, liver damage, and behavioral alterations due to exposure to solvents have been reported in workers in the shoe industry [3].

6. Occupational chemical risk in shoe industry in Sfax

6.1. Methods

The prevention of occupational risks, particularly chemicals, is based on the risk assessment (RA) according to the specified procedures. Thus, we realized an exposure assessment in making shoes in a sample of volunteer companies (**Table 4**).

This chemical risk assessment (CRA) aims to validate the hypothesis of decreasing gradient of RC craft enterprises to industrial enterprises in passing by semi-industrial. The simplified assessment methodology for chemical risk (SAMCR) of National Institute of Research and Safety (INRS) [15] was applied only for strand "Health." For each chemical, a hazard class (HC) is assigned based on the sign, and then the potential exposure class is determined based on the parameters "amount and frequency of use."

Process	Industrial	Semi-industrial	Artisanal
Number of companies	4	5	9
Equipment used	Mechanization *(workflow)*	Mechanization and manual	Manual
Number of employees	>20	[10–20]	<10
Installation location	IZ	IZ or Medina	Medina

IZ: industrial zone.

Table 4. Business definition.

Hazard Class	1	2	3	4	4 ou 3
Pictogram	None	Xi-Irritant	Xn-Harmful	T-Toxic	Not identified Not identified
Unidentified product: Not identified Glue			Xn-Harmful		
Unidentified product: Not identified Solvent			Xn-Harmful	T-Toxic	

Table 5. Hazard classes depending on the labeling [13–15].

The HC was determined only on the labeling. The allocation of HC to a preparation based on the analysis of the pictogram on the packaging (**Table 5**). For preparations of glues or solvents "unidentified" (no information on the composition and/or the manufacturing company), it was decided that the assigned pictogram would be one that represents the greatest health risk and consequently the HC."

http://dx.doi.org/10.5772/64936

Figure 2. Atmospheric sampling (a) "individual" and (b) "stationary."

		la France (EU[2])		USA (ACGIH[3])		Germany (MAK[4])		Our Study (Sfax—Tunisia)	
Solvent	**CAS[1]**	**VEA[5]**		**TLV-TWA[6]**				**EAV adopted**	
		ppm[7]	**mg. m^{-3}**	**ppm**	**mg. m^{-3}**	**ppm**	**mg. m^{-3}**	**ppm**	**mg. m^{-3}**
Acetone	67-64-1	500	1210	500	–	500	1200	**500**	**1200**
Cyclohexane	110-82-7	200	700	300	–	200	700	**200**	**700**
n-Hexane	110-54-3	20	72	50	–	50	180	**20**	**72**
Methylethylketone (MEK, 2-butanone)	78-93-3	200	600	200	–	200	600	**200**	**600**
Toluene	108-88-3	50	192	50	–	50	190	**50**	**190**

[1]CAS: Chemical abstract service.
[2]EU: European Union.
[3]ACGIH: American Conference of Governmental Industrial Hygienists.
[4]MAK: Maximum Arbeitsplatz-konzentration.
[5]EAV: Exposure average value calculated over a reference period of 8 hours.
[6]TLV-TWA: Time-weighted average (weighted average values of 8 hours per day and 40 hours per week).
[7]ppm: Parts per million per volume of air.

Table 6. Values of atmospheric exposure limits [17].

- *Unidentified glue:* The glue most dangerous to a symbol "Xn-Harmful" [5, 11] for the health and attributed HC is "3." Thus, any unidentified glue the symbol "Xn-Harmful" and "3" class was assigned.
- *Unidentified solvent (solvents, thinners, and strippers):* Two hypotheses "HC" were individualized [11, 13–15]:

- Let the symbol "T-Toxic" and class "4," the highest existing level;
- Let the symbol "Xn-Harmful" and "3" class more realistic for the majority of solvents.

Solvents	Identification	Chemical substance	Use	Nature	Pictogram	Hazard class
Acetone		Acetone		Ketones	Xi-Irritant	2
Alcohol	Isopropyl alcohol (IPA)	Isopropyl alcohol	Paint solvent	Alcohol	Xn-Harmful	3
Asouplex	Stripper	Acetone MEK +	Soften the "strong against Calorex"	Ketones	Xi-Irritant	2
Stripper		Methylene chloride (2%) in MEK	Washing, sole preparation (PU, PVC)		Xn-Harmful	3
Desmodur	DESMODUR RC	TDI (5%) in Ethyl acetate	Diluent PU glue		Xi-Irritant	2
Durcisseur	DESMODUR RC				Xi-Irritant	2
Diluent	Solvent mixture	Toluene (100%)	Thinner neoprene glue	Aromatic hydrocarbons	Xn-Harmful	3
PA	IPA			Alcohol	Xi-Irritant	2
Halogenation product		98% Ethyl acetate	Surface preparation sole	Acetates	Xi-Irritant	2
Flexibility	ASOUPLEX			Ketones	Xi-Irritant	2
Supersouple	ASOUPLEX			Ketones	Xi-Irritant	2
Tucosolve	Stripper	MEK, AE, Cyclohexane			Xn-Harmful	3
Bidonplaste *Elaste* *Special ink* *Solvent not defined*	**Not identified**	Solvent mixture			**Xn-Harmful** or **T-Toxic**	3 or 4

Table 7. Inventory preparations "solvent" used in the footwear industry and hazard class.

Our approach to RA is complete, and ambient measurements [16] (**Figure 2**) for solvents (acetone, cyclohexane, n-hexane, MEK, toluene) to which employees are more frequently exposed are taken.

Adhesives being constituted as solvent mixture, an exposure index (I.exp) was used and compared with the limit values (**Table 6**) [17, 18]:

$$I.exp = C1/EAV1 + C2/EAV2 \ldots + Cn/EAVn \quad (1)$$

Cn and EAVn are respectively the concentration and the average value of n pollutants exposure.

If I.exp is greater than 1, the limit value is considered outdated. Calculated from stationary sampling, this index is called pollution index (I.pol).

6.2. CRA

However much the manufacturing process, the inventory of solvent borne formulations showed HC "2" and "3". When the unidentified preparations, they were classified "3" or "4" (**Table 7**). For the 23 identified glue preparations, HC ranges from "1" to "3" (**Table 8**).

The potential risk score (PRS) process "industrial" 2 in case of danger class puts 13 chemical preparations to examine with high priority, including five of solvent preparations. For the HC "3," PRS do not exceed "10000," while with the hazard class "4" three of five solvent borne preparations pass a PRS 100000.

The situation of distinction for a process "semi-industrial" does not arise since we did not detect unidentified preparations. The preparations to consider with high priority are 2 solvents and 5 adhesives unidentified.

In the "artisanal" process, we have identified eight chemical preparations to examine with high priority: two solvents borne preparations and six adhesives.

Glues	Identification	Chemical substances	Use	Nature	Pictogram	Hazard class
110/Textile glue	Rubber 110F	Natural rubber, hexane, petroleum resin	Provisional or final bonding of leather fabric (handcrafted process)	Polychloroprene	Xn-Harmful	3
420	Ceylanprène 420	Polycholoroprène, cyclohexane, hexane, MEK, Resin (phenol, phenol terpene), Toluene	Bonding the leather	Polychloroprene	Xn-Harmful	3
425S	Ceylanprène 425	Polycholoroprène, cyclohexane, hexane, MEK, Resin (phenol, phenol terpene), Toluene	Bonding the leather	Polychloroprene	Xn-Harmful	3
715			Bonding the	Polychloroprene	Xn-Harmful	3

Glues	Identification	Chemical substances	Use	Nature	Pictogram	Hazard class
			leather			
945			Bonding the leather	Polychloroprene	Xn-Harmful	**3**
C3			Bonding the leather	Polychloroprene	Xn-Harmful	**3**
C4			Bonding the leather	Polychloroprene	Xn-Harmful	**3**
Calorex	Calorex	Polycholoroprène, cyclohexane, hexane, MEK, Resin (phenol, phenol terpene) on a ribbon	Polyester fabric coated with an adhesive	Polychloroprene	No	**1**
CD11			Gluing soles	Polyurethane	Xi-Irritant	**2**
CN13			Bonding the leather	Polychloroprene	Xn-Harmful	**3**
White glue	Glue "Latex"	Synthetic or natural rubber in an aqueous phase	For bonding wood heel	Aqueous emulsion	No	**1**
Yellow glue	Neoprene glue			Polychloroprene	Xn-Harmful	**3**
Colle neoprene	Neoprene glue			Polychloroprene	Xn-Harmful	**3**
Vinyl glue	Ex: R10	Polymer of vinyl acetate, acrylic polymer, water	Manufacture of the shoe	Aqueous emulsion	No	**1**
D1			Gluing soles	Polyurethane	Xi-Irritant	**2**
Forestali/ICF			Gluing soles	Polyurethane	Xi-Irritant	**2**
Plastipole	Polyurethane glue		Gluing soles	Polyurethane	Xi-Irritant	**2**
Polyurethane	Polyurethane glue		Gluing soles	Polyurethane	Xi-Irritant	**2**
Could	Polyurethane glue		Gluing soles	Polyurethane	Xi-Irritant	**2**
Tucoplast	Polyurethane glue		Gluing soles	Polyurethane	Xi-Irritant	**2**
UF43	Plestiprene	Polyurethane,	Gluing	Polyurethane	Xi-Irritant	**2**

http://dx.doi.org/10.5772/64936

Glues	Identification	Chemical substances	Use	Nature	Pictogram	Hazard class
	UF43	ethyl acetate, acetone, MEK, Toluene	soles			
Colle geant *Precision without glue*	**NOT IDENTIFIED**	**Solvent mixture**			**Xn-Harmful**	**3**

Table 8. Inventory preparations "glues" used in the footwear industry and hazard class.

6.3. Atmospheric measurements

Measurements of atmospheric concentrations (**Table 9**) show high levels especially in companies with semi-industrial process and type 1 craft.

Process		**Business**	**Number of workstation**	**Personal sampling**			**Atmosphere sampling**			
				Exposure index*			**Pollution index****			
				Not	**Scope**	**Average**	**Not**	**Scope**	**Average**	
Industrial		Industrial 1	5	3	0.3–1.1	0.7	2	0.5–0.8	0.6	
		Industrial 2	2	1	–	1.8	1	–	0.3	
		Industrial 3	5	4	0.9–9.4	3.7	1	–	0.8	
		Industrial 4	5	4	0.8–19.8	6.8	1	–	0.0	
Semi-industrial		Semi-industrial 1	5	3	0.4–1.2	0.7	2	0.0–0.3	0.1	
		Semi-industrial 2	5	4	0.5–2.0	1.0	1	–	0.4	
		Semi-industrial 3	4	2	0.2–2.5	1.3	2	0.2–0.4	0.3	
		Semi-industrial 4	5	4	0.9–2.4	1.7	1	–	0.1	
		Semi-industrial 5	4	1	–	5.6	3	4.3–8.8	6.8	
Artisanal	**Type 1**	Craft 1	2	1	–	0.1	1	–	0.4	
		Craft 2	1	/			1	–	0.7	
		Craft 3	1	/			1	–	0.8	
		Craft 4	3	2	0.4–1.7	1.1	1	–	1.6	
		Craft 5	2	1	–	1.5	1	–	0.9	
		Craft 6	2	2	1.6–3.4	2.5	/			
		Craft 7	2	1	–	6.4	1	–	5.1	
		Type 2		Craft 8	1	/			1	–0.5
Craft 9	1	/			1	–	0.6			
Total = 33						**Total = 22**				

Table 9. Atmospheric exposure of employees by industrial trial.

Stages of production	Sources of pollution
1. Individual sizing	Solvent vapor:
Post "UPPER PART MAKER"	– Glue container in use
	– Piece glued
	– Materials used for coating (pre-sizing operation)
2. The preparation of surfaces prior	– Clean and degrease the parts to be assembled (brush, cloth)
to bonding	– Washing of soles
Post "SOLE MAKER"	– Halogenation: Treatment of some soles (rubber)
3. Drying	Objects are placed in general on fixed or movable shelves near collages positions
4. Finish	– Cleaning after bonding to remove excess glue
	(solvent-soaked rag)
	– Coloring and application of polish (usually by spraying)
5. Cleaning equipment	Brushes and accessories: With solvents
6. Fractionation and storage	Adhesives wholesale packaging: A transfer in small containers used in gluing
"Decanting" preparations	station

Table 10. The steps of the shoe manufacturing (the welded Technique) and sources of pollution [11].

Exceedances of I.exp and/or I.pol are noted for exhibitors' positions including the upper part makers, smelters and finishing (**Table 10**) within three manufacturing processes (**Table 11**).

All measurements in different companies show that exposure to OS varied from one process to another and from one position to another (**Table 12**). The average atmospheric concentrations of hexane are particularly high, especially with the exposure average value (EAV) which overruns the position of the upper part maker, the sole maker, and finishing in all production processes except the process artisan type 2. The other solvents (acetone, cyclohexane, MEK, and toluene) are relatively high without exceeding the EAV.

6.4. Comments and discussion

Despite significant exposures to solvent-based preparations, an almost total lack of use of personal protective equipment such as gloves, masks, and goggles by the manipulators of the adhesives formulations and/or solvents was noted. The equipment of rooms by adequate ventilation is not rated very often, while the Tunisian regulations require companies to preserve human capital and implement a preventive political.

http://dx.doi.org/10.5772/64936

Process		Post working	Number of workstation	Personal sampling Exposure Index			Atmosphere sampling Pollution Index		
				levies	scope	average	levies	scope	average
Industrial		Chopped off	1	/			1	-	0.0
		Quilting	1	/			1	-	0.8
		Upper part maker	4	3	0.9 to 19.8	8.4	1	-	0.5
		Sole maker	7	5	0.3 to 3.5	1.5	2	0,3 - 0,8	0.5
		Finish	3	3	0.9 to 1.8	1.2	/		
		Serigraphy *(Woman)*	1	1	-	9.4	/		
Semi-industrial		Upper part maker	8	4	0.5 to 2.5	1.6	4	0.1 to 7.2	2.5
		Sole makers	12	10	0.2 to 5.6	1.8	2	0.2 to 8.8	4.5
		Finish	3	/			3	0.3 to 4.3	1.7
Artisanal	Type 1	Upper part maker	5	2	0.1 to 1.6	0.9	3	0.9 to 5.1	2.5
		Sole makers	8	5	0.4 to 6.4	2.7	3	0,4 - 0,7	0.6
	Type 2	Upper part maker / Sole makers	2	/			2	0.5 to 06	0.6
			Total = 33				Total = 22		

Type 1: The positions of Upper part maker and Sole maker are separated (made by different people)
Type 2: One confused post, in two Upper part maker and Sole maker

Table 11. Average atmospheric concentrations of solvents assayed.

Process		Post working	Sampling	Acetone (mg/m³)		n-Hexane (mg/m³)		methylethylketone (mg/m³)		Cyclohaxane (mg/m³)		Toluene (mg/m³)	
				scope	average	scope	average	scope	average	scope	average	scope	average
Industrial		Chopped off	1	-	0.0	-	0.0	-	0.0	-	0.0	-	0.0
		Quilting	1	-	0.0	-	36.8	-	0.0	-	160.5	-	6.8
		Upper part makers	4	0.0 to 1652.0	413.0	28.1 to 975.0	300.1	0.0 to 2656.9	776.6	0.0 to 183.8	66.8	13.6 to 321.0	103.5
		Sole makers	7	0.0 to 3077.3	439.6	14.3 to 139.3	51.9	0.0 to 121.4	17.3	0.0 to 72.1	21.2	00 to 71.9	21.3
		Finish	3	-	0.0	53.8 to 86.6	66.8	0.0 to 165.3	108.7	0.0 to 34.2	12.9	56 to 43.7	20.8
		Serigraphy	1	-	0.0	-	401.8	-	1090.8	-	1281.1	-	38.6
Semi-industrial		Upper part makers	8	0.0 to 114.3	16.1	0.0 to 204.0	57.4	0.0 to 509.5	113.4	-	118.5	4.4 to 500.0	106.6
		Sole makers	12	0.0 to 989.1	222.0	0.0 to 214.1	58.6	0.0 to 622.5	134.2	-	94.0	4.0 to 615.9	121.6
		Finish	3	0.0 to 474.3	158.1	0.0 to 116.0	43.8	0.0 to 350.3	136.2	-	3.9	10.3 to 329.0	130.4
Artisanal	Type 1	Upper part makers	5	0.0 to 217.6	43.5	0.0 to 112.8	52.9	0.0 to 334.5	124.0	-	111.5	7.8 to 480.0	138.5
		Sole makers	8	-	0.0	3.3 to 147.3	58.6	0.0 to 442.5	150.8	-	112.2	0.4 to 578.1	115.0
	Type 2	Ipper part makers /Sole makers	2	0.0 to 43.4	21.7	21.6 to 27.2	24.4	-	0.0	-	26.6	27.7 to 35.9	31.8
			Total = 55										

Type 1: The positions of Upper part maker and Sole maker are separated (made by different people)
Type 2: One confused post, in two Upper part maker and Sole maker

Table 12. Average atmospheric concentrations of solvents assayed.

The natural ventilation has been the main alternative ventilation: Some kind of artisans 2 working doors open because they have no other coming from vents (without windows), while others worked closed doors because of their irregular position (no statement to the National Case Social Security and/or cohabitation a group of craftsmen in a building intended for habitation and not commercial and industrial travailing each for its own account).

Semi-industrial footwear companies are craft enterprises working on a large scale with some mechanization, which paid their employees the number of parts produced per day (upper part makers and sole makers). They correspond to industrial companies that should be in IZ [19]. Another feature in this type of business is the presence of women in age mount positions rods and finishing [20]. Shoes manufacturing companies are increasing in the region with an extension to the new IZ south of the city of Sfax. The semi-industrial companies continue to invade Medina and especially the houses which are in most cases protected historical monuments.

The SAMCR of INRS was chosen because it is one of the reference methods in RA established from the Kinney model (mathematical model) [21].

If these atmospheric samples and toxicology tests were made for the first time in the shoe industry in Sfax, they concerned a sample of companies of the three shoes manufacturing processes (industrial, semi-industrial, and artisanal) preceded by a preliminary RA with a job and a study inventory of products handled.

Interest carcinogenic characteristics of certain preparations and especially their potential toxicity for reproduction must be taken, as a female presence in the shoe manufacturing sector increasingly important and cases of couple infertility [22] have been highlighted.

7. Conclusion

Identification and CRA are important steps in this process of prevention in the shoe industry. This detection was to (i) highlight the chemical preparations that are really harmful. This could be achieved by the inventory and prioritization of chemical preparations; and (ii) identify the nature of the manipulated chemical preparations. This identification was made by labeling of preparations. After this step, the semi-industrial process CRA is most at risk followed the industrial process and finally that of the artisanal process. When the study of exposure to workplace needs to know not only the average exposure over the EAV, but also to identify the polluting phases to determine the short-term exposure over the TLV.

Apart from the strengthening of health and safety measures in the shoe industry, two actions are to be undertaken in parallel: (i) the substitution of certain solvents by others in the manufacture of glues and derivatives seems necessary while maintaining the preparations use properties and (ii) a change in the Tunisian regulations on hygiene and safety, in particular, is necessary to mandate the RA in business.

Acknowledgements

This study was partly funded and supported by Sfax University. We are grateful to the director and staff of the adhesive company SIFCOL for their cooperation. We are also indebted to the workers who participated voluntarily in the study.

http://dx.doi.org/10.5772/64936

Author details

Imed Gargouri[1,2*], Moncef Khadhraoui[1] and Boubaker Elleuch[1]

*Address all correspondence to: Imed.Gargouri@fmsf.rnu.tn

1 National School of Engineers, Laboratory of Water, Energy and Environment-AD 10-02, Sfax University, Sfax, Tunisia

2 Medical School, Department of Occupational Medicine and Professional Pathology, Sfax University, Sfax, Tunisia

References

[1] National Institute for Research and Security (NIRS). Breakdown of gluing small items workshops (shoes). Guide pratique de ventilation n° 5. ED 672, 1987: 28.

[2] McCann M. The Leather, the fur and the shoe. Encycl Security Health Labour (ILO) 2000, 3rd French edition (Translation 4th English edition), 3, 88: 13 Available from: www.ilo.org/public/french/protection/safework/cis/products/encyclo/pdf [Accessed: 2007.04.06].

[3] Leather Shoe Technical Centre (CTC). Shoe manufacturing secrets in comics. Available from: www.ctc.fr/faq/questions.php3?theme=1 [Accessed: 2008.02.29].

[4] Governorate of Sfax. Portal of Tunisian Industry (Industry Promotion Agency). Available from: http://www.tunisieindustrie.nat.tn [Accessed: 2008.02.25].

[5] Dor F, Bonvallot N. Hazard identification: a step in the evaluation of health risks deepen. Environnement, Risque et Santé 2005,6:279–87.

[6] Margossian N. Chemical risk. 1st Edition. Dunod. Paris, France 2002: 235 ISBN: 2-10-006562-9

[7] Martel B. Guide chemical risk: identification, assessment, control. 3rd Edition. Dunod. Paris, France 2002: 388 ISBN : 2-10-005585-2

[8] Shoe glue. Keck Chimie S.A. Available from: www.keck-chimie.fr [Accessed: 2008.02.29].

[9] The different adhesives. In collage technique. Art of collage montage. Available from: www.art-pjm.com/collage-technique/technique-collage-les-colles [Accessed: 2008.02.29].

[10] Folder laminating adhesives. The developments reinforce for without solvents and aqueous bases. ED Emballage Digest. Available from: www.emballagedigest.fr/blog.php?2005/08/30/2230-dossier-colles [Accessed: 2008.02.29].

[11] NIRS. Breakdown of gluing small items workshops (shoes). Practical Guide of ventilation (n° 5) 1987, ED 672:28.

[12] Gérin M. Industrial solvents: Health, Safety, substitution. 1st Edition. Masson. Paris, France 2002 : 257 ISBN 2-294-00297-0

[13] NIRS. Database "Toxicological card index" Available from: www.inrs.fr [Accessed: 2005. 09.10].

[14] NIRS. Database "Solvex". Available from: www.inrs.fr/solvants/bsolvants.nsf [Accessed: 2006.03.20].

[15] Vincent R, Bonthoux F, Mallet G, Iparraguirre J-F, Rio S. Simplified assessment methodology of chemical risk. Cahier Note Doc 2005, 200 (ND 2233):39–62.

[16] SKC Inc. Air sample pumps, calibrators & accessories. Sorbent sample tubes and media collection. Available from: http://www.skcinc.com/ [Accessed: 2006.03.15].

[17] Triolet J. Occupational exposure limits for hazardous substances. ACGIH values (USA) and the MAK Commission (Germany). ND 2114-176-99. Cahier Note Doc 1999,176:59–90.

[18] Armstrong TW, Caldwell DJ, Verma DK. A Proposed methodology for setting occupational exposure limits for hydrocarbon solvents. Occup Hyg day Approximately 2005,2:600–7.

[19] Poirot P, Hubert G. Pelle-exposure profiles to solvents and compared to the values short-term limits. ND 2235-200-05. Cahier Note Doc 2005,200:83–93.

[20] Pagès million Falcy Mr. Risk Assessment solvents pregnancy. DMT 1999,80:335–53.

[21] Lieve P, Van De Laer M, JM Lamotte, Emelen JV. The risk analysis. Federal Ministry of Employment and Labour (Belgium) 2002 September: 58.

[22] Cho SI, Damokosh AI, Ryan AI, Chen D, Hu YA, Smith T. Effects of exposure to organic solvents on menstrual cycle length J Occup Environ Med 2001,43:567–75.

Chapter 10

Fracture Toughening Mechanisms in Epoxy Adhesives

Aldobenedetto Zotti, Simona Zuppolini,
Mauro Zarrelli and Anna Borriello

Additional information is available at the end of the chapter

http://dx.doi.org/10.5772/65250

Abstract

Fracture toughness is generally considered as the main properties of a polymer or a polymer adhesive system for measuring the material resistance to the extension of cracks. Epoxy adhesives are generally brittle in nature; however, the addition of a second dispersed phase could induce a remarkable increase of damage tolerance performance by an enhancement of the material fracture toughness. The fracture behavior of a filled epoxy resin is strongly affected by the dimensions, the shape, and the chemical nature of the considered filler. The chapter describes the different toughening mechanisms for polymer adhesives with special attention toward innovative nanofiller such as graphene nanoplatelets and hyperbranched polymer nanoparticles.

Keywords: fracture mechanisms, epoxy adhesives, filled epoxy resins, inorganic fillers, nanocomposites, hyperbranched polymer fillers

1. Introduction

Epoxy resins are among the most commonly used polymeric materials in structural applications, such as adhesives and fiber-reinforced composites. Moreover, they are widely employed in other nonstructural applications, i.e., as a coating in electronic and textile industries due to their great thermal, electronic, and mechanical properties [1, 2].

Although their wide use in the cited fields, highly cross-linked epoxy resins are rigid and brittle in nature and lose much of their structural integrity when damaged, thus limiting their use as structural materials (adhesives and composites).

In the last three decades, also due to the successful development of nano- and microparticles with different shapes, dimensions, and chemical features, nano-/microparticles have been extensively studied as effective tougheners to overcome the natural fragility of epoxy resins. As reported by Kinloch [3], the addition of 15–20 vol% fractions of dispersed rubbery particles (diameter of ~0.5–2 μm) can lead to an increase of the strain energy release rate G_{IC} by a factor of 10–15. Quan et al. [4] have increased the DGEBA epoxy resin fracture energy from 343 J/m^2 for the neat matrix to 2671 J/m^2 for epoxy modified by 30 vol% of core-shell rubber particles. Carbon nanotubes are also used as tougheners in epoxy resin and, according to the work by Gojny et al. [5], a nanotube content of only 1 wt% increases the critical stress intensity factor K_{IC} from 0.65 to 0.82 MPa m$^{1/2}$. Zotti et al. [6] report an increase of the G_{IC} value of about 350% with the addition of metallic iron core-inorganic shell nanoparticles (0.2 wt%). Ternary systems are extensively used in order to exploit the synergistic effect of the single elements on the fracture toughness; this is explained by Wang et al. [7], who have obtained an increase of about 108% in K_{IC} using a rubber/graphene/epoxy system.

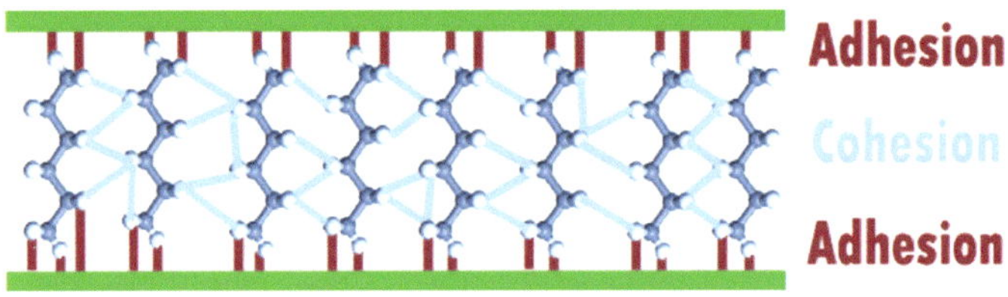

Figure 1. Comparison between cohesion and adhesion in an adhesive joint.

In an adhesive two main forces could be distinguished, i.e., cohesive and adhesive forces. The first could be considered as the inherent strength of the material related only to the chemical nature in terms of morphology and chemical structure. The second defines the complex interaction system between the adhesive and substrate therefore strongly affected by the contact surface morphology. **Figure 1** shows the difference between cohesive and adhesive forces.

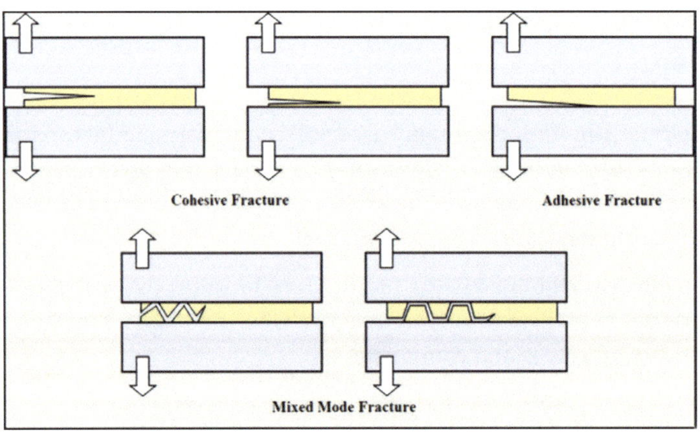

Figure 2. Adhesive joint fracture mode illustration.

http://dx.doi.org/10.5772/65250

Two adherent surfaces can fail according different fracture types (see **Figure 2**):

1. Cohesive fracture: the fracture propagates through the bulk resin that constitutes the adhesive layer. The fractured adhesive will cover both adherent surfaces. This kind of fracture generally occurs when the adhesion strength is stronger than the cohesion one.

2. Adhesive fracture: the fracture propagates through the adherent/adhesive surface. Only one of the adherent surfaces will be covered with the adhesive. This kind of fracture generally occurs when the adhesion strength is weaker than the cohesion one.

3. Mixed mode fracture: the fracture jumps from one interface to the other. It is possible to observe traces of adhesive on both surfaces of the substrate.

In materials science, fracture toughness describes the ability of a material containing a crack to resist fracture, and is one of the most important properties of any material system for many design applications. There are two alternative approaches to fracture analysis: the energy criterion and the stress-intensity approach [8].

According to the energy approach, for brittle materials, the crack propagates when the available energy is higher than the energy needed to create two new fracture surfaces. For ductile materials also, other types of energy dissipation should be taken into account (such as plastic deformation) in the global energy balance to follow correctly the crack onset and growth. Based on the Griffith energy criterion, Irwin developed the present version of this approach [8]; for a linear elastic, it is possible to define the energy release rate G as the rate of change in potential energy with the crack area material. When a material fails a critical value of the energy release rate, $G = G_C$, can be defined as the measure of material fracture toughness [8]. According to the Irwin theory, assuming a loaded infinite plate containing a crack whose measure in length is 2a (**Figure 3**), the energy release rate is given by:

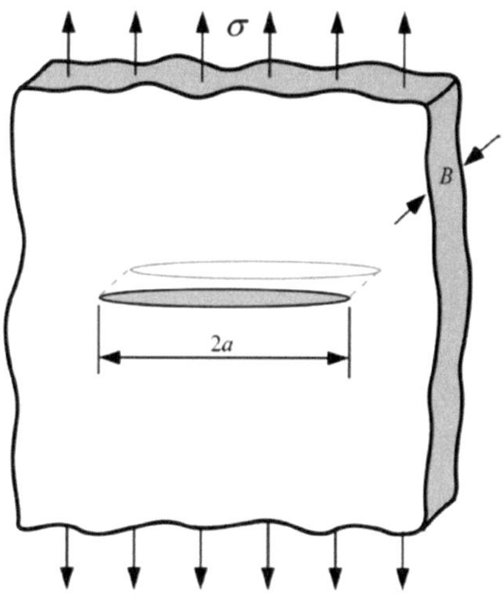

Figure 3. Through-thickness crack in an infinite plate subject to a remote tensile stress. In practical terms, "infinite" means that the width of the plate is >> 2a [8].

$$G = \frac{\pi \sigma^2 a}{E} \tag{1}$$

where E is Young's modulus, σ is the remotely applied stress, and a is the half-crack length.

Since, $G = G_C$ when the crack starts to propagate, the relation between stress and crack size for failure can be expressed as:

$$G_C = \frac{\pi \sigma_f^2 a}{E} \tag{2}$$

Considering an elastic material, each stress component of an elemental unit near the tip of the crack can be written as reported in **Figure 4**. It is interesting to note that in each equation, a proportional factor K_I, called stress intensity factor, appears. Knowing the value for this parameter, the entire stress state near the crack tip can be defined. Similar to the energy approach, failure occurs in correspondence to a critical stress intensity factor, $K_I = K_{IC}$. K_{IC} is an alternate measure of fracture toughness, and in the case depicted in **Figure 3**, the stress intensity factor can be written as follows:

$$K_I = \sigma \sqrt{\pi a} \tag{3}$$

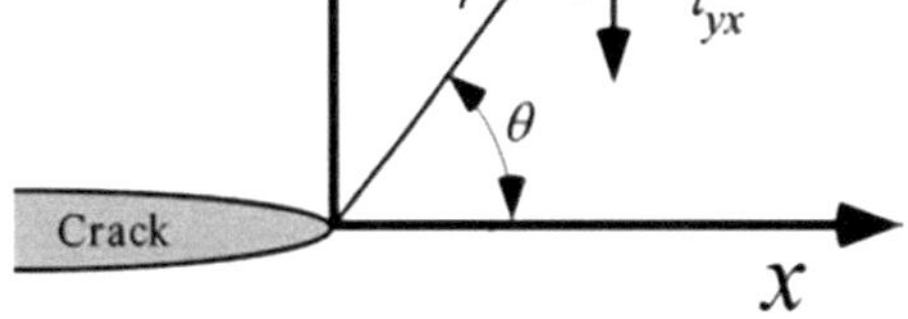

$$\sigma_{xx} = \frac{K_I}{\sqrt{2\pi r}} \cos\left(\frac{\theta}{2}\right)\left[1 - \sin\left(\frac{\theta}{2}\right)\sin\left(\frac{3\theta}{2}\right)\right]$$

$$\sigma_{yy} = \frac{K_I}{\sqrt{2\pi r}} \cos\left(\frac{\theta}{2}\right)\left[1 + \sin\left(\frac{\theta}{2}\right)\sin\left(\frac{3\theta}{2}\right)\right]$$

$$\tau_{xy} = \frac{K_I}{\sqrt{2\pi r}} \cos\left(\frac{\theta}{2}\right)\sin\left(\frac{\theta}{2}\right)\cos\left(\frac{3\theta}{2}\right)$$

Figure 4. Stresses near the tip of a crack in an elastic material [8].

Failure occurs when $K_I = K_{IC}$. In this case, K_I is the driving force for fracture and K_{IC} is a measure of material resistance. As with G_C, the property of similitude should apply to K_{IC}. That is, K_{IC} is assumed to be a size-independent material property [8].

The comparison of Eqs. (1) and (3) results in a relationship between K_I and G:

$$G = \frac{K_I^2}{E} \quad (4)$$

This same relationship obviously holds for G_C and K_{IC}. Thus, the energy and stress-intensity approaches to fracture mechanics are essentially equivalent for linear elastic materials [8]. As previously stated, epoxy resins are brittle in nature, and in order to improve the cohesive fracture toughness, a second nano-/microphase is generally added. The fracture behavior of toughened polymers may involve several mechanisms, each one contributing toward the total fracture toughness of the material [9].

The chapter aims to describe the different mechanisms that cause the epoxy adhesive fracture toughening, with special attention to the correlation between fracture surface morphology and fracture toughening mechanisms. The chapter is divided into two main parts: (i) an overview on the possible fracture toughening mechanisms that may occur in epoxy adhesives and (ii) a study of toughened epoxy adhesives based on the filler nature.

2. Epoxy resins and failure modes

The epoxy group, when bonded chemically with other molecules, forms a large three-dimensional network. The process by which chemical bonding is achieved is called curing, where the fluid resin changes to a solid form. Among curing agents, various amines [10], anhydrides [11] and Lewis acids [12] are employed to cure the epoxies. The properties of the resulting epoxide resins (**Figure 5**) depend upon the epoxy, the curing agent, and the curing process. For example, when curing agents such as aromatic amines [13], acid anhydrides [14], and BF_3MEA [15] are used with proper epoxy resins and fully cured, the epoxies attain high glass transition temperature (T_g) and are thus suitable for high-temperature applications.

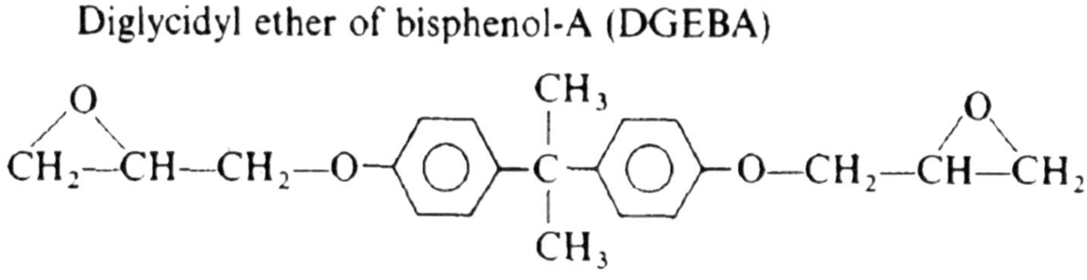

Figure 5. Chemical structure of DGEBA, an epoxy resin [16].

The cure process of an epoxy resin leads to the formation of a network consisting of epoxy molecules cross-linked to the reactive groups of the curing agent following the chemical stoichiometry of the compound. The mechanical properties of the cured resin are strictly dependent on the level of cross-linking achieved during the processing stage; in fact, a lower cross-link density will reduce the glass transition temperature and the storage modulus improving the fracture toughness allowing greater elongation before failure. In the case of fully cross-linked system, the higher complexity of the formed network structure will induce an increase of the T_g with a lowering of the maximum deformation allowed before failure.

The topic of improving the mechanical properties of polymers by the addition of second-phase fillers has received a great attention in the scientific and industrial community throughout the last few decades [16]. Rubber and inorganic particles are the two main classes of fillers used to increase the fracture toughness of epoxy resins. Over the decades, starting from these two classes, new filler typologies were developed, and very recently nanomodification has become a suitable tool to optimize properties of polymers by designing their structure at a nanoscale. That has led to considerable improvements of mechanical properties of polymers, especially in terms of stiffness, strength, and toughness [17, 18].

The fracture behavior of toughened polymers may involve several mechanisms, each one contributing toward the total fracture toughness of the material. Such possible mechanisms are schematically depicted in **Figures 6** and **7**, and are the following: (1) shear band formation, (2) fracture of rubber particles, (3) stretching, (4) debonding and (5) tearing of rubber particles, (6) transparticle fracture, (7) debonding of hard particles, (8) crack deflection by hard particles, (9) cavitated or voided rubber particles, (10) crazing, (11) craze tip plastic deformation, (12) diffuse shear yielding, (13) shear band/craze interaction, and (14) pinning of crack front (**Figure 6**).

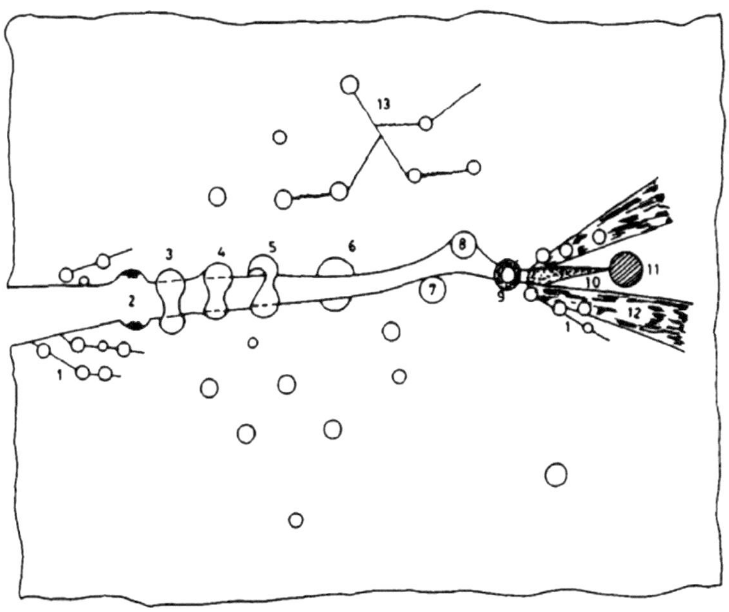

Figure 6. Toughening mechanisms in particle filled epoxy adhesive: (1) shear band formation, (2) fracture of rubber particles, (3) stretching, (4) debonding and (5) tearing of rubber particles, (6) transparticle fracture, (7) debonding of hard particles, (8) crack deflection by hard particles, (9) cavitated rubber particles, (10) crazing, (11) craze tip plastic deformation, (12) diffuse shear yielding, and (13) shear band/craze interaction [19].

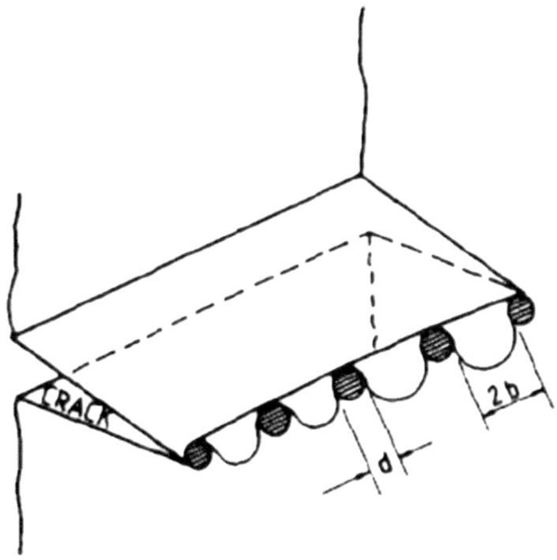

Figure 7. Crack pinning mechanism. The bowed crack front is at the verge of breaking away from pinning [19].

Energy dissipation mechanisms can occur in a toughened polymer at the same time, depending upon the nature of particles and matrix. The modeling of absorbed energy associated with each mechanism represents an important issue in the designing as well as in the verification stage. The main relationship reported in the current literature is reported below.

2.1. Rubber particle mechanisms: stretching, tearing, and debonding

Kunz-Douglass et al. [20] have developed a simplified model to estimate the energy dissipation (ΔG_{IC}) associated with the stretching, tearing, and debonding of rubber particles. According to this approach, rubber particles store elastic energy, equal to ΔG_{IC}, which is irreversibly dissipated (e.g., as heat) when the particle fails either by debonding from the matrix or by tearing, and is given by:

$$\Delta G_{IC} = 4\gamma V_p[1 - \frac{6}{(\lambda^2 + \lambda + 4)}] \quad (5)$$

where γ is either the specific energy of interface required to debond rubber or the rubber tearing energy, V_p is the rubber particles volume fraction, and λ is either the extension ratio at the time of debonding or rubber tearing.

2.2. Transparticle fracture

Considering an elastic particle of mean radius r, its specific fracture energy γ_p is given by:

$$\gamma_p = \frac{2r}{E} \eta^2 \sigma_f^2 \quad (6)$$

where η is the stress concentration factor at the particle equator and σ_f is the fracture stress. Considering that the number of particles N of mean radius r, per unit area of surface is:

$$N = \frac{3}{2}\frac{V_p}{\pi r^2} \tag{7}$$

Thus, the increase in toughness due to the fracture of particles is:

$$\Delta G_{IC} = 3\eta^2 r E V_p \varepsilon_f^2 \tag{8}$$

where ε_f ($=\sigma_f/E$) is the fracture strain of the particle. For a uniaxial stress field $\eta = 2$.

2.3. Crack deflection

The crack is diverted by the particle, resulting in an increase of the crack surface area. This causes an increase in fracture energy given by:

$$\Delta G_{IC} = \frac{3\gamma_m V_p}{2} \tag{9}$$

where γ_m is the matrix specific fracture energy. According to this equation, the crack deviation leads to an increase in the fracture surface area, which is equivalent to half the surface area of the particle (i.e., $2\pi r^2$) minus the matrix area in absence of the particle (i.e., πr^2). Generally, the crack does not propagate along the whole particle surface, and then the deviation may be less. Consequently, Eq. (9) could overestimate the fracture toughness increase.

Deflection of cracks is commonly distinguished into:

1. Tilting: in-plane deflection, resulting in a mixed-mode (I/II) crack tip stress state.
2. Twisting: out-of-plane deflection, resulting in a mixed-mode (I/III) crack tip stress state.

Slender rods or fibers with high aspect ratios are more effective than disc-shaped particles or spheres in deflecting and twisting the propagating crack and hence in toughening [21, 22].

2.4. Crack-pinning

The obstacles (hard filler particles) create the obstruction to the propagation of the crack front and cause an increase in toughness by bowing out the crack front between the particles (**Figure 7**). Lange [23] has given a relation for the increase in fracture energy due to pinning as:

http://dx.doi.org/10.5772/65250

$$\Delta G_{IC} = \frac{T}{2b} \tag{10}$$

where, T is the line energy of the crack front and $2b$ is the interparticle spacing. The interparticle spacing can be obtained from:

$$b = d\frac{(1-V_p)}{3V_p} \tag{11}$$

where $d = 2r$. For a penny-shaped crack [24], Lange showed that the line energy is:

$$T = \frac{2r}{3}\gamma_m \tag{12}$$

Thus,

$$\Delta G_{IC} = \frac{r\gamma_m}{3b} \tag{13}$$

This equation predicts a linear relationship between the increase in fracture energy and the ratio r/b. However, this is not observed in practice except in the case of a glass-filled alumina [25].

The increase in G_{IC} required to bow out the crack front was then calculated by Evans assuming that the derivative of strain energy with respect to the interparticle spacing is equal to the increase in toughness. This model was confirmed by the experimental results of Moloney et al. [26] relative to epoxy resin filled with alumina and silica particles at low volume fractions. The agreement between theory and experimental results fails when high volume fractions are considered, likely because the theory assumes a semicircular "bow" crack front, while in these conditions it is semielliptical [26].

2.5. Crack tip yielding

The cited mechanisms can induce yielding and flow of the polymer near the crack tip. This effect hinders the crack propagation phenomenon, increasing the fracture toughness of the resin. It is well known that a polymer is characterized by covalent bonding along the polymeric chains, but chains can interact among themselves through weak van der Waals forces. When a local stress is able to overcome these forces, a narrow gap is formed. This forming gap, called craze, is not able to grow further, because it is held together by the covalent bonding of the chains. Crazes scatter the light due to their microscopic dimensions. The polymeric chains that bridge the craze are grouped to form fine filaments called fibrils, visible only with an electron microscope (see **Figure 8**).

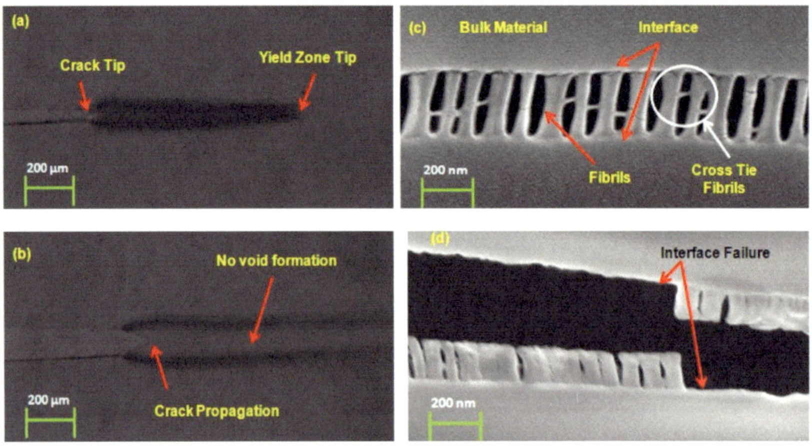

Figure 8. Shear yielding in 0.5 mm thick sheet of polycarbonate (a) before crack propagation, (b) after crack propagation and crazing in 800 nm thick polystyrene film, (c) fibril formation, and (d) fibril breakage [27].

3. Toughening mechanisms in modified epoxy adhesives

All possible failure mechanisms for filled polymer, has been reported in Section 2. Which of the cited mechanisms occur in the polymer depends on different factors: filler nature, matrix typology, curing agent, temperature, and load rate [19]. In the following paragraphs, the plausible mechanisms that can occur in epoxy adhesives filled with different typologies of filler are reported and discussed.

3.1. Rubber-toughened epoxies

The toughening of thermoplastics by rubber has been in existence for over 70 years but the addition of rubber to thermoset resins (epoxies) is only about 45 years old. The work by Sultan et al. [28] showed that worthwhile improvements could be effected by adding certain liquid rubbers to epoxide formulations.

Carboxyl-terminated butadiene-acrylonitrile rubber (CTBN) (**Figure 9**) is widely employed as a toughening agent in epoxy adhesives because it satisfies both the requirements needed to maximize the toughening effect [19, 29, 30]. The first requirement is the compatibility with the matrix, i.e., the liquid rubber must be soluble in the uncured matrix, but must precipitate out during the curing process. The second requirement is the reactivity between epoxy group and rubber chains.

http://dx.doi.org/10.5772/65250

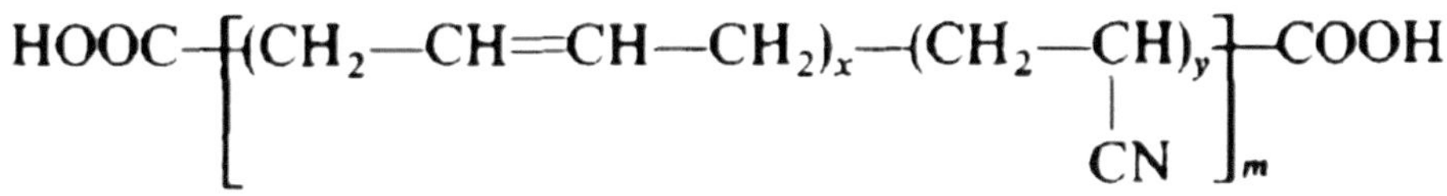

Figure 9. Chemical formula of CTBN rubber [19].

The addition of rubber to a brittle resin modifies its various characteristics. For example, it causes a reduction in stiffness, lowers T_g, plasticizes the matrix, reduces the yield strength, and increases the linear thermal expansion coefficient [31]. However, it also significantly increases the fracture resistance [32]. Because of such a gain in toughness, rubber-toughened epoxies are being used for various engineering applications.

Generally, the fracture energy, G_{IC}, increases with increasing elastomer concentration up to about 16–17% and then decreases (**Figure 10**). Such behavior is seen because of the change of the elastomer from a dispersion phase to a blend at high CTBN concentration. Using this filler typology, the maximum fracture energy of the rubber-modified epoxy can also be 30 times that of the unmodified epoxy [33].

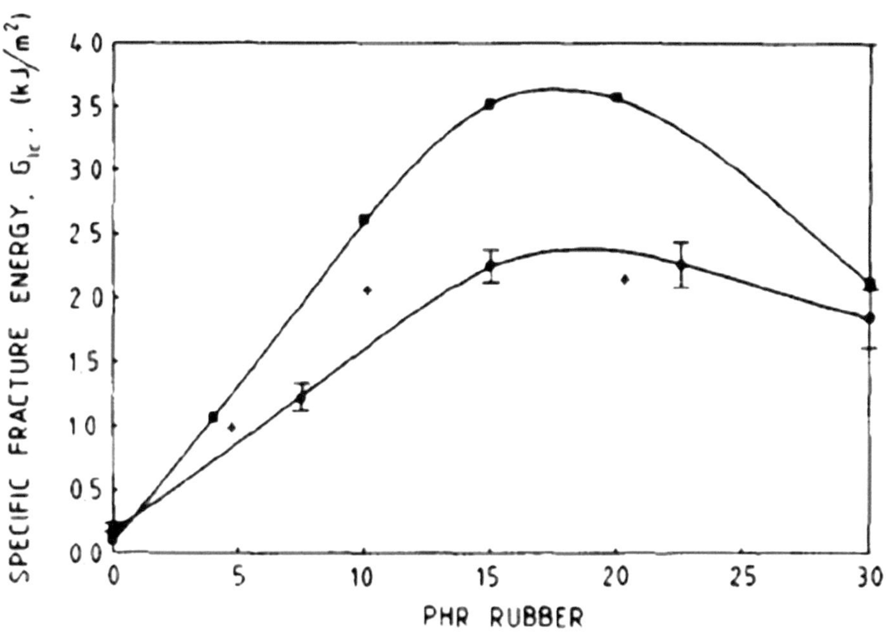

Figure 10. Variation of specific fracture energy, G_{IC}, with rubber content. Different symbols are relative to different works [19].

Besides the reactive rubbers, several insoluble, unreactive rubber modifiers, such as core-shell rubbers (CSR) and dispersed acrylic rubbers (DAR) were used to modify the epoxy systems [34]. Dispersion of epoxy insoluble rubbers has resulted particularly useful in many adhesive and composite applications as the dispersed phase volume is not remarkably affected by curing condition variations [35], and above a certain level of content, CTBN rubbers do not affect glass transition temperature of the final system. Moreover, the modulus of these modified epoxy systems can be varied independently of T_g.

Many theories have been proposed to explain the toughening mechanisms occurring in rubber-filled epoxy resins mainly based on the mechanisms identified in rubber-toughened thermoplastics. These theories involve deformation mechanisms such as shear yielding, crazing, and interaction among crazes and shear bands to explain toughening effect. According to these theories, rubber particles are considered initiation sites for crazes and shear bands and they also act as obstructions for the growing crack, reducing the intrinsic flaw size. The effect of rubber particles on crazes and crack propagation leading to reduced flaw size implies an increase in fracture toughness of the final system [10, 36].

The remarkable increase in volume during tensile creep tests on rubber-filled epoxy resins suggests the occurrence of a massive crazing of the specimen. It is noteworthy that this approach is verified for thermoplastic materials, but other contributions are to be considered to explain the toughening mechanisms in epoxide resins.

Stretching and tearing of the rubber particles can be considered as a different toughening mechanism. This mechanism is proposed by Kunz-Douglass et al. [20] by following the propagation of a crack using an optical microscope, have found that stretching and tearing of particles could be considered as the major source of energy absorption also in toughened epoxies. In their model, contribution of shear yielding to the fracture toughening was completely ignored as they did not notice this contribution neither in unmodified or modified epoxy.

Detailed microscopic analysis of the stress whitening region in fractured rubber-toughened epoxy resin showed that shear yielding cannot be totally neglected. In fact, these surfaces appear to be characterized by many small holes larger than the original particle size and yielding of the matrix (observed as tear marking) near the particles. The holes are the consequence of the dilatation of the original rubber particles, while the shear yielding is caused by the intense tri-axial stress state around the particles, which induce a noteworthy deformation into the matrix.

A similar model was proposed by Yee and Pearson [37], to account for the increased toughness of the rubber-filled epoxies. According these authors, the large dissipation of energy near the crack tip is responsible for the increased crack growth resistance and this deformation involves mainly two processes: (1) the cavitation in the rubber particles and in the surrounding matrix and (2) the plastic shear yielding in the epoxy matrix. According to this model, shear yielding is one of the main source of energy absorption for rubber-toughened epoxy resin, which causes a significant tip deformation that induce a remarkably reduction of the local stress concentration enhancing the fracture toughness of the final system.

Another remarkable mechanism in rubber-modified epoxy resins is the voiding. The presence of the rubber particles ahead the crack tip induces a tri-axial stress field when tensile load is applied to a notched specimen that causes a high hydrostatic tension around particles. The consequent dilatation promotes their cavitation and the enlargement of the resultant voids

(**Figure 11**). These voids formed are further enhanced by thermal stresses induced during the cooling from the curing to ambient temperature.

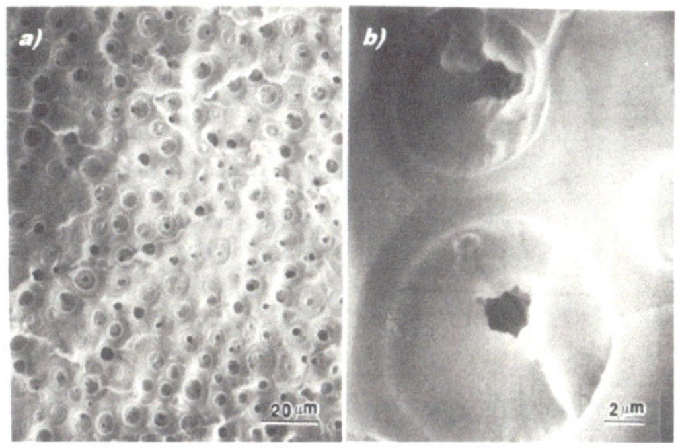

Figure 11. Fracture surface of a rubber-toughened epoxy resin: (a) low and (b) high magnification [38].

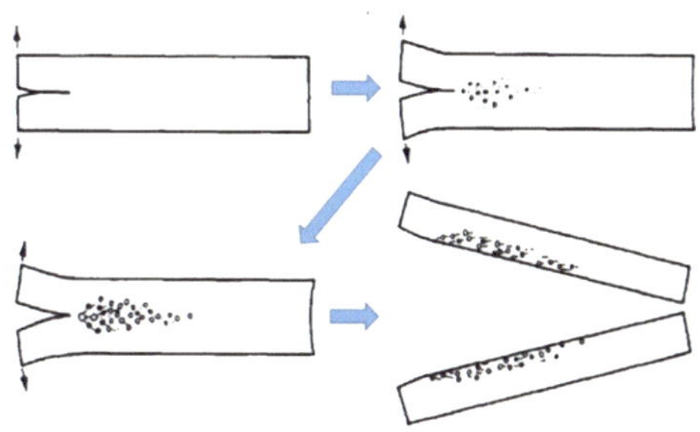

Figure 12. Schematic representation of the deformation processes ahead of a crack tip [37].

In the work by Yee et al. [37] two epoxide resin systems toughened with CTBN elastomers were studied by uniaxial tensile dilatometry to investigate the involved fracture mechanisms. Results have shown that the presence of the rubber particles has a twofold effect: (1) inducing shear flow at lower strain rates and (2) promoting cavitation at higher strain rates. In both cases, there is the formation of a shear yielded region ahead the crack tip, which hints the crack propagation. By analyzing tensile dilatometry results, it was showed that the formation of a region characterized by voids and shear bands ahead the crack tip occurs. This region blunts the crack and this implies that the crack tip behaves as it was larger than reality, increasing the fracture toughness level. On increasing the load, voids continue to grow and the region characterized by shear band extends its dimensions. Moreover, the crack propagating through the voided plane increases the fracture path, and then the energy is dissipated. The fracture process described above is schematically depicted in **Figure 12**. As shown in **Figure 11**, the matrix surrounding the voids exhibits a notable plastic deformation as evidenced by the beveled edges of the voids. The plastic deformation becomes larger as the particle-particle distance decreases; this is an evidence for particle-particle interaction.

The interaction between particles can also be seen by comparing the diameters of the fracture surface cavities (SEM) with the diameters of the undeformed rubber particles at various levels of rubber content. It appears clear that many cavities are larger than the corresponding rubber particles with the same filler content, indicating a strong interaction among near particles. At high rubber content, the expansion in the diameter from the undeformed to the fracture cavity becomes more relevant due to the increase of the particle interaction [38].

The critical strain-energy release rate, G_{IC}, is plotted against the rubber content for three rubber-toughened epoxies as given in **Figure 13**. These results show that fracture toughness is a strong function of rubber content, but not significantly dependent on the rubber particle size. The Hycar CTBN 1300X8 modified epoxies (828-8), containing ~1 μm diameter rubber particles, are not significantly tougher than the Hycar CTBN 1300X15 modified epoxies (828-15), which have ~5–10 μm diameter rubber particles [37].

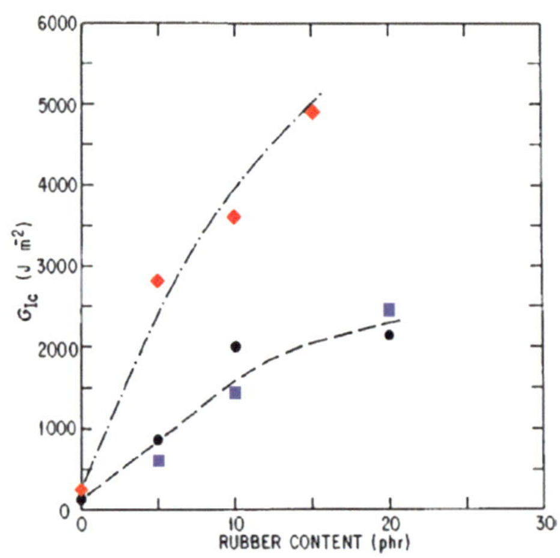

Figure 13. The strain-energy release rate G_{IC} as functions of rubber content, for three types of modifier: (•) 828-8, (■) 828-15, (♦) 828-BPA-8 [37].

In summary, in rubber-toughened epoxy resin, the main fracture mechanism recognized are the following: shear band formation near rubber particles, fracture of rubber particles after cavitation, stretching, debonding and tearing of rubber particles, voiding, crazing, diffuse shear yielding, and shear band/craze interaction.

3.2. Particle-filled epoxies

Over the decades, besides using rubber as a toughening agent for polymers, several inorganic fillers such as alumina [39], silica [40], glass beads [41], and aluminum hydroxide [42] have been considered. Generally, the preparation of inorganic particle-filled resins is relatively simple: filler and resin are mixed and cured, imposing vacuum before the cure process to allow the spilling out of the trapped air bubbles. To obtain polymers with high fracture toughness, high modulus, and T_g, rubber particles are often added to the inorganic particles/epoxy resin system [43, 44].

Inorganic particles are able to provide high physical and mechanical properties to the hosting matrix and therefore they could be an ideal candidate in the case of advanced engineering applications. These fillers have many advantages, such as low cost, reduce matrix shrinkage and polymerization heat during the cure process, and increase the elastic modulus, T_g, and thermal conductivity [10, 23, 39, 40]. Also fracture toughness is affected by the presence of the particle filler: in fact it is generally higher than those of pure epoxy resin [40, 41]. The entity of the gain in mechanical properties depends on different factors: chemical nature of the filler, particle size, filler content, functionalization of the particles surface, and strain rate. Many works on fracture toughness of the particle-filled epoxy during fast [45–47] and slow [48] crack propagation have reported an increase of the K_{IC} value of the composite, together with an improvement in Young's modulus and yield stress (see **Table 1**).

Volume fraction, V_p (%)	Mean particle size, d (μm)	K_{IC} (MPa m$^{1/2}$)	E (GPa)	σ_y (Mpa)
40	300	1.76	–	–
	160	1.74	–	–
	100	1.87	–	–
	60	1.83	–	–
0	100	0.60	3.5	100
20	100	1.30	6.1	108
30	100	1.62	7.7	121
40	100	1.87	9.8	133
50	100	2.21	12.5	150
40 (silane-treated)	100	1.90	–	–

Table 1. Epoxy/silica composite with different filler content and surface treatment [19].

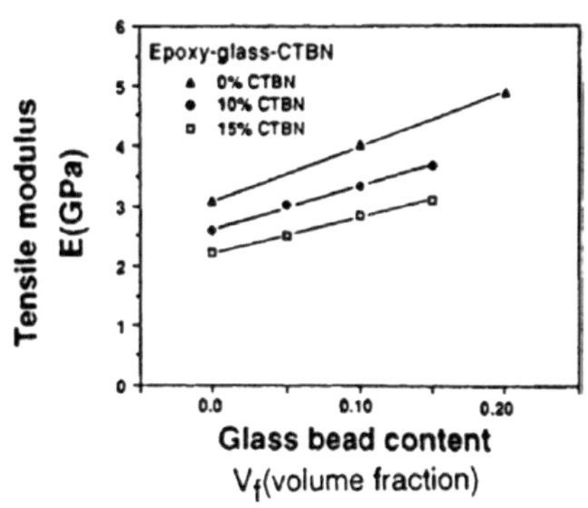

Figure 14. Tensile modulus as a function of glass bead content for different epoxy-CTBN compositions [44].

A simple and effective way to improve fracture energy without loss of strength and elastic modulus is to incorporate inorganic fillers in rubber-toughened epoxies. As reported in **Figure 14**, the presence of the CTBN rubber lowers the mechanical properties (tensile modulus, E), but the addition of small amount of inorganic filler restore the modulus to the initial value. The G_{IC} value for the hybrid glass rubber-modified epoxy is much larger than for the glass-unmodified epoxy for a given temperature and volume fraction of glass [49].

It is well known that the presence of rigid (micro) fillers may induce several toughening mechanisms in epoxy matrices, for example, (1) crack deflection [21, 50], (2) plastic deformation [26, 51], and (3) crack front pinning [36, 52]. The first mechanism is always present, but minimally contributes to the total increase of fracture toughness in particulate-filled epoxy resins. Similar to rubber filled epoxy resins, the presence of inorganic particles induces a triaxial stress state in matrix around it, promoting plastic deformation in the polymeric phase and then crack tip blunting. However, the main source of toughening in particulate-filled epoxy resins is the crack pinning.

This statement is supported by the work of Moloney et al. [26], in which the effect of alumina and silica particles on the fracture toughness of a filled epoxy resin was studied. It is noteworthy that in case of weak filler particles, such as aluminum hydroxide and dolomite [26], crack pinning cannot completely explain the fracture toughening, in fact, up to a certain filler content crack pinning is applicable, but over a critical content (~20%), the crack tends to propagate through particles, promoting a transparticle-fracture mechanism.

As reported by Kinloch et al. [36], who studied a hybrid system consisting of glass beads and rubber particles in an epoxy matrix, crack pinning is the main fracture toughness mechanism in systems operating at low temperature. Fracture can be described by the critical crack opening displacement criterion. At higher temperatures the crack pinning mechanism is overshadowed by the large plastic flow processes at the crack tip.

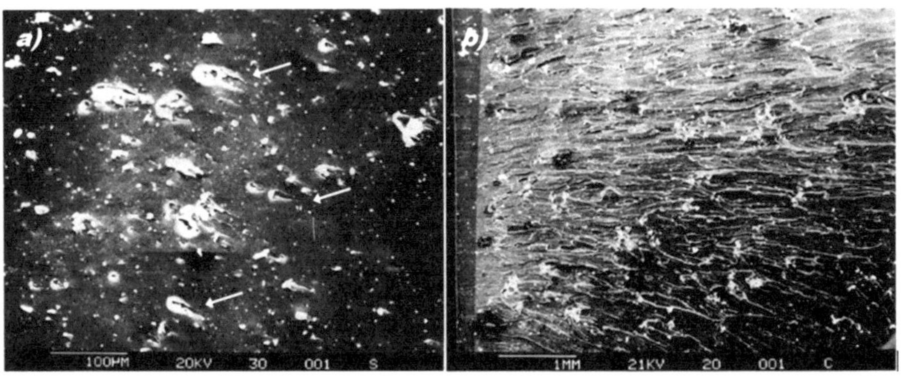

Figure 15. Fracture surfaces for (a) zirconia-epoxy showing crack pinning "tails" (white arrows) and (b) zirconia-rubber-epoxy showing crack initiations and arrests in stress whitened region [53].

The occurrence of crack pinning in zirconia epoxy and zirconia-rubber epoxy systems was also confirmed by the work of Low et al. [52], according to which crack pinning is the main fracture toughening mechanism in zirconia epoxy system, while in the hybrid resin there are evidence of cavitation and shear flow near the crack tip. Crack pinning is clearly indicated by the presence of the characteristic tails (white arrows) as given in **Figure 15**. **Figure 15(b)** shows the

http://dx.doi.org/10.5772/65250

hybrid system fracture surface, which is evident from the presence of many sites of crack initiation and arrest, with consequent remarkable fracture toughness increase (rubber particles debond/tear in correspondence of each initiation site, rising the dissipated energy).

Maxwell et al. [53] developed a hybrid glass-filled epoxy-polymer containing both rubbery and rigid particles. The values of G_{IC} for the unmodified and rubber-modified epoxy materials as a function of volume fraction of glass particles are shown in **Figure 16**. It is interesting to note that in systems without glass bead ($V_f = 0$), the value of G_{IC} continuously increases with the temperature, and in particular, in a rubber-modified system, the rise is of about an order of magnitude. At higher glass beads content, the G_{IC} levels up to a certain value and then decreases; this effect is more evident for the hybrid system. The optimum value of the filler content for the glass-epoxy system is about 0.3 with a G_{IC} increase of about 400%. Conversely, in the hybrid system, the optimum value is dependent on the temperature.

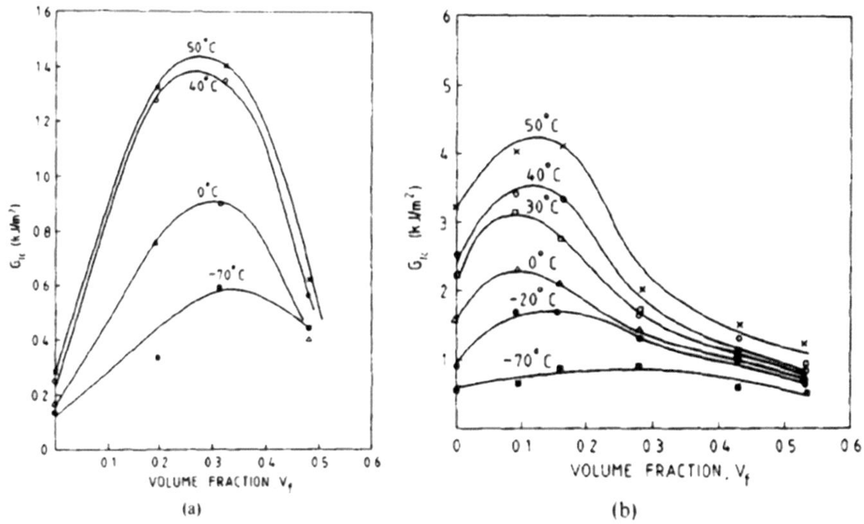

Figure 16. Fracture energy, G_{IC} , against volume fraction, V_f, of glass particles for (a) an unmodified epoxy and (b) a rubber-modified epoxy [49].

SEM observation can help to understand the fracture mechanisms that occur in a filled epoxy adhesive. A typical micrograph of a glass particle-filled resin is shown in **Figure 17(a)**, in which the tails behind the particles typically associated with the crack pinning mechanism are clearly visible. These tails are associated with the bifurcation of the crack front when it encounters the pinning position (i.e., the rigid particle) and to the subsequently meeting again of the two new crack fronts. **Figure 17(b)** shows the micrograph of the hybrid glass/rubber system. It is interesting to note the presence of many small holes, larger than the rubber particles, and a very rough surface, due to the massive shear flow during the crack propagation. The presence of few small tails behind the glass particles indicates a little contribution of the crack pinning to the global fracture toughening mechanism. However, the presence of few crack pinning characteristic tails in the hybrid system was predictable considering the curve of G_{IC} as a function of glass bead content, in fact, the increase of G_{IC} for the presence of inorganic filler is very high (400%) in the rubber-free system, but is minimum (~40%) in the rubber toughened system.

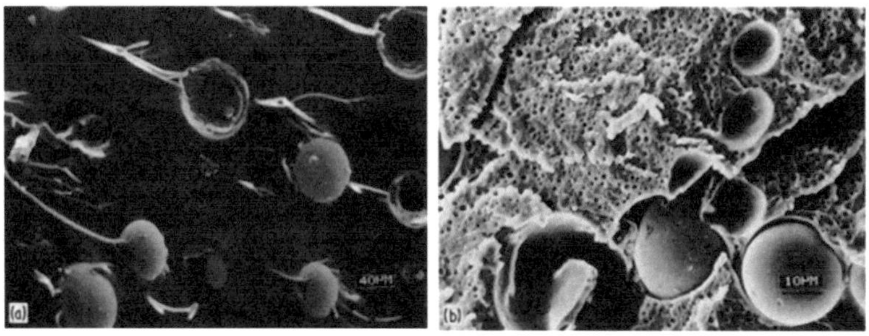

Figure 17. Scanning electron micrographs of fracture surfaces. (a) Glass particle filled unmodified epoxy (arrow indicates direction of crack growth). (b) Hybrid particulate-filled epoxy-polymer [19].

3.3. Thermoplastic-modified epoxies

Thermoplastic particles can be used as toughening agent in brittle epoxy resins, as they promote delocalized microcracking and crack bridging [54]. This typology of particles is extensively used to avoid a decline in modulus of the hosting matrix.

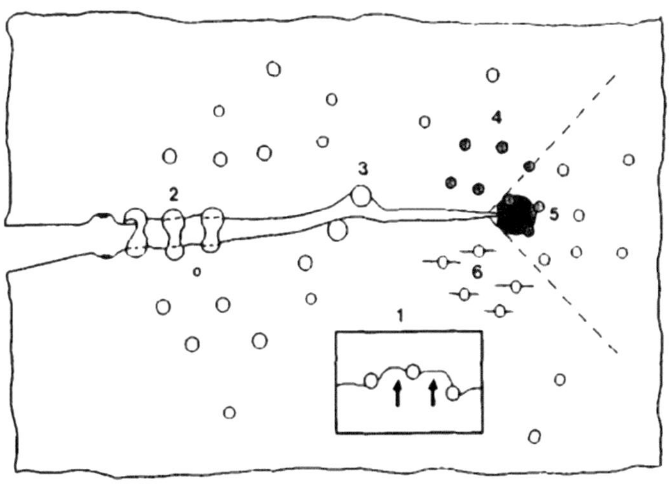

Figure 18. Schematic diagram of toughening mechanisms proposed for thermoplastic-modified epoxies: (1) crack pinning, (2) particle bridging, (3) crack path deflection, (4) particle yielding, (5) particle-yielding induced shear banding, and (6) microcracking [57].

Thermoplastic-modified epoxies are toughened by a mechanism called crack bridging (**Figure 18**) [55, 56]. The increase in K_{IC} observed in this toughened resin is due to the effect of the rigid plastic particles that bridge the two crack surfaces limiting the opening of the crack tip. Crack bridging can also be explained considering that the energy is dissipated during the plastic deformation and consequent tearing of the plastic particles.

The effect of the crack bridging on the morphology of the fracture surfaces is simple to observe; generally, the surface of broken particles appears rich, yielded after a large plastic deformation.

Microscopic evidence for crack bridging can be seen in **Figure 19**, in which highly drawn particles of polyamide-12 [58] in the process zone ahead of the precrack are shown. It can be seen that the particles adhere well to the matrix and are capable of large plastic strains before failure. It is also apparent from the large deformations that the particles failed well after the crack had advanced through the matrix.

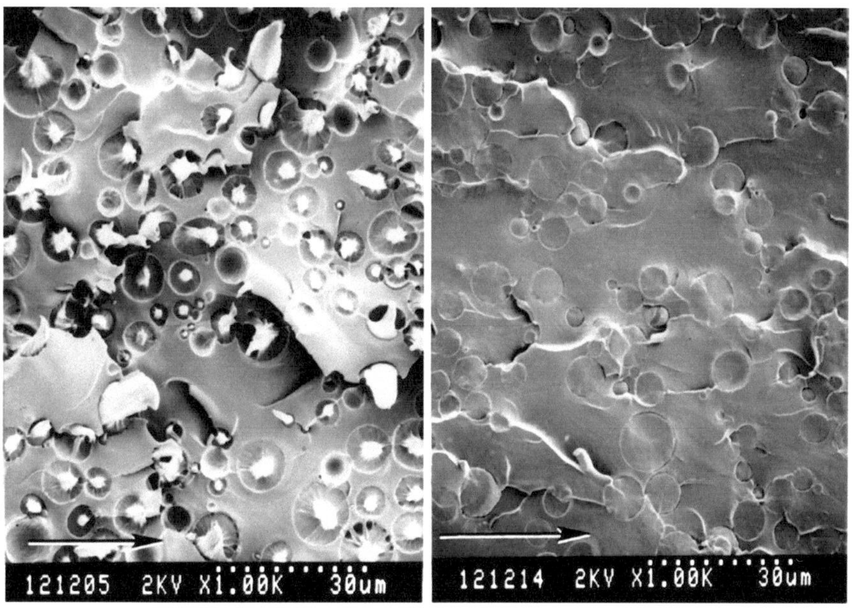

Figure 19. SEM micrograph showing the deformed polyamide-12 particles on the fracture surface a polyamide-12 modified epoxy from within the process zone (arrow indicates direction of crack propagation) [54].

Many works [57, 59] also mention the crack pinning as one of the mechanisms contributing to the fracture toughening effect of thermoplastic particles. These particles behave like impenetrable objects, similar to inorganic particulate filler previously cited. Indeed, the higher brittleness of the epoxy matrix compared to the ductile thermoplastic phase allows the consideration of these particles as relatively impenetrable.

Crack path deflection [58] is another cited mechanism responsible for the toughening of thermoplastic particle-filled epoxy resins. As described previously, in this mechanism, the particles cause the deviation of the crack front from the original path, increasing the fracture surface area and consequently the energy dissipation. Occurring of this toughening mechanism can be verified observing the fracture surfaces with SEM.

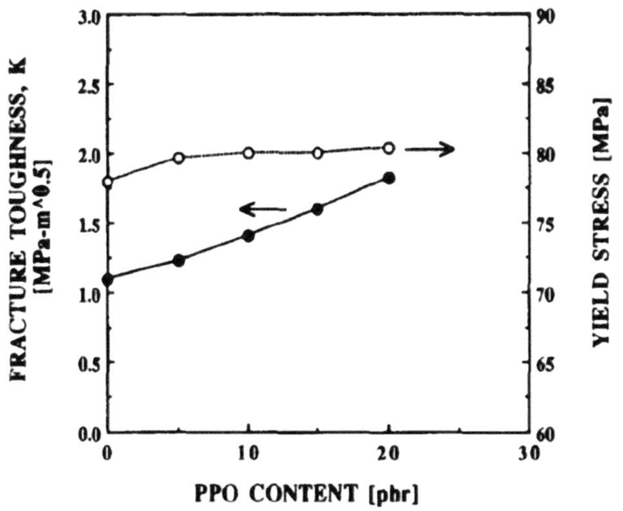

Figure 20. Fracture toughness increases with PPO content while the yield strength remains constant [54].

Thermoplastic particles also induce shear banding in the matrix, similar to rubber particles. In fact, during the yielding of the rigid particles, their modulus reduces up to that of the rubber, and this mismatch in modulus between matrix and yielded particles induces a remarkable stress concentration, initiating shear banding in the surrounding matrix. Pearson et al. [54] have modified an epoxy based on the diglycidyl ether of bisphenol A (DGEBA) with poly(phenylene oxide) (PPO). The fracture toughness of PPO-modified epoxies increases almost linearly with PPO content. These results are presented graphically in **Figure 20**. The simple linear relation between PPO content and fracture toughness is surprising since the morphology undergoes a dramatic change from a particulate morphology to the one that consists of continuous domains. Pearson et al. stated that the toughening mechanism is not massive shear banding as has been found for rubber-modified epoxies; instead, the PPO-modified epoxies are toughened by a microcracking mechanism and branching (**Figure 21**).

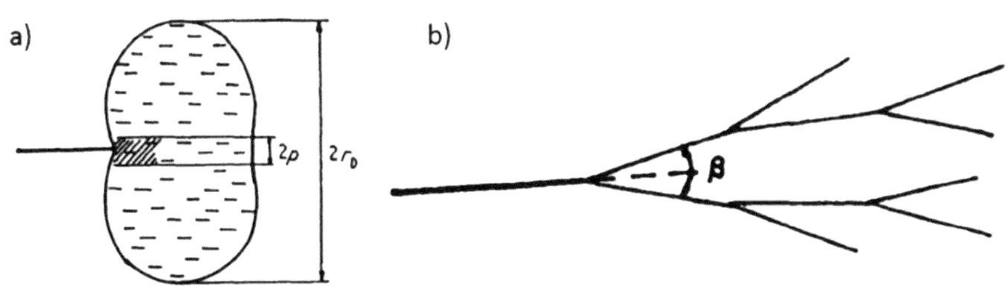

Figure 21. (a) Microcracking and (b) Branching mechanism at the crack tip [54].

According Hodgkin et al. [60], the criteria for achieving good thermoplastic toughening could be listed as follows:

1. Thermoplastic backbone: this should have high thermal stability, should be soluble in the uncured epoxy, but must be phase separated during cure to form a multiphase morphology (as previously stated for the rubber particles).
2. Morphology: increase in optimum toughness is generally achieved through attainment of a cocontinuous or phase inverted morphology.
3. Reactive endgroups: it does not appear necessary to use functionalized endgroups in the toughening phase.
4. Cross-link amount: increasing the crosslinking density of the epoxy resins increases the efficiency of the toughening with thermoplastic particles (conversely to rubbers).
5. Molecular weight: high molecular weights of the employed thermoplastic filler promote the toughening of the matrix.

3.4. Layered nanofiller-modified epoxies

One of the most promising approaches to synthesize materials with barrier properties [61, 62], fire resistance [44], and increase of mechanical properties [10, 44] consists in dispersing an inorganic clay mineral in an organic polymer on a nanometer scale. Crack deflection is likely the major energy dissipation mechanism that occurs in nanoclay-filled polymers, where 1 nm thick clay platelets often tend to be stacked into microsized tactoids, readily able to perturb the growing crack front. Moreover, also debonding of nanoclay (often regarded as nanoclay delamination) and matrix shear banding contribute to the overall fracture toughness improvements [62].

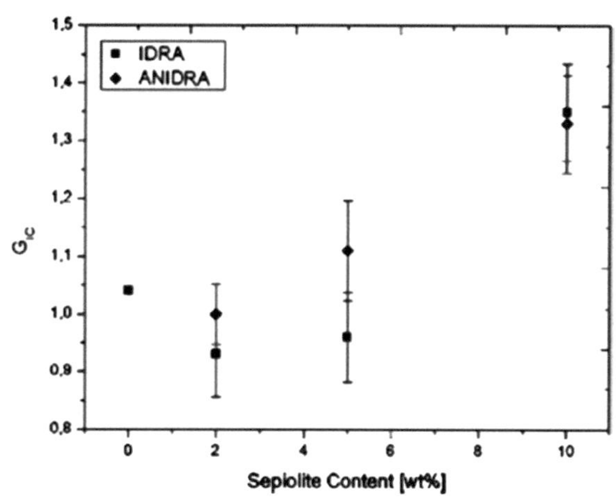

Figure 22. Critical strain energy release rate, G_{IC} , versus sepiolite content [10].

Evidence of the cited mechanisms can be seen in the work by Zotti et al. [10]. In this work the addition of sepiolite clay induces an increase of the strain energy release rate up to 30% (**Figure 22**). These authors pointed out that sepiolite clay particles might act as stress concentration sites promoting clay-matrix debonding (orange arrow), which produces micro- and nanovoids and matrix shear yielding in the regions close to the crack tip (**Figure 23**). The high roughness of the fracture surface indicates the presence of a crack deflection mechanism.

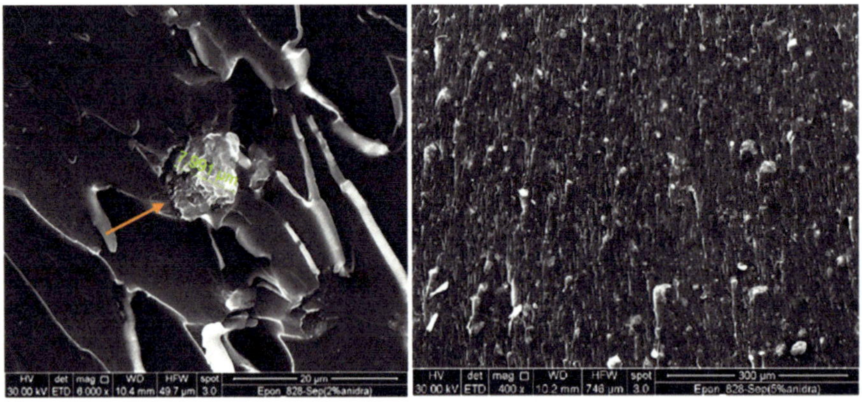

Figure 23. Fracture surface of a sepiolite/epoxy; (left) high and (right) low magnification [10].

To the group of the layered nanofiller belong graphene and its derivatives. This innovative material is constituted by a single sheet of sp^2 hybridized carbon atoms and therefore shows a honeycomb structure. This bidimensional structure shows excellent thermal and electrical conductivity, similar to carbon nanotubes (CNTs). Graphene also shares with CNTs a very high elastic modulus [55, 63]. Similar to clays, graphene is characterized by a layered structure, but owns higher mechanical properties thanks to the carbon back bone.

Graphene is able to remarkably increase the fracture toughness of an epoxy resin also with low content: in fact in the work of Raffiee et al. [56] a 65% increment in fracture toughness for 0.125 wt% of filler amount was reported. Fracture surface analysis indicates a doubled surface area for the graphene/epoxy system compared to the pure epoxy. This strong increase in the fracture surface area is indicative of a crack deflection mechanism. However, the fracture toughness contribution of this mechanism reduces increasing the filler content.

Chandrasekaran et al. [64] report similar observations in their work, where graphite nanoplatelets (GNPs) and thermally reduced graphene oxide (TRGO) were dispersed in epoxy matrix.

The fractured surface of graphene/epoxy nanocomposite, observable with SEM, shows higher roughness than the pure epoxy as presented in **Figure 24**. This again indicates that crack deflection is one of the toughening mechanisms in these nanocomposites. The difference between the two graphene types GNPs and TRGO is that, TRGO is produced via chemical exfoliation and hence contains some remnant oxide functionalities and a fewer number of layers. These factors improve the graphene dispersion in the epoxy matrix compared to the GNP/epoxy system. In fact, GNPs are prepared by metal intercalation of graphite and have less oxide functionalities and more number of layers compared with TRGO. The better dispersion explains the higher surface roughness for the TRGO/epoxy composite when compared with GNP/epoxy (**Figure 24**).

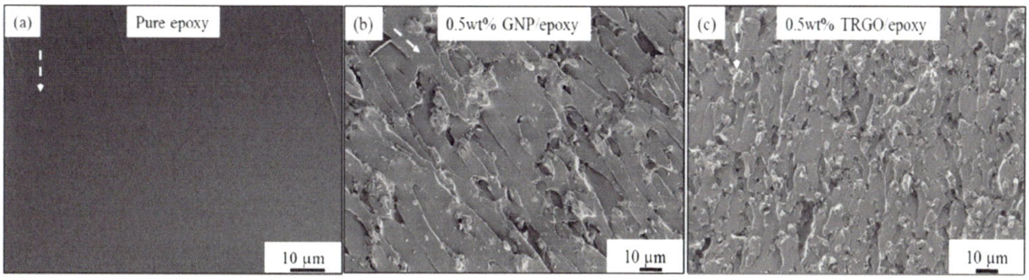

Figure 24. Increase in surface roughness of fracture surface (b and c) graphene/epoxy nanocomposite compared with (a) pure epoxy (white dotted line arrow indicate crack propagation direction) [65].

A 2D schematic of crack bifurcation and propagation at different heights showing the interaction of the crack front with the TRGO/GNP particle is shown in **Figure 25**. During the interaction between propagating crack front and graphene nanoplatelet there are two possible situations that can happen: (1) the crack front is deflected, taking a torturous path or (2) the crack front bifurcates around the platelet (**Figure 25a** and **b**) [65].

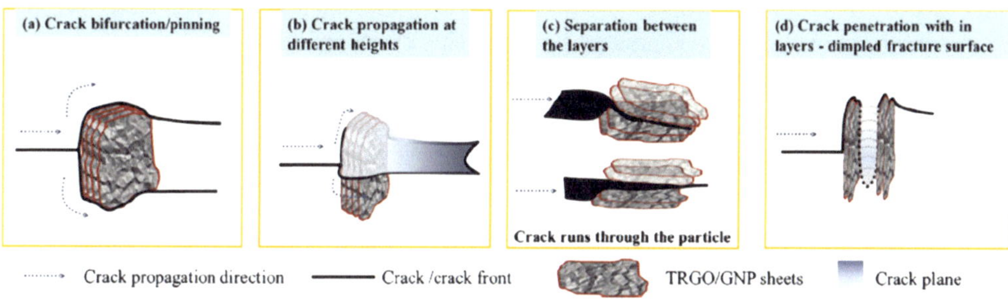

Figure 25. Schematic on the interaction of crack front with GNP/TRGO particles (white dotted line arrow indicates crack propagation direction): (a) crack bifurcation/pinning, (b) crack propagation at different heights, (c) separation between layers and (d) crack penetration with in layers - dimpled fracture surface [64, 65].

In **Figure 26**, the fracture surface of both the sides (for the ease of comparison, side B was electronically tilted by 180°) of the specimen is presented. A feature that was observed in the fracture surfaces is that there are places where there are no flow patterns as pointed out by the solid arrows in **Figure 26**. This fact indicates that this face is not the fracture surface of epoxy resin but the surface of TRGO/GNP particle. In this case, the crack runs along the surface of TRGO or GNP/epoxy interface. This failure mechanism occurs when the crack meets the edges of the graphene sheets. The crack continues to propagate in between the graphene sheets and finally passes through by separating the layers. Since the force between the sheets is a secondary force (van der Waals), the separation of sheets is facilitated with easy crack propagation (**Figure 25c**).

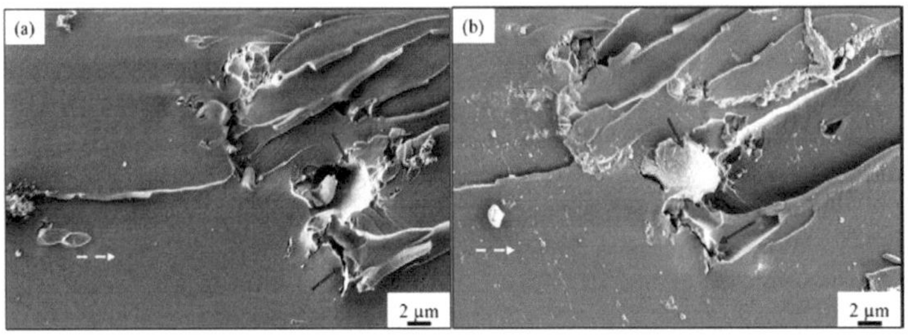

Figure 26. Scanning electron micrograph of both the sides of fracture surface for 0.1 wt% TRGO/epoxy (a) Side A and (b) Side B showing separation in-between the graphene layers that are oriented parallel and perpendicular to the propagating crack [64, 65].

Combination of crack deflection and separation between the graphene layers is another mechanism that can occur in graphene-toughened epoxy resins (**Figure 24d**). When the crack front is deflected by the graphene platelet, it goes around the particle until it meets the graphene edges. In some cases, the crack front can run between the graphene sheets, producing a characteristic "dimple type" fracture surface [65].

On increasing the graphene content, the surface roughness remarkably increases, as reported in **Figure 27**. It is interesting to note that at lower filler content (**Figure 25a**), bifurcation and pull-out are the main failure mechanisms. By increasing the filler weight percent, crack pinning and crack deflection became the predominant failure modes (**Figure 25b** and **c**). Finally, higher filler content promotes the separation between graphene sheets, reducing the surface roughness and consequently the K_{IC} value (**Figure 25d**).

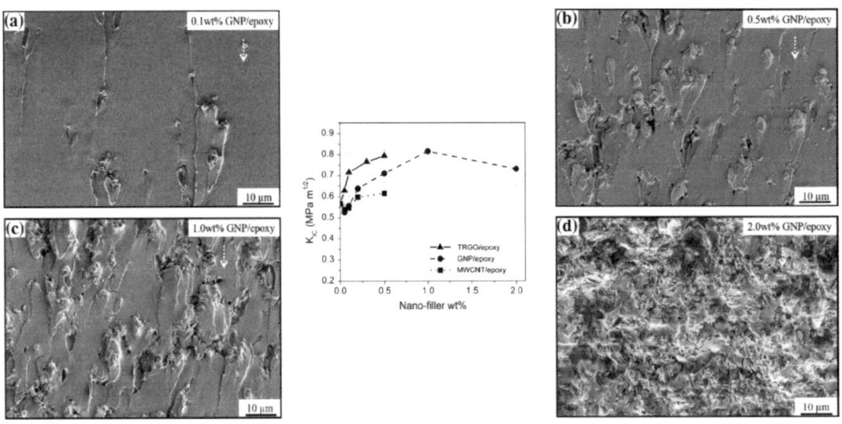

Figure 27. Fracture toughness as a function of weight percent of filler and corresponding SEM micrographs of the fracture surface of GNP/epoxy: (a) 0.1 wt%; (b) 0.5 wt%; (c) 1.0 wt%, and (d) 2.0 wt% [65].

http://dx.doi.org/10.5772/65250

3.5. Hyperbranched-modified epoxies

Hyperbranched polymers are characterized by different properties that stimulate their use as toughening agent in epoxy resins such as low melt and solution viscosities, large number of functional groups in the peripheral region, and small dimensions compared to polymer of equal molecular weight [66, 67]. Hyperbranched polymers have also been explored as novel, high potential, and low viscosity tougheners for epoxy resins that increase fracture toughness properties without deleterious effects on other properties [68].

SEM observations of fracture surface, performed by Fu et al. [69], showed cavitations at center and fibrous yielding phenomenon at edges, which indicated that the particle cavitations, shear yield deformation, and *in situ* toughness mechanism are the main toughening mechanisms. In this work, Fu et al. have synthesized an epoxy-terminated low viscosity liquid thermosetting aromatic polyester hyperbranched epoxy resin (named as HTTE) for the modification of DGEBA epoxy resins. The SEM of the *XY* plane (central region of the sample—**Figure 28a**) shows "sea-island" structures, "dimple-like" structures, and macroscopically sized stress-whitened zones surrounding the crack, which are generally the characteristics of rubber-modified epoxy resins and are called "toughness dimples." The rubbery HBP particles cavitating generate voids, which scatter the light and induce the stress whitening phenomenon. As previously reported, cavitation is an important source of energy dissipation in rubber toughened epoxies, since cavitation absorbs energy inducing shear yielding in the surrounding matrix (the last mechanism is favored at high filler content, where the surroundings of the particles can efficiently interact with that of the others).

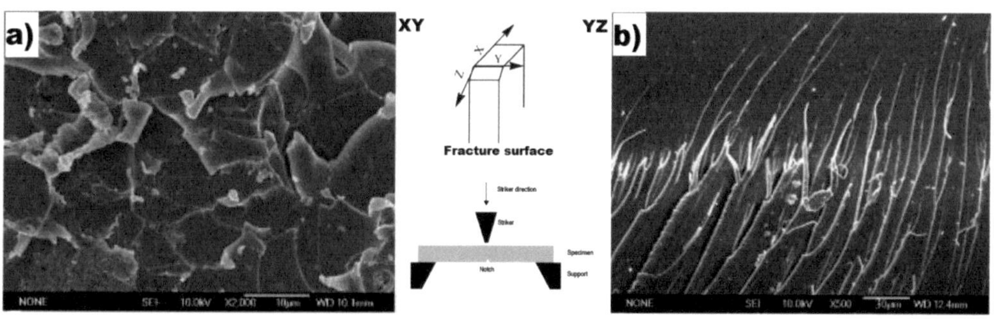

Figure 28. SEM micrographs of the fracture surface in the (a) *XY* and (b) *YZ* plane of the HTTE/Epoxy blends [69].

Figure 28(b) reports the SEM micrograph relative to the *YZ* plane (the region where the crack growth terminated). In this micrograph, the ductile yielding and the consequent fibrillation are evident, which indicate the transition between brittle and ductile states. This transition is in agreement with the large increases of impact strength. The fibril formation mechanism probably consists of the following: (1) voids generated by the debonding between HBP particles and matrix, induce the formation of crazes; (2) after initiation of crazes and voids, the epoxy matrix is elongated under a high tensile stress at the crack-tip region, producing the characteristic fibrils.

Also, Varley et al. [70] demonstrated that the toughening mechanism operating in the HBP/ epoxy system is based upon a particle cavitation process. **Figure 29(a)** shows the substantial improvement in fracture toughness (in terms of K_{IC}) with increasing HBP addition.

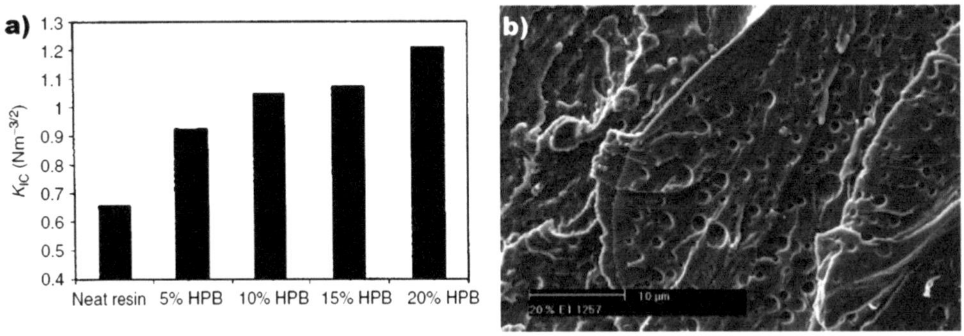

Figure 29. (a) Increase in fracture toughness of the cured resin samples as a function of increasing HBP content and (b) SEM of the 20% HBP/epoxy system [70].

It is noteworthy that this result has been obtained with some decrease in both the T_g and flexural properties. This behavior can be explained in terms of the phase-separated morphology evident from scanning electron micrograph (**Figure 29b**). The size of the HBP particles (~1–5 μm) is of the order suitable to induce toughening mechanisms such as particle cavitation. Interestingly, at the higher levels of HBP concentration the epoxy matrix shows considerable evidence of ductile flow in the continuous epoxy network.

4. Concluding remark

Epoxy adhesives are widely employed in structural applications and the addition of filler is a common method to overcome the brittle nature of this class of polymers. The aim of this contribution was to identify the different mechanisms responsible of the fracture toughening in filled epoxy resins, with special attention to innovative fillers such as graphene nanoplatelets and hyperbranched polymers.

We have reported a classification of the different toughening mechanisms that can occur in a filled epoxy adhesive, based on the filler characteristics. Each filler contributes to the matrix toughening by different mechanisms. In most of the cases, the main occurring mechanisms for filled epoxy systems are crack deflection, matrix deformation, and crack pinning. As the contribution to fracture toughening for each mechanism is indivisible to that of the others, it is not possible to identify the fracture mechanisms only on the basis of the fracture toughness variation. For this purpose, electron microscopy is extensively employed, which allows the identification of the typical features of the searched mechanism in the fracture surface of the toughened epoxy resin. For example, in rubber filled epoxy resins the SEM analysis highlights the presence of small holes, generally larger than the original particles, due to the filler and

http://dx.doi.org/10.5772/65250

matrix cavitation followed by the growth of these voids (cavitation and voiding mechanisms). Crack pinning, common in rigid and thermoplastic particle-toughened epoxy resin, is characterized by tails beyond the particles, caused by the bifurcation of the crack propagation path in the presence of the filler surface, and subsequent meeting of the previously separated crack fronts.

The knowledge of the fracture toughness mechanisms that can occur in toughened epoxy adhesives (and, in general, epoxy resins) allows the development of novel materials able to resist extreme loading conditions, increasing the reliability of the designed final element.

Acknowledgements

This study was financially supported from European Project: EXTREME dynamic loading—pushing the boundaries of aerospace composite material structures—GA 636549.

Author details

Aldobenedetto Zotti, Simona Zuppolini, Mauro Zarrelli* and Anna Borriello

*Address all correspondence to: mauro.zarrelli@cnr.it

Institute of Polymers, Composites and Biomaterials, National Research Council of Italy, Portici, Naples, Italy

References

[1] Bauer SR, editor. Epoxy Resin Chemistry. Washington, DC: American Chemical Society; 1979. DOI: 10.1021/bk-1979-0114

[2] Cho YS, Lee HK, Shim MJ, Kim SW. Characteristics of polymer insulator materials: voltage-lifetime characteristics of DGEBA/MDA/SN system. Materials Chemistry and Physics. 2000;66(1):70–76. DOI: 10.1016/S0254-0584(00)00272-8

[3] Kinloch AJ. Toughening epoxy adhesives to meet today's challenges. MRS Bulletin. 2003;28(6):445–448. DOI: 10.1557/mrs2003.126

[4] Quan D, Ivankovic A. Effect of core-shell rubber (CSR) nano-particles on mechanical properties and fracture toughness of an epoxy polymer. Polymer. 2015;66(1):16–28. DOI: 10.1016/j.polymer.2015.04.002

[5] Gojny FH, Wichmann MHG, Kopke UB, Fiedler K. Carbon nanotubereinforced epoxy-composites: enhanced stiffness and fracture toughness at low nanotube content.

Composites Science and Technology. 2004;64(15):2363–2371. DOI: 10.1016/j.compscitech.2004.04.002

[6] Zotti A, Borriello A, Zuppolini S, Pomogalio AD, Lesnichaya VA, Zarrelli M. Fabrication and characterization of metal-core carbon-shell nanoparticles filling an aeronautical composite matrix. European Polymer Journal. 2015;71:140–151. DOI: 10.1016/j.eurpolymj.2015.07.052

[7] Wang F, Drzal LT, Qin Y, Huang Z. Enhancement of fracture toughness, mechanical and thermal properties of rubber/epoxy composites by incorporation of graphene nanoplatelets. Composites: Part A. 2016;87:10–22. DOI: 10.1016/j.compositesa.2016.04.009

[8] Anderson TL, Anderson TL. Fracture Mechanics: Fundamentals and Applications. 3rd ed. Boca Raton (Florida): Taylor and Francis; 2005. 640 p.

[9] Selby K, Miller LE. Fracture toughness and mechanical behaviour of an epoxy resin. Journal of Materials Science. 1975;10(1):12–24. DOI: 10.1007/BF00541027

[10] Zotti A, Borriello A, Martone A, Antonucci V, Giordano M, Zarrelli M. Effect of sepiolite filler on mechanical behaviour of a bisphenol A-based epoxy system. Composites Part B: Engineering. 2014;67:400–409. DOI: 10.1016/j.compositesb.2014.07.017

[11] Xavier FF, Xavier R, Àngels S. From curing kinetics to network structure: a novel approach to the modeling of the network buildup of epoxy-anhydride thermosets. Journal of Polymer Science Part A: Polymer Chemistry. 2014;52(1):61–75. DOI: 10.1002/pola.26972

[12] Ma H, Spychaj T, Adamus J. Lewis acid type deep eutectic solvents as catalysts for epoxy resin crosslinking. RSC Advances. 2015;5:82813–82821. DOI: 10.1039/C5RA12664A

[13] Oleinik EF. Epoxy-Aromatic Amine Networks in the Glassy State Structure and Properties. In: Dušek K, editor. Epoxy Resins and Composites. 4th ed. Berlin, Heidelberg: Springer; 2005. pp. 49–99. DOI: 10.1007/3-540-16423-5_12

[14] Shechter L, Wynstra J. Glycidyl ether reactions with alcohols, phenols, carboxylic acids and acid anhydrides. Industrial and Engineering Chemistry. 1956;48(1):86–93. DOI: 10.1021/ie50553a028

[15] Tokizawa M, Okada H, Wakabayashi N. Novel cycloaliphatic epoxy resins. II. Curing reaction with BF3MEA and its cured properties. Journal of Applied Polymer Science. 1993;50:875–884. DOI: 10.1002/app.1993.070500515

[16] McClintock RM, Hiza MJ. Epoxy resins as cryogenic structural adhesives. Advances in Cryogenic Engineering. 1960;3:305–315. DOI: 10.1007/978-1-4684-3105-6_34

[17] Azeez AA, Rhee KY, Park SJ, Hui D. Epoxy clay nanocomposites-processing, properties and applications: A review. Composites Part B: Engineering. 2013;45(1):308–320. DOI: 10.1016/j.compositesb.2012.04.012

[18] Jajam KJ, Rahman MM, Hosur MV, Tippur HV. Fracture behavior of epoxy nanocomposites modified with polyol diluent and amino-functionalized multi-walled carbon

nanotubes: a loading rate study. Composites Part A: Applied Science and Manufacturing. 2014;59:57–69. DOI: 10.1016/j.compositesa.2013.12.014

[19] Garg AC, Mai YW. Failure mechanisms in toughened epoxy resins: A review. Composites Science and Technology. 1988;31(3):179–223. DOI: 10.1016/0266-3538(88)90009-7

[20] Kunz-Douglass S, Beaumont PWR, Ashby ME. A model for the toughness of epoxy-rubber particulate composites. Journal of Materials Science. 1980;15(5):1109–1123. DOI: 10.1007/BF00551799

[21] Faber KT, Evans AG. Crack deflection processes—II experiment. Acta Metallurgica. 1983;31(4):577–584. DOI: 10.1016/0001-6160(83)90047-0

[22] Wetzel B, Haupert F, Zhang MQ. Epoxy nanocomposites with high mechanical and tribological performance. Composites Science and Technology. 2003;63(14):2055–2067. DOI: 10.1016/S0266-3538(03)00115-5

[23] Lange FF. Fracture of Brittle Matrix Particulate Composite. In: Broutman LJ, editor. Composite Materials, Vol. 5: Fracture and Fatigue. New York: Accademic Press; 1974. pp. 2–44.

[24] Newman JC, Raju IS. An empirical stress-intensity factor equation for the surface crack. Engineering Fracture Mechanics. 1981;15(1–2):185–192. DOI: 10.1016/0013-7944(81)90116-8

[25] Lange FF. Fracture energy and strength behavior of a sodium borosilicate glass-Al2O3 composite system. Journal of the American Ceramic Society. 1971;54(12):614–620. DOI: 10.1111/j.1151-2916.1971.tb16016.x

[26] Moloney AC, Kausch HH, Stieger HR. The fracture of particulate-filled epoxide resins. Journal of Materials Science. 1983;18(1):208–216. DOI: 10.1007/BF00543827

[27] Measurement of Cohesive Parameters of Crazes in Polystyrene Films (Experimental and Applied Mechanics) [Internet]. Available from: http://what-when-how.com/experimental-and-applied-mechanics/measurement-of-cohesive-parameters-of-crazes-in-polystyrene-films-experimental-and-applied-mechanics/

[28] Sultan JN, McGarry FJ. Effect of rubber particle size on deformation mechanisms in glassy epoxy. Polymer Engineering and Science. 1973;13(1):29–34. DOI: 10.1002/pen.760130105

[29] Ismail H, Galpaya D, Ahmad Z. The compatibilizing effect of epoxy resin (EP) on polypropylene (PP)/recycled acrylonitrile butadiene rubber (NBRr) blends. Polymer Testing. 2009;28(4):363–370. DOI: 10.1016/j.polymertesting.2008.11.007

[30] Pearson RA, Yee AF. Influence of particle size and particle size distribution on toughening mechanisms in rubber-modified epoxies. Journal of Materials Science. 1991;26(14):3828–3844. DOI: 10.1007/BF01184979

[31] Chikhi N, Fellahi S, Bakar M. Modification of epoxy resin using reactive liquid (ATBN) rubber. European Polymer Journal. 2002;38(2):251–264. DOI: 10.1016/S0014-3057(01)00194-X

[32] Liu HY, Wang GT, Mai YW, Zeng Y. On fracture toughness of nano-particle modified epoxy. Composites Part B: Engineering. 2011;42(8):2170–2175. DOI: 10.1016/j.compositesb.2011.05.014

[33] Bascom WD, Cottington RL, Jones RL, Peyser P. The fracture of epoxy- and elastomer-modified epoxy polymers in bulk and as adhesives. Journal of Applied Polymer Science. 1975;19(9):2545–2562. DOI: 10.1002/app.1975.070190917

[34] Sue HJ, Garcia-Meitin EI, Pickelman DM. Literature Review of Epoxy Toughening. In: Arends C, editor. Polymer Toughening. New York: CRC Press; 1996. pp. 154–168.

[35] Arends CB. Design of Tough Epoxy Thermosets. In: Arends CB, editor. Polymer Toughening. New York: CRC Press; 1996. pp. 339–380.

[36] Kinloch AJ, Shaw SJ, Tod DA. Deformation and fracture behaviour of a rubber-toughened epoxy: 1. Microstructure and fracture studies. Polymer. 1983;24(10):1341–1354. DOI: 10.1016/0032-3861(83)90070-8

[37] Yee AF, Pearson RA. Toughening mechanisms in elastomer-modified epoxies: Part 1. Journal of Materials Science. 1986;21(7):2462–2474. DOI: 10.1007/BF01114293

[38] Yee AF, Pearson RA. Toughening mechanisms in elastomer-modified epoxies: Part 2. Journal of Materials Science. 1986;21(7):2475–2488. DOI: 10.1007/BF01114294

[39] Young RJ, Beaumont PWR. Failure of brittle polymers by slow crack growth. Journal of Materials Science. 1977;12(4):684–692. DOI: 10.1007/BF00548158

[40] Lee J, Yee AF. Fracture of glass bead/epoxy composites: on micro-mechanical deformations. Polymer. 2000;41(23):8363–8373. DOI: 10.1016/S0032-3861(00)00187-7

[41] Dillingham RG, Boer FJ. Interphase composition in aluminum/epoxy adhesive joints. The Journal of Adhesion.1987;24(1–2):315–335. DOI: 10.1080/00218468708075434

[42] Tsai JL, Huang BH, Cheng YL. Enhancing fracture toughness of glass/epoxy composites by using rubber particles together with silica nanoparticles. Procedia Engineering. 2011;14:1982–1987. DOI: 10.1016/j.proeng.2011.07.249

[43] Zhang H, Berglund LA. Deformation and fracture of glass bead/CTBN-rubber/epoxy composites. Polymer Engineering and Science. 1993;33(2):100–107. DOI: 10.1002/pen.760330208

[44] Zotti A, Borriello A, Ricciardi M, Antonucci V, Giordano M, Zarrelli M. Effects of sepiolite clay on degradation and fire behaviour of

a bisphenol A-based epoxy. Composites: Part B. 2015;73:139–148. DOI: 10.1016/j.compositesb.2014.12.019

[45] Spanoudakis J, Young RJ. Crack propagation in a glass particle-filled epoxy resin. Part 1 effect of particle volume fraction and size. Journal of Materials Science. 1984;19(2): 473–486. DOI: 10.1007/BF02403234

[46] Evans AG. The strength of brittle materials containing second phase dispersions. Philosophical Magazine. 1972;26(6):1327–1344. DOI: 10.1080/14786437208220346

[47] Young RJ, Beaumont PWR. Failure of brittle polymers by slow crack growth. Journal of Materials Science. 1975;10(8):1343–1350. DOI: 10.1007/BF00540824

[48] Kinloch AJ, Maxwell DL, Young RJ. The fracture of hybrid-particulate composites. Journal of Materials Science. 1985;20(11):4169–4184. DOI: 10.1007/BF00552413

[49] Faber KT, Evans AG. Crack deflection processes-I Theory. Acta Metallurgica. 1983;31(4): 565–576. DOI: 10.1016/0001-6160(83)90046-9

[50] Kawaguchi T, Pearson RA. The moisture effect on the fatigue crack growth of glass particle and fiber reinforced epoxies with strong and weak bonding conditions. Part 2. A microscopic study on toughening mechanism. Composites Science and Technology. 2004;64(13–14):1991–2007. DOI: 10.1016/j.compscitech. 2004.02.017

[51] Kinloch AJ, Maxwell D, Young RJ. Micromechanisms of crack propagation in hybrid-particulate composites. Journal of Materials Science Letters. 1985;4(10):1276–1279. DOI: 10.1007/BF00723480

[52] Low IM, Bandyopadhway S, Silva VM. On hybrid toughened DGEBA epoxy resins. Polymer International. 1992;27:131–137.

[53] Maxwell D, Young RJ, Kinloch AJ. Hybrid particulate-filled epoxy-polymers. Journal of Materials Science Letters. 1984;3(1):9–12. DOI: 10.1007/BF00720061

[54] Pearson RA, Yee AF. Toughening mechanisms in thermoplastic-modified epoxies: 1. Modification using poly(phenylene oxide). Polymer. 1993;34(17):3658–3670. DOI: 10.1016/0032-3861(93)90051-B

[55] Monti M, Rallini M, Puglia D, Peponi L, Torre L, Kenny JM. Morphology and electrical properties of graphene-epoxy nanocomposites obtained by different solvent assisted processing methods. Composites Part A: Applied Science and Manufacturing. 2013;46:166–172. DOI: 10.1016/j.compositesa.2012.11.005

[56] Rafiee MA, Rafiee J, Wang Z, Song H, Yu Z, Koratkar N. Enhanced mechanical properties of nanocomposites at low graphene content. ACS Nano. 2009;3(12):3884–3890. DOI: 10.1021/nn9010472

[57] Hendrick JL, Yilgor I, Wilkes GL, McGrath JE. Chemical modification of matrix resin networks with engineering thermoplastics. Polymer Bulletin. 1985;13(3):201–208. DOI: 10.1007/BF00254652

[58] Cardwel BJ, Yee AF. Toughening of epoxies through thermoplastic crack bridging. Journal of Materials Science. 1998;33(22):5473–5484. DOI: 10.1023/A:1004427123388

[59] Hendrick JL,Yilgor I, Jurek M, Hendrick JC, Wilkes GL, McGrath JE. Chemical modification of matrix resin networks with engineering thermoplastics: 1. Synthesis, morphology, physical behaviour and toughening mechanisms of poly(arylene ether sulphone) modified epoxy networks. Polymer. 1991;32(11):2020–2032. DOI: 10.1016/0032-3861(91)90168-I

[60] Hodgkin JH, Simon GP, Varley RJ. Thermoplastic toughening of epoxy resins: a critical review. Polymers for Advanced Technologies. 1998;9(1):3–10. DOI: 10.1002/(SICI)1099-1581(199801)9:1<3::AID-PAT727>3.0.CO;2-I

[61] Kojima Y, Usuki A, Kawasumi M, Okada A, Kurauchi T, Kamigaito O. Sorption of water in nylon 6-clay hybrid. Applied Polymer Science. 1993;49(7):1259–1264. DOI: 10.1002/app.1993.070490715

[62] Yano K, Usuki A, Okada A. Synthesis and properties of polyimide-clay hybrid films. Journal of Polymer Science Part A: Polymer Chemistry. 1997;35(11):2289–2294. DOI: 10.1002/(SICI)1099-0518(199708)35:11<2289::AID-POLA20>3.0.CO;2-9

[63] Shiu SC, Tsai JL. Characterizing thermal and mechanical properties of graphene/epoxy nanocomposites. Composites Part B: Engineering. 2014;56:691–697. DOI: 10.1016/j.compositesb.2013.09.007

[64] Chandrasekaran S, Sato N, Tolle F, Mülhaupt R, Fiedler B, Schulte K. Fracture toughness and failure mechanism of graphene based epoxy composites. Composites Science and Technology. 2014;97:90–99. DOI: 10.1016/j.compscitech.2014.03.014

[65] Quaresimin M, Schulte K, Zappalorto M, Chandrasekaran S. Toughening mechanisms in polymer nanocomposites: from experiments to modelling. Composites Science and Technology. 2016;123:187–204. DOI: 10.1016/j.compscitech.2015.11.027

[66] Guo QP, Habrard A, Park Y, Halley PJ, Simon GP. Phase separation, porous structure, and cure kinetics in aliphatic epoxy resin containing hyperbranched polyester. Journal of Polymer Science Part B: Polymer Physics. 2006;44(6):889–899. DOI: 10.1002/polb.20757

[67] Ratna D, Varley R, Raman RKS, Simon GP. Studies on blends of epoxy-functionalized hyperbranched polymer and epoxy resin. Journal of Materials Science. 2003;38(1):147–154. DOI: 10.1023/A:1021182320285

[68] Luciani A, Plummer CJG, Nguyen T, Garamszegi L, Manson JAE. Rheological and physical properties of aliphatic hyperbranched polyesters. Journal of Polymer Science Part B: Polymer Physics. 2004;42(7):1218–1225. DOI: 10.1002/polb.10671

[69] Fu JF, Shi LY, Yuan S, Zhang DS, Chen Y, Wu J. Morphology, toughness mechanism, and thermal properties of hyperbranched epoxy modified diglycidyl ether of bisphenol A (DGEBA) interpenetrating polymer networks. Polymers for Advanced Technologies. 2008;19(11):1597–1607. DOI: 10.1002/pat.1175

[70] Varley RJ, Tian W. Toughening of an epoxy anhydride resin system using an epoxidized hyperbranched polymer. Polymer International. 2004;53(1):69–77. DOI: 10.1002/pi.1324

Chapter 11

Wood Adhesives and Bonding Theory

Onur Ülker

Additional information is available at the end of the chapter

http://dx.doi.org/10.5772/65759

Abstract

In this last century, world had grown faster than before; now people need more furniture than in the past century. More furniture means, more particleboards and more adhesives. Wood adhesives are used in every step of furniture manufacturing. Wood adhesives aim to bond wooden materials with each other or with different materials. Today, production with a faster pace is more important. Furniture production lines could be more productive with fast curing glues. Wood adhesives are used in more than 70% of wood products today in the world. The main reason is their use in gluing furniture joints and wood composite materials. In this chapter, readers can find four different topics: (1) technical properties of wood adhesives, (2) environment friendly adhesives, (3) semisynthetic adhesives, and (4) synthetic adhesives.

Keywords: wood adhesives, bonding theory, penetration of adhesives, measures of wetting, organic adhesives

1. Introduction

Wood is an anisotropic and porous material with many inherited anatomical features. Major features are longitudinal tracheids in softwood species, and vessel elements and longitudinal fibers in hardwood species. The lumens of their cells are large enough to provide a good pathway for flow of liquid resin. Interconnecting pits are often adequate to permit resin flow. However, high-molecular weight resins or occlusions in the pits or lumens may inhibit flow. This conglomeration of resin and wood substance is called the "interphase region." Two substrates, each with its own interphase, and the interface between the substrates, comprise the "bond line." The geometry of the interphase region varies as a result of many factors, such as wood anatomy, permeability, porosity, resin viscosity, surface energy, consolidation pressure, and others [1].

A cursory glance through the literature shows the creation of a new or existing wood adhesives, which are seen as research and development activities to improve. Glue in all industrial sectors is a modern need; it is also important in the furniture industry. Now in the new world "green living, green thinking" is the order. It needed a lot of research on environmental and human friendly adhesives for the new chorus. For example, there are a lot of research works done for the elimination of formaldehyde emissions from particleboard adhesive. Wood glue is a major area of importance for industrial and commercial activities. This particular study was made to explain the importance to improve the adhesion strength.

2. Technical properties of wood adhesives

It is more important to know the use of wood adhesives and their technical properties. These properties are given below.

2.1. What is bonding? (Science of adhesion)

In furniture and forest product industry, "wood adhesives" have played an important role in the development and efficient use of wood. In wood products, the most commonly used material is glue. If we check different wood products (plywood, MDF, particleboard, OSB, structural frame and wooden architectural doors, windows, and frames), adhesives are more important to retain their structure. Significant amounts of adhesives are used in floor coverings, kitchen counters and for ceiling and wall tiles. They are also used in nonstructural applications, within car upholstery and accessories. Adhesives increase the resistant strength and stiffness of the composite sheet. The adhesion of the glue depends on the wood-adhesive bonding chain.

Adhesive bonding performance between wood elements is presumed to be significantly influenced by the degree of penetration of the adhesive into the porous network of interconnected cells. Research on the bonding performance has been conducted through microscopic examination and associated techniques, in an effort to establish relationships with the bond performance. Variation between wood species, the wide variety of adhesive application and curing processes, and many types of adhesive chemistries and formulations make sweeping generalities difficult. However, troubleshooting bonding problems and designing new adhesive systems and processes may be facilitated by understanding the fundamentals of adhesive penetration [1].

The interphase region is an uneven layer, as illustrated in **Figure 1**. The geometry of the interphase is assumed to affect bond performance. Adhesive joints under load must transfer stress from component to component through the interphase region. The structural makeup of the interphase, its volume and shape, will dictate the magnitude of stress concentrations and ultimately have a significant impact on the performance of the bond [1].

http://dx.doi.org/10.5772/65759

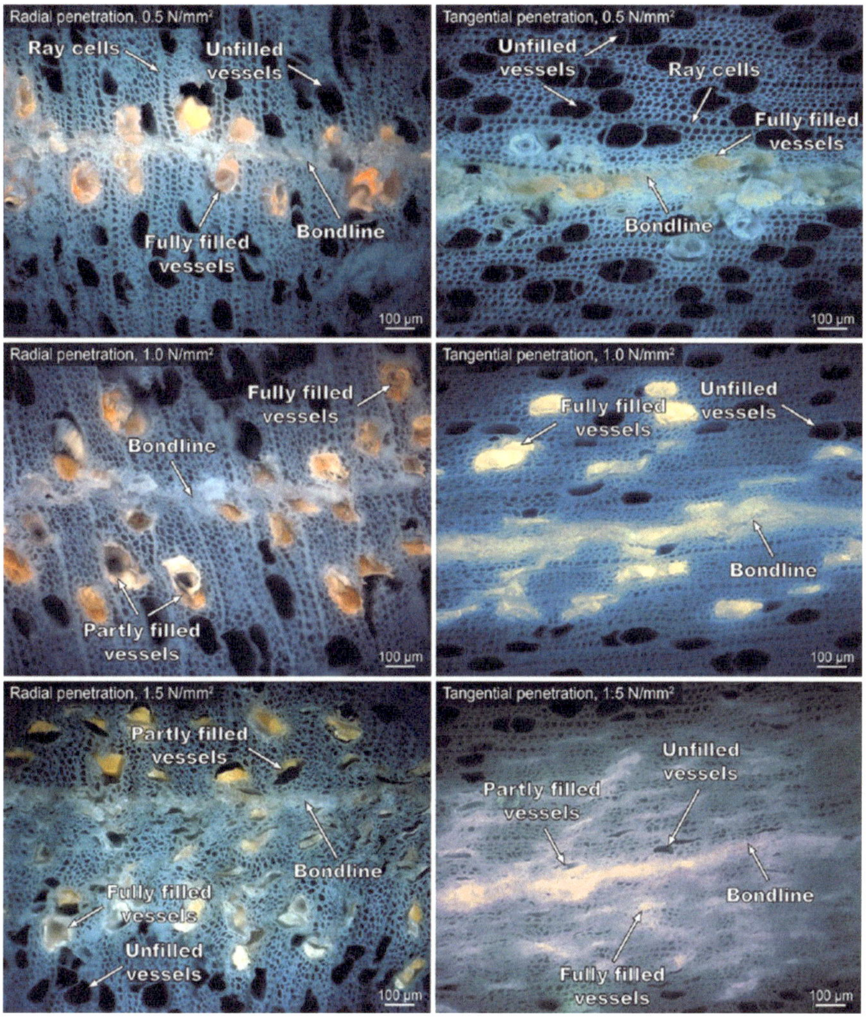

Figure 1. Example of epi-fluorescence microphotograph with the penetration of UF resin into poplar at three different pressures applied during the press cycle: 0.5 N/mm^2, 1 N/mm^2, and 1.5 N/mm^2 for radial and tangential penetration [2].

2.2. Penetration of wood adhesives

There were many researches made on penetration of wood adhesives. The penetration of adhesive into wood can be categorized into two groups:

- Gross penetration
- Cell wall penetration.

Gross penetration results from the flow of liquid resin into the porous structure of wood, mostly filling cell lumens. Hydrodynamic flow and capillary action could be explained as gross penetration. Cell-wall penetration occurs when resin diffuses into the cell wall or flows into microfissures.

In wood, the least resistance to hydrodynamic flow is in the longitudinal direction, following the lumens in the long and slender tracheid of softwood, or through the vessels of hardwoods. Since vessels are connected end-to-end with perforation plates and there is no pit membrane, this cell type dominates the penetration of adhesives in hardwoods. Using optical microscopy, the author has observed resin in pit chambers of both hardwood and softwood species and in cell lumens in which the only entry pathway for the resin was through the pit.

Adhesive penetration influences link 4 through 7 in reference [3]. All of the potential adhesion mechanisms are influenced by penetration. The concept of mechanical interlocking is obviously dependent on penetration of the adhesive phase beyond the external wood surface. In addition, the combined adhesion force due to covalent bonding and formation of secondary chemical bonds is directly related to the area of surface in contact between the adhesive and the cell wall.

In reference [3], a chain-link analogy for an adhesive bond is proposed as shown in **Figure 2** and inferred that the bond is only as good as the weakest link in the chain. Adhesive penetration plays a vital role in this analogy. Link 1 is the pure adhesive phase, unaffected by the substrates. Links 2 and 3 represent the adhesive boundary layer that may have cured under the influence of the substrates and is no longer homogeneous. Links 4 and 5 represent the interface between the boundary layer and the substrate and constitute the "adhesion" mechanism. That mechanism may be mechanical interlocking, covalent bonding, or secondary chemical bonds due to electrostatic forces. Links 6 and 7 represent wood cells that have been modified by the process of preparing the wood surface or the bonding process itself.

Figure 2. Chain link analogy for an adhesive bond in wood.

For example, rotary peeling of veneer causes fractures that initiate in the radial-longitudinal plane. The cells in the region may have been weakened, and thus increase the potential for failure of the bond. Planning, flaking, sanding, and other mechanical surface preparation techniques will also cause minute fractures in the wood cells. Finally, links 8 and 9 represent the unadulterated wood. A properly designed adhesive bond would have the lower limit of structural integrity located at links 8 and 9. In other words, the wood should be the weakest link [1].

http://dx.doi.org/10.5772/65759

2.3. Overview of adhesion and cohesion

Adhesion is the tendency of dissimilar particles or surfaces to bond to one another. The internal forces between molecules that are responsible for adhesion are chemical bonding, dispersive bonding, and diffusive bonding. These intermolecular forces can make cumulative bonding and bring certain emergent mechanical effects.

Cohesion word (cohaerere in Latin language) means "stick or stay together." Cohesive force is the tendency of similar molecules to stick together. They attract mutually. Cohesive force caused by the shape and structure of molecules, which makes the distribution of orbiting electrons irregular when molecules get close to one another, creating electrical attraction that can maintain a microscopic structure such as a water drop.

Chain link analogy for adhesion and cohesion is shown in **Figure 3**. The adhesive and cohesive definition refers to the forces that keep together the adhesive with the substrate (adhesion) and the adhesive to itself (cohesion). These forces correspond to:

Figure 3. Chain link for adhesion and cohesion.

a. chemical bonds

b. intermolecular forces

2.4. Explaining the wetting

The meaning of "wetting" is always misunderstood because there are many explanations for wetting in the literature. Some examples are given below:

a. *"The liquids which spread over a solid surface, the adhesion of these liquids to the solid could explain the wetting properties of the solid* [4]."

b. *"Good wetting will assist spreading and penetration, but it is not identical with them, good wetting is a zero contact angle* [5]."

c. *"Wetting is that phenomena of a liquid spreading out over and intimately contacting a solid surface* [6]."

d. *"If the molecules at the interfaces of the liquid and the solid are attracted more strongly by the solid than by the liquid, the liquid wets the surfaces and tends to creep outward along them, on a non-wetted surface the liquid molecules are still influenced by their own attraction for one another* [7]."

"In reference [8], *wetting is the term that has come into common usage to describe what happens when a liquid comes in contact with a solid surface."*

Because several phenomena occur when this contact is made, it seems logical to assume that the term wetting is best used in a generic sense. Putting this concept in perspective, the term wetting is used to cover the processes of adhesion, penetration, and spreading. All these cases

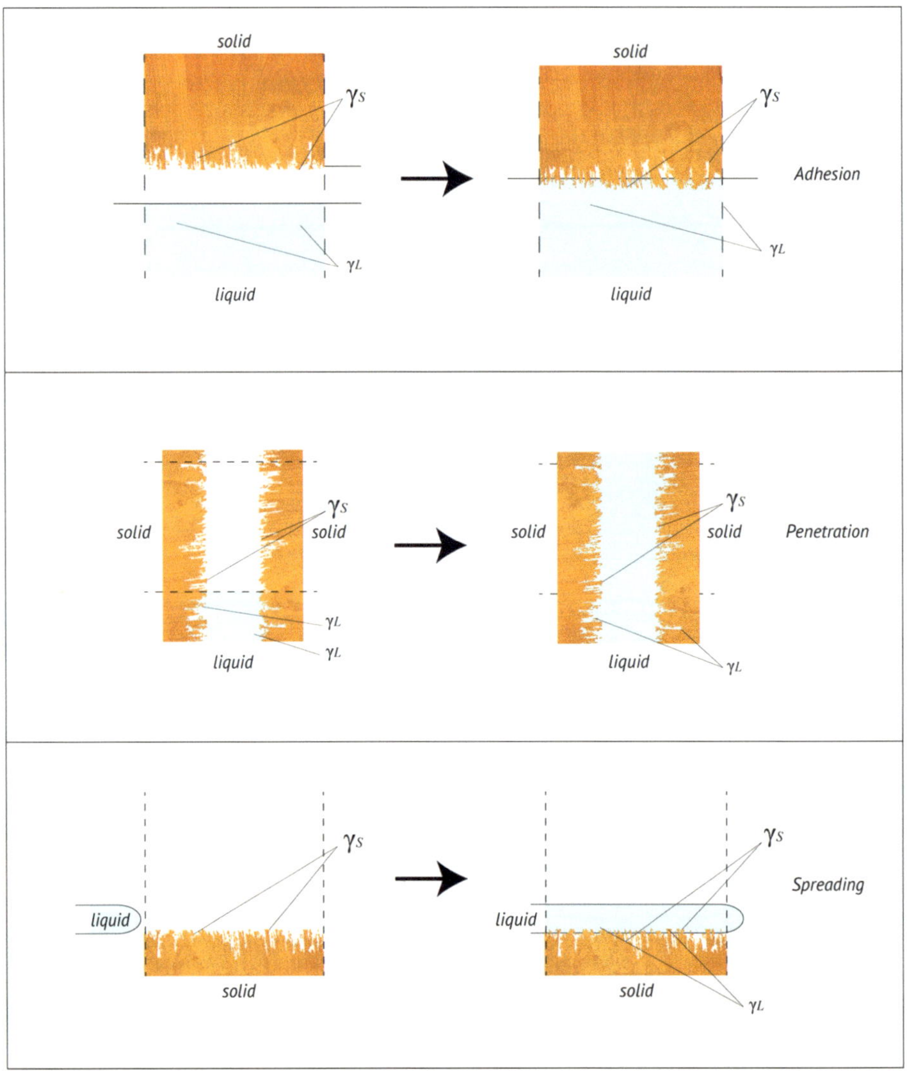

Figure 4. Explaining the wetting.

http://dx.doi.org/10.5772/65759

explain definitely different types of wetting. Setting adhesion as a subgroup of the wetting is a result of its surface energetics. It is identified simply as the wetting conditions that apply during "face to face contact" [9].

We can clarify this by stating that using the surface energetics approach. The word adhesion implies strictly an interracial phenomenon, while in practical, two materials bond together. Wetting phenomenon is shown in **Figure 4**.

"Penetration" refers to wetting conditions when a liquid works its way up along the walls of a solid material, and "spreading" refers to wetting conditions involved when a liquid flow out over a surface. In **Figure 4**, this theory is explained. In reference [10], wetting is explained as a process that takes place when a liquid contacts a solid surface. Usually the process of adhesion takes place in air; thus, three phases are involved: liquid, solid, and vapor.

2.5. Measures of wetting

The contact angle of a liquid with a solid surface is a convenient measure of wettability; it is an indicator of the affinity of a liquid for a solid. In reference [11], contact angle measurements are made in various ways, the balance of a liquid drop which is lean on a plane solid surface is under the movement of three surface tensions. In **Figure 5**, basic conditions are explained.

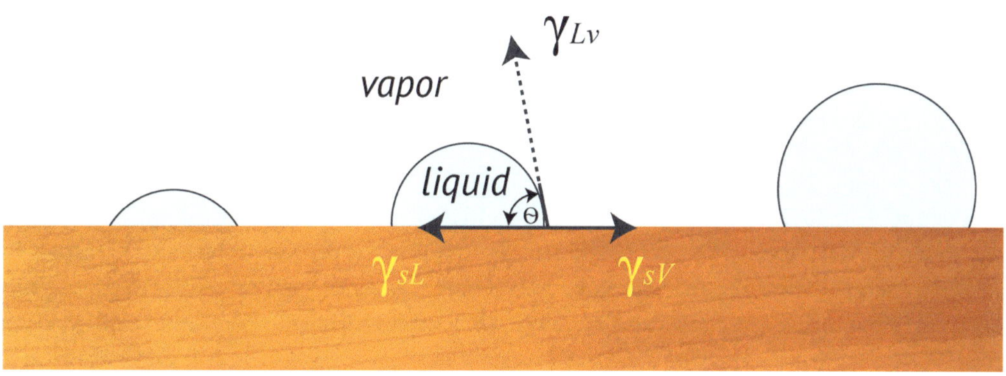

Figure 5. Finite contact angle with liquid resting on solid surface.

The three types of surface stress are explained below:

a. at the interface of the liquid and vapor phase, γ_{Lv};

b. at the interface of the solid and the liquid phase, γ_{sL};

c. at the interface of the solid and vapor phase, γ_{sv}.

Using of contact angle in wettability reduces the fact that the tendency for a given mass of liquid to spread and adhere to a solid surface increases as the angle "θ" decreases.

If the contact angle is "θ", it will be inverse measure of wettability, while the cosine of "θ" is an apparent direct measure. In Young's equation for the classical case of the three phase line, get in touch with between a smooth, rigid, and solid phase "S", a liquid "L", and vapor "V', expresses the relationship between the equilibrium contact angle "θ ", and the three-surface tensions γ_{Lv}, γ_{sL}, and γ_{sv} .

As mentioned in reference [12], "Young equation" is given below,

$$\gamma_{sv} = \gamma_{sL} - \gamma_{Lv} \cdot \cos\theta \tag{1}$$

This equation contains two solid-surface tensions, which are extremely difficult, if possible at all, to measure, it is usually combined with in reference [12], and this relationship relates the work of adhesion, W_{sL}, between solid and liquid:

As mentioned in reference [13], ''Dupre equation" is given below,

$$W_{sL} = \gamma_{sv} + \gamma_{Lv} - \gamma_{sL} \tag{2}$$

The combination of Eqs. (1) and (2) yields the original Young-Dupre equation, which has been one of the most useful tools in the experimental approach to study surface behavior:

As mentioned in reference [10], ''Young and Dupre's equation" is given below,

$$\mathrm{W}_{SL} = \gamma_{Lv}(1 + \cos\theta) \tag{3}$$

In reference [14], many of the main derivations and arguments concerning the validity of these equations are described. Their conclusion, after detailing the results of the various thrusts to this end, seems worthy of particular note: '' *The present reviewers lack sympathy with much of this work, finding it difficult to understand why it should have been thought necessary to raise so many objections to, or to devise so many derivations of, what is, in their opinion, a set of self-evident equations." We lay ourselves open to the charge of enjoying the advantages of hindsight, and perhaps also to that of being naive in outlook, but we do not see why it should have taken so long for workers to appreciate, for example, that we must be the work required to separate the liquid from the solid to give both solid and liquid phases in equilibrium with the vapor phase. If the three-phase system is truly in equilibrium, both before and after separation, then (these) equations necessarily follow* [14]."

Although in reference [14], the results of various derivations of Young's and Dupre's equations have been discussed, it is important here to note briefly the major results because of their relationship and bearing on other methods of measuring wettability. W_{SL}, the reversible work of adhesion per unit area, is not the same value when measured in a vacuum on the one hand and in the saturated vapor of the wetting liquid on the other [15, 16]. The two are distinguished, as W_{SL}^0 or W_A^0 or the work under vacuum, and W_{sL} or W_A as the work in the saturated vapor of the wetting liquid [10].

http://dx.doi.org/10.5772/65759

$W_{SL}{}^{0}$ has been shown to always be greater than W_{sL} [11]. In essence, what this means is that on adsorption of a gas or vapor the surface free energy (surface tension) of the solid is reduced. This phenomenon is expressed by the following equation:

As mentioned in reference [11], ''Boyd and Livingston equation'' is given below,

$$\pi_{sv} = \gamma_s - \gamma_{sv} \tag{4}$$

where :

π_{sv}: the change in the surface free energy upon adsorption of the vapor of the contacting liquid.

γ_s: the solid surface free energy in a vacuum.

γ_{sv}: the solid surface free energy in saturated vapor.

In reference [10], Eq. (3) includes the factor π_{sv} in cases other than vacuum, and more correctly assumes the following form:

$$W_{SL} = \pi_{sv} + \gamma_{Lv}(1+\cos\theta) \tag{5}$$

3. Environmental friendly adhesives (organic adhesives)

Glue is the most important raw material coming after wood in furniture industries. Especially after World War II, glue ameliorated its time and bonding techniques and developments occurring in plywood block board and in the production of chipboard. It has led to developments in a positive direction. Results of the physical and mechanical properties of these materials have been borne and assessed in a wide variety of places where there are opportunities for its usage. In past, plant and animal glues are used, later they left the place for the synthetic resin. Animal glue has to fight against instability and resistance to hot water and microorganisms and the lack of block board limit their usage, though the use and application of synthetic resin in plywood production eliminate these drawbacks. Thus, these materials are now in different atmospheric conditions, letting in water, as well as in direct contact with the water on the concrete breakdown, in mold making, etc., as these are used widely in very different fields. Adhesives used in the furniture industry prior to 1930s were obtained from crop and animals, and those obtained from animals are classified as:

a. Animal glue or gelatin—obtained from skin, bone, and fish residues.

b. Blood—obtained from the raw with blood from slaughterhouses.

c. Casein—derived from an animal milk protein.

3.1. Environmentally friendly adhesives obtained from animals

Adhesives obtained from animals are gelatin-type adhesives. They are obtained from waste and by-products of the animal industries. Raw materials of animal adhesives are hides, sinews, and bones of cattle and other animals. The wastes of leather industry (from tanned hides) are also utilized. Adhesives that are made from hides are of higher grade than glue obtained from bones and tendons [17].

Adhesives obtained from animals are of gluten origin and are obtained by boiling collagens in water agent. To prepare such a solution, adhesive solution is left in cold water which induces swelling (15-30 minutes for the powder and 2 hours for beads), and then if necessary, heat at temperatures not exceeding 60°C in bain-marie. Animal glues can be wetted with water and pasted on any support material. During withdrawal, dry animal glue is proportional to the water used to prepare the solution. The viscosity of the binder solution varies with the change in pH. At low viscosities, pH is between 4.5 and 5.

Adhesives obtained from animals are treated one or three times with cold and clean water depending on the part of the animal, glue weight, and size keeping it for a while in this aqueous gel and by heating at 60°C temperature to bring it to the suitable condition. If the glue is not suitable for boiling, it will weaken its gluing property. In plywood and veneer industries, the glue is applied with a gluing machine. It is more important not to use excess glue unnecessarily. Because, in this case, the deterioration of the balance between the amount of water present in the glue and moisture content of the wood material will cause distortion, such as a crack up, swelling, and corrugated formation of drawbacks. In this regard, the inconveniences of glue application are much more than the minimum driving. These issues suggest that when applied, the amount of glue from animals should be held in 10 per 9.29 m^2.

Plywood unit is prepared by applying hot melted adhesive between the layers placed in the cold press, uses the fluid state glue, and is subsequently heated to 60°C in order to bring it by a press. Later heat up to 25°C for the purpose of heated press 60°C so. Subsequently, 25°C > a is cooled. Unfavorable side of animal glues are that they comes off when contacted with water or at 80°C temperature and due to higher relative humidity of animal glue bonding there is the loss of precision. Furthermore, animal glues, with the action of microorganisms can easily undergo deterioration.

3.2. Environmentally friendly adhesives obtained from botanical plants

Botanical adhesives are obtained mostly by processing starchier plants. As well as some of the resins of wood species fall into this group. Both in animal glue and vegetable glue, formaldehyde (an organic compound with the formula CH_2O) and so on are used to prevent microorganisms.

Plant-derived adhesives are divided into two groups as starch- and cellulose-derived. The first group is adhesives derived from starch, extracted from plants, such as corn, rice, potatoes, and wheat and is generally used in bookbinding, paper bags, and cardboard boxes.

http://dx.doi.org/10.5772/65759

The second group of adhesives is made from cellulose, derived from trees, shrubs, or fruits such as bananas, which are used more in stickers affixed to glass.

4. Semisynthetic adhesives

Cellulosic derivatives are thermoplastic adhesives. Thermoplastics are in solid phase at normal temperature. On heating, they soften and melt; this property helps them to protect the floor when a chemical alteration occurs. They occur as an adhesive solution, dispersion, and in solid form. Once heated, they become solid ones. A liquefied thermoplastic material can provide adhesion when allowed to cool. In solution and dispersion they remain as a film by the evaporation of the liquid solvent. They are classified into two groups:

Cellulose nitrates are cellulose, which are obtained from cotton by reaction with nitric acid and sulfuric acid. The color of cellulose nitrate normally is opaque or transparent, but under the day light, it will be dark. It is resistant to water and oils, and it biodegrades at moderate temperature in weak acids, alkalis, and organic solvents. But it is a strong, flexible adhesive. There is even aged adhesive reversibility.

Cellulose ethers are prepared by reactions of suitable alcohol sulfate or chloride salts in an alkaline environment. They are sold in the market as organic solvents or by dissolving the gel powder in water and a vaporizable solvent. After the evaporation of the solvent, they remain as thermosetting polymers. Upon evaporation, there is a degree of shrinkage and decrease in volume. In these cases, the adhesive is not suitable to be used for structural strength.

5. Synthetic adhesives (inorganic adhesives)

In 1930s, synthetic resin adhesives were used in the woodworking industry. They have many advantages for use in the woodworking industry. In the outdoor furniture, synthetic resin adhesives can be used in joints that remain as strong as the wood even in unprotected exposure to the weather. Most of the "animal" adhesives can be used in furniture joints for interior use only [18].

Inorganic adhesives are based on typical compounds, such as sodium silicate, magnesium oxychloride, lead oxide (litharge), sulfur, and various metallic phosphates. These materials form strong resistant bonds for special applications, and are still widely used. The advent of synthetic organic polymer adhesives during the last two decades has led to a decline in the use of many of the older inorganic adhesives laboratory recipes [19].

The more important adhesives for wood are currently produced by chemical synthesis. Chemical synthesis, usually converts synthetic adhesives from liquid to solid by a hardener or a setting agent. These agents may be furnished separately for addition to the resin before use, or they may be present (particularly with spray-dried powdered resins) in the resin as supplied [18, 20].

The advantages and disadvantages of synthetic resin adhesives are given below.

Advantages

a. low cost materials,

b. easy to use and to scale-up for industrial use, and

c. high temperature resistant joints.

Disadvantages

a. cannot cope with large CTE mismatch,

b. are only good for low stress applications,

c. joints not leak-tight, and

d. surface preparation is critical.

5.1. PVA adhesives

Polyvinyl resin emulsions are thermoplastics, softening when the temperature is increased to a particular level and hardening again when cooled. PVA adhesives are copolymer based. PVA adhesive solidifies by evaporation or by absorption of water by the gluing material. Solidification time is relatively short, about 45 seconds. The best temperature for using PVA adhesive is around 20°C. PVA resins have long storage and working lives at normal room temperatures. Users must avoid in-storage evaporation or freezing. Packaging should be modified to keep them cool. They are diluted to last longer. PVA adhesives are milky-white fluids to be used at room temperature in the form supplied by the manufacturer. Emulsion films are resistant, waterproof, usually applied quickly, does not smell, and they do not change the taste. They are durable to machine use and are oil-resistant [17, 18, 20].

5.2. PVAC adhesives

PVAC (the term embraces both homopolymer and copolymers) was in the forefront of the transition of adhesives from natural to synthetic adhesives. While overall consumption of homopolymer plus copolymer doubled during the 1975–1987 period, the copolymer adhesives increased by a dramatic 250%. The preferred comonomers are ethylene acrylate esters. Three uses for PVAC in packaging, construction, and textiles account for 80% of the polymer and 95% of compound usage. PVAC adhesives are found to be used in a dozen construction applications; the largest is ready-mix joint cement for gypsum board. These are highly filled formulations with a polymer content of 3%. Concrete adhesives containing PVAC serve to bond new concrete to old. Vinyl acetate-ethylene is the material of choice for vinyl and paper lamination to hardboard gypsum board and other substrates [21].

As a furniture adhesive, PVAC is used for general assembly applications, film overlay and high-pressure lamination, edge gluing, wood veneer, and edge bonding. The demand has increased by 400% in the last 20 years. In furniture market "white glue" continues to be a staple for both home and shop.

5.3. Hot-melt adhesives

To spread and for adhesion of substrates, adhesives should have the "soaking" feature. In most adhesives used in packaging, easily evaporating agent such as water flow (water as solvent) is used for dissolving the adhesive. It evaporates leaving behind a sticky substance and it easily glues. On the other hand, a solvent is used in Hot-melt before applying instead of dissolving it. The unifying feature can be caused by the solidification of hot melt adhesives. Heat dissipates more quickly in the case of a volatile liquid even for a waterproof layer. Fast heat dissipation provides quick bond formation. In the furniture industry, hot-melt adhesives are furnished in solid form.

Therefore, hot melt is ideal for applications, especially where high speed, early adhesion is important. The lack of a liquid solvent, frost, weathering, and decay do not affect the shipment and storage of hot-melt adhesives. The biggest disadvantage of hot-melt adhesives is their limited heat resistance. In furniture market, they usually are sold as chunks, granules and in cord form on reels.

Hot-melt adhesives are thermoplastic materials. Even at mildly elevated temperature, a major part of the glue is lost by melt at the end. The bond is formed very quickly depending upon the temperature difference between the glue and the parts being joined. The setting time as brief as a fraction of a second has been reported. The durability of hot melt used in the packaging decrease at 60°C and deteriorate at 70°C. The most common use of hot melt is in packaging material, joining a low molecular weight resin and a waxy ethylene-vinyl acetate copolymer (EVA). According to the attached, EVA additives integrate power, product durability, and resistance to hot seam and heat. (If the resin is useful in the hot seam accelerates adhesion largely determines the color and odor). Most crystallized substances as waxy material softening point, resistance to heat and set the time. Most frequent uses of hot melts in wood gluing have been for edge banding of panel boards [22–24].

5.4. Contact adhesives

Contact adhesives are generally based on synthetic rubber, which is obtained by dissolving in suitable liquids. Contact adhesives are sold under a variety of names in the market. As a result of evaporation of the flux contained in the glue, the glue dries. Bonding of large surfaces with contact adhesive is difficult.

Contact adhesives are usually used to bond:

a. plastic laminates to plywood or particleboard for counter-tops,

b. restaurant and kitchen tables,

c. edge of particle boards and MDF,

d. upholstering of sofas, and

e. PVC edge to boards.

Contact adhesives should be applied to both surfaces, and allowed to dry, depending on the ambient temperature and the chemical structure of glue. The wait and airtime (10 minutes, 15 minutes) varies. Contact adhesives are unique in that they develop considerable strength immediately upon contacting surfaces [16].

5.5. Urea formaldehyde resin adhesives

Urea formaldehyde resins are widely used in chipboard or plywood production. UF resin is produced by heating suitable urea and formaldehyde at 115°C for 5 hours. They are usually produced during the production of high quality glue E3 formaldehyde emissions. UF resins came into market in 1930s. UF resin can be formulated either for hot pressing or for room temperature curing by different types and amounts of catalyst. UF resins are compatible with various low cost extenders or fillers, thus permitting variation in both quality and cost. They are available with solid contents from about 40–70% percent. They are also marketed as dry powders, with or without incorporating the catalyst [18].

UF resins are used as adhesives provide a number of advantages to manufacturers in the wood industry. The use of UF resins adhesives make it superior to other chips and boards them, providing the reasons for the choice of MDF and plywood production as follows:

a. low cost,

b. a very different baking (curing) conditions,

c. easy to use,

d. low energy consumption during production with low firing temperature,

e. ability to easily dissolve in water,

f. microorganisms and abrasion resistance,

g. hardness,

h. great thermal properties, and

i. colorless.

UF resin has some disadvantages despite an outstanding advantage. UF resins have a high durability, especially in low humidity and high temperature. Hence UF resin produced from forest industry products are suitable for indoor use only. The combination of temperature with humidity reduces adhesive property of the urea formaldehyde and melamine urea formaldehyde adhesives in the product and leads to formaldehyde gas emission.

5.6. Melamine resin adhesives

Melamine-formaldehyde resin is obtained from polycondensation of melamine and formaldehyde. In the reaction between formaldehyde and melamine, the melamine (2, 4, 6 - triamino-1, 3, 5- triazine) gives derivatives containing different numbers of methyl groups participating amine group. The number of methyl groups may be up to six. The water-soluble methyl melamine cross-linking at elevated temperatures over a methylene or ether bridges are converted to the formaldehyde resin.

MF resins are usually used in the impregnated decor paper, barrier lining the balance and preservation, post-forming craft, and overlays. They are also used in production for hardwood Kraft paper impregnated overlay and coatings for the tray. MF resin adhesives are sold in furniture market as powders. MF resin adhesives are prepared by mixing with water or used with a MF hardener. The color of MF adhesive is almost white, but the addition of filler usually gives them a light tan color similar to the urea resins. MF resins are considerably more expensive than PF or UF resins. Uncatalyzed MF resin adhesives also have been investigated for gluing heavy laminated ship timbers at curing temperatures of 140–190°F [18, 20].

5.7. Silicone adhesives

Silicone adhesives are known as polysiloxanes. In the chemical structure of silicones, silicon (*Si*) and oxygen (*O*) atoms are sequentially arranged instead of carbon (*C*) contained polymer is the common name. The most methyl or phenyl groups are located bound to silicon atoms of the silicone molecules. Silicones are the most fluidic produced in the form of a resin. Silicone fluids are quite stable substances, and they are not affected by water or influenced by rising heat.

They are very good electric insulators as well as hydraulic fluids and emulsion-breaking agents, and they are also used to reduce water permeability of various materials, such as paper. Silicone rubbers are also electrically insulating and chemically resistant and maintain flexibility in a wide temperature range. These are the important features. It is most commonly used in protective sheath and insulating varnishes.

In furniture industry, silicone adhesives are usually used to:

a. finish material at counter-tops,

b. bond edges and finish points at parquets,

c. bond kitchen and bath cupboards, and

d. bond upholstering fabric of sofas.

Single component silicone adhesives require a humidity of 5–95% to cure. Besides the presence of humidity, a temperature between 5 and 40°C is required to cure the adhesive [25].

6. Conclusions

The aim of this chapter has been to present a selective review of the literature of wood adhesives. Different kinds of wood adhesives were explained. Today wood products and wood adhesive industry have many aspects in common. Furniture designers should know wood adhesives and their using tips. Bonding theory and wetting phenomenon were explained in this chapter. An effort was made to tie the status of research in wood adhesion with main adhesive types. The principal results of this section can be summarized as follows:

a. The positive relationships of glue-bond quality and adhesive penetration into the wood structure.

b. The positive relationships glue-bond quality and wettability of the wood structure.

c. Explaining bonding theory with samples.

d. Chain link analogy for adhesion and cohesion has a strong influence on optimum conditions for good bonding.

e. Organic (animal, plant), semisynthetic, and synthetic adhesives were explained with their usage.

On the forefront of adhesion research, types and using tips of adhesives are important. Wood adhesives and adhesion theory depends on wood surface, contact angles, and adhesive type.

Author details

Onur Ülker

Address all correspondence to: ulker79o@hotmail.com

Department of Interior Architecture and Environmental Design, Fine Arts Faculty, Kırıkkale University, Kırıkkale, Turkey

References

[1] Kamke FA, Lee JN. Adhesive penetration in wood – A review. Wood and Fiber Science. 2007; 39: 205–220.

[2] Gavrilović-Grmuša I, Dunky M, Djiporović-Momčilović M, Popović M, Popović J. Influence of pressure on the radial and tangential penetration of adhesive resin into poplar wood and on the shear strength of adhesive joints. BioRES. 2016; 11: 2238–2255. DOI: 10.15376/biores.11.1.2238-2255

[3] Marra AA. Technology of wood bonding: Principles in practice. 1st ed. Van Nostrand Reinhold. New York; 1992. 454 p.

[4] Baier RE, Shafrin EG, Zisman WA. Adhesion mechanisms that assist or impede it. Science. 1968; 162: 1360–1368.

[5] Marian JE, Stumbo DA. Adhesion in wood. Part I: Physical factors. Holzforsch. 1962; 16: 134–148.

[6] Schneberger GL. Part II. Physical principles. Adhesive Age. 1970; 18: 28–31.

[7] De Bruyne NA. The action of adhesives. Scientific American. 1962; 206: 114–126.

[8] Patton TC. A simplified review of adhesion theory based on surface energetics. Tappi. 1970; 53: 421–429.

[9] Sharpe LH, Schonhorn H. Surface energetics, adhesion, and adhesive joints. Contact angle, wettability, and adhesion. Advances in Chemistry Series. 1964; 43: 189–201.

[10] Collet BM. A review of surface and interfacial adhesion in wood science and related fields. Wood Science and Technology. 1972; 6: 1–42.

[11] Boyd GE, Livingston IL. Adsorption and the energy changes at crystalline solid surfaces. The Journal of the American Chemical Society. 1942; 64: 2383–2388.

[12] Young T. Cohesion of fluids. Philosophical Transactions of the Royal Society. 1805; 95: 65–72.

[13] Dupre, A.M. Dupre, P. Theorie Mecanique de La Chaleur. France: Gauthier Villan; 1869; 517 p.

[14] Elliott G. E. P., Riddiford, A. C. Contact angles. Recent progress in surface science. New York: Academic Press; 1964; Vol. 2, pp. 111–128. DOI:10.1016/0021-9797(67)90183-X

[15] Bangham D. H. The Gibbs adsorption equation and adsorption on solids. Transactions of the Faraday Society. 1937; 33.194: p. 805–811. DOI: 10.1039/TF9373300805

[16] Bangham R. E., Razouk, R. I. Adsorption and the wettability of solid surfaces. Transactions of the Faraday Society. 1937; 33.199: p. 1459–1463. DOI: 10.1039/TF9373301459

[17] Carroll MN, Bergin EG. Catalyzed PVA emulsions as wood adhesives. Journal of Natural Products. 1967; 17: 45–50.

[18] Selbo ML. Adhesive bonding of wood. Technical Report. United States Department of Agriculture Technical Bulletin No. 15 12. U.S. Department of Agriculture. Drake, New York. 1975. 124 p.

[19] Vick CB. Adhesive bonding of wood materials. In: Hand book of wood, wood as an engineering material. General Technical Report FPL-GTR-113. USDA. Madison; 1999. 24 p.

[20] Kardashov DA. Synthetic adhesives. No. FSTC-HT-23-1168-73. Army Foreign Science and Technology Center. Charlottesville, VA. 1973. 273 p (in Russian).

[21] Skeist I. Handbook of adhesives. New Jersey: Springer Science & Business Media; 2012; 800 p. DOI: 10.1007/978-1-4613-0671-9

[22] Cass SB. A comparison of hot press interior plywood adhesives. Journal of Natural Products. 1961; 11: 285–287.

[23] Selbo ML. Adhesives for structural laminated lumber. Adhesive Age. 1961; 4: 22–25.

[24] Snider RF. What to look for in glues for furniture production. Adhesive Age. 1960; 3: 38–40.

[25] Silicone Adhesives [Internet]. 2016. Available from: http://www.adhesives.org/adhesives-sealants/adhesives-sealants-overview/adhesive-technologies/chemically-curing/two-component-(2-c)/silicone-adhesives [Accessed: 2016-06-17].

Chapter 12

Dynamic Characterization of Adhesive Materials for Vibration Control

Jon García-Barruetabeña and
Fernando Cortés Martínez

Additional information is available at the end of the chapter

http://dx.doi.org/10.5772/66104

Abstract

This chapter focuses on the dynamic characterization of adhesive materials for vibration control proposes. First, the experimental characterization and modelization of the relaxation and complex moduli of the flexible adhesive ISR 70-03 by means of a dynamic mechanical thermal analysis technique (DMTA) are presented. Then, the interconversion path between the relaxation modulus $E(t)$ and the corresponding complex modulus $E^*(\omega)$ for linear viscoelastic solid materials is explored. In contrast to other approximate methods, in this work the fast Fourier transform (FFT) algorithm is directly applied on relaxation functions. Finally, an experimental study for the structural noise and vibration reduction in a cabin elevator by means of adhesive-bonded joints of panels is presented.

Keywords: elastomers and rubber, nondestructive testing, relaxation modulus, complex modulus, viscoelasticity, material functions interconversion, noise and vibration reduction, adhesive-bonded joints

1. Introduction

Regarding damping, joint procedures by means of screws, rivets, or by joining do not introduce relevant damping outside of some specific frequency ranges [1–5]. Therefore, they are not relevant for vibration control porpoises. However, adhesive joint can be designed for structural noise control because they are able tointroduce effective modal damping below 1 kHz [6].

In particular, viscoelastic adhesives are widely employed in engineering applications and also found their widespread application in many sectors such as the automotive industry, aerospace, wind power, and human transportation. The mechanical properties of general viscoelastic materials depend on temperature, frequency and amplitude, prestress, dynamic load level, relative humidity, and among others. Service temperature, frequency, and amplitude of deformation are the most relevant ones. Thus, suitable mechanical characterization is essential in order to obtain reliable predictions.

2. Experimental characterization of a flexible adhesive

The experimental characterization by means of adynamic mechanical thermal analysis technique (DMTA) and the modelization of the relaxation and complex moduli of the flexible adhesive ISR 70-03 [7] are presented. The manufacturing procedure of the test experiments is detailed. Then, the influence of the strain level and specimen thickness into material behavior is studied. Next, using a procedure based on the time-temperature superposition principle, relaxation and dynamic master curves under tension strain are obtained. Finally, a generalized Maxwell model and a fractional derivative model of these master curves are implemented. As a conclusion, models capable of describing together the influence of time, temperature, and strain level are proposed.

Flexible adhesive shows viscoelastic behavior [8, 9]. In these viscoelastic materials (VEMs), the energy dissipation is a consequence of the phase difference between the stress σ and the strain ε. In frequency domain, this behavior is represented by the complex modulus approximation, which can be obtained from the relationship between the harmonic stress $\sigma(t)$

$$\sigma(t) = \sigma_0 e^{i\omega t} \tag{1}$$

and the stationary harmonic strain $\varepsilon(t)$ given by

$$\varepsilon(t) = \varepsilon_0 e^{i(\omega t - \varphi)} \tag{2}$$

σ_0 is the stress amplitude, ε_0 is the strain amplitude, ω is the excitation frequency, and φ is the phase delay. Therefore, the frequency domain stress-strain relationship $\tilde{\sigma}(\omega)$–$\tilde{\varepsilon}(\omega)$ results in

$$\tilde{\sigma}(\omega) = E^*(\omega)\tilde{\varepsilon}(\omega) \tag{3}$$

where the complex modulus $E^*(\omega)$ can be written as

$$E^*(\omega) = E^{'}(\omega) + iE^{''}(\omega) = E^{'}(\omega)[1 + i\eta(\omega)] \tag{4}$$

$E^{'}(\omega)$ is the storage modulus, $E^{''}(\omega)$ is the loss modulus, and $\eta(\omega)$ is the loss factor calculated as

$$\eta(\omega) = \frac{E^{''}(\omega)}{E^{'}(\omega)} \tag{5}$$

The complex modulus $E^*(\omega)$ behavior is influenced by multiple factors (excitation conditions, amplitude, frequency, temperature, prestress, relative humidity, and among others) where temperature, frequency, and amplitude are the most relevant ones [9]. Involving the frequency influence, the ASTM E 756–04 "standard test method for measuring vibration-damping properties of materials" [10] details the methodology needed to characterize the mechanical behavior of nonself-supporting viscoelastic materials in the frequency range of 50–5 kHz, implying the use of multimaterial Oberst beam specimens. Nevertheless, the main inconvenience of ASTM E 756–04 standard consists in introducing additional damping or mass through the excitation or through the measurement devices. Many authors investigated these inconveniences [11], which lead to generate alternative methods with improved accuracy [12–20]. Others techniques based on forced vibrations as the Weissenberg rheogoniometer [21], the dynamic mechanical analysis (DMA) technique [22], and the dynamic mechanical thermal analysis (DMTA) technique have also been implemented [23].

In particular, DMTA technique considers together time and temperature by applying the superposition (TTS) principle [24]. The superposition principle relates the material response at a given time t and at a given temperature T under different conditions assuming a relation

$$E(t, T) = E(t_0, \alpha_T T_0) \tag{6}$$

where T and T_0 represent the reference temperature and the reference time, respectively, and α_T is the shift factor relating T to T_0. Accordingly, this principle can also be applied to frequency domain by means of

$$E^*(f, T) = E^*(f_0, \alpha_T T_0) \tag{7}$$

where f and f_0 represent the frequency to be shifted and the reference one, respectively, whereas the shift factor α_T relates f to f_0 in the frequency domain. Arrhenius and William-Landel-Ferry (WLF) [25] are the commonly applied models, where the former model is given by

$$\log\alpha_T = C\left[\frac{1}{T} - \frac{1}{T_0}\right] \tag{8}$$

where C is a constant, the William-Landel-Ferry model can be expressed by

$$\log\alpha_T = \frac{-C_1 \Delta T}{C_2 + \Delta T} \tag{9}$$

where C_1 and C_2 are constants and $\Delta T = T - T_0$.

Hence, if the objective is to describe the material behavior in frequency or temperature ranges outside of the tested ones [25], the TTS principle can be used to derive the master curves (MC).

It should be noted that to apply DMTA techniques for characterizing viscoelastic materials, defect free adhesive specimens are required. The required quality of the specimens can be obtained by applying nondestructive evaluation (NDE) techniques (such as ultrasonic, acoustic emission, radiography, thermography, shearography, holography, and vibration analysis [26], the neutron radiography and the ultrasonic scan [27, 28]). Through these NDE techniques allow the detection of voids such as cracks or other regions of uncured adhesive [29–33]. Most of the research studies [31–33] are focused on the NDE of adhesive joints, but pay attention to adhesion failures.

As a resume, the experimental characterization of the material response to both time and frequency domain is described and also focusedon the specimen preparation and validation. Therefore, the objective of this research is to characterize, under tension strain, the relaxation and the dynamic behavior of the flexible adhesive ISR 70-03:

- The materials and the experimental techniques employed are presented.
- The influence of the strain level and test specimen thickness into specimen behavior is analyzed.
- Relaxation and dynamic master curves are developed using the time-temperature superposition principle. Related shift factors α_T are fitted to the Arrhenius model (Eq. (8)).
- The constructed master curves are fitted to a generalized Maxwell and to a fractional derivative model.

Hence, the proposed models are capable of describing the influence of time, temperature, and strain level over the mechanical properties of the flexible adhesive ISR 70-03.

2.1. Experiments

The mechanical behavior of the analyzed material was characterized by means of relaxation and dynamic tests at different temperatures. A DMTA equipment (RSA3 of TA Instruments) equipped with climate chamber, the Faculty of Engineering of the University of Oviedo, was employed.

2.1.1. Materials and experiments

The studied flexible adhesive is modified silane, commercially named ISR 70-03 produced by Bostik. Regarding the test specimens that were obtained from plates of the cured adhesive produced using casts of 50 × 70 mm × h, where h is the nominal thickness. Three casts with different thickness h of 0.5 mm, 1.0 and 1.5 mm were manufactured. They were manufactured using Teflon to guarantee that a plate of solid material can be demoldedwithout degradation after the curing. The cure time was 48 h for all plates; at room temperature no specific equipment was employed [34]. Consequently, the proposed procedure can be outlines as follows:

- First, the nozzle is drawn along the length of the cast in a zig-zag motion without removing the tip. It should be empathized that the nozzle does not touch with the cast surface in order to ensure an adequate lower surface finish for the adhesive plate.
- Second, when enough amount of adhesive is spread, it is forced to fill the cast, in a single uniform motion by means of a spatula made of Teflon. The spatula has round corners in order to obtain a uniform upper surface finish.

No chemical products are used to prepare the specimens. From the uniform obtained plates, rectangular specimens were cut and measured using an optical microscope. The obtained width and thickness values of each sample are presented in **Table 1**.

Identification sample	Width (mm)	Thickness (mm)
P1	5.45	1.38
P2	5.94	1.50
P3	6.58	1.47
P4	5.26	0.85
P5	6.24	0.78
P6	6.21	0.92
P7	5.25	0.75
P8	5.91	0.45
P9	5.66	0.48
P10	5.69	0.47

Table 1. Specimen dimensions.

To ensure the quality of the produced specimens, tomography techniques by means of neutron radiographies were used. Two defects were found: internal voids and superficial flaws. Accordingly, defect-free samples were identified and selected for testing and denoted as P1, P3, P6, and P10. **Figure 1** shows some of the manufactured samples with defects (P2 and P9) and without (P6 and P10), respectively.

Finally, a DSC test (see Ref. [35] for details) was carried out to determine the glass transition temperature, $T_g = -64°C$.

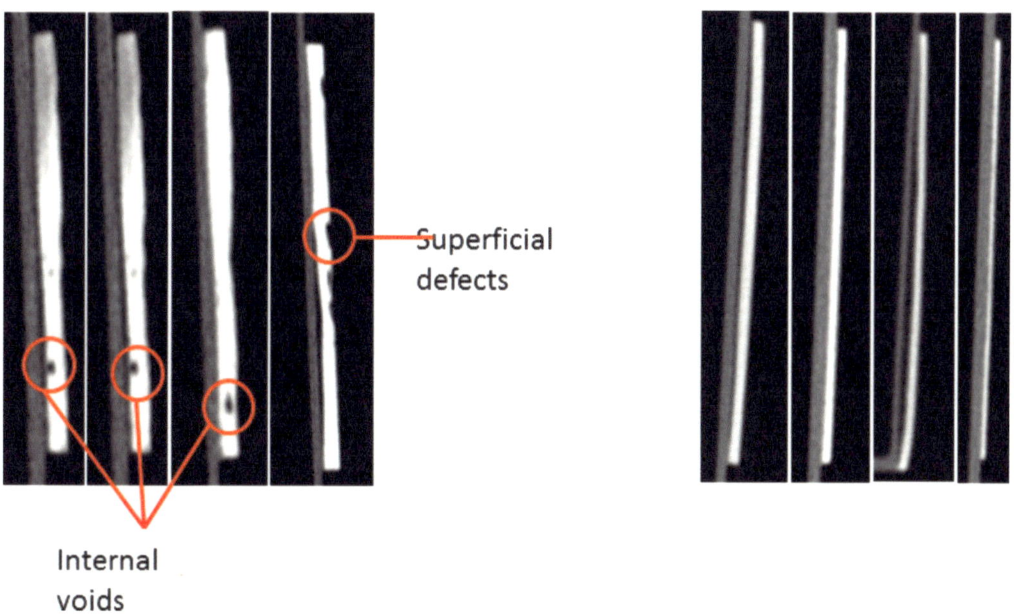

Figure 1. Tomography analysis: specimens with defects, defect-free specimens.

2.1.2. DMTA test

The specimens were studied using a tensile fitting tool shown in **Figure 2**. For the relaxation and dynamic tests, the specimens were prestressed to avoid the buckling effect of tightening the tool screws. Hence, it was verified that the measured force was higher than the prestress at any time all over the experiments. Under these conditions, two different tests were performed in the RSA3 DMTA. First, linearity was studied and second the experiments to obtain the material master curve (MC) were performed. In the latter group, the equipment climate chamber was used and the stabilization time at each temperature was about 10–15 min. These master curves represent the relation between the stress and the strain. Therefore, the relaxation master curve represents the relaxation modulus $E(t)$ while the dynamic one represents the complex modulus $E^*(\omega)$.

Concerning the linearity, the tests were performed at a reference temperature of 20°C. Both specimen thickness and the strain level parameters were analyzed:

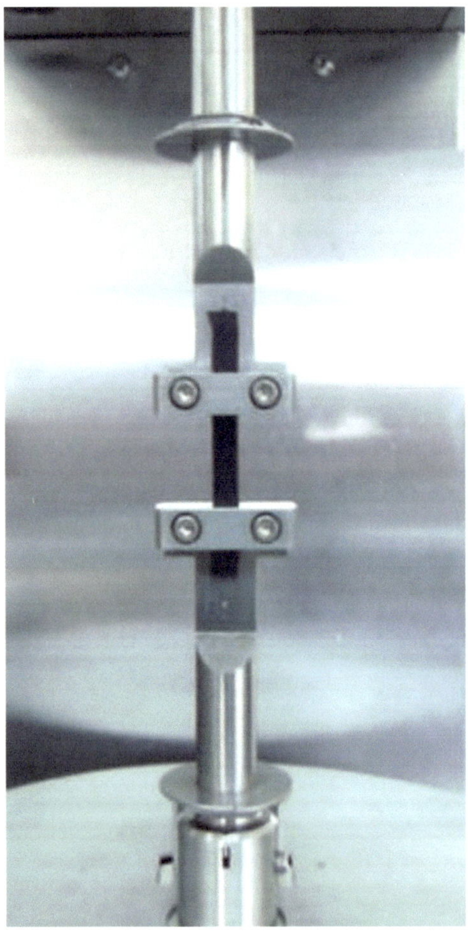

Figure 2. Detail of tension supporting tool.

First, relaxation tests were performed using three samples (P3, P6, and P10) with dimensions shown in **Table 1**. Two test series differing in the induced strain level were completed. Strain levels of $\varepsilon = 0.5$ and $\varepsilon = 2\%$ were, respectively, applied in a time range of 10^{-2} s–10^{2} s.

Following, seven relaxation tests were carried out using only the specimen P1 (see **Table 1**). In these tests, the analyzed time range was 10^{-2} s–4×10^{3} s for induced strain levels of $\varepsilon = 0.2$, $\varepsilon = 0.5$, $\varepsilon = 1$, $\varepsilon = 2$, $\varepsilon = 5$, $\varepsilon = 7$, and $\varepsilon = 8\%$.

For the MC, relaxation and dynamic tests were conducted over sample P3. In both cases, the strain level induced was 0.5%. For the relaxation MC, the analyzed temperature range was −40 to 50°C, where 10 different relaxation tests were carried out. For the dynamic MC, the temperature range was −10 to 20°C and four tests were deformed.

2.2. Results and discussion

2.2.1. *Linearity analysis of the adhesive behavior*

Next, the linearity regarding the material behavior is analyzed involving two test conditions, sample thickness and strain level influence, both by means of relaxation tests.

First, the thickness influence is studied. **Figure 3** shows relaxation test series for $\varepsilon = 0.5\%$ and **Figure 4** shows relaxation test series for $\varepsilon = 2\%$.

From **Figures 3** and **4**, it should be remarked that the relaxation modulus $E(t)$ for the thickness $h = 1.5$ mm is slightly higher than the others, anyway less than 5% in both cases. Considering that the results were derived from different specimens, certain dispersion between the obtained relaxation modulus $E(t)$ can be expected. Therefore, it can be concluded that the specimen thickness has a negligible influence on these test results.

Next, the strain influence on the range of $0.2\% < \varepsilon < 8\%$ over the relaxation modulus $E(t)$ is analyzed through seven tests. The results are illustrated in **Figure 5**.

From **Figure 5**, it can be noted that the $\varepsilon = 0.2\%$ and $\varepsilon = 0.5\%$ curves show slight fluctuations that are not visible for the other curves. These inaccuracies can be expected with the strain decrement, because the force may be lesser than the machine resolution. Consequently, as the strain increases, this effect is less significant. Hence, only the curve $\varepsilon = 0.2\%$ could be rejected due to the fluctuations.

Figure 5 shows that the adhesive material behavior depends on the imposed strain level where the higher the strain, the lower the experimental relaxation modulus $E(t)$. This implies that the material softens when the strain level grows up. With the aim of comparing the relaxation moduli $E(t)$ obtained for the different strain levels, **Figure 6** shows the corresponding ratios taking $\varepsilon = 0.5\%$ as a reference.

Assuming inherent scatter (see **Figure 6**), a nearly constant $E(t)$ratio can be verified for the range of strains analyzed, even for $\varepsilon = 0.2\%$. Consequently, the strain influence on the relaxation modulus can be modeled using Eq. (10)

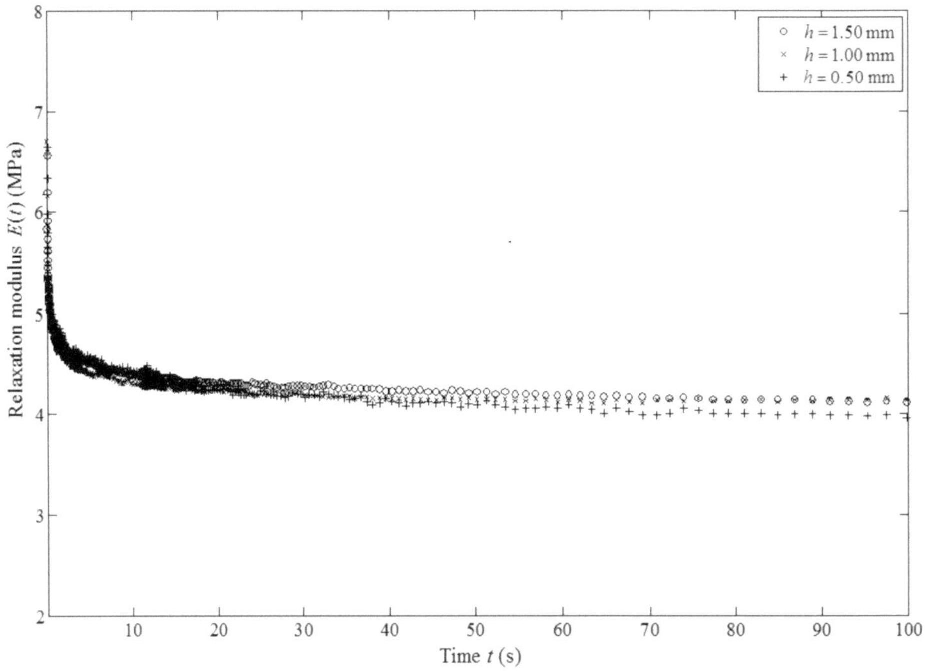

Figure 3. Relaxation modulus tests for thickness influence evaluation for $\varepsilon = 0.5\%$.

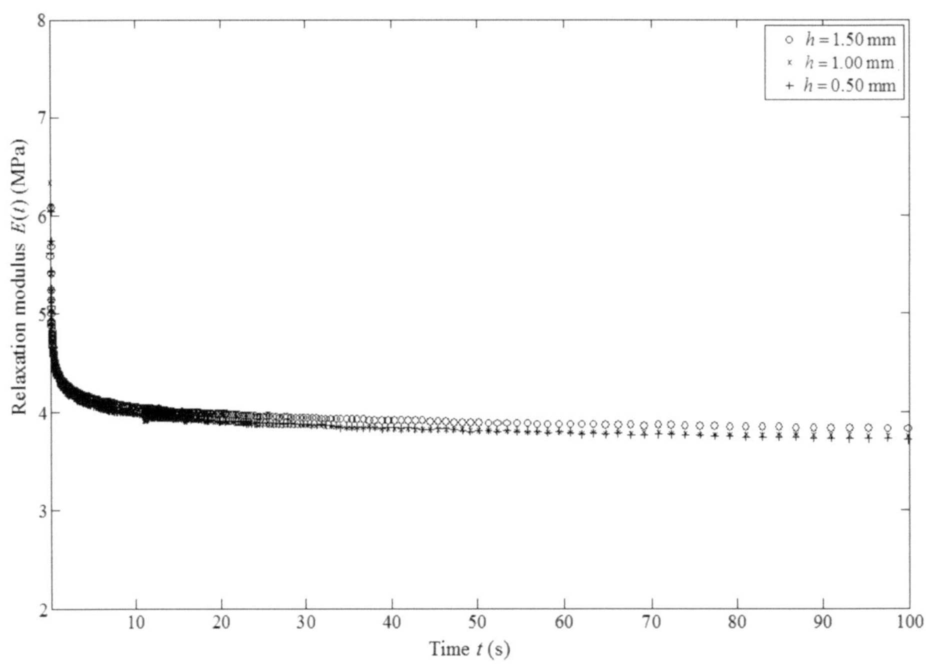

Figure 4. Relaxation modulus tests for thickness influence evaluation for $\varepsilon = 0.5\%$.

http://dx.doi.org/10.5772/66104

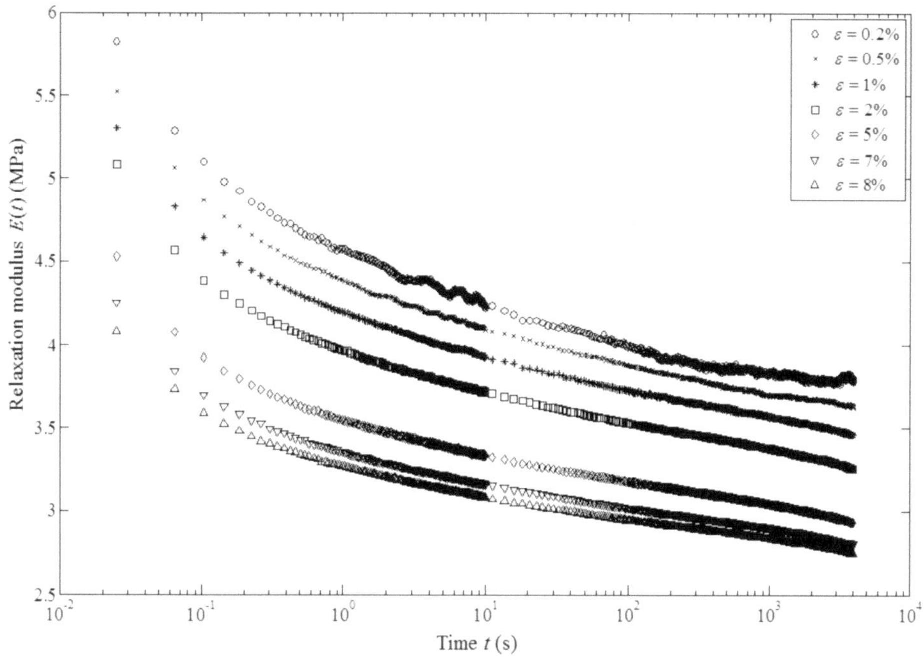

Figure 5. Relaxation modulus tests for strain influence evaluation.

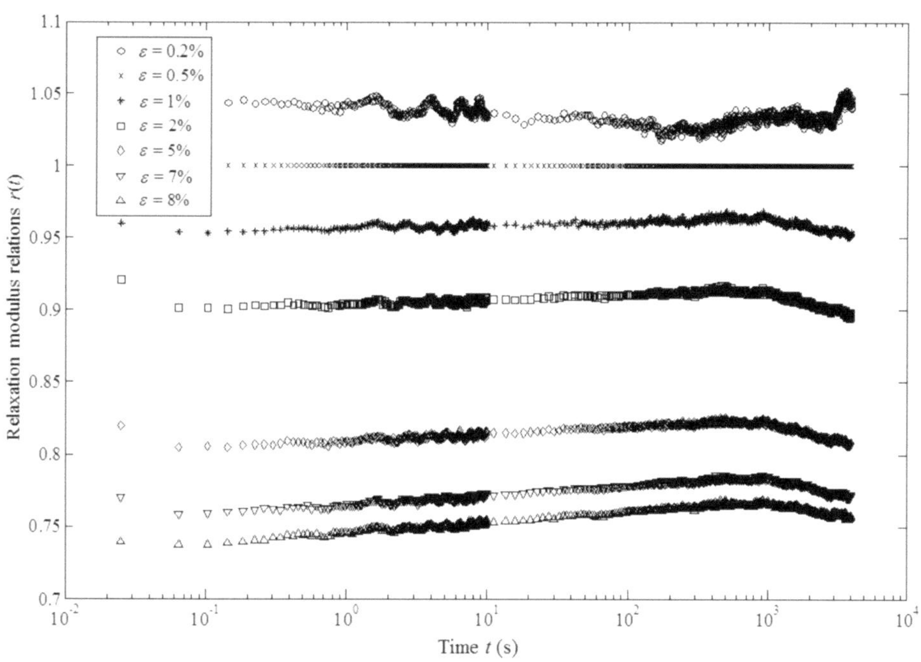

Figure 6. Ratio of relaxation moduli $r(t)$ for $\varepsilon = 0.5\%$ as a reference.

$$E(\varepsilon, t) = r(\varepsilon) \times E_{\text{ref}}(t) \tag{10}$$

where $r(\varepsilon)$ is the ratio between the relaxation modulus $E(\varepsilon, t)$ for the strain ε and that taken as a reference $E_{\text{ref}}(t)$. In this context, **Figure 7** shows the evolution of the mean value of the ratio $r(\varepsilon)$ of different strains as a function of the strain ratio, taking $\varepsilon_{\text{ref}} = 0.5\%$ as a reference.

The decay of the mean value of these ratios with the strain presented in **Figure 7** can be fitted by an experimental function according to

$$r(\varepsilon) = \gamma_1 + \gamma_2 e^{-\gamma_3 \frac{\varepsilon}{\varepsilon_{\text{ref}}}} \tag{11}$$

where the parameters γ_1, γ_2, and γ_3 were estimated by least squares, as $\gamma_1 = 0.73780$, $\gamma_2 = 0.30504$, and $\gamma_3 = 0.14792$, and where ε_0 represents the reference strain level, being $\varepsilon_{\text{ref}} = 0.5\%$ for the present cases. The function $r(\varepsilon)$ of Eq. (11) is represented in **Figure 7** by the discontinuous trace.

In order to verify the proposed model, the results provided by Eqs. (10) and (11) are compared with the experimental data obtained for the strain $\varepsilon = 2\%$, and as **Figure 8** illustrates, Eqs. (10) and (11) provide an accurate prediction of the material relaxation in the 10^{-2} s–4×10^3 s range of time.

Based on the results, it can be observed that mechanical properties of the flexible adhesive ISR 70-03 do not depend on sample thickness. In contrast, they do depend on the strain level. Thus,

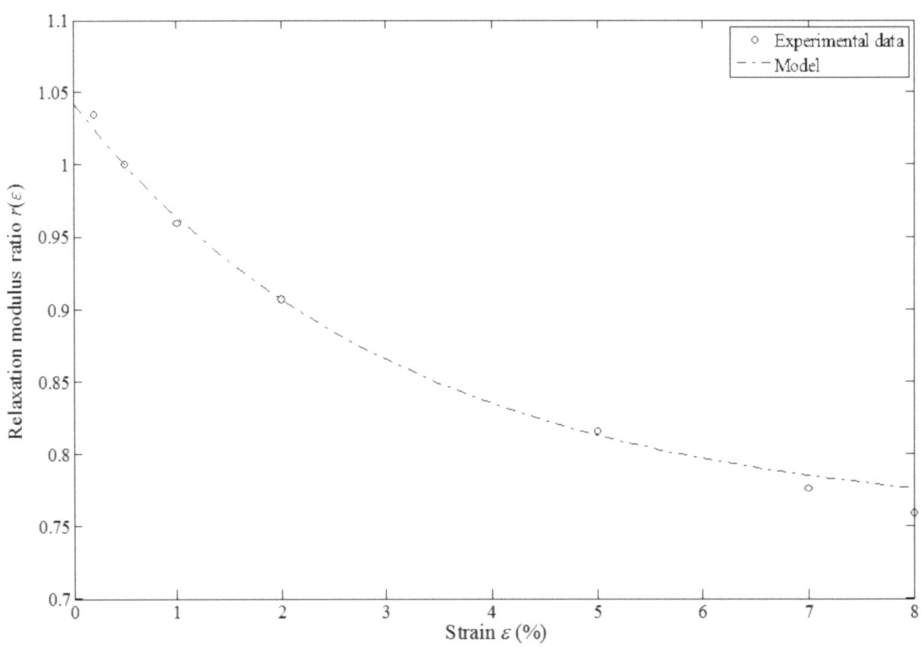

Figure 7. Ratio of the relaxation modulus relations $r(\varepsilon)$ by Eq. (11).

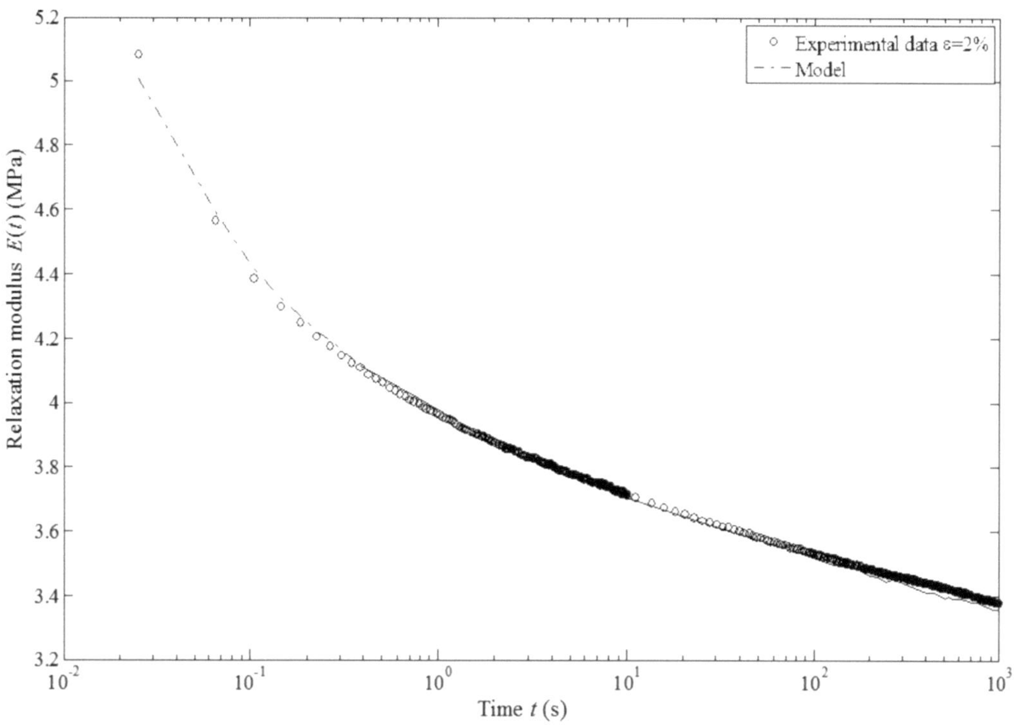

Figure 8. Relaxation modulus given by Eq. (**10**).

the lower the strain, the higher the relaxation modulus, leading to the conclusion that the material softens as the strain level grows up. Consequently, the exponential model given by Eq. (11) represents the influence of the strain level and Eq. (10) allows computing the relaxation modulus function $E(t)$ for any strain level, once the relaxation modulus for a reference strain level is obtained.

This increase in the relaxation modulus $E(t)$ as a consequence of the decrease in the strain level is in accordance to the observed behavior in other viscoelastic materials [36] so as other rubber like compounds [37] used for vibration control.

2.2.2. Relaxation master curve

Following, the time domain master curve (MC) is obtained applying the time-temperature superposition (TTS) principle [25]. The MC is built-up from 10 relaxation curves, obtained for 10 different temperatures ranging from −40°C–50 °C all of them in the 10^{-2} s–10^{2} s range, as shown in **Figure 9**.

From **Figure 9**, it should be remarked that the higher the temperature, the lower the relaxation modulus. It should be noted also that the test carried out at $T = 30°C$ intersects the one at $T = -20°C$ due to an error during the test. Analogous situation can be verified for the curves$T = -10°C$ and $T = -20°C$. Reasons for removing these curves from the analysis will be discussedlater.

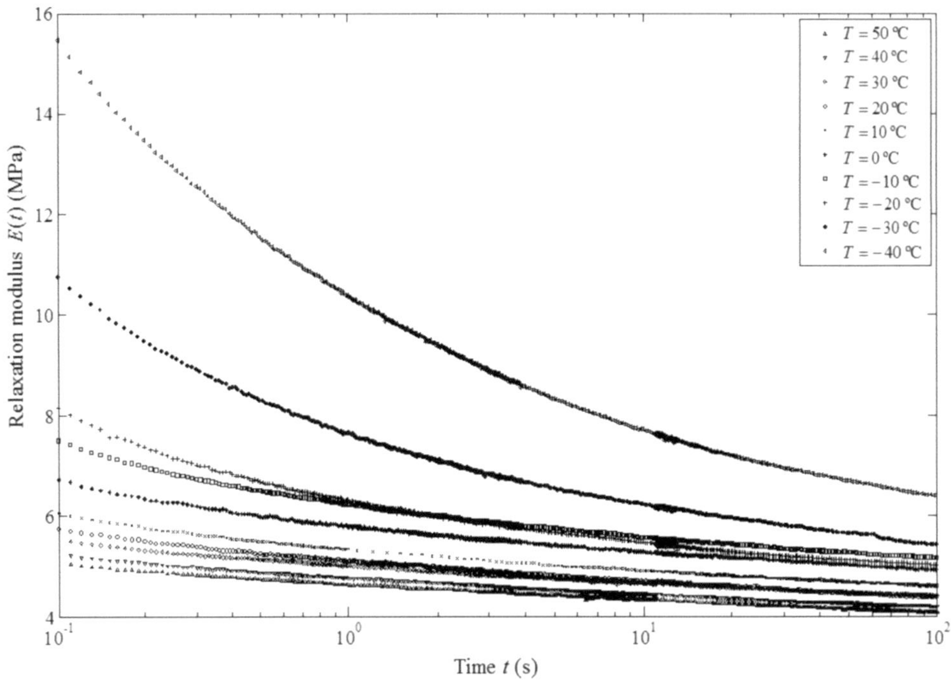

Figure 9. Relaxation curves resulting from a broad range of temperatures.

Based on the industrial application, the temperature $T_0 = 20°\text{C}$ is chosen as a reference to derive the MC. First, all the possible shift factors $\alpha_T(t_0)$ between the reference curve and that for $T = 10°\text{C}$ are computed. Thereafter, the times t and t_0 are determined for which the relaxation modulus curves coincide, $E(t, 10°\text{C}) = E(t_0, 20°\text{C})$. Accordingly, the shift factor $\alpha_T(t_0)$ at any time t_0 is computed as

$$\alpha_T(t_0) = t/t_0. \tag{12}$$

Based on the TTS principle, singe shift factor α_T should exist for each temperature. Therefore, the optimum shift factor $\alpha_T(t_0)$ is computed from all the possible factors. This is done by minimizing the error function defined as the difference between the original and the shifted curves. Then, a preliminary MC is reached, for which the procedure is repeated for any curve represented in **Figure 9**.

Figures 10 and **11** show the range of possible factors $\alpha_T(t_0)$ obtained from each preliminary MC and the one to be shifted below and above $T_0 = 20°\text{C}$, respectively.

From **Figures 10** and **11**, significant conclusions can be obtained. Involving the curves for temperatures $T = -40, -20, -10, 0, 10, 40$, and $50°\text{C}$ nearly constant $\alpha_T(t_0)$ factors are computed in accordance to the TTS principle. Hence, a reasonable optimum value can be determined at each temperature. On the contrary, the curve for $T = -30°\text{C}$ exhibits high scatter and should, therefore, be disregarded from the analysis.

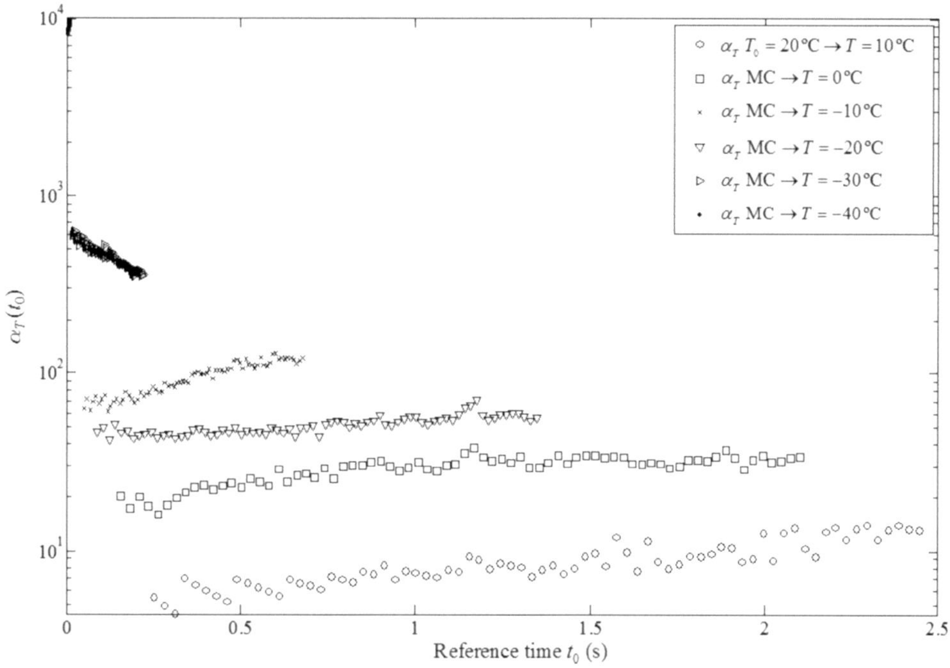

Figure 10. Range of possible factors $\alpha_T(t_0)$ in time domain below T_0.

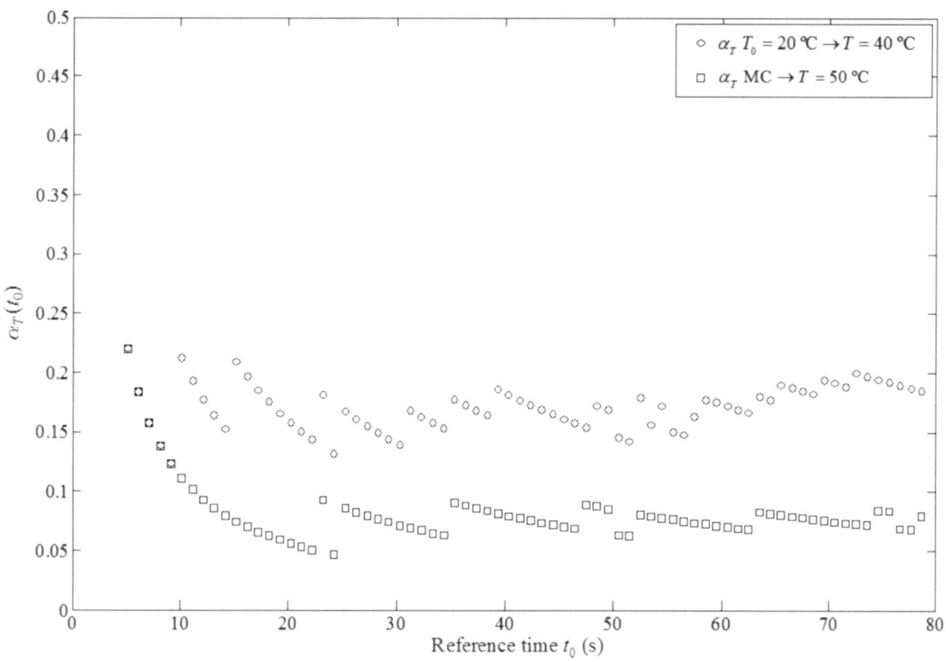

Figure 11. Range of possible factors $\alpha_T(t_0)$ in time domain above T_0.

It should be mentioned that the position for the curves resulting for $T = -10$ °C is shifted and, therefore, at least one of them must be removed from the analysis. Decision is taken based on the optimum shift factors α_T represented as a function of temperature in **Figure 12**.

It follows from **Figure 12** that, in order to obtain a monotonically decreasing $\log\alpha_T - T$ curve without irregularities, it is convenient to discard the information provided for $T = -20$°C. The observed behavior can be modelized using the Arrhenius model (Eq. (8)), the results of which is represented in **Figure 12** by the discontinuous trace. The parameter C was estimated by linear regression, giving $C = 4419$ K, corresponding to a regression coefficient $R^2 = 0.999$.

Therefore, applying the presented procedure and computing the optimums α_T values shown in **Figure 12**, each single curve has been shifted to built-up the desired relaxation master curve, as shown in **Figure 11**. The constructed MC decays from 16 MPa down to 4 MPa in the covered range 10^{-5} s–1.6×10^3 s.

Based on the results, it can be said that the studied material can be used for vibration control proposes. It should be remarked also that based on the mechanical strength, the ISR 70-03 can be employed not only for non-structural applications but also more demanding structural designs that can be proposed [38, 39].

2.2.3. Dynamic master curve

The dynamic MC can be constructed using an analogous procedure to that followed in the derivation of the master curve for the relaxation modulus. Hence, **Figure 12** represents four

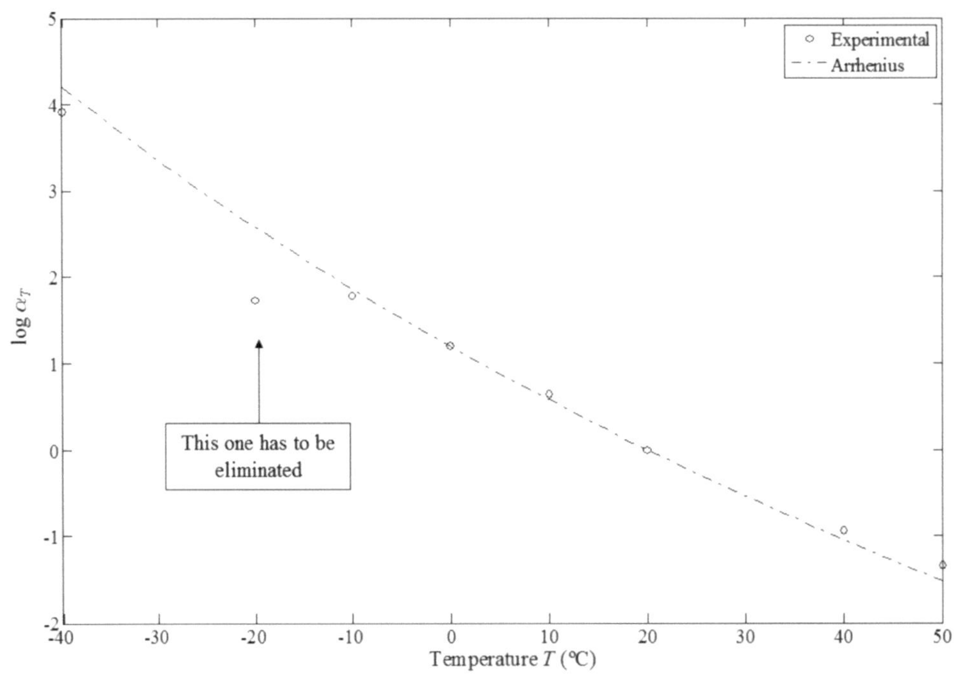

Figure 12. Optimum shift factor α_T values for time domain and Arrhenius model fit.

curves of the storage modulus E' and loss factor η, respectively, in the 10^{-1} Hz–2×10^{1} Hz frequency range, for four different temperatures, $T = -10, 0, 10,$ and 20°C. In this case, taking as well into account the industrial application, the same reference temperature $T_0 = 20$°C has been adopted.

Figure 13 shows that the higher the temperature, the lower both the storage modulus E' and the loss factor η. Obviously, this in accordance with the results obtained in the time-domain case.

Therefore, all the possible shift factors $\alpha_T(f_0)$ between the preliminary master curve and the one to be shifted are computed using the storage modulus $E'(f, T) = E'(f_0, T_0)$, and the shift factors $\alpha_T(f_0)$. Thus, the shift factor $\alpha_T(f_0)$ represented in **Figure 14** satisfying

$$\alpha_T(f_0) = f/f_0. \tag{13}$$

Taking into account the moderate scatter in the results assignable to the experimental nature of the data, the $\alpha_T(f_0)$ values of the three curves shown in **Figure 15** may be taken as a constant. To compute a single value for each temperature, an analogous minimization procedure to that followed in the relaxation case was applied to the complex modulus $E^*(f, T)$. The resulting optimums shift factors α_T are computed and represented in **Figure 16**.

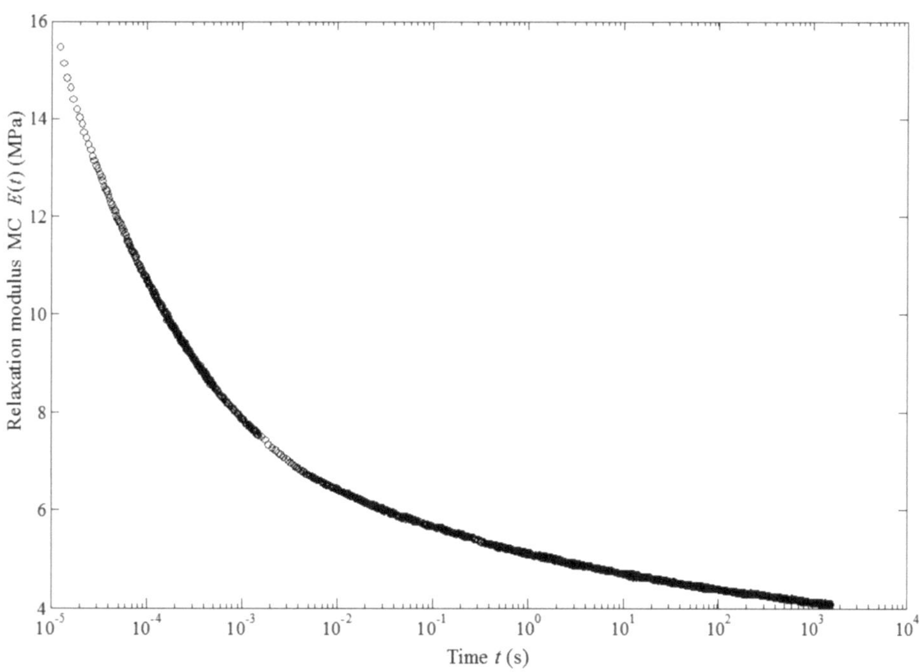

Figure 13. Storage modulus and loss factor curves for four different temperatures.

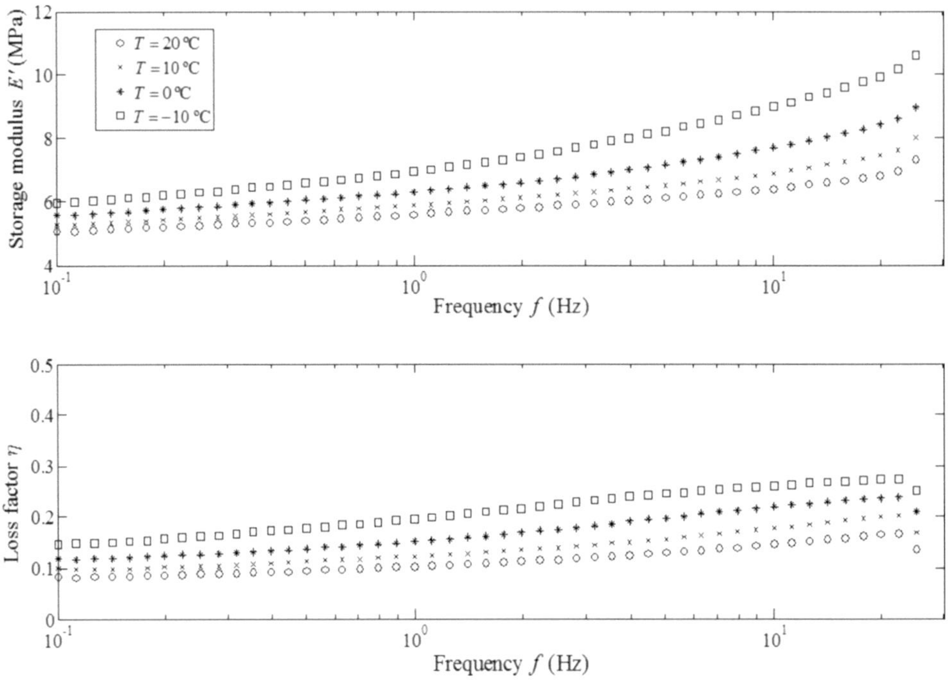

Figure 14. Relaxation master curve in time domain.

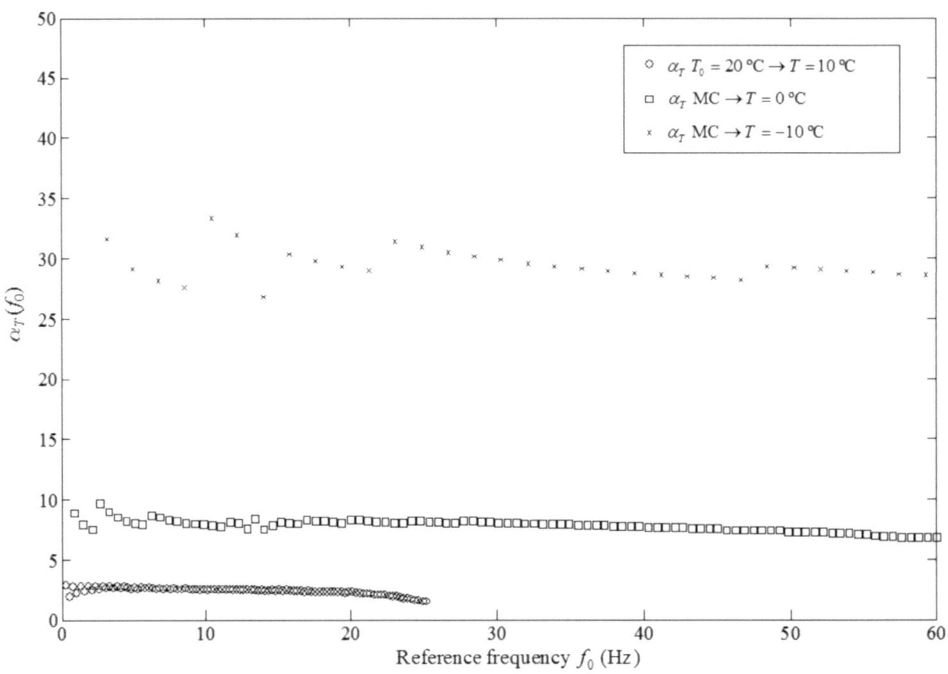

Figure 15. Shift factors $\alpha_T(f_0)$ in frequency domain below T_0.

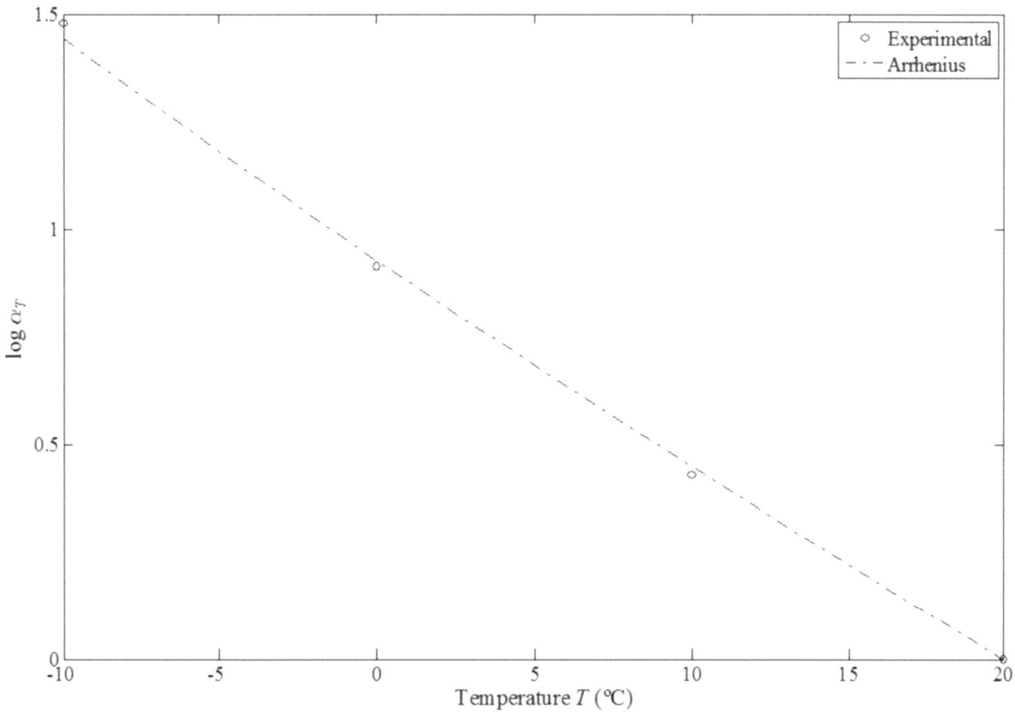

Figure 16. Optimum shift factor α_T values for frequency domain and Arrhenius model fit.

It can be seen from **Figure 16** that the determinedvalues decay with the temperature. Similarly as in the time-domain case, it follows that the Arrhenius model (Eq. (8)), which is represented in **Figure 13** by the discontinuous trace, is a good candidate for describing the dynamic behaviorobserved in the material. The parameter C became $C = 3710$ K for a regression coefficient $R^2 = 0.999$. Consequently, the corresponding MC is constructed as shown in **Figure 17**.

In **Figure 17** the MCs for the storage modulus $E^{'}$ and loss factor η in the range 0.1 Hz–700 Hz are illustrated where the rubbery behavior and the beginning of the transition zone are present. As it was concluded in the time-domain analysis, the ISR 70-03 is applicable for vibroacoustic control of structures even for low frequency applications [36].

2.3. Models

Next, a relaxation modulus model capable of taking into account together the influence of strain, time and temperature as

$$E(\varepsilon, t, T) = r(\varepsilon) \times E_{\text{ref}}(t, T). \tag{14}$$

is proposed. Concretely, a generalized Maxwell model (also known as the Prony series model) [40] and a fractional model [41] are fitted to the previously obtained master curve $E_{\text{ref}}(t)$, in the

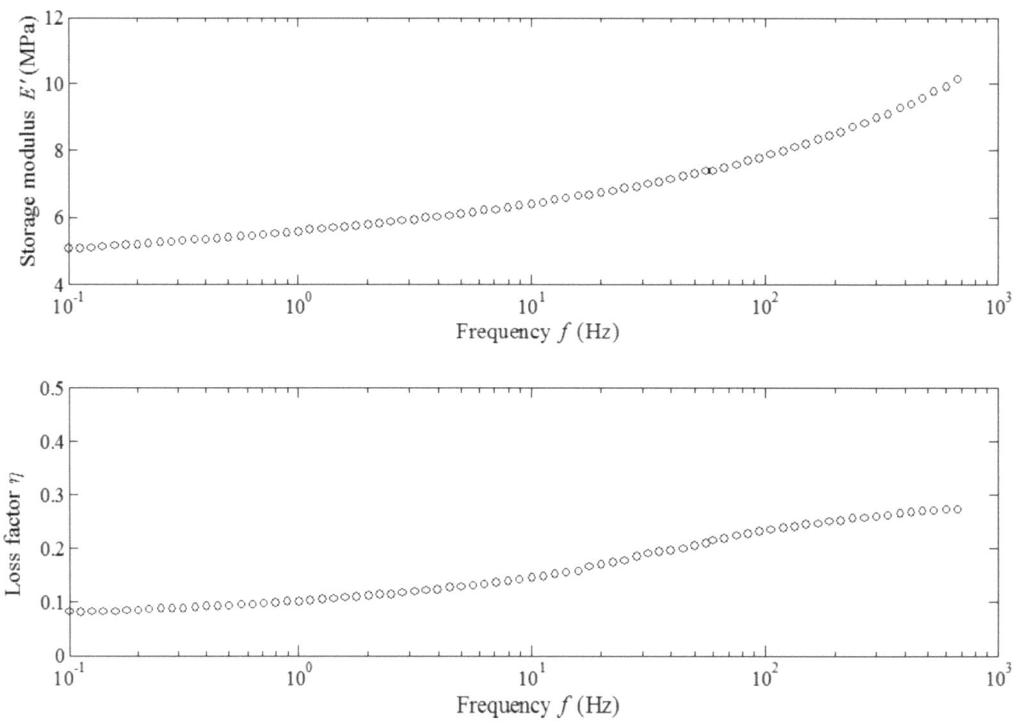

Figure 17. Master curves for the storage modulus and loss factor in frequency domain.

time range of 1×10^{-5} s–1.6×10^{3} s. Also, a complex modulus model is suggested to modelize the dynamic master curve $E^*(f)$ in the frequency range of 0.1 –700 Hz.

2.3.1. Relaxation models

Next, two relaxation models are presented. Concerning the generalized Maxwell (or Prony series) model, $E_{\text{ref}}(t)$ yields

$$E_{\text{ref}}(t) = E_0 + \sum_{i=1}^{N} E_i e^{-t/\tau_i}. \quad (15)$$

where E_0 is the relaxed modulus, E_i represents stiffness parameter and τ_i denotes relaxation time. $N = 9$ is the number of terms chosen for the generalized Maxwell model to accurately represent the experimental results. The fitting has been carried out by least squares, corresponding to a regression coefficient of $R^2 = 0.999$. The obtained values for E_0, E_i and τ_i are presented in **Table 2**.

Concerning the fractional derivative model, the four-parameter derivative model [42]

$$\sigma(t) + \tau^{\alpha} D^{\alpha} \sigma(t) = E_r \varepsilon(t) + \tau^{\alpha} (E_u - E_r) D^{\alpha} \varepsilon(t). \quad (16)$$

is employed, where σ denotes the stress, E_r and E_u are relaxed and unrelaxed modulus, τ is the relaxation time, and D^{α} is the α order fractional derivative operator. The G1 numerical

Stiffness parameters (MPa)	Relaxation time (s)
$E_0 = 4.101$	
$E_1 = 0.378$	$\tau_1 = 1.60\times10^{-3}$
$E_2 = 0.427$	$\tau_2 = 2.91\times10^{2}$
$E_3 = 0.523$	$\tau_3 = 9.965$
$E_4 = 0.773$	$\tau_4 = 4.14\times10^{-1}$
$E_5 = 1.014$	$\tau_5 = 2.01\times10^{-2}$
$E_6 = 1.521$	$\tau_6 = 1.68\times10^{-3}$
$E_7 = 2.501$	$\tau_7 = 2.44\times10^{-4}$
$E_8 = 3.364$	$\tau_8 = 4.82\times10^{-5}$
$E_9 = 5.904$	$\tau_9 = 9.83\times10^{-6}$

Table 2. Coefficients for the Prony series fitted to the relaxation master curve.

approximation [43] of the fractional derivative of a generic function $f(t)$ at the instant $t_n$is used, given by

$$\mathrm{D}^{\alpha} f(t)\big|_{t=t_n} = \frac{1}{(\Delta t)^{\alpha}} \sum_{j=0}^{n-1} A_{j+1} f(t_{n-j}). \tag{17}$$

where $A_{j+1} = A_j(j-\alpha-1)/j$ with $A_1 = 1$, and Δt is the time step. Thus, applying a strain step as $\varepsilon(t) = \mathrm{H}(t)$ where $\mathrm{H}(t)$ is the Heaviside function, the relaxation modulus $E(t_n) = E_n$ where $t_n = n\times\Delta t$ can be calculated as

$$E_n = \frac{E_{\mathrm{r}} + (E_{\mathrm{u}}-E_{\mathrm{r}})\left(\frac{\tau}{\Delta t}\right)^{\alpha} \sum_{j=0}^{n-1} A_{j+1} - \left(\frac{\tau}{\Delta t}\right)^{\alpha} \sum_{j=1}^{n-1} A_{j+1} E_{n-j}}{1 + \left(\frac{\tau}{\Delta t}\right)^{\alpha}}. \tag{18}$$

An error minimization procedure has been applied for the curve fitting, and $E_{\mathrm{r}} = 3.271$ MPa, $E_{\mathrm{u}} = 20.147$ MPa, $\tau = 1.589\times10^{-7}$ s and $\alpha = 0.116$ have been determined.

In **Figure 18**, both models are compared with the experimental data.

From **Figure 18**, it should be noted that the fractional model has been fitted using a time step of $\Delta t = 1\times10^{-2}$ s. It should be pointed out that both models are able to reproduce the experimental relaxation master curve. Nevertheless, the curve provided by the fractional model is smoother. Besides, it should be pointed out that the fractional derivative model needs only four parameters whereas the generalized Maxwell model needs 19 parameters. On the contrary, the computation of Eq. (18) is much larger than that of Eq. (15). As a conclusion, it should be verified that relaxed and unrelaxed moduli provided by both models are coherent. Hence, it should be highlighted that involving the generalized Maxwell model, the unrelaxed modulus E_{u} can be calculated as

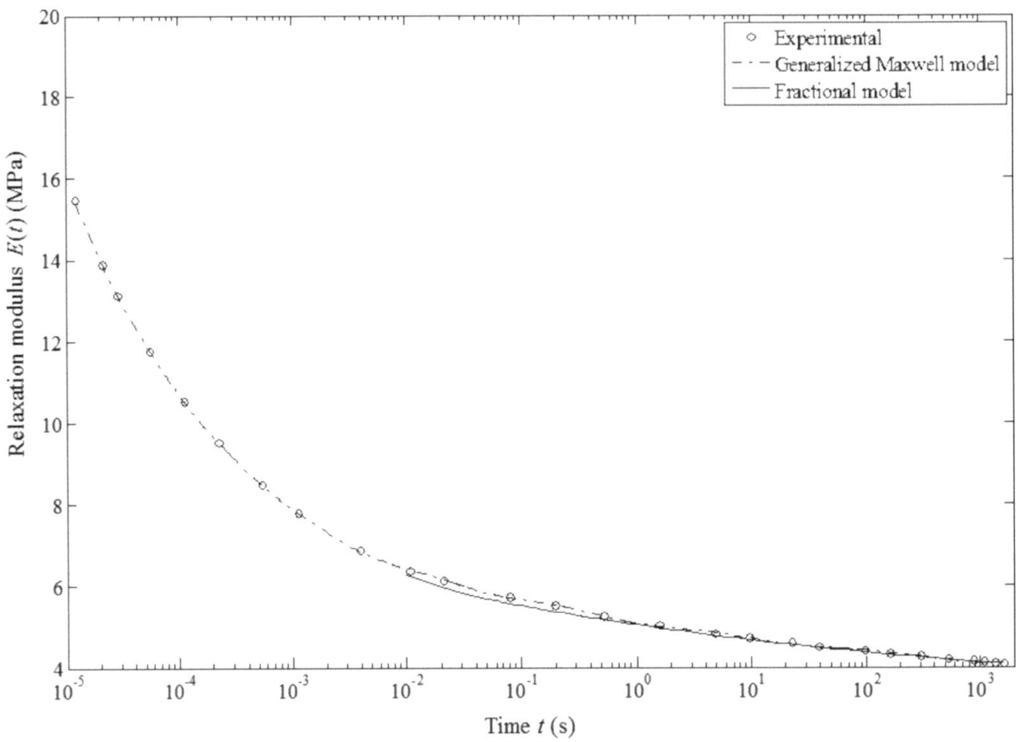

Figure 18. Relaxation models in time domain.

$$E_u = E_0 + \sum_{i=1}^{N} E_i. \tag{19}$$

Hence, the obtained unrelaxed modulus E_u for the generalized Maxwell model is $E_u = 20.806$ MPa while that for the fractional model is $E_u = 20.147$ MPa.

Apart from the computational cost, fractional models describe precisely the viscoelastic behavior even in time domain despite the low number of parameters needed to describe wide time ranges [44–47].

2.3.2. *Dynamic models*

Following, the complex modulus $E^*(\omega)$ for the generalized Maxwell and fractional derivative models is derived from the Fourier transform of Eqs. (15) and (16), respectively. The one for the generalized Maxwell model yields

$$E^*(\omega) = E_0 + i\omega \sum_{i=1}^{N} \frac{\tau_i E_i}{1 + i\omega\tau_i}. \tag{20}$$

http://dx.doi.org/10.5772/66104

Stiffness parameters (MPa)	Relaxation time (s)
$E_0 = 5.049$	
$E_1 = 0.032$	$\tau_1 = 2.42\times10^{-7}$
$E_2 = 0.070$	$\tau_2 = 4.31\times10^{-3}$
$E_3 = 0.097$	$\tau_3 = 5.89\times10^{-3}$
$E_4 = 0.211$	$\tau_4 = 5.82\times10^{-3}$
$E_5 = 0.690$	$\tau_5 = 3.29\times10^{-1}$
$E_6 = 0.924$	$\tau_6 = 2.40\times10^{-2}$
$E_7 = 1.442$	$\tau_7 = 1.70\times10^{-3}$
$E_8 = 2.637$	$\tau_8 = 2.80\times10^{-4}$
$E_9 = 18.17$	$\tau_9 = 2.16\times10^{-5}$

Table 3. Coefficients for the Maxwell model fitted to the dynamic master curve.

where $\omega = 2\pi f$, f is the excitation frequency. The curve fitting is carried out by least squares with $N = 9$. The obtained numerical values are presented in **Table 3**.

Concerning the one based on fractional derivatives, the corresponding complex modulusresults in

$$E^*(\omega) = \frac{E_r + (E_u - E_r)(i\omega\tau)^\alpha}{1 + (i\omega\tau)^\alpha}. \tag{21}$$

Accordingly, fitting by least squares, $E_r = 4.419$ MPa, $E_u = 31.66$ MPa, $\tau = 1.17\times10^{-4}$ s and $\alpha = 0.353$ have been found, the regression coefficient satisfying $R^2 = 0.998$.

Both models, Eqs. (20) and (21), are contrasted to the experimental dynamic master curve shown in **Figure 19**.

From **Figure 19**, it should be highlighted that the generalized Maxwell model fits the experimental storage modulus E'. However, the fractional derivative model reproduces better the experimental loss factor η. It should be remarked also that the fractional derivative model needs only four parameters. Besides, the fractional model parameters extraction Eq. (21) is faster than that of Eq. (20).

Involving the curve fitting, it should be noted that for time domain the difference between the unrelaxed modulus E_u provided by both models is 1.71%. Regarding frequency domain, the difference is 7.39%. However, for the relaxed modulus, the differences between the generalized and the fractional models for time and frequency domains are 20.24 and 12.47%, respectively. Consequently, the results provided by these models differs for $t \rightarrow \infty$ and for $f = 0$ Hz.

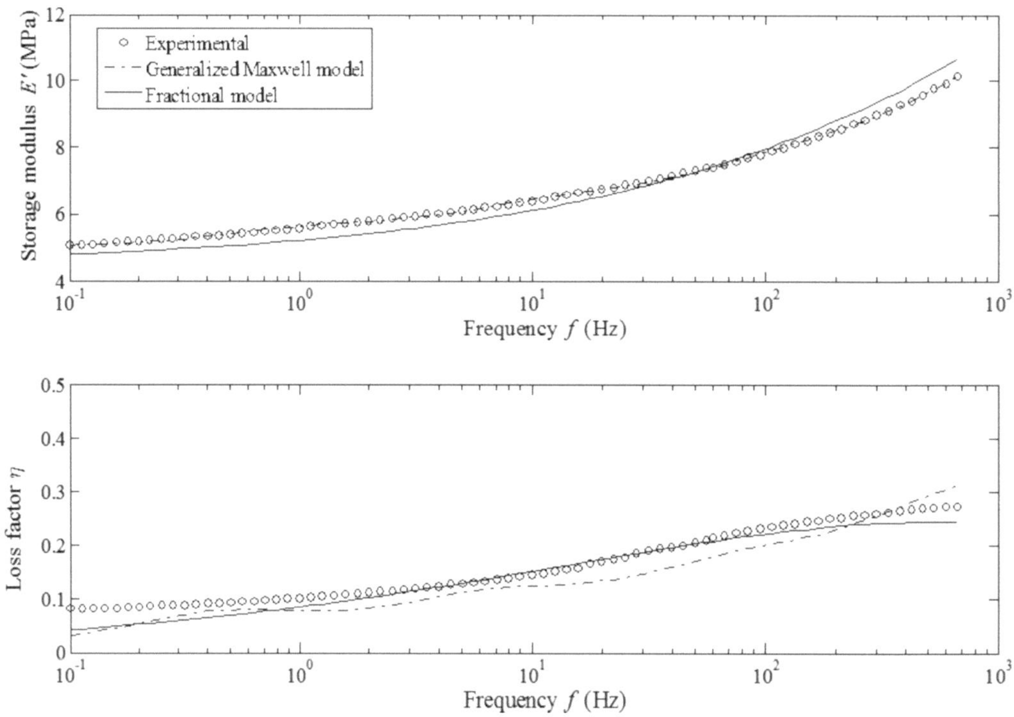

Figure 19. Dynamic models in frequency domain.

Hence, it can be stated that fractional derivative models are really valuable tools for describing the viscoelastic behavior principally in frequency domain but also in time domain [44–47].

2.4. Conclusions

The experimental characterization and modelization of the relaxation and complex moduli of the flexible adhesive ISR 70-03 havebeen performed using dynamic mechanical thermal analysis (DMTA).

- To conduct the experiments, defect-free samples have been manufactured using Teflon™ casts. Regarding validation, it can be concluded that tomography techniques by means of neutron radiography are able to identify internal and external defects in cured adhesives.
- Regarding the linearity, it can be stated that the relaxation test results are not influenced by sample thickness. On the contrary, the strain influence has been verified. Consequently, it has been modeled using an exponential model. Therefore, it can be noted that the material stiffens when the strain level decreases.
- Involving the master curves, the relaxation modulus $E(t)$ and the complex modulus $E^*(\omega)$ have been derived by means of a procedure based on the TTS principle. Besides, the temperature dependence has been modelized by the Arrhenius model.

http://dx.doi.org/10.5772/66104

- Finally, the generalized Maxwell model and fractional derivative models have been fitted to the time and frequency domains master curves. The generalized Maxwell one has been fitted using 19 parameters whereas the fractional derivative model has four parameters. Involving time domain, the fitting is accurate enough for both models. Regarding frequency domain, the storage modulus $E^{'}$ and loss factor η may be represented also by the generalized Maxwell and fractional derivative models. However, the fractional model fits the experimental response in a wide time or frequency ranges with a lower number of parameters. Hence, the fractional derivative model should be used for both time and frequency domains [44–47].

As a conclusion, a model capable of representing the influence of time, temperature and strain level over the mechanical properties of the flexible adhesive ISR 70-03 has been presented.

3. Relaxation modulus: complex modulus interconversion for linear viscoelastic adhesives

The interconversion path between the relaxation modulus $E(t)$ and the corresponding complex modulus $E^{*}(\omega)$ for linear viscoelastic solid materials is explored. The key difference with other approximate methods relies on the fact that in the presented procedure, the fast Fourier transform (FFT) algorithm is directly applied on the time-dependent part of the viscoelastic response $R(t)$. First, method foundations are outlined. Next, a theoretical example is developed using the generalized Maxwell model. Using this example, influence of sampling conditions and experimental error and data dispersion is studied. Finally, the accuracy of the method is proved by an application example using experimental data. As a conclusion, the proposed procedure is able to compute the complex modulus by means of relaxation tests and vice versa.

Concerning VEM behavior modeling, the memory of viscoelastic materials such as viscoelastic adhesives can be properly represented using the Boltzmann superposition principle [48]. Therefore, time evolution of stress $\sigma(t)$ can be evaluated using relaxation functions $R(t)$ through convolution integrals given by

$$\sigma(t) = E_{\mathrm{r}}\, \varepsilon(t) + \int_{0}^{t} R(t-\lambda)\dot{\varepsilon}(\lambda)\mathrm{d}\lambda. \tag{22}$$

where $\varepsilon(t)$ is the strain, E_{r} represents the viscoelastic constant, λ denotes the integration variable and $(\dot{.})$ represents the time derivative. In frequency domain, viscoelastic behavior can be represented by the complex modulus approximation [49], as shown in Eq. (4).

Concerning experimental characterization of viscoelastic adhesives, ASTM E 756-04 [10] details the methodology to characterize the mechanical behavior of non-self-supporting viscoelastic materials, implying the use of multimaterial Oberst beam specimens.

In this context, the DMTA technique allows to take into account together temperature and time (or temperature and frequency) by means of the time-temperature superposition (TTS) principle [24, 50] introducing no extra mass or damping.

Thus, frequency-time interconversion methods are valuable and useful tools [51–53] due to the fact that they can also be applied to overcome the inherent difficulties of relaxation or dynamic characterization [54–57], depending on the tested material. The most widely applied methods [51] for material functions conversion from time to frequency domains are those based on the Prony series model [40], and the opposed conversion can be achieved through the algorithms proposed by Ninomiya and Ferry [58]. The former can be obtained by fitting the experimental data by means of the generalized Maxwell model [59], whereas the latter is based on experimental data fitting.

As summary, the objective of this section is to propose an interconversion method between time and frequency domains capable of obtaining the complex modulus $E^*(\omega)$ by means of relaxation tests, and vice versa. The main advantage of this procedure is the direct application of the fast Fourier transform (FFT) algorithm on experimental data. On the contrary, other existing methods [55–65] are based on fitting models or theoretical functions. In particular, the proposed method is relevant when a Prony series cannot be accurately fitted to the experimental data in time or frequency domains. This section is structured as follows:

- Method foundations are outlined.
- A theoretical example is developed using a generalized Maxwell model. Using this example, influence of sampling conditions and experimental error and data dispersion are studied.
- The accuracy of the method is proved by an application example using experimental data.

3.1. Method foundation

An experimental relaxation test consists on applying a strain step as $\varepsilon(t) = \varepsilon_0 \mathrm{H}(t)$, where ε_0 represents the magnitude of the strain and $\mathrm{H}(t)$ is the Heaviside function. Consequently, applying a strain step and substituting its time derivative into Eq. (22), it yields

$$\sigma(t) = E_\mathrm{r}\, \varepsilon_0 \mathrm{H}(t) + \int_0^t R(t\text{–}\lambda)\, \varepsilon_0 \delta(\lambda) \mathrm{d}\lambda = [E_\mathrm{r} + R(t)]\varepsilon_0. \tag{23}$$

where $\delta(t)$ is the Dirac function. Then, the relaxation modulus $E(t)$ can be deduced as

$$E(t) = \frac{\sigma(t)}{\varepsilon_0} = E_\mathrm{r} + R(t). \tag{24}$$

where the long-term part of the relaxation modulus is represented by the viscoelastic constant E_r and where the time-dependent component is represented by $R(t)$.

Then, applying the Fourier transform, the complex modulus $E^*(\omega)$ is derived. On the one hand, applying the Fourier transform over Eq. (24), it results in

$$\tilde{E}(\omega) = \frac{E_r}{i\omega} + \tilde{R}(\omega) \tag{25}$$

where $\tilde{(.)}$ represents the Fourier transform. On the other hand, applying the Fourier transform over Eq. (1) it yields

$$\tilde{\sigma}(\omega) = E_r\, \tilde{\varepsilon}(\omega) + i\omega\, \tilde{R}(\omega)\, \tilde{\varepsilon}(\omega) \tag{26}$$

where from the complex modulus $E^*(\omega)$ can be derived, yielding

$$E^*(\omega) = E_r\ + i\omega\, \tilde{R}(\omega) \tag{27}$$

Therefore, by substituting the Fourier transform of the time-dependent part of the viscoelastic response $\tilde{R}(\omega)$ into Eq. (25), a relationship between complex modulus $E^*(\omega)$ and the Fourier transform of the relaxation modulus $\tilde{E}(\omega)$ is obtained,

$$E^*(\omega) = i\omega\, \tilde{E}(\omega) \tag{28}$$

As a result, the complex modulus $E^*(\omega)$ of a linear viscoelastic material can be obtained from the Fourier transform of its relaxation modulus $E(t)$. Nevertheless, applying the fast Fourier transform (FFT) algorithm, the resulting complex modulus $E^*(\omega)$ will suffer from leakage because $E(t)$ is not periodic [66, 67] and $E(t) \underset{t\to\infty}{=} E_r \neq 0$. Therefore, to avoid leakage, it should be remarked that the time-dependent part of the viscoelastic response $R(t)$ representing viscoelastic component disappears with time, $R(t) \underset{t\to\infty}{=} 0$. Hence, FFT algorithm does not produce leakage on complex modulus $E^*(\omega)$ approximation if Eq. (27) is used instead of Eq. (28). The drawback of the procedure is that the viscoelastic constant E_r must be extracted from the experimental data.

3.2. Theoretical example

Next, the influence of sampling conditions and experimental error and data dispersion is studied using an exponential material model. The study analyses some aspects related to the FFT algorithm, which are: leakage, signal discretization and the analyzed ranges. In this analysis, the FFT algorithm proposed by Cooley and Tukey [64] to compute the discrete Fourier transform is employed [65] despite the documented drawbacks [69]. There are other methods, as those presented by Dutt and Rokhlin [66] that have been used in several of applications [67–70]. However, as stated, the algorithm proposed by Cooley and Tukey [64] will be used.

An exponential damping model is widely used in the literature [71, 72], because it is capable of modeling damping mechanisms arising from viscoelastic nature of materials. Its time-dependent part of the viscoelastic response $R(t)$ is given by

$$R(t) = c\mu\, e^{-\mu t} \tag{29}$$

resulting in a relaxation modulus as

$$E(t) = E_r + E_1\, e^{-\mu t} \tag{30}$$

where $\mu = \tau_m^{-1} = E_1/c$ represents the material relaxation parameter, τ_m is the relaxation time, and E_1 and c are the stiffness and damping coefficients, respectively. The Fourier transform $\tilde{R}(\omega)$ of the time-dependent part of the viscoelastic response $R(t)$ is given by

$$\tilde{R}(\omega) = E_1 \frac{1}{\mu + i\omega} \tag{31}$$

Accordingly, the complex modulus $E^*(\omega)$ yields

$$E^*(\omega) = E_r + E_1 \frac{i\omega}{\mu + i\omega} \tag{32}$$

wherefrom storage modulus $E^{'}$ and loss factor $\tan\,\delta$ can be directly obtained as

$$E^{'}(\omega) = E_r + E_1 \frac{\omega^2}{\mu^2 + \omega^2} \tag{33}$$

and

$$\tan \delta(\omega) = \frac{E_1 \mu \omega}{E_r \mu^2 + E_1(\mu^2 + \omega^2)} \tag{34}$$

respectively. For the numerical application, it is considered that $E_r = 3$ MPa, $E_1 = 6$ MPa and $c = 0.1$ MPa s.

3.2.1. Leakage

Now, the leakage influence is studied. The conversion from time to frequency is achieved using the procedures described in Section 2 where the validation is done correlating the exact complex modulus (Eq. (32)) with that provided by Eqs. (28) and (27). All these complex modulus are represented in **Figure 20** as storage modulus $E^{'}$ and loss factor $\tan\,\delta$.

From **Figure 20**, it should be remarked that the direct use of Eq. (28) derives in erroneous results due to leakage, while through Eq. (27), the complex modulus $E^*(\omega)$ can be precisely computed from the relaxation modulus.

For the transformation from frequency to time domain, the exact relaxation modulus given by Eq. (30) is compared with those computed by the inverse FFT applied on Eqs. (28) and (27). Unfortunately, the leakage resulting from Eq. (28) provides a numerical instability, the relaxation modulus being infinity for every time. Thus, **Figure 21** illustrates only two curves instead of three: the analytic response given by Eq. (30) and the estimation for $E(t)$ by means of Eq. (27).

http://dx.doi.org/10.5772/66104

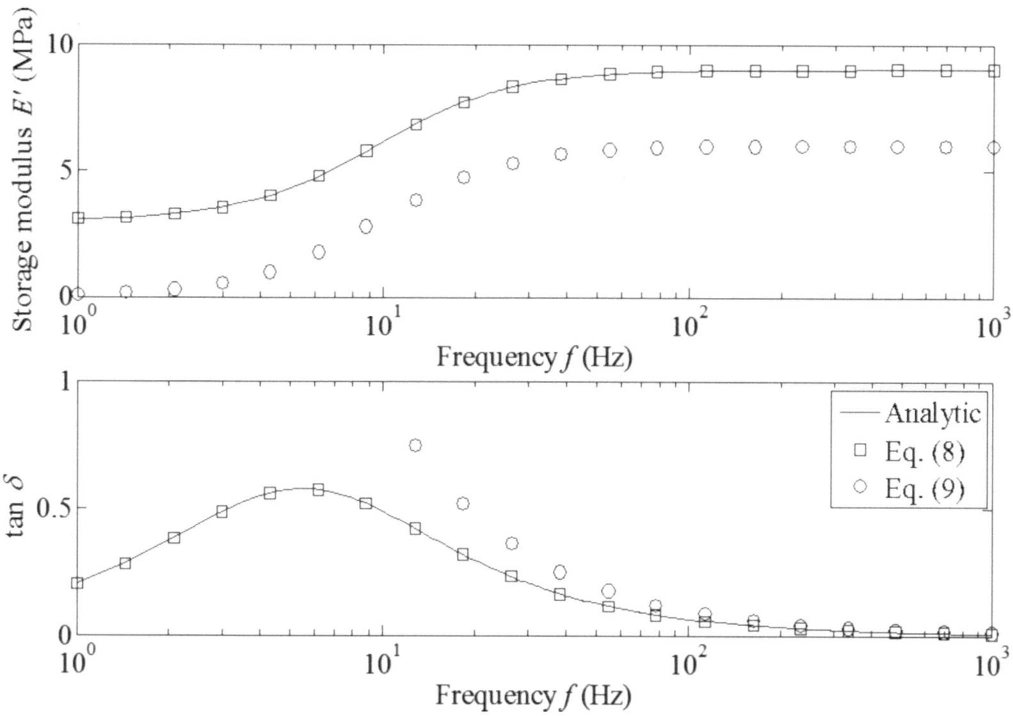

Figure 20. Influence of leakage, conversion from time to frequency. Comparison among the analytic generalized Maxwell model complex modulus $E^*(\omega)$ provided by Eq. (**32**), the one computed by means of Eq. (**28**) and therefore suffering leakage, and the one computed by means of Eq. (**27**) and therefore avoiding leakage.

From **Figure 21**, it should be pointed out that the proposed procedure is capable ofaccurately-computing the relaxation modulus $E(t)$ from the corresponding complex modulus $E^*(\omega)$.

3.2.2. Influence of time and frequency sampling

In this section, the influence of the time and frequency sampling is analyzed. It should be remarked that involving the conversion from time to frequency of a function defined up to a maximum time t_{max}, the discretization time Δt determines the Nyquist frequency f_{max}, according to

$$f_{max} = \frac{1}{2\Delta t} \tag{35}$$

the resulting discretized frequency being

$$\Delta f_{max} = \frac{1}{t_{max}} \tag{36}$$

having

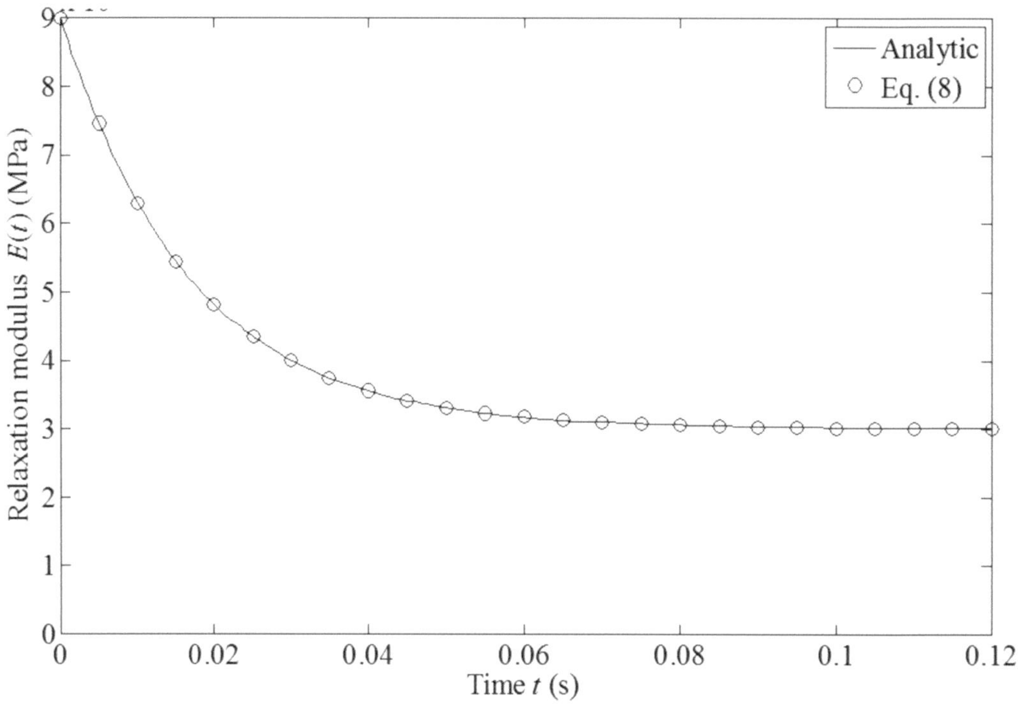

Figure 21. Influence of leakage, conversion from frequency to time. Comparison between the analytic generalized Maxwell model relaxation modulus $E(t)$ provided by Eq. (**30**), and the one computed through Eq. (**27**) avoiding leakage.

$$N = \frac{t_{\max}}{\Delta t} \tag{37}$$

discrete data.

For the conversion from frequency to time, these three equations can be inversely taken into account.

For the present analysis, $t_{\max} = 0.5$s is chosen, and five different discretization cases are analyzed: $\Delta t_1 = \tau_m/2 = 0.0083$ s, $\Delta t_2 = \tau_m/4 = 0.0042$ s, $\Delta t_3 = \tau_m/8 = 0.0021$ s, $\Delta t_4 = \tau_m/16 = 0.0010$ s and $\Delta t_5 = \tau_m/32 = 0.0005$ s. Thus, **Figure 22** shows six curves; the five analyzed cases plus the analytic response given by Eq. (30).

From **Figure 22**, it should be pointed out that the higher the Δt, the lower the $f_{\max}$ and better the accuracy. Thus, $\Delta t_1 = \tau_m/2$ is only able to represent the low-frequency range, representing the rubbery and the beginning of the transition zones of the viscoelastic material [54]. On the contrary, $\Delta t_5 = \tau_m/32$ is enough to accurately represent the complex modulus $E^*(\omega)$ in the whole frequency range, including the vitreous one [54].

For frequency to time domain transformation, a maximum frequency $f_{\max} = 1$ kHz is considered, and four discretization cases are studied: $\Delta f_1 = \tau_m^{-1}/2 = 29.94$ Hz, $\Delta f_2 = \tau_m^{-1}/4 = 14.97$ Hz,

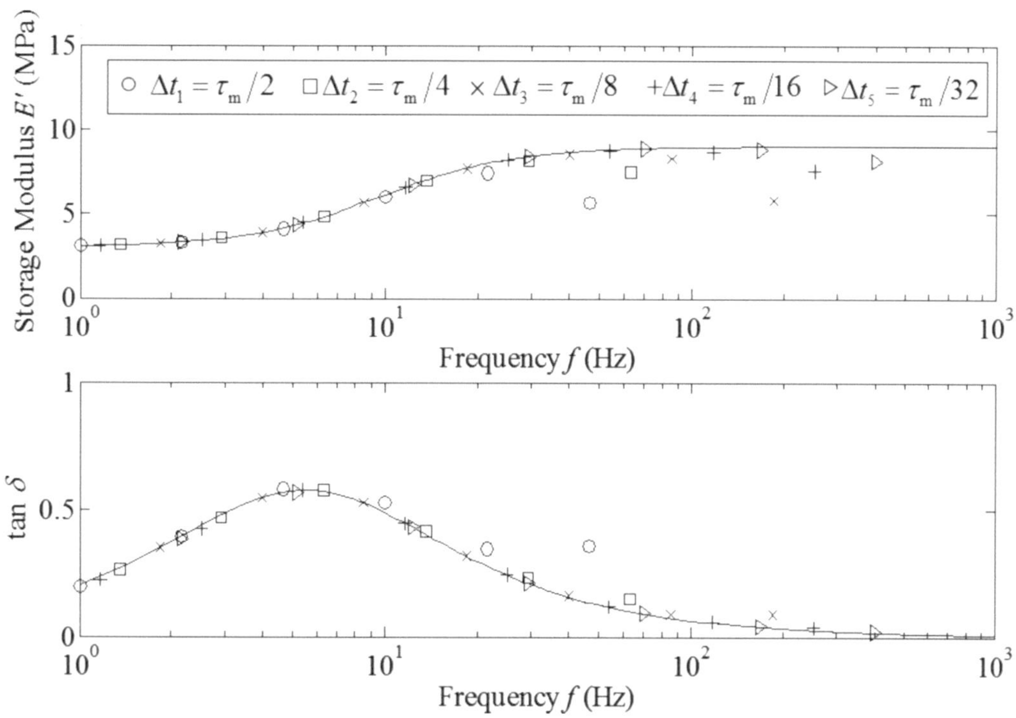

Figure 22. Influence of the analyzed time range, conversion from time to frequency. Comparison between generalized Maxwell model complex modulus $E^*(\omega)$ provided by Eq. (**32**) and the result provided by the proposed interconversion method for the different truncation times.

$\Delta f_3 = \tau_m^{-1}/8$ Hz $= 7.48$ Hz and $\Delta f_4 = \tau_m^{-1}/16$ Hz $= 3.74$ Hz. Hence, **Figure 4** shows five curves; the four analyzed cases and the analytic response given by Eq. (30).

From **Figure 23**, it should be noted that in two of the considered cases, $\Delta f_1 = \tau_m^{-1}/2$ and $\Delta f_2 = \tau_m^{-1}/4$, the relaxation is not properly represented. Thus, differences are verified for $t < 0.02$ s. Consequently, they are not useful to compute the relaxation modulus $E(t)$. Considering the cases $\Delta f_3 = \tau_m^{-1}/8$ and $\Delta f_4 = \tau_m^{-1}/16$, the relaxation is reached, providing analogous accuracy. Hence, for the present case, a $\Delta f_3 = \tau_m^{-1}/8$ is small enough to accurately compute the relaxation modulus $E(t)$.

3.2.3. *Influence of the maximum time and frequency*

Next, the influence of t_{max} and f_{max} is analyzed. First, the conversion from time to frequency is analyzed. The previously defined function discretization parameter Δt_5 is employed. Five truncated signals are considered, as $t_{max,1} = 2\tau_m = 0.0334$ s, $t_{max,2} = 4\tau_m = 0.0668$ s, $t_{max,3} = 8\tau_m = 0.1336$ s, $t_{max,4} = 16\tau_m = 0.2672$ s and $t_{max,5} = 32\tau_m = 0.5344$ s. On the one hand, **Figure 24** presents the exact $E(t)$ given by Eq. (30), in which each employed truncation is represented. On the other hand, **Figure 25** shows six curves corresponding to the five analyzed cases plus the analytic response given by Eq. (32).

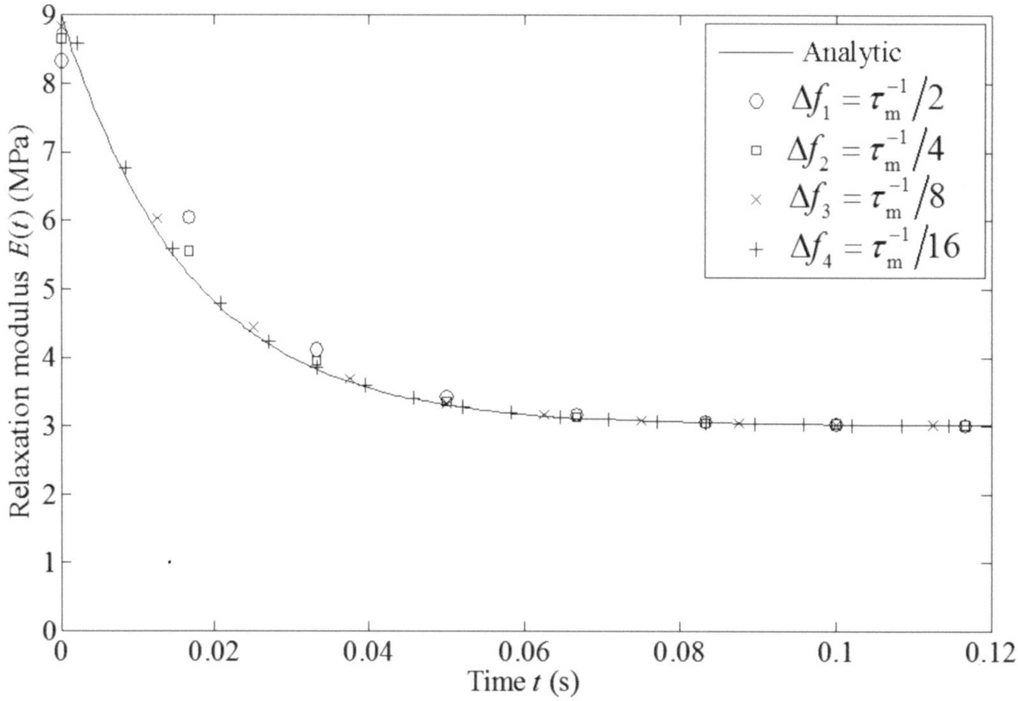

Figure 23. Influence of sampling frequency, conversion from frequency to time. Comparison between generalized Maxwell model relaxation modulus $E(t)$ provided by Eq. (**30**) and the result provided by the proposed interconversion method for different sampling frequencies.

From **Figure 24**, it should be remarked that in two of the considered cases, $t_{max,1}$ and $t_{max,2}$, the relaxation has not been reached, implying that only the vitreous zone can be represented, as **Figure 25** shows. Even if for $t_{max,3}$ and $t_{max,4}$ the relaxation has been reached, only the transition zone can be represented. In fact, to include the rubbery zone, a maximum spam $t_{max,5}$ has to be taken into account.

Next, for the conversion from the complex modulus to the relaxation modulus, a previously defined discretization frequency Δf_3 is chosen. Three cases of maximum frequency are analyzed: $f_{max,1} = 0.1\ \tau_m^{-1} \approx 6$ Hz, $f_{max,2} = \tau_m^{-1} \approx 60$ Hz and $f_{max,3} = 10\ \tau_m^{-1} \approx 600$ Hz. These frequency ranges are supposed to cover the rubbery, transition and vitreous zones, respectively, as shown in **Figure 26**. Therefore, **Figure 27** shows four curves matching to the three studied cases plus the analytic response given by Eq. (30).

From **Figure 27**, it should be noted that the lower the f_{max}, the worse the accuracy. Consequently, the $f_{max,1} = 0.1\ \tau_m^{-1} \approx 6$ Hz is not able to represent the relaxation modulus $E(t)$. Regarding $f_{max,2} = \tau_m^{-1} \approx 60$ Hz, differences are encountered during the relaxation until the viscoelastic constant E_r is reached. On the contrary, $f_{max,3} = 10\ \tau_m^{-1} \approx 600$ Hz is enough to accurately represent $E(t)$ in the whole time range.

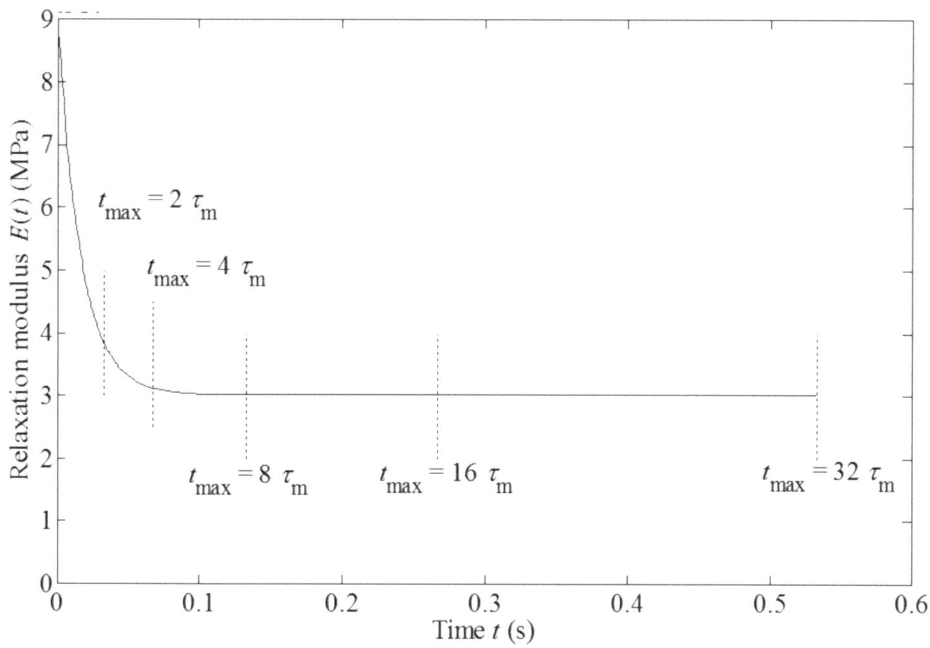

Figure 24. Influence of the analyzed time range, conversion from time to frequency. Analytic generalized Maxwell model relaxation modulus $E(t)$ provided by Eq. (**30**), in which different truncation times are illustrated.

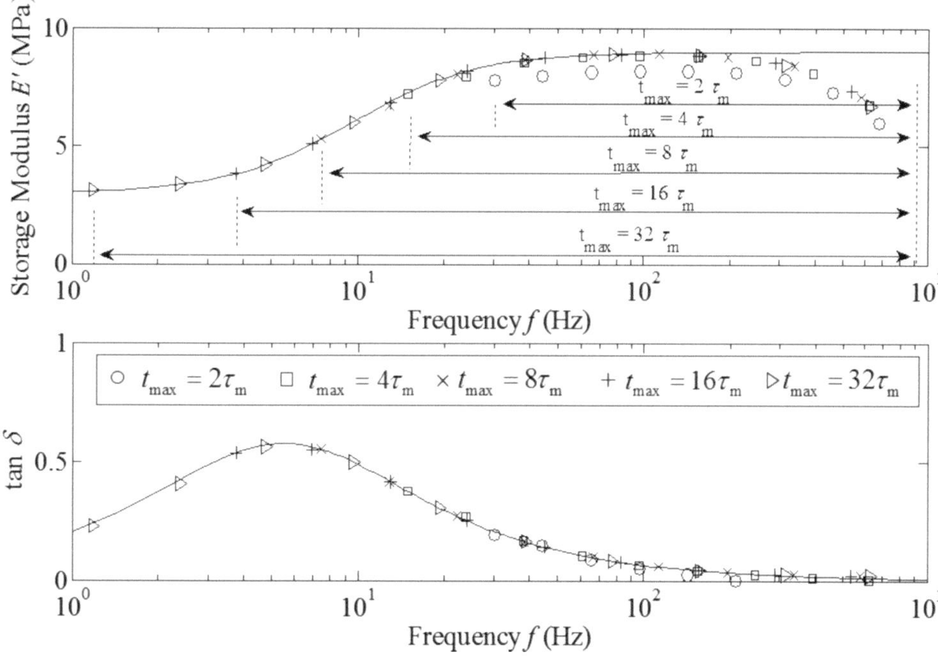

Figure 25. Influence of the analyzed time range, conversion from time to frequency. Comparison between the generalized Maxwell model complex modulus $E^*(\omega)$ provided by Eq. (**32**) and the result provided by the proposed interconversion method for the different truncation times.

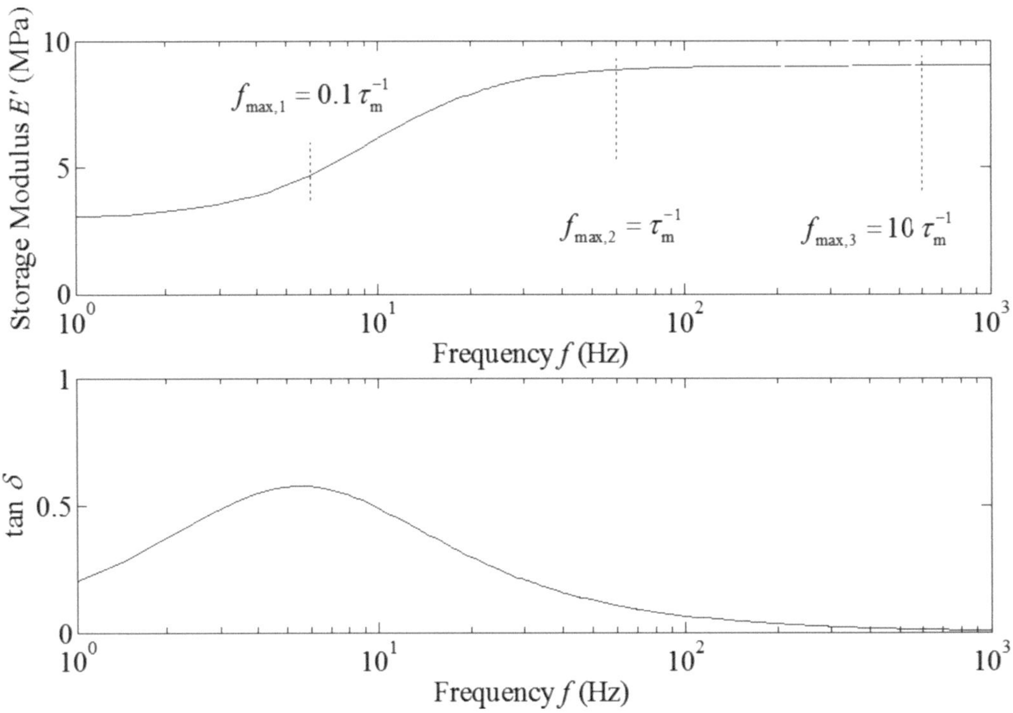

Figure 26. Influence of the analyzed frequency range, conversion from frequency to time. Analytic generalized Maxwell model complex modulus $E^*(\omega)$ provided by Eq. (**32**), in which different truncation frequencies are illustrated.

3.2.4. Influence of experimental error and data dispersion

Next, the precision of the interconversion is studied considering eventual data dispersion. Under this condition, some pseudo-experimental data for relaxation modulus$\overline{E}(t)$ and complex modulus $\overline{E}^*(\omega)$ have been generated evaluating Eqs. (30) and (32), respectively, in some unevenly spaced data points, in which random eventual error $\alpha(t)$ and $\alpha^*(\omega)$ have been introduced, as

$$\overline{E}(t) = E(t) + \alpha(t) \tag{38}$$

and

$$\overline{E}^*(\omega) = E^*(\omega) + \alpha^*(\omega) \tag{39}$$

Then, these generated data have been resampled in order to obtain evenly spaced data $\overline{E}_{es}(t)$ and $\overline{E}^*_{es}(\omega)$. For the present case, linear interpolation has been applied.

For the present numerical application, $\Delta t = 10^{-4}$s and $t_{max} = 1$ s are used. **Figures 28** and **29** show the conversion from relaxation modulus $E(t)$ to complex modulus $E^*(\omega)$. The former

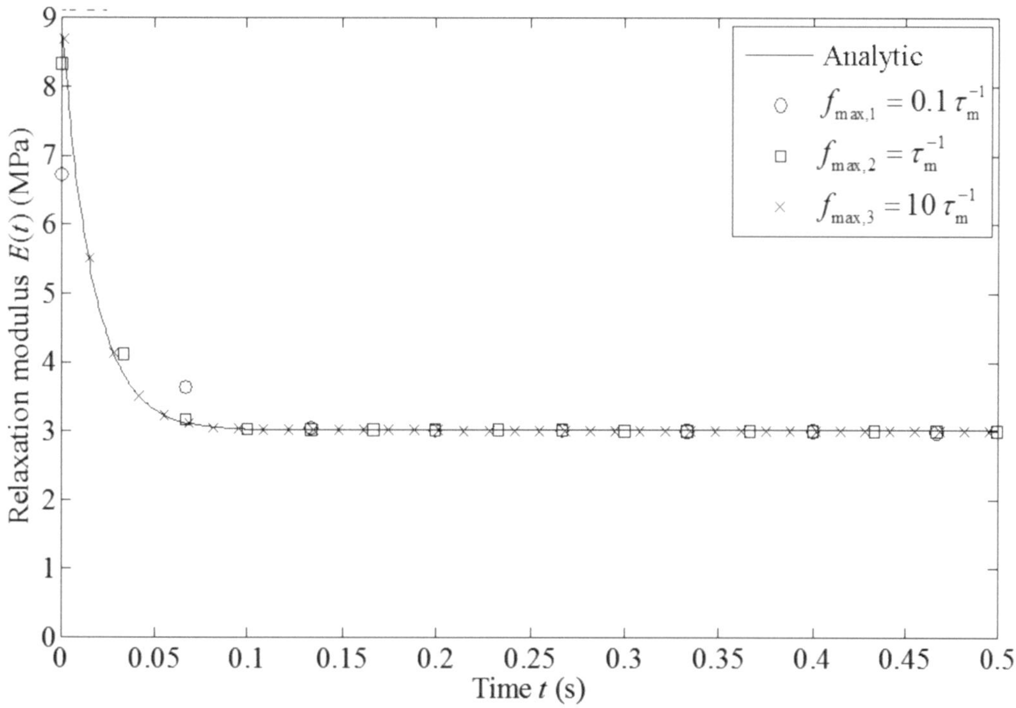

Figure 27. Influence of the analyzed frequency range, conversion from frequency to time. Comparison between the generalized Maxwell model relaxation modulus $E(t)$ provided by Eq. (**30**) and the result provided by the proposed interconversion method for different truncation frequencies.

illustrates Eq. (30) together with the pseudo-experimental data $\overline{E}(t)$ and the latter illustrates the converted modulus with the analytic solution for $E^*(\omega)$ given by Eq. (32).

From **Figure 29**, it should be pointed out that the low-frequency range is properly reproduced while the estimation of $E^*(\omega)$ for the higher frequencies differs from the analytic one (Eq. (32)). The reason is that not enough points were taken in $E(t)$ during the relaxation, and therefore, a linear interpolation technique is not enough to represent the employed model. Therefore, a higher number of data points are needed, especially during the relaxation. Besides, a higher order interpolation technique will provide better accuracy.

Regarding the inverse conversion, $\Delta f = 0.5$ Hz and $f_{max} = 1$ kHz are chosen to guarantee a wider time range. **Figures 30** and **31** show the conversion from complex modulus to relaxation modulus. **Figure 30** illustrates Eq. (32) with the pseudo-experimental data $\overline{E}^*(\omega)$, and **Figure 31** illustrates the converted modulus with the analytical solution for $E(t)$ given by Eq. (30).

From **Figure 31**, it should be noted that the converted relaxation modulus accurately reproduces the model provided by Eq. (30).

As a conclusion, it can be stated that the proposed procedure is able to provide an accurate approximation of the relaxation modulus $E(t)$ and of the complex modulus $E^*(\omega)$even though

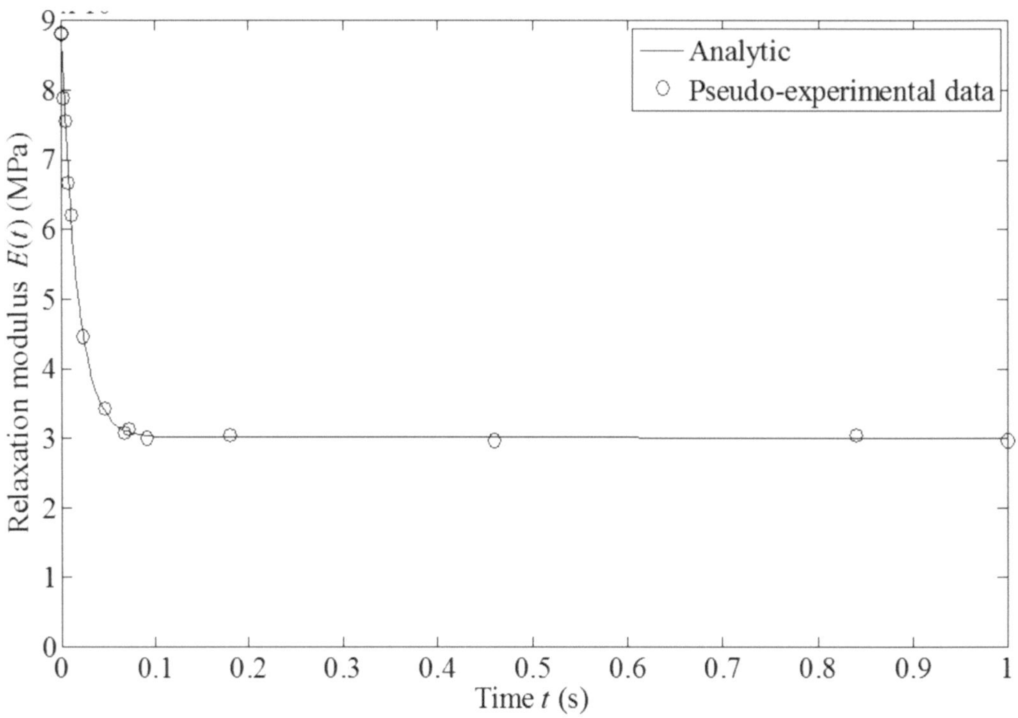

Figure 28. Influence of data dispersion, conversion from time to frequency. Analytic generalized Maxwell model relaxation modulus $E(t)$ provided by Eq. (**30**) together with the employed unevenly spaced data.

the original data does not match the exact response and even though data are not properly spaced.

3.3. Application example using experimental data

Finally, an application example is presented in which experimental data for DMTA obtained relaxation and complex moduli $E_{exp}(t)$ and $E^{*}_{exp}(\omega)$ of a flexible adhesive [73] are used to assess the present procedure. The employed flexible adhesive is a modified silane. Concretely, ISR 70-03 is employed [73]. It should be remarked that the behavior of the employed material was fitted to an exponential relaxation model [73] considering nine relaxation functions.

The experimental relaxation modulus $E_{exp}(t)$ covers the time range 10^{-5} s–3×10^{3} s while the complex modulus $E^{*}_{exp}(\omega)$ covers the frequency range 10^{-1} Hz–7×10^{2} Hz. It should be reminded that an interpolation technique is used to equally space the data. Also, this interpolation step is needed to reach the needed t_{max} and Δt where a cubic interpolation is employed. It should be remarked that, due to the fact that $E_{exp}(t)$ and $E^{*}_{exp}(\omega)$ are experimental data, there is no a relaxation time τ_m associated withthem as a result there is no underlying model. Thus, the desired Δt is estimated using the criteria $\tau_m = 0.66\, t_r$ where t_r is the elapsed time from the strain is applied until relaxation is reached. Therefore, a $\Delta t = 10^{-5}$ s is derived.

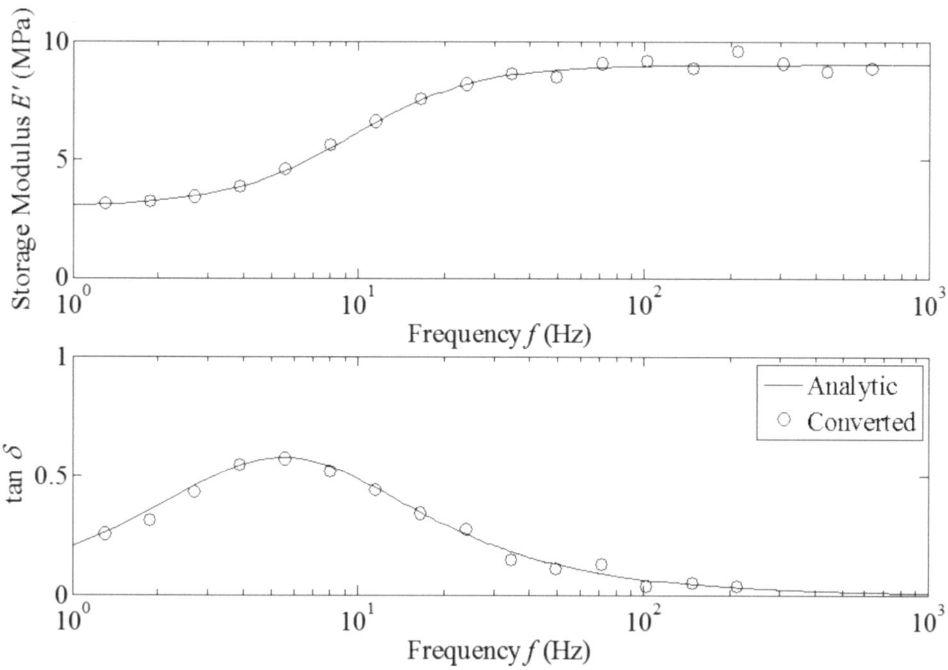

Figure 29. Influence of data dispersion, conversion from time to frequency. Comparison between analytic generalized Maxwell model complex modulus $E^*(\omega)$ provided by Eq. (**32**) and the converted one using data dispersion.

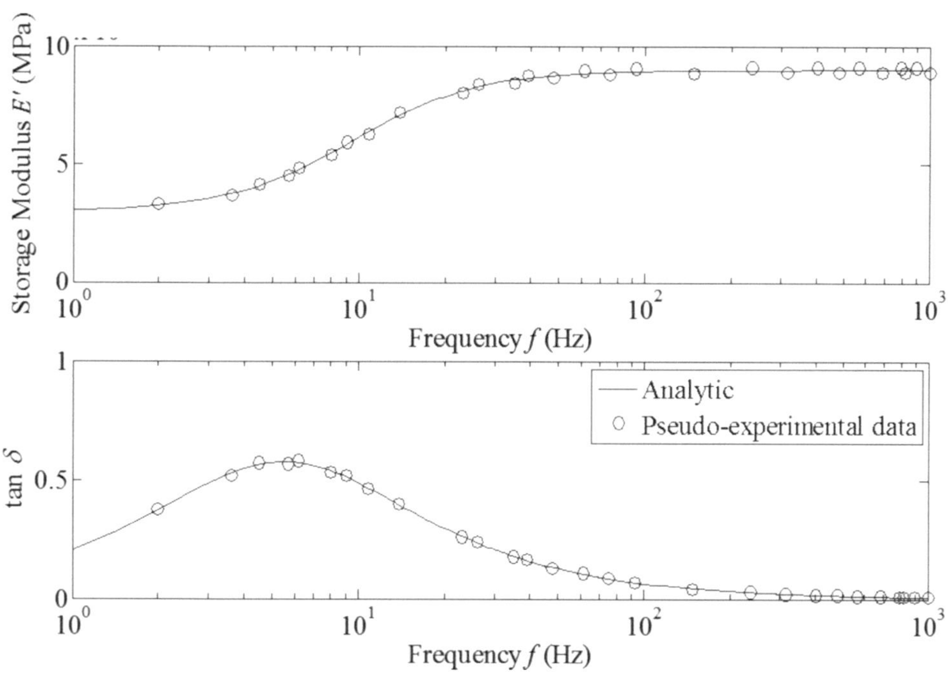

Figure 30. Influence of data dispersion, conversion from frequency to time. Analytic generalized Maxwell model complex modulus $E^*(\omega)$ provided by Eq. (**32**) together with the employed unevenly spaced data.

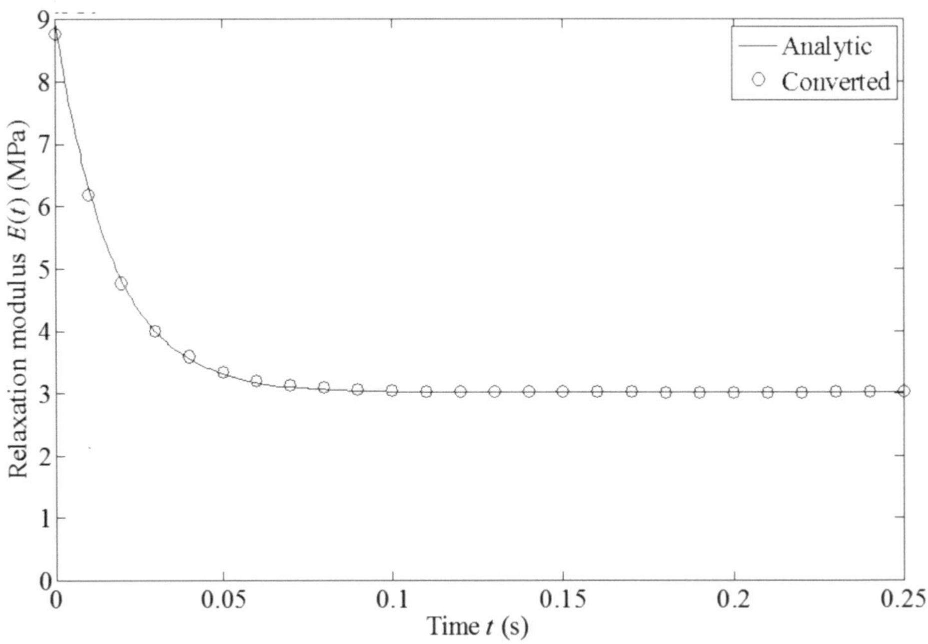

Figure 31. Influence of data dispersion, conversion from frequency to time. Comparison between analytic generalized Maxwell model relaxation modulus $E(t)$ provided by Eq. (**30**) and the converted one using data dispersion

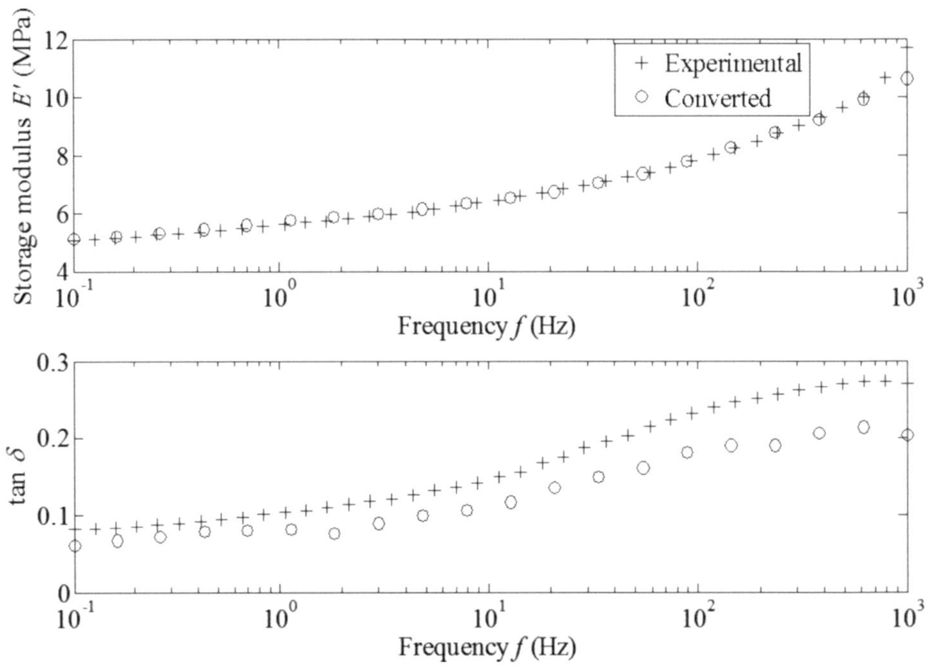

Figure 32. Application example using experimental results. Conversion from time to frequency: comparison between the experimental complex modulus $E^*(\omega)$ of a flexible adhesive and the converted one from its respective experimental relaxation modulus $E(t)$.

Regarding the dynamic case a $\Delta f = 0.01$ Hz is selected. Therefore, **Figure 32** shows the comparison of the conversion from relaxation modulus $E(t)$ to complex modulus $E^*(\omega)$ with the experimental data and **Figure 33** shows the inverse conversion.

From **Figure 32**, it should be pointed out that the described procedure achieves an accurate estimation for $E^*_{exp}(\omega)$. Regarding the storage modulus, the described procedure reproduces the experimental data in the whole frequency range. Concerning, the loss factor tan $\delta(\omega)$, the procedure matches the tendency of the experimental data but differences are encountered, being significant in the high-frequency range where these grow up to 30%. Nevertheless, the accuracy of the interconversion can be improved by reducing the Δt used. Regarding the interconversion from frequency to time, from **Figure 33**, it should be noted that the described methodology provides a precise approximation for the experimentally obtained relaxation modulus during the relaxation. However, an error of 7% is found for the upper time limit. As it was presented in Section 3, the accuracy can be improved by reducing the Δf for the interpolation step or by widening the frequency range, this is employing a higher f_{max}.

3.4. Concluding remarks

In this section, the interconversion between the complex modulus $E^*(\omega)$ and the corresponding relaxation modulus $E(t)$for linear viscoelastic materials has been analyzed. In

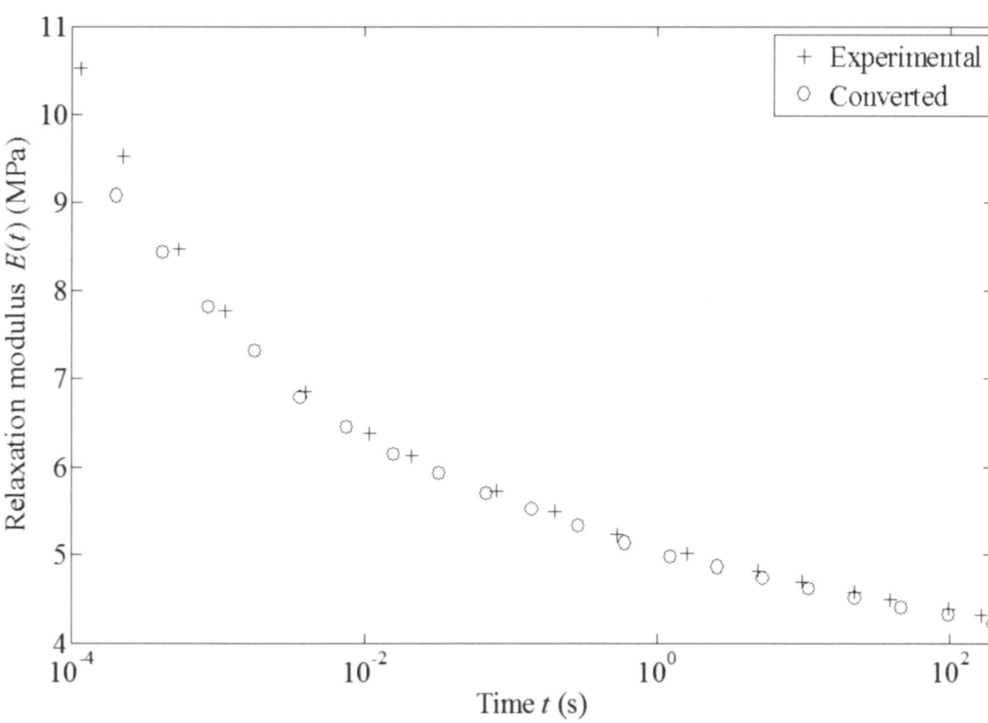

Figure 33. Application example using experimental results. Conversion from frequency to time: comparison between the experimental relaxation modulus $E(t)$ of a flexible adhesive and the converted one from its respective experimental complex modulus $E^*(\omega)$.

contrast to other approximate methods, in this research the FFT algorithm proposed by Cooley and Tukey has been applied on time-dependent part of the viscoelastic response $R(t)$. Together with the procedure itself, the influence of leakage, discretization and studied ranges so as the experimental error and data dispersion have been studied via an analytical material model. Also, an application case employing experimental data has been developed to prove the methodology. As a conclusion, the presented methodology is able to estimate the complex modulus $E^*(\omega)$ by means of relaxation tests, and vice versa.

4. Structural noise and vibration reduction in a cabin elevator prototype by means of adhesive-bonded joints of panels

This section presents an experimental study for the structural noise and vibration reduction in a cabin elevator by means of adhesive-bonded joints of panels. For that noise and vibration measurements are carried out on two prototypes: one of them built with classical panel joining technologies and the other one with adhesive joints. Through the experiments, the benefits from the vibroacoustic point of view of joining panels by means of adhesive-bonded joints in contrast to the traditional joining technologies are put into evidence.

4.1. Prototype description

Measurements are performed in order to prove adhesive joints benefits using the selected material. Therefore, in an elevator cabin property of ORONA S. Coop., the interior sound pressure level L_p was measured in conjunction with the vibration of a side panel $\ddot{s}(t)$. Besides, the vibrations of the floor of the cabin $\ddot{u}(t)$, $\ddot{v}(t)$, $\ddot{w}(t)$ in x, y and z directions, respectively, were also measured.

Hence, an elevator cabin prototype was built-up and two set of side panels were specifically manufactured. Thus, the results of the system with metallic joints are compared to those provided by the prototype system with adhesive joints. The experimental program was carried out under operational conditions and the analyzed frequency range is 5 Hz–1 kHz.

A scheme of both systems is shown in **Figures 34** and **35**, where **Figure 34** represents the position of the metallic joints, and **Figure 35** shows the system with the continuous adhesive single lap joints.

From **Figure 34**, it should be noted that two kinds of joints are employed in the original system. On the one hand, a joint typology is used between the basis and the side panel. On the other hand, another one is employed to join the metal sheets that constitute this side panel. From **Figure 35**, it should be pointed out that the adhesive joints are continuous; whereas the metallic ones are discrete joints. Besides, the joint thickness and overlapping length dimensions are $h_1 = 2$ mm and $l_1 = 50$ mm for the basis joint and $h_2 = 2$ mm and $l_2 = 20$ mm for the joint between the sheets.

Regarding the experiments, two kinds of responses were measured. On the one hand, the interior sound pressure level L_p was registered at the center of the cabin at a height of 1.5 m

http://dx.doi.org/10.5772/66104

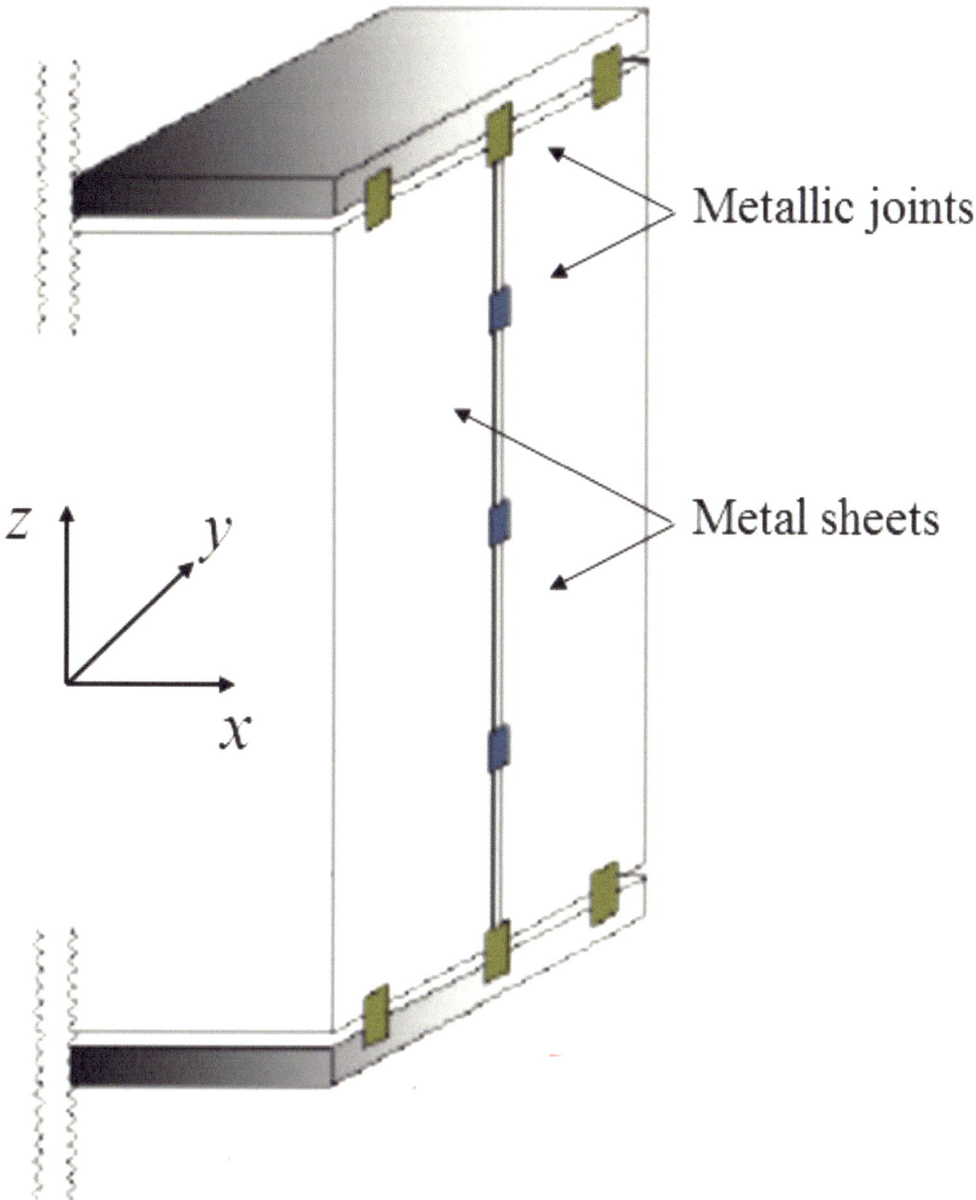

Figure 34. Original elevator scheme with metallic joints.

and facing to the side panel under study. On the other hand, the out of plane acceleration $\ddot{s}(t)$ of the side panel was measured. The floor accelerations $\ddot{u}(t)$, $\ddot{v}(t)$, $\ddot{w}(t)$ in x, y and z directions, respectively, were also registered.

Hence, **Figure 36** shows a scheme in which the placement of the accelerometers and the microphone is represented.

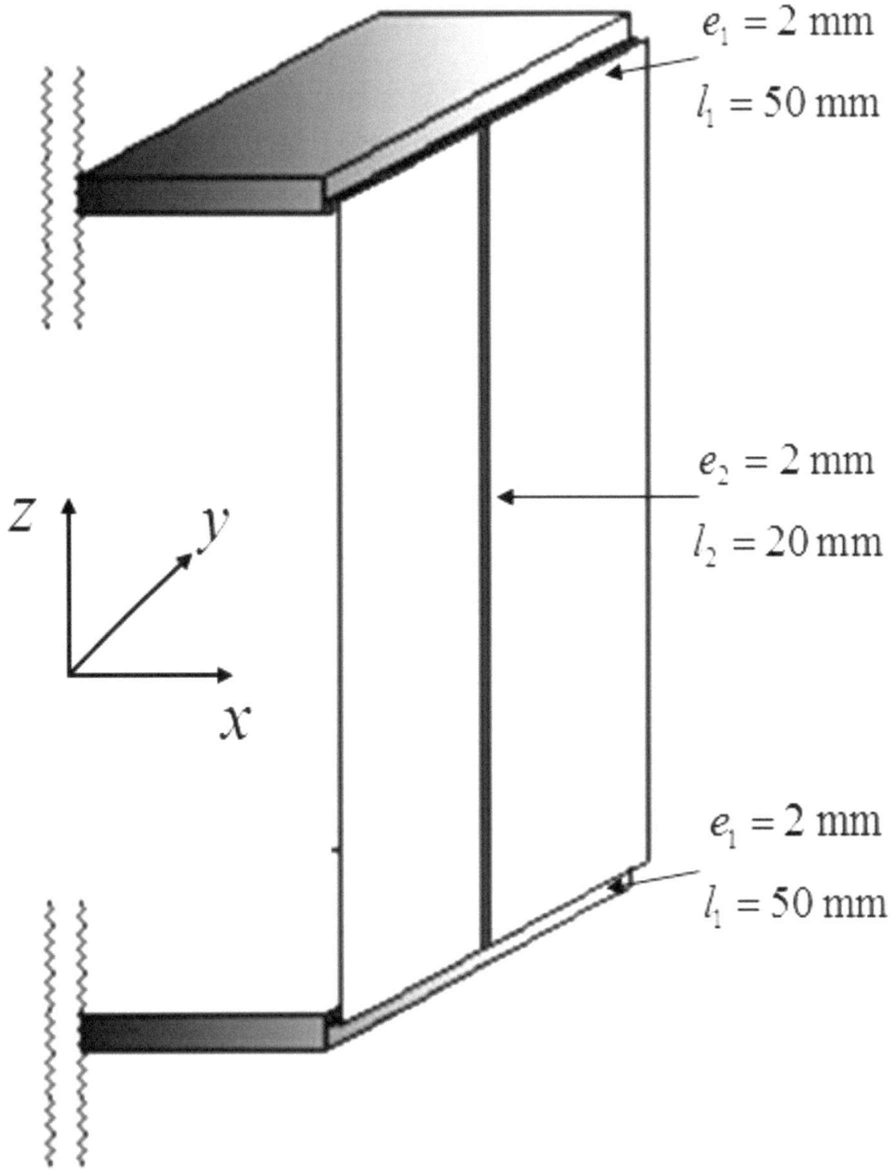

Figure 35. Prototype elevator scheme with adhesive joints.

Next, the responses of the elevator cabin with metallic joints are compared to those of the system with adhesive joints where the ISR 70-03 adhesive is employed. First, the sound pressure level L_p is analyzed together with the acceleration auto-spectrum of the side panel $\ddot{s}(t)$ (see **Figure 36**). Then, the floor acceleration auto-spectra in x, y and z directions are studied. The results are presented in third octave bands.

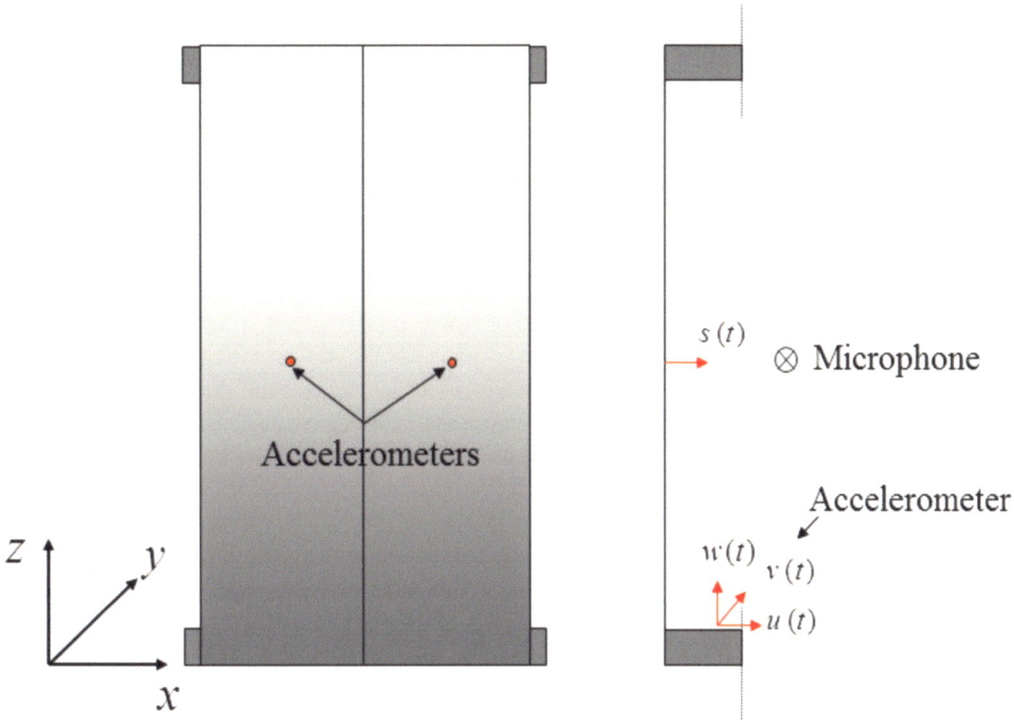

Figure 36. Representation of the elevator cabin measurement points.

4.2. Structural noise

Next, the sound pressure level inside the cabin is presented together with the vibration auto-spectrum of $\ddot{s}\ (t)$ for the right-hand side of the panel, that of the left-hand side one being analogous (see **Figure 36**). Hence, **Figure 37** shows the interior sound pressure level L_p for the system with metallic joints and the corresponding one for the system with adhesive joints. The results are presented using the A-weighting [74]. **Figure 38** shows both acceleration spectra of $\ddot{s}\ (t)$.

From **Figures 37** and **38**, it should be emphasized that the vibroacoustic response of an elevator cabin can be improved by adhesively bonding the side panels. Thus, according to **Figure 37**, the sound pressure level L_p has been reduced for even all frequency bands except for the one of 100 Hz. Small increments can be found also for the lowest frequency bands. Hence, the total sound pressure level has been determined, being $L_{p,\text{metal}} = 72.27$ dB for the system with metallic joints and $L_{p,\text{adh}} = 71.80$ dB for the one with adhesive joints. However, concerning the human ear, the A-weighting [74] is taken into account and the following values are reached: $L_{p,\text{metal}} = 50.89\ \text{dB(A)}$ and $L_{p,\text{adh}} = 49.59\ \text{dB(A)}$, respectively.

Hence, it can be concluded that the interior sound pressure level has been reduced by means of adhesive joints where a reduction of $\Delta L_p = 1.30\ \text{dB(A)}$ has been achieved. It should be

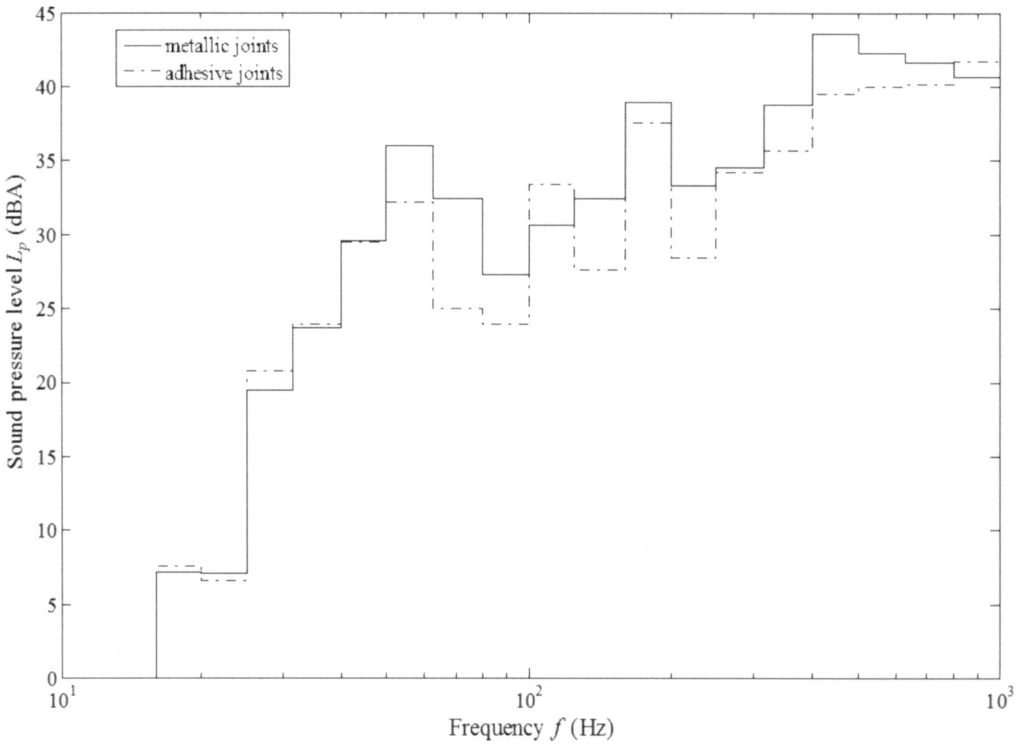

Figure 37. Structural noise response: sound pressure levels.

remarked that, for the present case only one of the three side panels has been adhesively bonded.

From **Figure 38**, it should be noted that the system with metallic joints contains the highest vibration levels in the low-frequency range, especially for frequencies smaller than 100 Hz. It is important to remark that these vibration levels are an order higher than those for higher frequencies.

In order to evaluate the effectiveness of the provided solution, the RMS acceleration $\ddot{s}_{RMS}$ has been computed, the result is shown in **Table 4**. The study is carried out taking into account two frequency groups: the first one made up by the frequency bands below 100 Hz and the second one by the frequency bands between 100 Hz and 1 kHz.

From **Table 4**, it should be pointed out that the level of vibration of the side panel has been reduced in 20 and 30% for the low and high frequency bands, respectively.

In short, it can be concluded that adhesive joints are able to reduce the interior structural noise of an elevator cabin, by means of introducing effective modal damping for the side panels.

http://dx.doi.org/10.5772/66104

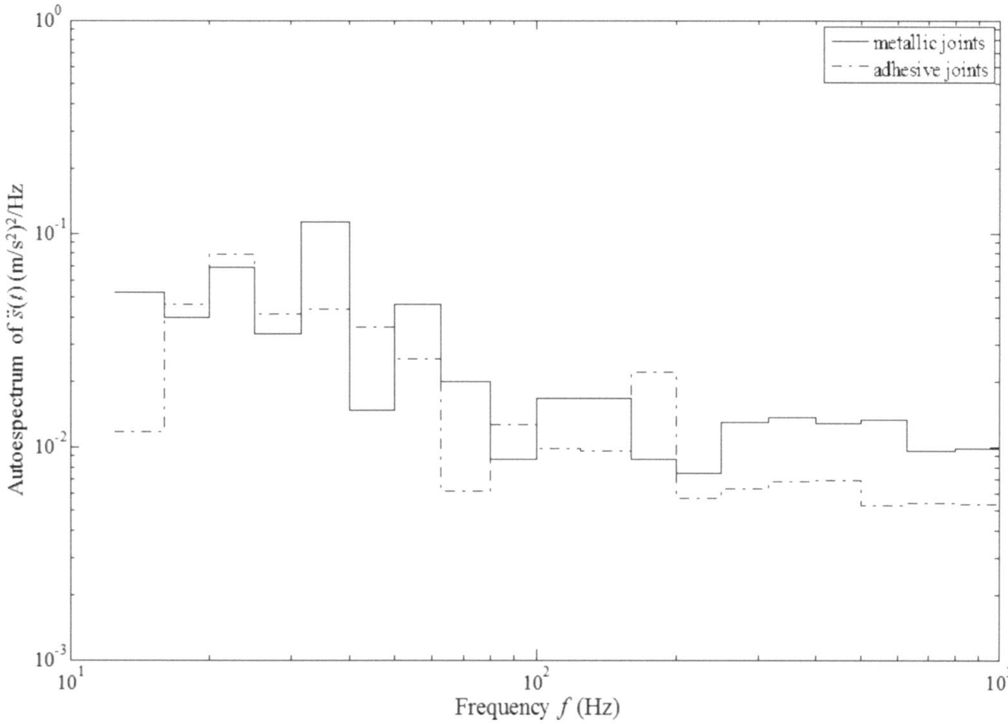

Figure 38. Structural noise response: auto-spectrum of $\ddot{s}(t)$.

		$\ddot{s}_{RMS}$ $(m/s^2)^2$	$\ddot{u}_{RMS}$ $(m/s^2)^2$	$\ddot{v}_{RMS}$ $(m/s^2)^2$	$\ddot{w}_{RMS}$ $(m/s^2)^2$
$f < 100$ Hz	Metallic	1.82	1.06	0.83	2.03
	Adhesive	1.48	0.86	0.76	1.72
100 Hz $< f <$ 1 kHz	Metallic	3.64	3.23	3.02	3.50
	Adhesive	2.53	2.00	1.89	2.44

Table 4. Vibrational results.

4.3. Vibration floor response

Next, the vibration spectra of the cabin floor are presented. The aim of this analysis is to verify that the vibrational ride comfort has not been decreased by the use of adhesive joints. **Figures 39–41** illustrate the acceleration auto-spectra for $\ddot{u}(t)$, $\ddot{v}(t)$ and $\ddot{w}(t)$, respectively, for the systems with metallic and adhesive joints.

From **Figures 39–41** it should be pointed out that, accordingly to the panel response, the highest vibration levels are found for frequencies below 100 Hz. Concerning **Figures 40** and **41**, it should be noted that the smallest vibration level is encountered for the medium-frequency

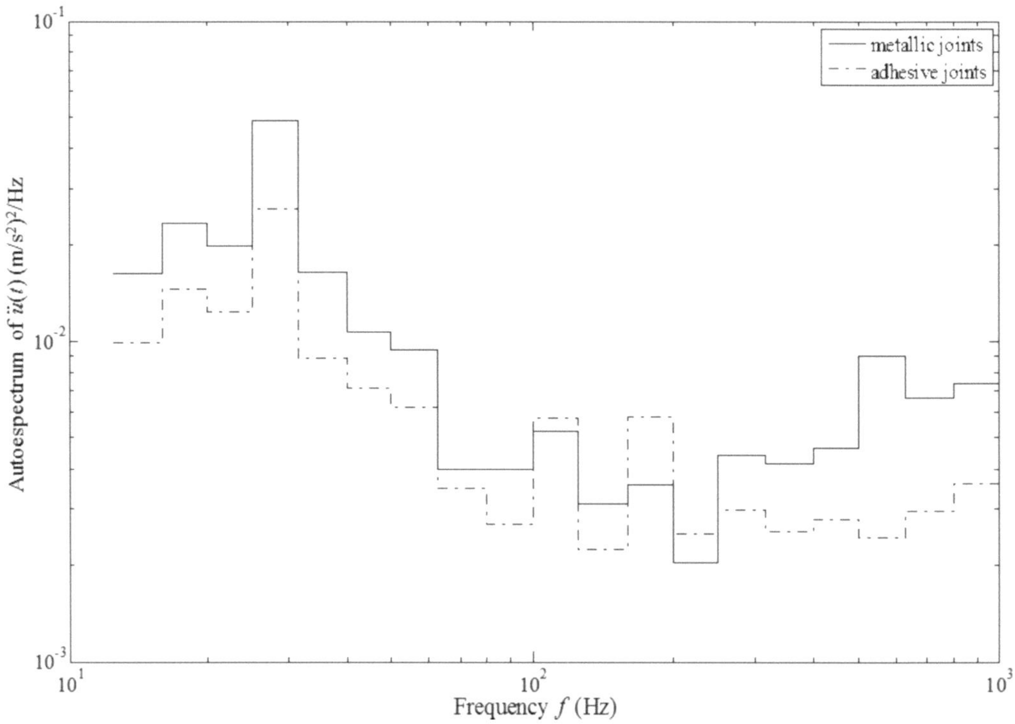

Figure 39. Cabin floor response acceleration auto-spectra in the x direction.

range. It is worth mentioning that the acceleration level on the z direction, see **Figure 41**, is one order of magnitude higher than those for x and y directions. The obtained RMS acceleration values for the systems with metallic and adhesive joints are shown in **Table 4**. The results are provided in two frequency bands as well.

From **Table 4**, it should be pointed out that in spite of just one side panel has been joined with the adhesive material its effect over the vibrational response of the floor is significant. Thus, for the horizontal vibration a reduction up to 40% is obtained while the vertical vibration level has been reduced to 30%.

4.4. Conclusions

In this section, a study for the noise and vibration reduction in an elevator cabin prototype by means of panel adhesive-bonded joints has been presented. The interior sound pressure level of the considered elevator cabin prototype has been reduced in $\Delta L_p = 1.30$ dB(A). The vibration level of the side panel has been significantly reduced in the low and high frequency bands. Also, the vibration level of the elevator cabin floor has been reduced also.

As conclusion, taking into account the presented results obtained with adhesive joints in only side panels, ride comfort in an elevator cabin can be notably enhanced by substituting traditional joining technologies by adhesive-bonded joints.

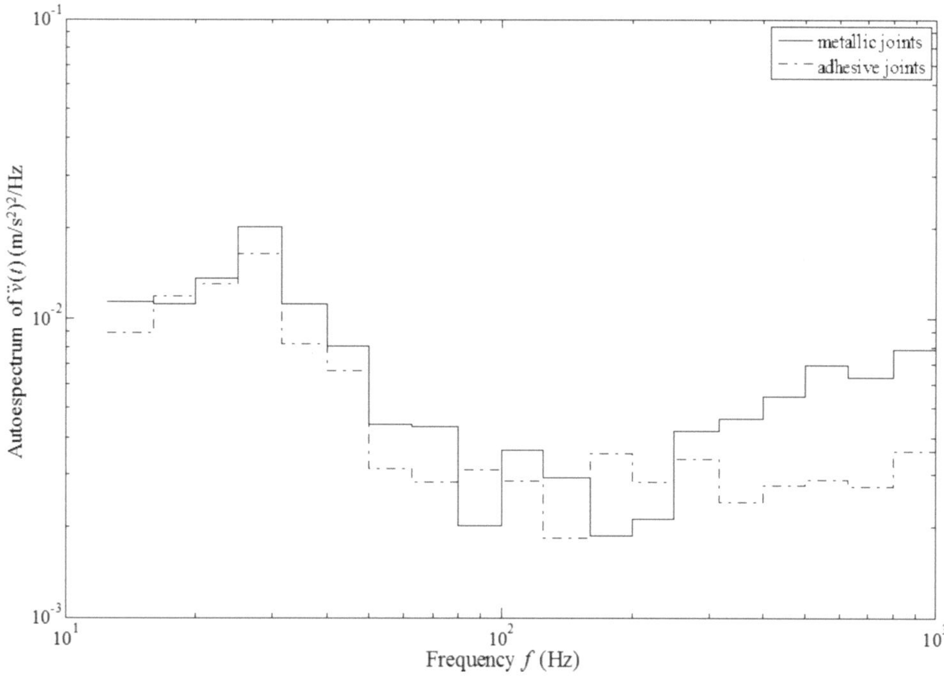

Figure 40. Cabin floor response acceleration auto-spectra in the y direction.

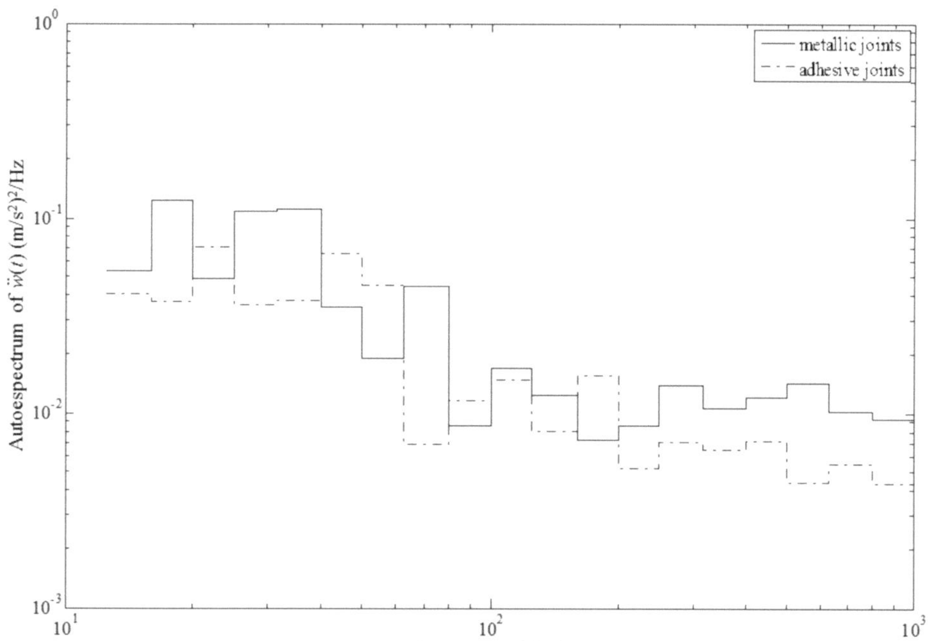

Figure 41. Cabin floor response acceleration auto-spectra in the z direction.

5. Chapter conclusions

This chapter has been focused on the dynamic characterization of adhesive materials for vibration control proposes. First, the experimental characterization and modelization of the relaxation and complex moduli of the flexible adhesive ISR 70-03 by means of a dynamic mechanical thermal analysis technique (DMTA) has been presented. Then, the interconversion path between the relaxation modulus $E(t)$ and the corresponding complex modulus $E^*(\omega)$ for linear viscoelastic solid materials has been explored. In contrast to other approximate methods, in this work the fast Fourier transform (FFT) algorithm has been directly applied on relaxation functions. Finally, an experimental study for the structural noise and vibration reduction in a cabin elevator by means of adhesive-bonded joints of panels has been presented to probe the benefits of adhesive joints on the vibroacoustic behavior of equipment subjected to dynamics loads.

Acknowledgements

Parts of this chapter are reproduced from authors' recent work [73, 75].

Author details

Jon García-Barruetabeña* and Fernando Cortés Martínez

*Address all correspondence to: jgarcia.barruetabena@deusto.es

Faculty of Engineering, University of Deusto, Bilbao, Spain

References

[1] Feng L. An experimental method to determine vibroacoustic properties of joints. In: Euro Noise International INCE Symposium; Washington, DC INCE/USA 1998. pp. 1–6.

[2] Feng L. Characterization methods and ranging of mechanical joints. In: ISMA05; San Antonio, Texas 2000. pp. 793–800.

[3] Feng L. An experimental method for vibration insertion loss of mechanical joints. ACUSTICA. 2001;**87**:191–198.

[4] Feng L., Liu M. Experimental study of structure-borne sound transmission loss of mechanical joints. Acoustical Society of America. 2001;**110**(3):1391–1397.

[5] Feng L. Vibration reduction through joint in finite systems. In: Proceedings InterNoise; Christchurch, New Zealand 1998. pp. 655–670.

[6] Zatarain M., Villasantea C. New light and highly-damped bonded structures for noise and vibration reduction. Cirp Annals. 2007;**53**(1):345–348.

[7] Bostik industry. Bostik Industry, Technical Sheet for Industrial Special Range (ISR) 70–03. English version [Internet]. Available from: http://www.bostikindustrie.nl [Accessed: 15 December 2010]

[8] Watrnaka G.M. Strain-frequency temperature relationships in polymers. Journal of Engineering for Industry. 1968;**90**(3):491–498.

[9] Sjöberg M. On Dynamic Properties of Rubber Isolators. [thesis]. KTH: 2002.

[10] ASTM E 756–04. Standard test method for measuring vibration damping properties of materials. American Society for Testing and Materials. 2004.

[11] Fasana A.G. Oberst beam test technique. Proceedings of the Society of Photo-Optical Instruments. 1998;**3327**:451–462.

[12] Wojtowickia J.L., LucJ. New approach for the measurement of damping properties of materials using the Oberst beam. Review of Scientific Instruments. 2004;75(8) 2569–2581.

[13] Caracciolo R.,Gasparetto A., Giovagnoni M. An experimental technique for complete dynamic characterization of a viscoelastic material. Journal of Sound and Vibration. 2004;**272**:1013–1032.

[14] Caracciolo R., Gasparetto A., Giovagnoni M. Measurement of the isotropic dynamic young's modulus in a seismically excited cantilever beam using a laser sensor. Journal of Sound and Vibration. 2000;**231**(5):1339–1353.

[15] Cortés F., Elejabarrieta M.J. Viscoelastic materials characterisation using the seismic response. Materials and Design. 2007;**28**:2054–2062.

[16] Liao Y., Wells V. Estimation of complex modulus using wave coefficients. Journal of Sound and Vibration. 2006;**295**:165–190.

[17] Beda T. Identification of viscoelastic fractional complex modulus. AIAA Journal. 2007;**42** (7):1450–1456.

[18] Kulik V.M., Semenov B.N., Boiko A.V., Seoudi B.M., Chun H.H., Lee I. Measurement of dynamic properties of viscoelastic materials. Experimental Mechanics. 2009;**49**:417–425.

[19] Maheri M.R., Adams R.D. Determination of dynamic shear modulus of structural adhesives in thick adherend shear test specimens. International Journal of Adhesion and Adhesives. 2002;**22**:119–127.

[20] Jahani K., Nobari A. Identification of dynamic (Young's and Shear) moduli of a structural adhesive using modal based direct model updating method. Experimental Mechanics. 2008;**48**:599–611.

[21] Hunston D.B. Viscoelastic characterization of structural adhesive via force oscillation experiments. Adhesion and Adsorption of Polymers. 1980;**12**:321–339.

[22] Deng S. Temperature-dependent elastic moduli of epoxies measured by DMA and their correlations to mechanical testing data. Polymer Testing. 2007;**26**:803–813.

[23] Ward I.M., Sweeney J., editors. An introduction to the mechanical properties of polymers. West Sussex: John Wiley and Sons; 2004.

[24] Ferry J.D., editor. Viscoelastic properties of polymers. 3rd ed. New York: John Wiley and Sons; 1980.

[25] Williams M., Landel R., Ferry J.D. The temperature dependence of relaxation mechanisms in amorphous polymers and other glass-forming liquids. Journal of the American Chemical Society. American Chemical Society. 1955;**77**:3701–3707.

[26] Segal E., Rosel J.L., editors. Nondestructive testing of adhesive bond joints. In: Sharpe RS, Research Techniques in Non-Destructive Testing, vol. IV. London: Academic Press; 1980.

[27] Fassbender R.H., Hagemaier D.J., Radeky R.L. Verification of defects found by nondestructive inspection in metallic adhesive bonded structures. International Journal of Adhesion and Adhesives. 1980;**1**(2):79–88.

[28] Biggiero G., Canella G. Ultrasonic testing for bond efficiency of carbon steel to stainless steel adhesive joints. International Journal of Adhesion and Adhesives. 1985;**5**(3):143–148.

[29] Assender H.E., Bowditch M.R., Gery N.F.C., Harris A.E., O'Gara P.M., Shaw S.J. A novel system for self-validating adhesive joints. International Journal of Adhesion and Adhesives. 2000;**20**:477–488.

[30] Chang D.J., Muki R. Stress distribution in a lap joint under tension-shear. International Journal of Solids and Structures. 1974;**10**:503–517.

[31] Michaloudaki M., Lehmann E., Kosteas D. Neutron imaging as a tool for the non-destructive evaluation of adhesive joints in aluminium. International Journal of Adhesion and Adhesives. 2005;**25**:257–267.

[32] Maev R.Gr., Titov S. Pulse-echo ultrasonic NDE of adhesive bonds in automotive assembly. In: ECNDT; Paris. France 2006.

[33] Chapman G.B., Sadler J., Maev R.Gr., Titov S., Maeva E.Y., Severina I., Severin F. Ultrasonic pulse-echo NDE of adhesive bonds in sheet-metal assemblies. In: IEEE Ultrasonics Symposium; Vancouver, Canada 2006. pp. 902–905.

[34] Da Silva L., Adams R., Gibbs M. Manufacture of adhesive joints and bulk specimens with high-temperature adhesives. International Journal of Adhesion and Adhesives. 2004;**24**:69–83.

[35] Wunderlich B., editor. Thermal analysis of polymeric materials. Berlin: Springer Verlag; 2005.

[36] Jones D., editor. Handbook of viscoelastic vibration damping. Wiley; New York, EEUU 2001.

[37] Rao M.D. Recent applications of viscoelastic damping for noise control in automobiles and commercial airplanes. Journal of Sound and Vibration. 2003;**262**:457–474.

[38] Banea M., Da Silva L. Mechanical characterization of flexible adhesives. The Journal of Adhesion. 2009;**85**:261–285.

[39] Kadioglu F., Adams R. Non-linear analysis of a ductile adhesive in the single lap joint under tensile loading. Journal of Reinforced Plastics and Composites. 2009;**28**:2931–2936.

[40] Chen T. Determining a Prony series for a viscoelastic material from time varying strain data. NASA Center for AeroSpace Information. Maryland, EEUU 2000.

[41] Bagley R.A. Theoretical basis for the application of fractional calculus to viscoelasticity. Journal of Rheology. 1983;**27**(3):201–210.

[42] Pritz T. Analysis of four-parameter fractional derivative model of real solid materials. Journal of Sound and Vibration. 1996;**195**(1):103–115.

[43] Oldham K.B., Spanier J., editors. The fractional calculus. New York: Academic Press; 1974.

[44] Lewandowski R., Chorazyczewski B. Identification of the parameters of the Kelvin-Voigt and the Maxwell fractional models, used to modelling of viscoelastic dampers. Computers & Structures. 2010;**88**:1–17.

[45] Haneczok G., Weller M. A fractional model of viscoelastic relaxation. Materials Science and Engineering. 2004;**370**:209–212.

[46] Alcoutlabi M., Martinez-Vega J. A modified fractional model to describe the viscoelastic behavior of solid amorphous polymers. The effect of physical aging Journal of Macromolecular Science, Part B. 1999;**38**:991–1007.

[47] Jia J., Shen X., Hua H. Viscoelastic behavior analysis and application of the fractional derivative Maxwell model. Journal of Vibration and Control. 2007;**13**:385–395.

[48] BoltzmannL. On the theory of the elastic aftereffect. Poggendorff's Annals of Physics and Chemistry. 1876;**7**:624–645.

[49] Nashif A.D., Jones D.I.G., Henderson J.P., editors. Vibration damping. New-York: Wiley Interscience; 1985.

[50] Roland C.M. Characteristic relaxation times and their invariance to thermodynamic conditions. Soft Matter. 2008;**4**:2316–2322.

[51] Emri I., Von Bernstorff B.S. Re-examination of the approximate methods for interconversion between frequency and time-dependent material functions. Journal of Non-Newtonian Fluid Mech. 2005;**129**:75–84.

[52] Park S.W., Schapery R.A. Methods of interconversion between linear viscoelastic material functions. Part I–A numerical method based on Prony series. International Journal of Solids and Structures. 1999;**36**:1653–1675.

[53] Park S.W., Schapery R.A. Methods of interconversion between linear viscoelastic material functions. Part II–An approximate analytical method. International Journal of Solids and Structures. 1999;**36**:1677–1699.

[54] Van Krevelen D.W., Te Nijenhuis K. Their numerical estimation and prediction from additive group contributions. In: Properties of Polymers: Their Correlation with Chemical Structure; Amsterdam, Netherlands 2009.

[55] Hatfield W.J., Quake Stephen R. Dynamic properties of an extended polymer in solution. Physical Review Letters. 1999;**82(**17) 3548–3560.

[56] Doi M., Edwards S.F., editors. The theory of polymer dynamics. Oxford: Clarendon; 1986.

[57] Mark E.J., editor. Physical properties of polymers handbook. New York, EEUU 2007.

[58] Ninomiya K., Ferry J.D. Some approximate equations useful in the phenomenological treatment of linear viscoelastic data. Journal of Colloid Science. 1959;**14**:36–48.

[59] Fernández P., Rodriguez D., Lamela M.J., Fernández-Canteli A. Study of the interconversion between viscoelastic behavior functions of PMMA. Mechanics of Time-Dependent Materials. 2010;**15**(2) 1–12.

[60] Schwarzl F. On the interconversion between viscoelastic material functions. Pure and Applied Chemistry. 1970;**23**:219–234.

[61] Sorvari J., Malinen M. Numerical interconversion between linear viscoelastic material functions with regularization. International Journal of Solids and Structures. 2007;**44**:1291–1303.

[62] Leblanc J. Transform rheometry: a new tool to investigate intrinsically non-linear viscoelastic materials. Annual Transactions of the Nordic Rheology Society. 2005;**13**:3–21.

[63] Bellanger M., editor. Digital processing of signals: theory and practice. New York: Wiley-Interscience; 1984.

[64] Cooley J., Tukey J. An algorithm for the machine calculation of complex Fourier series. Mathematics of computation. American Mathematical Society. 1965;**19**:297–301.

[65] Oran B., editor. The fast Fourier transform and its applications. Pretince Hall International; New Jersey, EEUU 1988.

[66] Dutt A., Rokhlin V. Fast Fourier transforms for nonequispaced data, II. Applied and Computational Harmonic Analysis. 1995;**2**:85–100.

[67] Fourmont K. Non-equispaced fast Fouriertransforms with applications to tomography. The Journal of Fourier Analysis and Applications. 2003;**9**:431–450.

[68] Greengard L., Lee J.Y. Accelerating the nonuniform fast Fourier transform. Siam review. Society for Industrial and Applied Mathematics. 2004;**46**(3):443–454.

[69] Marion D. Fast acquisition of NMR spectra using Fourier transform of non-equispaced data. Journal of Biomolecular NMR. 2005;**32**:141–150.

[70] Lee J.Y., Greengard L. The type 3 nonuniform FFt and its applications. Journal of Computational Physics. 2005;**206**:1–5.

[71] Adhikari S. Damping Models for Structural Vibration [thesis]. Cambridge University, Engineering Department: 2000.

[72] Adhikari S. Classical normal modes in non-viscously damped linear systems. AIAA Journal. 2001;**39**(5):1–3.

[73] García-Barruetabeña J., Cortés F., Abete J.M., Fernández P., Lamela M.J., Fernández-Cantelli A. Experimental characterization and modelization of the relaxation and complex moduli of a flexible adhesive. Materials and Design. 2011;**32**:2783–2796.

[74] Bies D., Hansen C., editors. Engineering noise control: theory and practice. London: Spon Pr; 2003.

[75] García-Barruetabeña J., Cortés F., Abete J.M., Fernández P., Lamela M.J., Fernadez-Cantelli A. Relaxation modulus-complex modulus interconversion for linear viscoelastic materials. Mech. Time-Depend Materials. 2013;**17**:465–479.

Chapter 13

Thermoplastic Adhesive for Automotive Applications

Giovanni Belingardi, Valentina Brunella,
Brunetto Martorana and Raffaele Ciardiello

Additional information is available at the end of the chapter

http://dx.doi.org/10.5772/65168

Abstract

The objective of this study is to give a general overview on the thermoplastic adhesives used in the automotive sector. Some of the main applications in which the hot-melt adhesives (HMAs) are used in automotive industries are indicated, together with the adhesive characteristics that explain the reasons for their adoption. The chemical and mechanical behavior of these adhesives and the generally used experimental characterization methods are presented and opportunely criticized. In this study, some of the main properties of thermoplastic adhesives are reviewed together with the standard tests used for their characterization. For what concerns the structural performance, single lap joint test is used to determine the shear strength of the adhesive joint. Thermogravimetric analysis and Fourier transform infrared spectroscopy are used to characterize the chemical properties of the adhesive. This study clarifies what are the potentialities of a thermoplastic adhesive in car industries compared to other adhesives.

Keywords: thermoplastic adhesive, hot-melt adhesive, mechanical strength, chemical characterization, automotive applications

1. Introduction

Adhesives can be categorized in terms of both their chemical composition or polymer type and their physical forms; the latter determines the preferred mode of application and this is the one we refer to in this chapter.

The adhesive joining of different substrates is rapidly diffusing in the manufacturing industries, and nowadays competes with the more traditional techniques of mechanical fastening.

This is due to several reasons. First, they permit to join parts made of materials that are difficult or even impossible to join in other ways; moreover, they permit the joining of substrates made of different materials (hybrid structures). Other advantages of adhesives include vibration and noise damping, insensitivity to corrosive environment, and direct and indirect reduction in joint weight.

The use of adhesives has progressively increased in the last decades in the automotive industry. The reasons are due to the already cited advantages but in particular for the possibility of making hybrid structures that represent one of the main paths toward lightweight vehicle design. Vehicles have to be designed increasingly lighter with the objective to reduce both their fuel consumption and environmental pollution. In the European Union, particularly, the CO_2 limit has been progressively reduced by the European Commission; the short-term normative objective is 95 g/km within 2021. In this contest, multimaterial structures and adhesives can offer great contributions to lightweight design.

Adhesives offer a good opportunity of weight reduction [1], in fact, the modern tendency is to replace threaded fasteners and welds with adhesives. Studies show that by reducing 100 kg of the vehicle weight, the CO_2 emission will be reduced by 3–5% depending on the vehicle size and powertrain type [2].

Further, Chang et al. [3] as well as Belingardi et al. [4] showed that adhesives can improve the stress distribution along the lap zone by a proper design of the joint geometry. Thus, the structural performance can be increased, and in this way, adhesives can offer at least the same stiffness and strength properties than welds. Furthermore, the lightweight trend leads to the design of new vehicle with light aluminum and magnesium alloys, composite materials, and polymers. Therefore, in many of these cases, the adhesive bonding is the only way to make a mechanical connection. Other advantages that can be pursued with the use of adhesives are the already mentioned lack of corrosion if compared to metal fasteners, reduction of noise and vibration, and also the possibility to be functionalized. For instance, adding some properly conceived particles to the base polymer both electrical conductivity and electrical isolation can be enhanced.

Initially, in order to manufacture strong and stiff structural joints thermosetting polymers have been preferred, as they guarantee higher strength and elastic modulus. However, in recent years, the use of thermoplastic adhesives has also been considered and progressively increased in automotive industry, initially for nonstructural applications and then also for structural applications.

Thermoplastic adhesives are plastic materials. The constituents of these adhesives are polymers that melt above a specific characteristic temperature and solidify upon cooling. The thermoplastic adhesive family also belongs to the hot-melt adhesives (HMAs). The HMA is a thermoplastic compound which is applied in the molten state at relatively elevated temperature [5]. The final properties of the adhesive are obtained by natural cooling that usually requires a few seconds. This characteristic of HMA is of particular interest for the automotive industry where the pace of production is very strict.

http://dx.doi.org/10.5772/65168

The possible constituents of the HMAs can be mixed correctly in order to obtain the required properties. This makes HMAs extremely versatile and various types of HMAs are available in the market.

The adhesive must be applied as a liquid, preferably of low viscosity, both to wet the adherent surface and to flow into the crevices and over the asperities that are universally found in solid surfaces. Usually, HMAs are provided in pellets of different dimensions (from few millimeters to 3–4 cm). In the use of HMA, it is important to mate the two surfaces to be bonded while the adhesive is still fluid. The most used method to spread the HMAs is by means of a hot melt gun.

HMAs are easy to apply. They can be applied by rolling, spreading with a gun, or with an extruder. For this reason, they are used in the manufacturing industry for a lot of different applications, such as:

- Fabrication of luggage, cartons, and trays
- Bookbinding
- Fabrication of furniture
- Fabrication of carpets
- Fabrication of bags, shoes, and other fashion pieces

The above-listed applications are only few of the very big number of applications in which HMAs are used. Since HMAs are generally considered nonstructural adhesives, their main use is limited to applications which do not require high level of adhesion strength. Accordingly, luggage, trays, cartons, and bookbinding industries are the main ones that commonly use HMAs. In other cases, in which higher levels of adhesion strength are required, industries use just a little percentage of HMAs, preferring other types of adhesives, such as the thermosetting ones. In furniture construction, for instance, only 11% of the total adhesives used are HMAs [6].

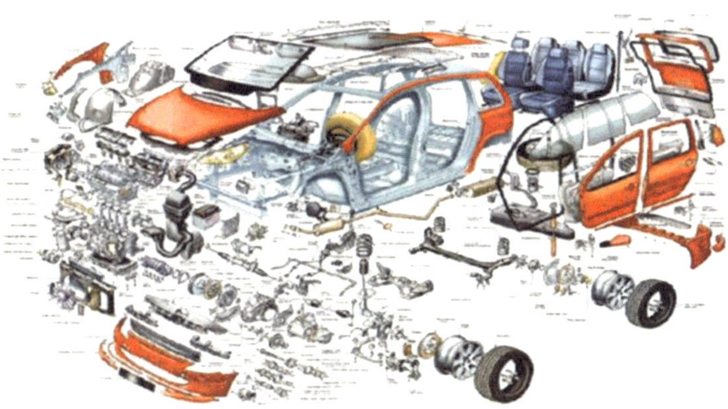

Figure 1. Exploded view of all car components.

However, in the last decades, due to their versatility, shortness of manufacturing time (compatible with the desired production pace), and good level of adhesion strength, the use of HMAs, but in general for all the adhesives, is also increasing in the automotive industry. The assembly process of the different parts that constitute a vehicle plays an important role in the vehicle construction. In this context, the adhesives represent a good way to simplify the vehicle assembly, in comparison with more traditional fasteners and welding. **Figure 1** shows an exploded view of all the vehicle components. This view helps to understand the grade of simplification introduced by the use of adhesives in the vehicle assembly process. For example, the possibility to join all the components of the door panel trims has a high impact on the reduction of cost and on the cycle time of a vehicle.

Nowadays, HMAs are used both for components inside the passenger compartment (generally called "interiors") and for exterior parts. Typical interior components, where HMAs are used to fix, are plastic components such as trim panels, instrument gauge springs, ducts and pipes for air conditioning, sun visor, and many other applications that require the fastening of nonstructural plastic, wooden, and fabric parts. For the exterior components, the environment, where the vehicle has to operate, imposes that the adhesives should overcome the possible degradation that can be consequent to the exposure to severe conditions and, in some cases, also relatively high loading conditions. Typical exterior applications for HMAs are plastic bumpers, back doors, car plates, external air conditioning ducts, lamps, lamp housings, car skirts, and glass roofs. Especially for external parts, the durability and the aging behavior of the HMAs are critical points that need to be assessed in a proper way.

As far as manufacturing process is concerned, HMAs represent a good opportunity to reduce the cycle time. In fact, they can be easily spread by a robotic gun and the solidification of this kind of adhesive occurs in few seconds. During this short time, adhesive reaches a high percentage of the maximum value of its final strength. HMA is also preferred over other adhesive for joining, repairing, and replacing plastic components.

However, HMAs have some disadvantages compared to traditional structural adhesives [7]. In fact, when compared to thermosetting adhesives, which are cross-linked polymers, they exhibit lower heat (i.e., lower operative maximum temperature) and strength resistance.

For what concerns the composition, the HMAs have a main polymer constituent which represents the backbone of the adhesive. The other constituents are generally intended to contribute to modify the main properties of the adhesive in order to increase the affinity with adherent materials and to lower the costs. The base of the HMAs corresponds to polyamides, polyurethane polyesters, EVA (ethylene co-vinyl acetate copolymers), and polyolefin [8]. Polyolefin-based HMAs are used in automotive industries both for interior and exterior components. Applications in the bumper subsystem, door panels, overhead system, seat subsystem, package trays, and instrument panels are only some of the typical uses of HMAs.

HMAs present different constituents, opportunely balanced, in order to give the best performance for the application they are prepared for. In the most typical composition polyethylene and polypropylene are blended together and then combined with a small amount of additives, such as tackifier resins and waxes.

This work provides a short resume on the types of HMAs used in the automotive industries in order to give a general overview on these kinds of adhesives. The chemical and mechanical behaviors of these adhesives are presented, together with the generally used experimental characterization methods. Some of the main applications in which the HMAs are used in the automotive industries are indicated, together with the adhesive characteristics that explain the reasons for their adoption. This study clarifies what are the potentialities of a thermoplastic adhesive in car industries compared to other type of adhesives.

2. Materials

A thermoplastic polymer that is heated to obtain a liquid of flowable viscosity, and, after application, cooled to obtain a solid is termed a hot-melt adhesive (HMA). Although many of these adhesives are polymers of reasonable molecular weight, it is a common practice to incorporate low-molecular weight additives for increasing the fluidity at the manufacturing application temperatures, and to modify properties and lower costs.

So a thermoplastic polymer and a diluent system represent the main composition of an HMA. The thermoplastic polymer constitutes the backbone of the adhesive, providing strength and toughness. The diluent system, represented by antioxidants, fillers, plasticizers, wax, and other resins, opportunely designed, contributes to enhance the specific properties of the adhesive. In this paragraph, the single contributions of the thermoplastic polymers and of the diluent system are discussed.

2.1. Polymers

The most popular HMAs are based on polyethylene, polyolefin or their mixtures, ethylene-vinyl acetate copolymers (EVAs), polyamides, polyurethane, and polyesters. These constituents need to be precisely designed in order to obtain the desired performance.

Polyethylene (PE) and other polyolefin (PO), such as polypropylene (PP), are adopted in the compound used with plastic materials that are difficult to bond. In general, it is possible to state that the molecular weight of the polymer determines the mechanical properties of the polymers [9]. These kinds of polymers are widely used in automotive applications. The use of this polymers in the HMAs contribute to give higher chemical resistance against acids, bases, and alcohols, good resistance to moisture, and a huge open time, compared to the EVA-based adhesive. The open time is the time between the laying of the adhesive and its solidification. It can be also defined as the time necessary to make the joint in order and to get its bonding quality. When this time is gone, i.e., the adhesive is solidified, it is not possible to constitute a joint. What makes the polyolefin-based HMAs of particular interest for the automotive sector is the high adhesion that this adhesive gives with polypropylene substrates, which is so common in automotive applications.

EVAs are used mainly for low-performance bonding [10–15]. This is one of the most used polymers for the HMAs because of its low cost and large production volume (over 100 million tons in 2006) [16]. The use of this polymer in a thermoplastic compound is limited to a maximum temperature of 80°C, and this polymer has the best mechanical strength between 30 and 50°C. Since EVA is a copolymer, the balance between the single constituent polymers gives different properties to the compound. In particular, high concentration of ethylene improves the adhesion to nonpolar materials, such as polyethylene, and gives benefit to the mechanical strength. On the other hand, an increase in the vinyl acetate concentration gives benefit to the adhesion with polar materials, such as paper. The prevalence of the vinyl acetate provides higher flexibility and better performance at lower temperatures [17].

Polyamides (PA) are produced in various molecular-weight ranges with consequent variation in melt viscosity, modulus, and softening point that range from 100 to 200°C [18]. Open times for these HMAs are typically in the range from 15 to 90 s. Blending of these resins is frequently practiced to obtain the desired properties. These HMAs offer high performance in severe chemical environment. They can bond many different substrates such as those made with metals, wood, ABS, and treated polyethylene and polypropylene.

Polyesters are very similar to polyamides. These polymers are used in the adhesives which require high performance. They have a high application temperature but lower strength and melting point. On the other hand, they present higher resistance to moisture. Polyesters are widely employed to bond fabric.

Polyurethanes promote good adhesion to different surfaces. They have good flexibility at low temperature, because of their low glass transition temperature. These adhesives have a lower application temperature, of the order of 50–70°C, compared to the others HMAs.

2.2. Tackifiers

The presence of tackifiers in the HMA formulation is essentially aimed to increase the adhesive strength between substrates. This is possible because tackifiers are able to increase the surface wetting of the substrates by reducing the viscosity of the adhesive. In addition to the chemical and mechanical characteristics, in most cases, they also give color to the adhesive. The selection of the tackifier involves the cost and the chemical stability of this constituent [6]. Typically, tackifiers have low molecular weight compared to the polymers. The glass transition and the melting temperature are usually above the room temperature. This constituent provides the desired viscoelastic properties to the adhesives. The tackifiers can be divided into three different groups: hydrocarbon resins, rosin esters, and waxes.

2.2.1. Hydrocarbon resins

Hydrocarbon resins are a sort of thermal plasticizer. Mainly they are C5 and C9 aromatic hydrocarbon resins. They have a low molecular weight. These resins are a product of petroleum cracking.

C5 aliphatic resins are byproduct of naphtha cracking. They are composed of diolefins, isoprene monomers, and olefins. These kinds of resins confer good heat stability properties and contribute to improve the aging of the adhesive.

http://dx.doi.org/10.5772/65168

C9 aromatic resins result from gasoline cracking and from the production of ethylene and propylene. Although these resins confer good adhesion strength, they undermine the heat stability and the aging properties [6].

2.2.2. Rosin esters

Rosin esters can be produced synthetically or naturally. Rosin resins come from pine tree and they are the oldest raw material that adhesive industries have used. There are three different kinds of rosin resins: gum rosin, wood rosin, and tall oil rosin. Nowadays, rosin esters are widely used because they provide compatibility for a huge variety of adhesive constituents.

The oldest are gum rosins. They are oleoresin (pine gum) of the living pine tree. The collection of resin takes place by wounding of tree and collecting the resin by cups.

Wood rosin is a different oleoresin. It is obtained through the burial of the pine tree stump. After about 10 years in the ground, the bark and sapwood degrade and leave the ground reach of this oleoresin.

Tall oil resin is obtained by industrial distillation of crude oil. Crude oil is composed of about 80% of acid and tall oil resin. The main characteristic of the tall oil resin is its tendency to crystallize. The use of gum and wood oil resin is preferred to tall oil resin because the last one contains some sulfur contaminants [19].

2.2.3. Waxes

The use of waxes in the HMA is mainly intended to reduce the viscosity of the hot melt compound but it is used also to lower the price of the adhesive. The waxes influence not only the viscosity but also other properties such as the softening point, blocking characteristic, and open time. The most used waxes in HMAs are microcrystalline waxes, synthetic waxes, and paraffin waxes. The first and the second ones are used in order to increase the adhesive strength. They contribute also to increase the chemical and mechanical behavior at high temperatures [20]. On the other hand, the paraffin waxes are used to increase antiblocking and barrier properties. The percentage of waxes in HMAs is generally about 20–30%. This quite high value is also consequent to the aim of reducing the adhesive cost [6].

3. Methods

In this section we give an overview of the main chemical and mechanical tests used to characterize HMA in the automotive industries. Of course, these characterization tests are applicable also in other research and industrial fields. Here, we explain which are the main targets of each test and how they are conducted. An example of an HMA currently being used by a carmaker is presented too.

3.1. The Fourier transform infrared spectroscopy (FTIR) analysis

The Fourier transform infrared spectroscopy is one of the most popular physical methods for the characterization and identification of polymers and of all the constituents including additives, such as stabilizers, filler plasticizers, and so on. In fact, FTIR spectra provide a pattern of frequencies corresponding to the "fingerprint" of the molecule.

The interferogram (intensity versus frequency) represents the output signal of this analysis [21–23]. The output diagram reports the material transmittance (or absorbance) as a function of the wave numbers (reciprocal of wavelength). This technique is also used to evaluate what is the degradation level of a material, by comparing the spectra after and before the aging [24].

Figure 2 shows the typical functional groups that are possible to identify in FTIR spectrum. The graph is presented as a function of wave numbers in cm^{-1}. As it is possible to see from the figure, in the left part we can find the single bond groups, in particular N-H and O-H which are located between 4000 and 3200 cm^{-1}. In the range between 3200 and 2800 cm^{-1} it is possible to find the C-H group. Triple bond groups are in the range between 2300 and 2100 cm^{-1}, while double bond groups range between 1800 and 1500 cm^{-1}. The right part of the spectra is called the fingerprint of the material. Every different compound has a different fingerprint and for this reason FTIR can be used to identify a compound [23].

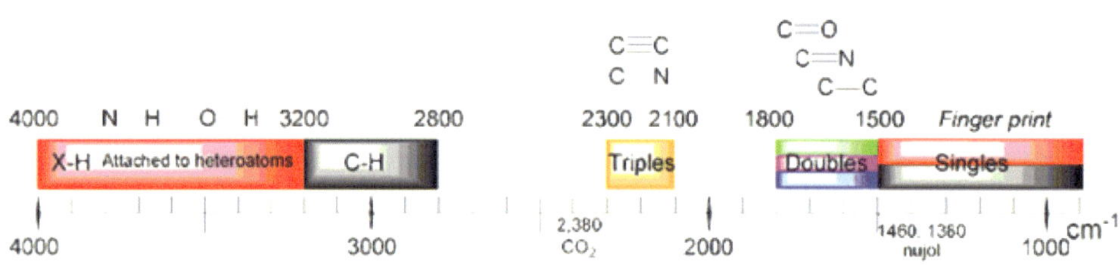

Figure 2. Typical functional groups of the Fourier transform infrared spectroscopy.

3.2. Thermal analysis

Thermal analysis embraces all methods in which measurements are made of a property that changes as the temperature changes

3.2.1. The differential scanning calorimetry (DSC) analysis

The differential scanning calorimetry (DSC) analysis is used to obtain the thermal behavior, thermodynamic data, and the phase transition temperatures of the considered material (in our case the hot-melt adhesive). With this technique the difference of the heat necessary to rise the temperature of a sample and a reference sample is measured.

The cell environment is purged usually with nitrogen gas, but sometimes it is also interesting to consider different atmosphere (air or oxygen). The test specimen is usually exposed to a heating-cooling-heating treatment cycle within a fixed temperature range at a fixed heating rate. During the test, the heat flow is continuously measured as a function of the temperature [25].

3.2.2. The thermogravimetric analysis (TGA)

The thermogravimetric analysis is a technique in which the mass of the sample is recorded continuously while the temperature is increased at a constant rate. As the temperature increases, the constituents of the sample start to decompose causing a loss of weight until the complete degradation of the sample.

The TGA instrumentation needs very precise components. These components typically consist of a controlled furnace, a precision balance (able to measure very small mass changes), and a thermocouple. The experiment can be conducted in nitrogen to know thermal stability or in air atmosphere to know the stability during thermal oxidation process.

3.3. Mechanical properties

3.3.1. The single lap joint (SLJ) test

The mechanical properties of the adhesives are evaluated using different types of joint specimen that have been designed to evaluate specific characteristics. The single lap joint (SLJ) test consists of the tensile test of a specimen made of two substrate parts joined by the adhesive layer of interest (see the specimen sketch in **Figure 3**). The adhesive layer is mainly submitted to shear loading. This is one of the most used tests in the automotive industry to assess the mechanical properties of the adhesive. This kind of joint is easy to make and it is able to evaluate the adhesive joint strength in its strongest direction (the adhesive layer is loaded in shear). The effect of eventual working temperature, different from the usual room temperature, can be evaluated and can be used with different substrate materials and thickness [26]. **Figure 3** shows a sketch of the assembly device that can be used to prepare SLJ specimens under controlled conditions, and a typical testing machine with a configuration of the test. The assembly device, **Figure 3(a)**, allows to fabricate SLJ specimens in the correct way, i.e., it is able to position properly the substrates and to regulate the thickness of the adhesive layer of the specimen. Usually, depending on the adhesive type, some weights are applied on the joint in order to allow the squeezing of the adhesive in excess. This process guarantees that the adherent surfaces are wet enough and in this way it is possible to obtain a great efficiency of the joint. **Figure 3(b)** shows the testing machine and the joint specimen in its fixtured condition. It can be noted that the intrinsic misalignment of the SLJ can be avoided using tabs at the extremities of the specimens.

The output of this test is the force-displacement or the average shear stress-displacement curve. This test is regulated by ASTM D1002, ASTM D3163, and ASTM D3164 standards that give general specifications for all types of adhesives and adherents. According to these regulations, SLJ is used to determine the shear strength of the adhesive. However, it should be reminded that, as it is well known in the literature, the shear stress distribution along the adherent overlap length is not uniform. This distribution is characterized by two relevant peaks at the two overlap extremities. Peaks values depend on several factors such as material modulus, overlapping length, and adhesive layer thickness. Thus, generally, from the performed test only the average value of the shear strength is evaluated.

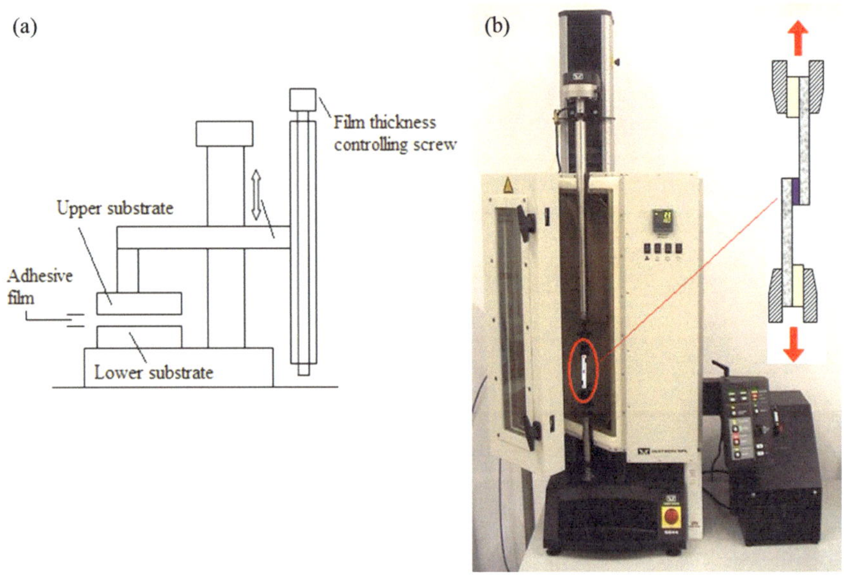

Figure 3. Experimental set-up: (a) assembly device for SLJ preparation, (b) uniaxial tensile testing machine, Instron 5544, with temperature chamber [7].

For what concerns the mechanical behavior, an important analysis for the tested mechanical structures, adhesive joints included, is their failure mode. In the case of adhesives, the analysis of the failure mode is extremely important to verify if the considered adhesive fits with the intended application and to understand the mechanical results. Visual inspection by means of an optical microscope, or a more precise equipment, is able to assess the failure mode. **Figure 4** shows the typical failures of adhesive-bonded structures. There are three base ways in which SLJ may fail:

a. Cohesive failure

b. Adhesive failure

c. Substrate failure

In the first one, the failure occurs in the adhesive layer, as shown in **Figure 4(a)**. In this case, it is possible to find the clear presence of the adhesive material on both the two adherent faces [7]. The second failure mode presents a complete separation between the adhesive layer and one of the substrates, as shown in **Figure 4(b)**. It should be mentioned that mixed modes also, i.e., with the presence of both adhesive and cohesive failures in the same joint, can be encountered. The last failure mode presents a failure in one of the two adherents, as shown in **Figure 4(c)**. **Figure 4(d)** shows a typical adhesive tensile failure of a HMA joint.

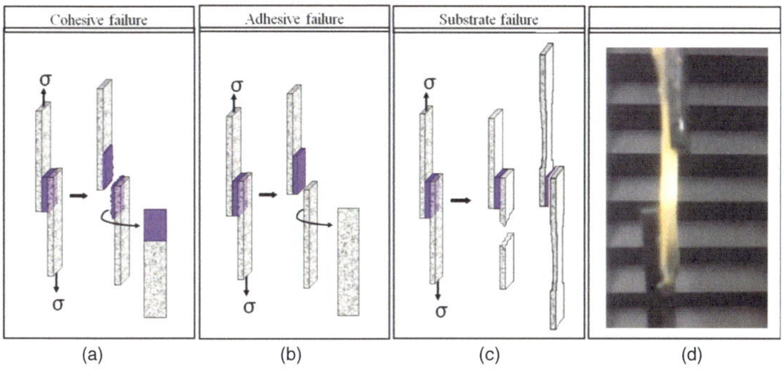

Figure 4. Failure modes of SLJ specimen [7]: a) cohesive failure, b) adhesive failure, c) substrate failure, d) adhesive tensile failure.

4. Results for one HMA sample experimental characterization

In this section, the chemical and mechanical characteristics of one typical HMA commonly used for interior and exterior automotive applications are presented. Polypropylene substrates have been used for the mechanical test (SLJ). First, the reason for the use of HMA is because of the good performance this adhesive gives with polypropylene substrates. The use of polypropylene in car components has increased over the time because of its low cost and good mechanical performance. Furthermore, both of them are easy to recycle and give the possibility to reduce the weight of the component in comparison to other plastic materials [27]. Another important characteristic of this particular adhesive is the possibility to apply without any surface treatment. As mentioned in the introduction, this adhesive is sold in a granular form. These pellets are melted in a portable furnace and sprayed on the components by a gun. This procedure fits perfectly with the typical manufacturing process of the automotive industries. In fact, this system can be easily mounted on a robot arm and thus adapted to perform joining process for different subassemblies.

4.1. Characterization of the hot hot-melt adhesive

First of all, we report the main results of the typical chemical-physical characterization process for an HMA used in the automotive field. In this section, FTIR, TGA, and DSC test results as well as the main result of the SLJ mechanical characterization tests are presented. The objective is to give a complete example of the procedure together with the main characteristics of a typical adhesive. These results are essential to understand whether the adhesive fits with the peculiar application requirements.

4.1.1. FTIR

Figure 5 shows FTIR spectrum of a commonly used HMA. It is possible to evidence the presence of the CH_2 and CH_3 group between the values of 2800 and 3000 cm^{-1} on the wave number axis, with a stretching peak at 2916 cm^{-1}. In this range, these are the highest peaks of the spectrum. Two other peaks are observed in the range between 1459 and 1376 cm^{-1} that can be interpreted as deformation of C-H links in CH_2 and CH_3 groups. Other absorption peaks are visible with lower wave number and lower intensities that are related to isotactic polypropylene.

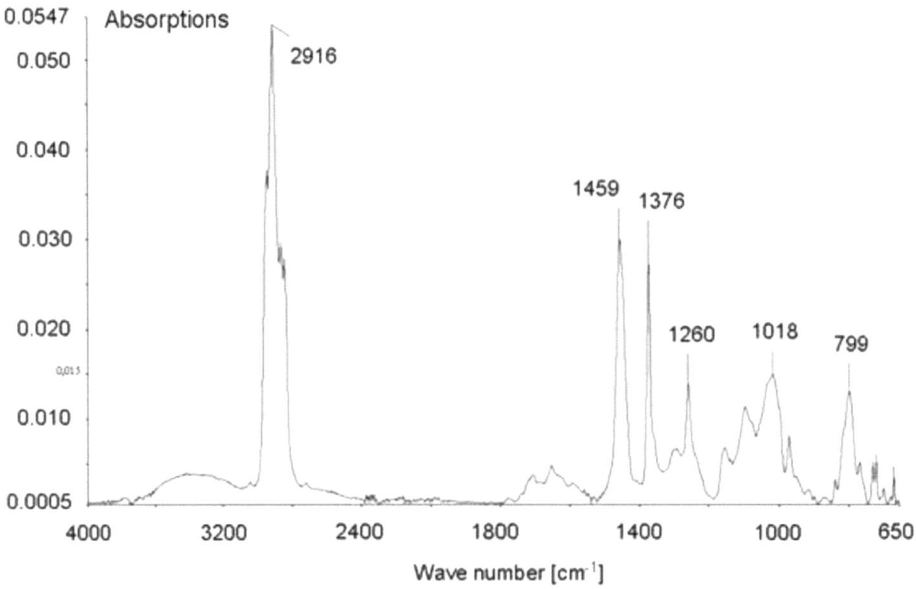

Figure 5. FT-IR spectrum for a polyolefin-based hot-melt adhesive [7].

This spectrum also shows two peaks at 1163 and 971 cm^{-1}, which are related to amorphous polypropylene. Moreover, other peaks related to crystalline polypropylene, in particular, the peaks at 997 and 838 cm^{-1} [28, 29], are visible. Polyethylene absorbance peaks at 720 and 730 cm^{-1} are presented [25]. More specifically, the peak at 720 cm^{-1} is related to both crystalline and amorphous polyethylene, while the peak at 730 cm^{-1} is the peak of the crystalline phase of the PE. The presence of the two phases of PE means that the copolymer has long chains of PE that can be crystallized.

The other region of interest in this spectrum is the region between 1800 and 1600 cm^{-1}. In this interval, the peak at 1778 cm^{-1} could be related to the presence of a partially hydrolyzed anhydride group. Another interesting peak is at 1710 cm^{-1} that refers to carbonyl acid C=O. The presence of these groups could be due to a treatment used in order to increase the polarity and consequently the adhesive properties of the HMA.

The recognition of these components and groups in the FTIR spectrum shows the potentiality of this analysis. The spectrum is able to show that this adhesive is a copolymer PP-PE HMA and the knowledge of the different groups is important to find the affinity with different adherents. This analysis can be also extremely useful to assess the aging of the HMA. For what concerns the studied adhesive, when the adhesive is aged, it oxidizes, especially in the ranges between 1800 and 1500 cm^{-1}. This behavior can be recognized by the presence of higher peaks in the cited range [23].

4.1.2. TGA results

Figure 6 shows the result of the TGA analysis with the chosen HMA. The green curve is relative to the degradation of the HMA with the increase of the temperature in nitrogen atmosphere. The thermal degradation of the polyolefin-based HMA takes place in one step. This behavior is typical of the polyethylene and polypropylene polymers that are the backbone of this adhesive. Degradation in inert atmosphere starts at about 250°C and ends almost at 480°C with a maximum degradation rate at 451°C. Higher temperatures do not cause any further degradation process. The carbon residue of the HMA is 0.68% of the original sample weight.

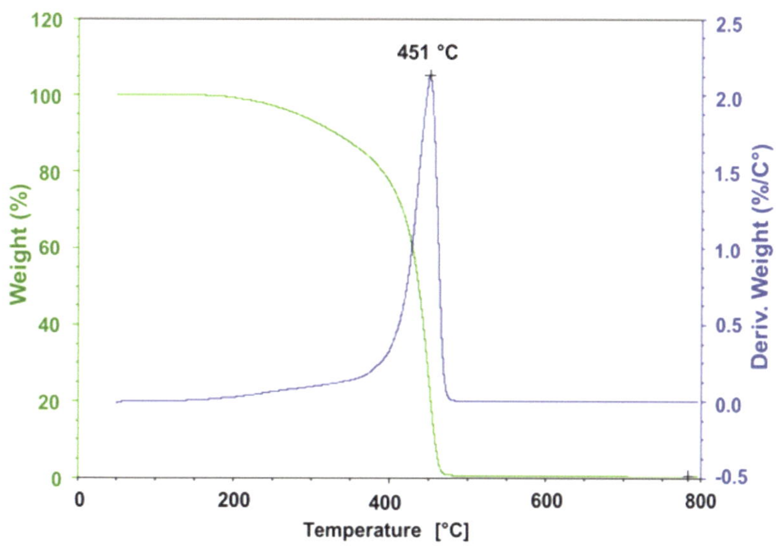

Figure 6. TGA result for a polyolefin-based hot-melt adhesive [7].

The blue curve, in the same figure, shows the derivative of the mass loss as a function of the temperature. This is the weight loss velocity and it is extremely significant to capture the processes that cause little variation of the mass (i.e., loss of volatile and nonvolatile constituents) and to evidence the temperature at which the degradation rate is maximum.

4.1.3. DSC

Figure 7 reports the relevant part of the DSC result curve. As it can clearly be seen in **Figure 7**, the considered polyolefin-based HMA exhibits a sharp phase change peak with a combination of shallow and broad peaks. In the heating curve, the endothermic peaks are located at about 124 and 154°C and could be associated to the melting point of polyethylene and polypropylene, respectively. In fact, as for FTIR results, a crystalline phase of PE and a crystalline phase of PP are present in HMAs.

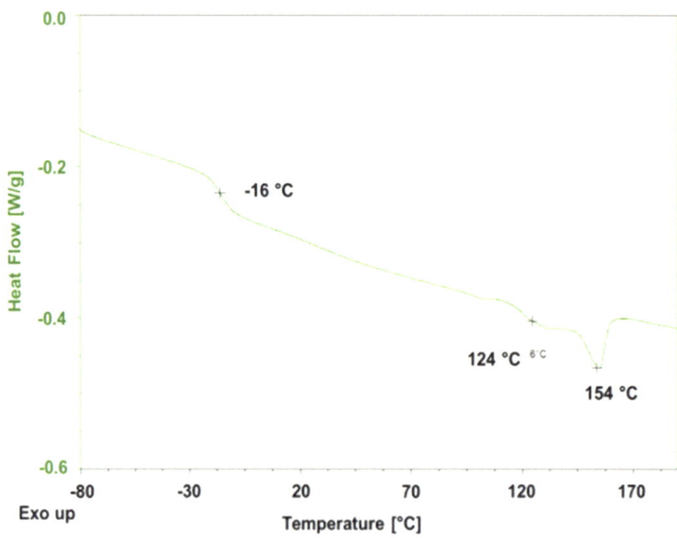

Figure 7. DSC results for a polyolefin-based hot-melt adhesive [7].

As it can be seen in **Figure 7**, only one glass transition temperature is exhibited at (−16°C) and this is correlated to a copolymer PE/PP. This means that the polyolefin-based HMA is more susceptible to become brittle and easily loses its toughness at or below this temperature.

4.1.4. SLJ tests

The main information about the modality of single lap joint test execution has been presented in a previous section. In addition, now it has to be specified that the test velocity for the results presented in this section is 100 mm/min. **Figure 8** shows the typical diagram of the applied load, displacement curves (**Figure 8a**), and the average shear stress, displacement curves (**Figure 8b**), in the case of specimen width of 20 mm, overlap length between the two substrates of 12.5 mm and three different values of the thickness of the adhesive layer, as specified in the diagram legend.

It is well visible that the first part of the curve is linear, and then there is a smooth passage to a nonlinear behavior that is connected to the plastic deformation of the adhesive layer. The plasticization starts from the overlap extremities, where, as mentioned, the stresses have their maximum values and move toward the overlap midspan. Then the maximum (peak or failure) load is reached and the sustained load decrease quite sharply.

The figure gives the possibility to make a comparison of the curves corresponding to three different values of the adhesive layer thickness, which is one important factor. It is well visible that, in accordance with the literature [30, 31], the smaller the adhesive layer thickness the larger the SLJ strength.

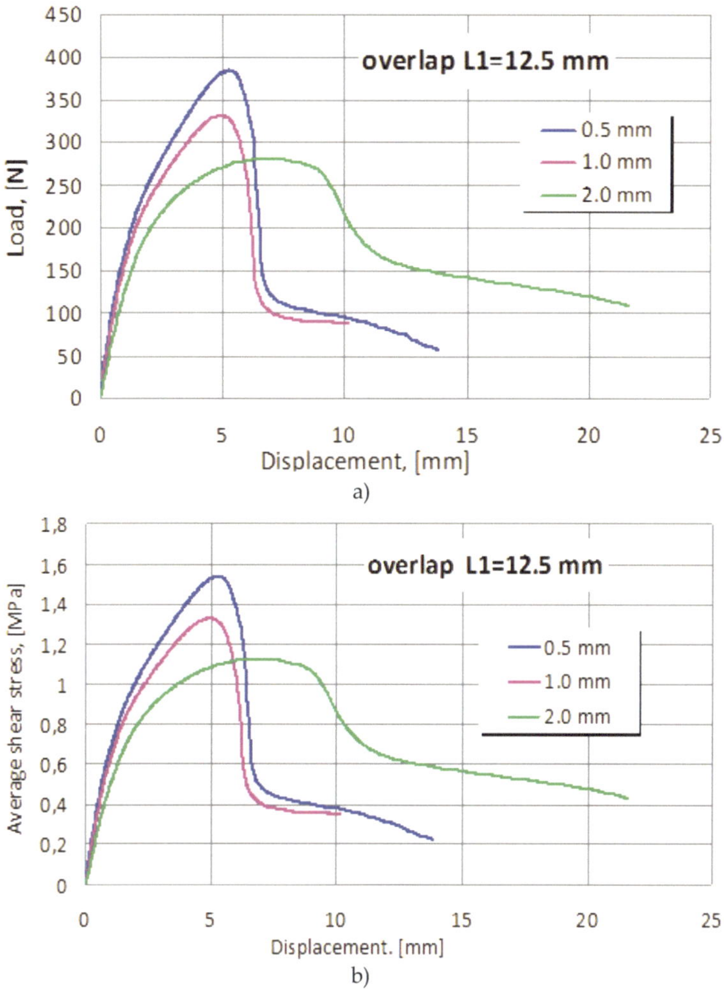

Figure 8. Typical force displacement (a) and average shear stress-displacement curve (b) obtained with SLJ tests for the considered HMA.

As anticipated in Section 1, HMAs are generally not considered structural adhesives. The mechanical characteristics of the polyolefin HMA are reported in this section. Having done several SLJ tests similar to that reported in **Figure 8**, by changing the overlap length, it is possible to develop an analysis of the effect of the two considered main factors (the overlap length and the adhesive layer thickness) on the structural performance. The reported results show the dependency of the failure loads of the SLJ test on different overlap lengths and adhesive thickness. **Figure 9** shows that larger adhesive layer thickness generally leads to lower failure loads, while longer overlap leads to higher failure loads. As it can be noticed, differently from the other two considered overlap lengths, the specimens with the overlap of 50 mm have the same failure loads. The reason for this has to be found in the different failure mode. In these latter cases, the failure zones are localized in the substrates and not in the adhesive layer.

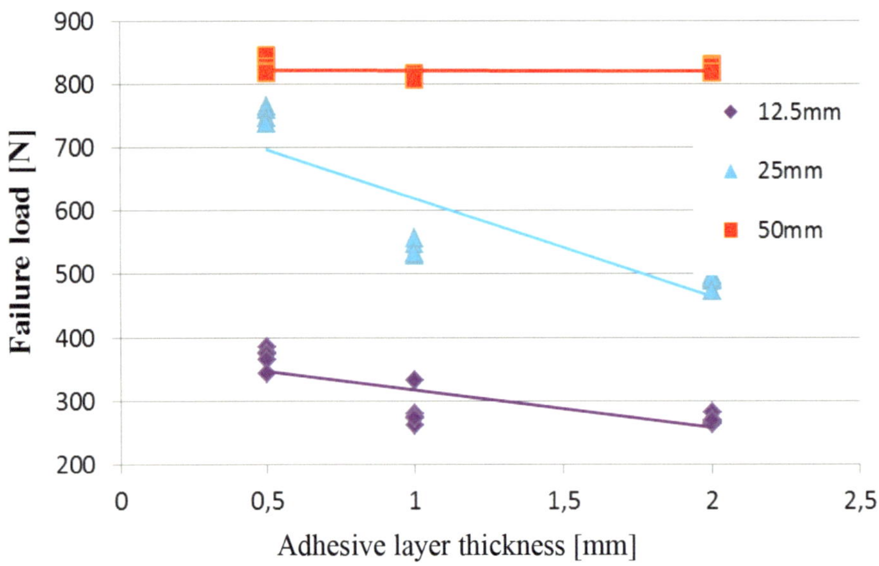

Figure 9. Failure load results from SLJ test for a Polyolefin based hot melt adhesive [21].

Figure 10 shows the same results but in terms of the average shear strength of the SLJ test, calculated by rating the failure load to the adhesion area. In this case both values and descending trends are very similar, if we look the overlap length of 12.5 and 25 mm, while the curve is once again flat for the 50 mm overlap. This last trend is different for the above-mentioned reason.

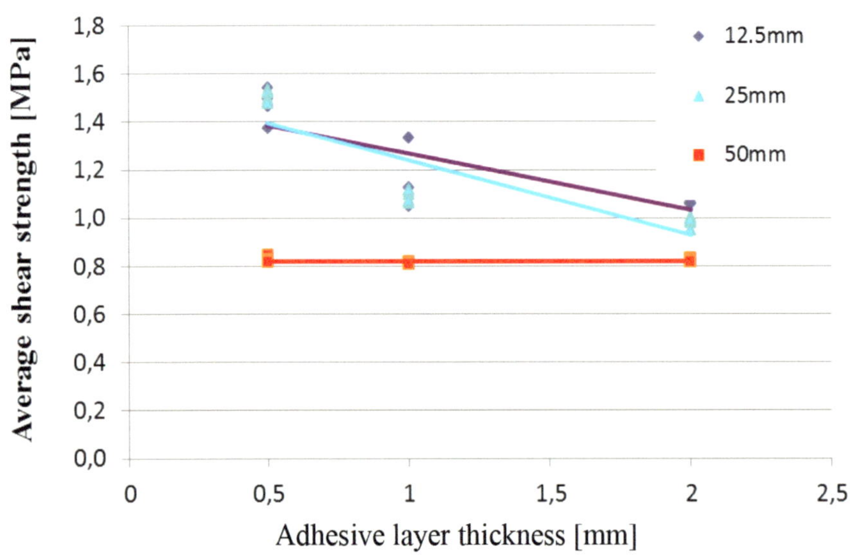

Figure 10. Average shear strength results from SLJ test for a Polyolefin based hot melt adhesive [21].

Visual inspection shows that the most dominant failure type is cohesive failure for overlap length of up to 25 mm, as it can be seen in **Figure 11**. Indeed, for the 0.5 and 1 mm adhesive layer thickness, 12.5 mm overlap SLJ exhibits cohesive failure and equivalent shear strength to 25 mm overlap sample. For 2 mm adhesive layer thickness, the observed dominant failure type is adhesive failure at one tip followed by adhesive layer shear failure. Finally, the 50 mm overlap length specimens show insignificant failure load variation with adhesive layer thickness and generally exhibit substrate failure.

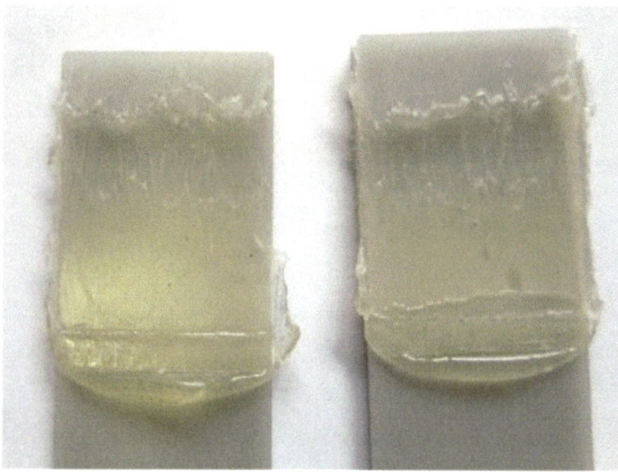

Figure 11. Failure surface of a SLJ test, overlap length 25 mm, adhesive thickness 1 mm.

The subsequent investigation of the load-displacement diagrams of the tested specimens leads to a better understanding of the substrate contribution and of the SLJ response to the type of failure mode for given geometry changes. The load-displacement curves expected from the SLJ shear tests are different in the case of 12.5 and 25 mm overlap length specimens with respect to 50 mm ones. This expectation is confirmed by the diagrams that are shown in **Figure 10**. The diagrams in **Figures 12(a)** and **(b)** have been obtained with two different values of the substrate overlap lengths and have similar trends, as previously described commenting **Figure 9**. On the contrary, curves in **Figure 12(c)** have completely different trends. After the elastic phase, a maximum is reached quite rapidly and then the curves have a slow decrease while the displacement reaches quite large values. For this latter type of specimens, the dominant plastic deformation of the PP substrate is responsible for this wide range of displacement after yielding. It is of interest to observe the curves reported in **Figure 13** where one of the characteristic curves of the SLJ and the tensile curve of the PP are superimposed. The SLJ curve is just a little bit below the PP curve, confirming that the structural behavior of this joint is dominated by the substrate (as depicted in **Figure 12(c)**). **Figure 12(d)** collects in one diagram the average shear stress–displacement curves for the three overlap lengths and one particular value of the adhesive layer thickness (0.5 mm). The shear stress curves for the two lower overlap lengths have similar trends and values. The shear stress-displacement curve for 50 mm overlap samples exhibits a completely different trend but, as explained before, this is related to the fact that it does not represent the actual adhesive curve, but the consequences of the PP substrate deformation.

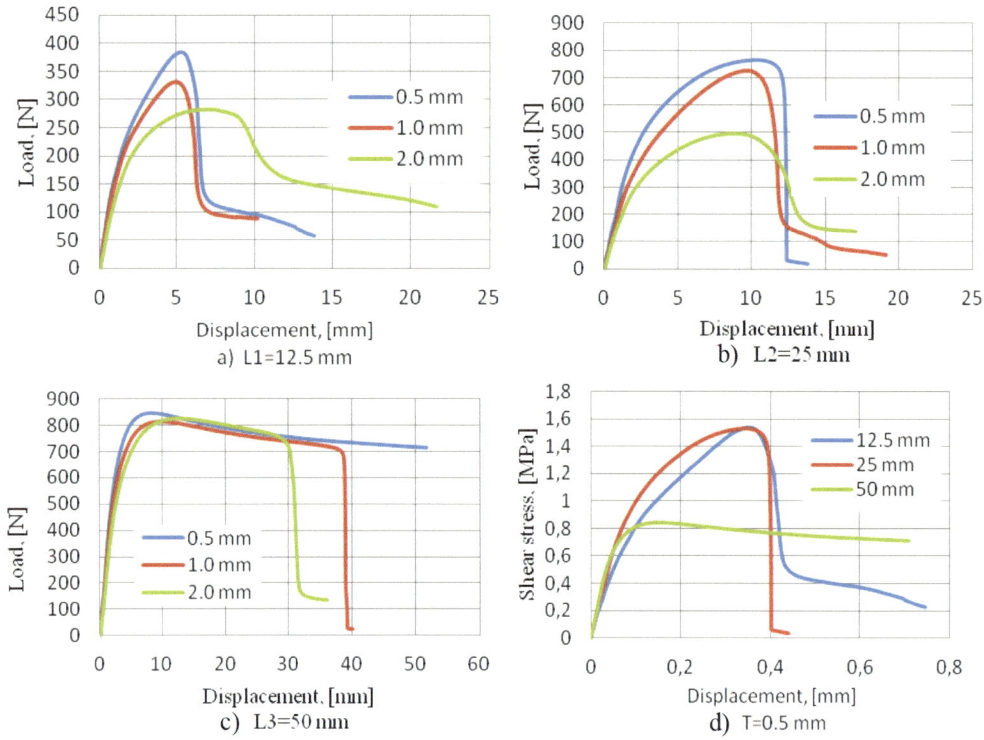

Figure 12. (a–c) Load versus displacement curves of SLJ for different bond lengths and adhesive layer thickness. (d) Average shear stress versus displacement curve [7].

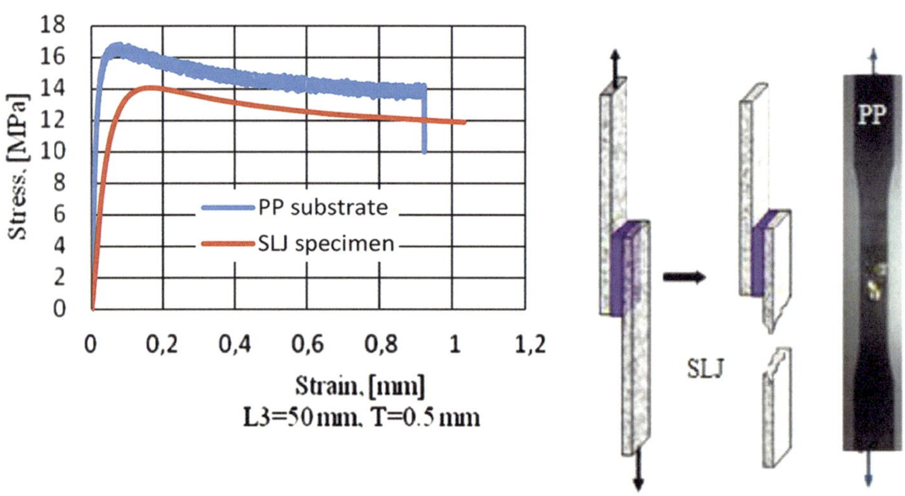

Figure 13. left - Typical stress-strain curve for SLJ and PP substrate test [7]; right - substrate failure during SLJ tensile test.

http://dx.doi.org/10.5772/65168

Finally, in order to explain the differences that are visible between the two curves reported in **Figure 13**, it is important to remember that the SLJ test introduces a slight but not negligible bending moment in the substrates. Due to the eccentricity of the applied load and uneven stress distribution near the joint extremities, the substrate experiences higher stress values than those induced by simple tensile tests.

5. Conclusion

This work represents a general overview of the current use of HMAs in the automotive industries. It also gives a comprehensive systematic summary of the ways adopted to characterize from the chemical-physical and mechanical points of view of an HMA. HMAs are a wide class of flexible thermoplastic adhesives, which can be employed for a wide range of uses both in the automotive field and in other mechanical applications. The motivations of this work are essentially due to the attempt of covering the lack of systematic information on this topic.

Some of the most relevant applications adopted by many carmakers have been listed. The adoption of HMAs has been progressively increased due to their particular properties: better recyclability, good mechanical performance, low cost, good resistance to moisture, good chemical stability, and lack of curing time. HMAs exhibit a lower strength and heat resistance when compared with structural thermosetting adhesive. Even though HMAs are traditionally associated with nonstructural applications (e.g., packaging, sealing, and clothing industries) in many literatures, the properties, that these materials give to the single lap joint loaded in shear, has resulted in an expansion of their use also in structural applications, such as in automotive vehicles, as well as in other nonstructural applications, such as furniture.

The results of some experimental tests have been reported, in relation to PP substrates, and duly commented, putting in evidence the role of two main factors: the overlap length and the adhesive layer thickness, in determining the strength of the SLJ. The comparison with the failure modes show that HMAs exhibit high plastic elongation compared to other structural adhesive (such as thermosetting adhesive), which generally present a brittle behavior.

Author details

Giovanni Belingardi[1*], Valentina Brunella[2], Brunetto Martorana[3] and Raffaele Ciardiello[1]

*Address all correspondence to: giovanni.belingardi@polito.it

1 Politecnico di Torino, Torino, Italy

2 Università di Torino, Torino, Italy

3 Centro Ricerche Fiat S.C.p.A., Torino, Italy

References

[1] Mallik P. Materials, Design and Manufacturing for Lightweight Vehicles. Woodhead publishing, Sawston, Cambridge (UK); 2010. 384 p. ISBN: 978-1845697822

[2] Structural bonding of lightweight cars [Internet]. 2010. Available from: http://www.dowautomotive.com/pdfs/Structural_Adhesives_Brochure.pdf [Accessed: 2016-06-01].

[3] Chang B, Shi Y, Dong S. Comparative studies on stresses in weld-bonded, spot-welded and adhesive-bonded joints. Journal of Materials Processing Technology. 1999; 87:230–236. DOI: 10.1016/S0924-0136(98)00355-0

[4] Belingardi G, Chiandussi G. Stress flow in thin walled box beams obtained by adhesive bonding joining technology. International Journal of Adhesion & Adhesives. 2004; 24:423–439. DOI:10.1016/j.ijadhadh.2003.11.009

[5] Pizzi A, Mittal K. Handbook of Adhesive Technology. 2nd ed. Marcel Dekker, New York (US), 2003, 999 p. ISBN: 0-8247-0986-1

[6] Skeist I. Handbook of adhesives. 3rd Van Nostrand Reinhold, New York (US). 1989; 931 p. ISBN: 978-0442276348

[7] Koricho E, Verna E, Belingardi G, Martorana B, Brunella V. Parametric study of hot-melt adhesive under accelerated ageing for automotive applications. International Journal of Adhesion and Adhesives. 2016; 68:164–181. DOI: 10.1016/j.ijadhadh.2016.03.006

[8] Shin H,Hamed G. Peel adhesion and viscoelasticity of poly (ethylene-co-vinyl acetate)-based hotmelt adhesives. The influence of wax. Journal of Applied Polymer Science. 1997; 63:333–342. DOI: 10.1002/(SICI)1097-4628(19970118)63:3<333::AID-APP8>3.0.CO;2-O

[9] Su W. Principles of polymer design and synthesis. Springer, Berlin (D). 2013; 306 p. DOI: 10.1007/978-3-642-38730-2

[10] Park Y, Joo H, Kim H, Lee Y. Adhesion and rheological properties of EVA-based hot-melt adhesives. International Journal of Adhesion & Adhesives. 2006; 22:571–576. DOI: 10.1016/j.ijadhadh.2005.09.004

[11] Kinloch A. Adhesion and adhesives-science and technology. Chapman and Hall, London (UK); 1994; 440 p. DOI: 10.1007/978-94-015-7764-9

[12] Petrie E. Reactive hot-melt adhesives for better structural bonding. Metal Finishing. 2008; 106:39–43. DOI: 10.1016/S0026-0576(08)80125-0

[13] Shin H, Hamed G. Peel adhesion and viscoelasticity of poly (ethylene-co-vinyl acetate)-based hot-melt adhesives. I. The effect of tackifier compatibility. Journal of

Applied Polymer Science. 1997; 63:323–331. DOI: 10.1002/(SICI)1097-4628(19970118)63:3<323::AID-APP7>3.0.CO;2-P

[14] Li W, Bouzidi L, Narine S. Current research and development status and prospect of hot-melt adhesives: a review. Industrial & Engineering Chemical Research. 2008; 47:7524–7532. DOI: 10.1021/ie800189b

[15] Park Y, Kim H. Hot-melt adhesive properties of EVA/aromatic hydrocarbon resin blend. International Journal of Adhesion & Adhesives. 2003; 23:383–392. DOI: 10.1016/S0143-7496(03)00069-1

[16] Anon. Production: grow is the norm. Chemical Engineering News. 2008; 84:59–68. DOI: 10.1021/cen-v084n028.p059

[17] Finley T. Secure devices for personal or article identification, such as driver's licenses and vehicle license plates, and methods of producing such improved devices. 2016. U.S. Patent 9272496.

[18] Frihart C. Specific adhesion model for bonding hot-melt polyamides to vinyl. International Journal of Adhesion & Adhesives. 2004; 24:415–422. DOI: 10.1016/j.ijadhadh.2003.11.008

[19] The chemistry of tackifying resins. [Internet] 1990. Available from: https://www.pstc.org/files/public/Donker.pdf [Accessed: 06-05-2016].

[20] Using Waxes in Hot Melt Adhesives. [Internet] 2009. Available from: http://adhesives.specialchem.com/tech-library/article/using-waxes-in-hot-melt-adhesives [Accessed: 06-05-2016].

[21] Colthup N, Doly L, Wiberley S. Introduction to infrared and Raman spectroscopy. 3rd ed. Academic Press Limited, London (UK); 1990. 547 p. ISBN: 978-0121825546

[22] Hummel D. Baum A. Atlas of Polymer and Plastic Analysis. Carl Hanser Verlag GmbH & Co., Munich (D), 1998. 1250 p. ISBN: 978-3446161269

[23] Ciardiello R, Belingardi G, Martorana B, Fondacaro D, Brunella V. A study of physical and mechanical properties of a nano-modified thermoplastic adhesive in normal and accelerated ageing conditions. European Conference on Composite Material. (ECCM17); 26–30 June, 2016.

[24] Mazurin O. Problems of compatibility of the values of glass transition temperatures-Glass Physics and Chemistry. 2007; 33:22–36. DOI: 10.1134/S108765960701004X). 2008. 363 p. ISBN: 978-0-8155-1533-3

[25] Ebnesajjad S. Adhesives technology handbook. 2nd ed. William Andrew, Norwich (NY-US). 2008. 363 p. ISBN: 978-0-8155-1533-3

[26] Polypropylene Compounds for Automotive Applications. [Internet] 2010. Available from: http://www.sumitomochem.co.jp/english/rd/report/theses/docs/20100100_a2g.pdf [Accessed: 06-05-2016].

[27] Park Y, Kim H, Kim H, Rafailovich M, Sokolov J. Viscoelastic properties and lap shear strength of EVA/aromatic hydrocarbon resins as hot-melt adhesives. Journal of Adhesion Science and Technology. 2003; 17:1831–1845. DOI: 10.1163/156856103322538714

[28] Hummel S. Atlas of Polymer and Plastic Analysis. Carl Hanser, Weinheim. 1991; 1250 p. ISBN: 9783527257980

[29] Gallo R, Brambilla L, Castiglioni C, Ipsale S, Severini F, Quasso F, Consolati G. Environmental degradation of a novel ethylene-propylene copolymer in thick sheets. European Polymer Journal. 2005; 41:359–366. DOI: 10.1016/j.eurpolymj.2004.09.014

[30] Adams R, Peppiatt N. Stress analysis of adhesively bonded lap joints. The Journal of Strain Analysis for Engineering Design. 1974; 9:185–196. DOI: 10.1243/03093247V093185

[31] da Silva L, Rodrigues T, Figueiredo M, de Moura M, Chousal J. Effect of adhesive type and thickness on the lap shear strength. The Journal of Adhesion. 2006; 82:1091–1115. DOI:10.1080/00218460600948511

Chapter 14

Ab-Initio Modeling of Adhesive Behaviors at Material Interfaces

Jun Zhong

Additional information is available at the end of the chapter

http://dx.doi.org/10.5772/64904

Abstract

In this chapter, density functional theory (DFT) is employed to identify the essentials of adhesive formation at the Al/Ceramic interfaces. Analyses of electronic structure in the DFT used for atomic bond evolution at interfaces, indicate that, adhesion energy mostly dominated by surface energy, may lead to formation of adhesives, which bind two type bulk-materials strongly together, especially for those polar systems involving cubic oxides, carbides and nitrides. In addition, ab-initio molecular dynamics (AIMD) based upon the DFT is adopted, for example, in the simulation of chemical reaction between two contacting reactive slabs: pure aluminum and iron-oxide. This may provide an insight into the dynamical formation of an adhesive (amorphous Al_2O_3-texture) occurring between two nascent material surfaces if lubricants are not present or are insufficient. Such a texture may bond onto a hard-roller surface as a protective thin film to resist the elevated temperature.

Keywords: density functional theory, ab-initio molecular dynamics, Al-Ceramic adhesives, amorphous Al_2O_3-texture, adhesion energy, surface and interface energies, adhesive transfer of materials

1. Introduction

Adhesives have been widely used to improve mechanical properties of materials such as hardening, deformation stability of surfaces or thin films for a variety of technological applications [1–6]. A significant effort focused on exploring nanomechanisms at interfaces where two solid surfaces contacted. For examples, adhesives were critical to successful manufacture for semiconductor circuits. Each deposited layer must adhere strongly to that beneath it to survive thermal stresses. Similarly, a recent report told that, wear of materials usually resulted in a very high economic loss to every country, for instance, economic losses due to the wear would amount to about 1.0–2.5% of the gross national product (GDP) in the USA [7]. Therefore, adhesives may have a tremendous effect on the US industry because it determines the lifetime of many working parts, from components of automotive engines to airplane brakes. A particular example is the aluminum-alloys processing which is also relevant to many other metallic alloys, that is, the Al-alloy products are manufactured through a series of processing steps that begin with casting an aluminum ingot. In order to reduce ingot thickness, it passes through a series of rolling operations under high loads at speeds up to 4500 ft/min. In this case, there will take place adhesives (the melting aluminum bonds to steel surfaces as chemical additives). On the positive side, many machining operations rely upon adhesive wear processes (from oil drilling to polishing a Si-wafer). The processing optimization is critical to commercial success. While on the negative side, an interest in dry machining is rising because oils used in the machining are likely evaporated due to frictional heating, and such a vapor impact on the long-term health of machine operators is increasingly concerned. Therefore, the better understanding of adhesion mechanisms may result in positive impacts on many industrial manufacturing.

In this chapter, we will introduce our previous studies of adhesives on aluminum and other alloy surfaces: (1) Adhesion is of particular interest. Although a work has reported the adhesion character at an interface between two pure materials (aluminum-diamond) [8], we will discuss an application of electronic structure to this field between aluminum and various ceramic compounds, to determine factors (e.g., an adhesion energy) that may qualitatively control the adhesion strengths at interfaces because this discussion may approach more to the reality even if such computed factors often show too large deviations from some continuum-scale (finite element) calculations. In addition, some particular adhesives, which bind two type bulk-materials strongly at the Al/Ceramic interfaces, are discussed. (2) Another simulation from ab-initio molecular dynamics (AIMD) based upon density functional theory (DFT) is present to bond breaking and formation, and adhesive transfer of materials over an interfacial area where two different material slabs (with pure aluminum and iron oxide) are compressed and then stretched out. In this study, contact of two reactive slabs may lead to a strong exothermic reaction to form a dynamical thin film bonding onto a hard-roller surface to prevent the surface from the melting.

In the following discussions, although we mainly focus on aluminum/aluminum-oxide samples, without loss of generality, we believe what we discuss here is also relevant to many other materials in which adhesion may occur.

2. Methodology

2.1. Density functional theory

Macroscopic materials consist of an enormous huge number (~1023 per cm^3) of nuclei and electrons, and these particles interact in a complex fashion. With the development of quantum mechanics in the early 1900s, the law governing those interactions was made clear, that is, the so-called many-particle Schrödinger equation. In order to solve this equation efficiently, the Born-Oppenheimer (BO) approximation [9] was usually invoked to separate the ionic degrees of the freedom from electronic ones. Even so, an exact solution for this many-particle problem was still impossible. In the early 1960s, density functional theory (DFT) was proposed as an alternative to the wavefunction-based description of quantum mechanics. The core concept of the DFT was to describe the state of a system not with the complex *all-electron wave function* (AE-WF), but with the much simpler *charge density*. Consequently, the DFT may drastically reduce the degrees of the freedom needed to describe the system, yet in principle maintain an accurate depiction of the physics. In recent years, the DFT has become the method of choice for describing large molecular clusters and solid systems.

DFT was based upon theorems developed by Hohenberg, Kohn and Sham [10, 11], which stated that the ground-state properties of a system were functionals of the charge density $n(\boldsymbol{r})$ alone. This charge density, which minimizes the energy functional under a given external potential: $V_{ext}(\mathbf{r})$ – the interaction between electrons and nuclei, is the ground-state charge density. So the total energy of a system can be expressed as a functional of the charge density, that is,

$$E_{tot}[n(\boldsymbol{r})] = <\Psi^{el} \mid \hat{H}^{el} \mid \Psi^{el}> = \overline{T[n(\boldsymbol{r})]} + \overline{V_{ee}[n(\boldsymbol{r})]} + \int V_{ext}(\boldsymbol{r}) n(\boldsymbol{r}) \mathrm{d}\boldsymbol{r}. \tag{1}$$

According to the Kohn-Sham ansatz [12], the kinetic term $\overline{T[n(\mathbf{r})]}$ is replaced with a kinetic energy of non-interacting electron gas: $E_{KE}[n(\mathbf{r})]$; the electron-electron interaction potential $\overline{V_{ee}[n(\mathbf{r})]}$ is replaced with a classical Coulombic interaction: $E_H[n(\mathbf{r})]$ – the Hartree energy, plus an additional non-classical term: $E_{XC}[n(\mathbf{r})]$ – the exchange-correlation (XC) energy. Therefore, Eq. (1) is modified as follows:

$$E_{tot}[n(\boldsymbol{r})] = E_{KE}[n(\boldsymbol{r})] + E_H[n(\boldsymbol{r})] + E_{XC}[n(\boldsymbol{r})] + \int V_{ext}(\boldsymbol{r}) n(\boldsymbol{r}) \mathrm{d}\boldsymbol{r}. \tag{2}$$

The advantage of this modification is that the first two and the last terms in Eq. (2) can be dealt with a traditional manner; while the third term the exchange-correlation energy, whose form is in principle unknown, can be simplified through several approximations. Eqs. (1) and (2) are called the N single-particle Kohn-Sham (KS) equations.

In order to solve the total energy $E_{tot}[n(\mathbf{r})]$ in Eq. (2) more accurately, a set of Lagrange multipliers $\{\varepsilon_i\}$, which correspond to the orthonormality of the N single-particle orbitals $\{\varphi_i\}$, are adopted in an effective Kohn-Sham Hamiltonian operator H_e^{KS}, so the ground-state energy $E_{tot}[n(\mathbf{r})]$ can be obtained through minimizing the charge density $n(\boldsymbol{r})$, see [11, 12]. One advantage of this method is that the number of the N single-particle KS equations increases linearly with the system size and number of electrons.

Even if the Kohn-Sham ansatz is exact in principle for ground-state systems, in practice the approximations are necessary because the exact functional form $E_{XC}[n(\mathbf{r})]$ in Eq. (2) is still unknown. Consequently, an accurate, yet efficient expression for the XC energy is essential. Even though some scientists were devoting themselves to extending the DFT to treat excited states in materials [13], most of current works mainly focus on systems at the ground state.

2.1.1. The exchange-correlation functional

Several approximations for the XC energy functionals have been developed. The most commonly used approximation in the DFT calculations was the so-called local density approximation (LDA) [14], which was based upon the so-called uniform electron gas. The LDA was an exact approximation for a uniform electron gas in a system with a slowly varying charge density [15]. Actually, the charge density $n(r)$ usually behaved like a rapidly varying function of position in some systems. Thus, the corrections to the XC energy based upon the gradient of the charge density were required to reduce the LDA error.

One famous commonly used extension to the LDA was the generalized-gradient approximation (GGA). An example of a GGA functional (PW91) predominantly used in solid-state calculations was that of Perdew and Wang [16]. Another modified expression known as the PBE functional has been suggested by Perdew, Burke and Ernzehof [17]. Comparing the GGA to the LDA, the former has demonstrated to tremendously improve the accuracy of calculations on solids and molecular clusters [15].

2.1.2. Plane waves and supercell

The method of Plane Waves (PW) basis-sets (orbitals), which was more suitable for calculations of periodic solids in the reciprocal space than others on extending its accuracy to any desired convergence, has been widely used for analyzing adhesion properties through calculations of various electronic structures. To simplify, they can be handled by sampling some special *k*-points in the reduced Brillouin-zone to determine the electronic wave function and the electric potential through the filled electronic bands. Two widely used methods were Chadi-Cohen [18] and Monkhorst-Pack [19] samplings.

In a larger supercell, for the wave vector selection, it is often sufficient to only consider a particular value-point—the Γ point because this point is the original position in *k*-space, and such a choice would confer significant savings with data storage and computational cost.

2.1.3. Pseudopotential approximation

The Plane Waves (PW) method cannot directly be carried out in the KS theorem since the wave functions of ionic-cores are rapidly varying with many nodes and require enormous PW basis-sets to reach the calculation accuracy. Since chemical reactions are solely related to valence electrons while ionic-core electrons are approximately "frozen" in their original configurations, in practice, to handle the external potential: $V_{ext}(\mathbf{r})$, a scheme of pseudopotential approximation (PP) is provided to just reproduce identical curves of valence wave functions in the place sufficiently far from the ionic-core region (R_C). So in the PP framework, the constructed charge densities outside the ionic-core region (R_C) must be identical to the true ones, which is called "the norm-conserving condition" [20].

To compromise the computational demanding in practice, a successful concept of "the ultrasoft-pseudopotential (US-PP)" was introduced by Vanderbilt [21]. Blöchl further expanded the US-PP concept by combining ideas from the PP and the LAPW (linearized augmented-plane-wave) methods in a conceptually elegant framework, he called it "the projector augmented-wave method (PAW)" [22]. And then, Kresse established an exact formal relationship between the US-PP and the PAW methods [23].

2.2. Molecular dynamics

Molecular dynamics (MD) is a methodology which depicts motions of a many-particle system using classical Newton's equations. It has been over 40 years since the first application of MD method created by Alder and Wainwright to a system with hard spheres [24]. The MD method is particularly useful for the study of dynamical properties of materials. More important, this study may help people extract physical insights from the many-particles system.

2.2.1. Equations of motion

In MD simulations, classical trajectories of particles in the real space for a finite size system are generated by solving the Lagrange equations numerically. When choosing the Cartesian coordinates, these equations will become Newton's equations, that is,

$$m_i \ddot{\mathrm{R}}_i = \mathrm{F}_i = \dot{\mathrm{P}}_i = -\frac{\partial V_{tot}(\mathrm{R}^N)}{\partial \mathrm{R}_i} \tag{3}$$

where $V_{tot}(\mathbf{R}^N)$ is the total potential energy of a system, m_i is the mass of the particle i, $\boldsymbol{P}_i$ is the momentum of the particle i, $\boldsymbol{F}_i$ is the total inter-particle force acting on the particle i, and " · " and " ·· " are *the first* and *the second* orders of the time derivative, respectively. The convenience of solving Eq. (3) should depend on some practical algorithms.

2.2.2. Algorithms

A standard way of solving Eq. (3) is the finite-difference method: Under a given configuration (positions and velocities of all particles) and other dynamical information in a system at the time t, a numerical integration would determine the new configuration of the system at the later time $t+\Delta t$ (Δt is a time step). Commonly used methods in MD are the Verlet algorithm [25], the Leapfrog algorithm [26] and the Gear predictor-corrector (GPC) algorithm [27]. An ideal algorithm should be simple, run fast and require little memory. It should also permit using a longer time step Δt to hold the whole trajectory of the system as close as possible and preserve conservation laws of momentum and energy. According to these, the Verlet algorithm can be the most widely used one. The leapfrog algorithm is essentially identical to the Verlet algorithm. The GPC method is often more accurate as well as more complicated than those two algorithms.

2.2.3. Statistical ensembles

The widely used statistical ensembles in MD simulations are often the microcanonical ensemble (NVE), the canonical ensemble (NVT), the isothermal-isobaric ensemble (NPT) and the grand canonical ensemble (μVT). Thermodynamic variables in above parenthesis for each ensemble are fixed in relevant ensembles [28]. In MD simulations, the (NVE) ensemble is the best to be realized since all equations of the motion conserve the total energy E and particle's number N. The constant volume V can be fixed using the free or the periodic boundary conditions; The (NVT) ensemble is very commonly used for practical calculations. Therefore, to control the temperature T in this ensemble, there are a number of ways to simulate it: (i) the simplest one is the velocity scaling method [28], but is crude in MD simulations. (ii) the Nosé-Hoover's method is a more conceptually fundamental, but has an adjustable parameter of the thermal inertia in practice [29].

For the (NPT) ensemble, Anderson provided an "extended system" method for a constant pressure MD [30]. In this method, a term of "external piston" was added into the Lagrange functional for adjusting a value of external pressure. In another way, Parrinello and Rahman [31] proposed a scheme of "variable cell constant MD" which allowed the simulation box to change the shape as well as the size [31]. This method, so far, is the most general constant pressure method in MD simulations.

The (μVT) is very useful for an open system: In this ensemble, the total particle's number N will not be conserved, see [32] for details.

2.2.4. Ab-initio molecular dynamics

The MD simulation based upon the force calculations through empirical potential [33] is called the *classical* molecular dynamics simulation, while that based upon the force calculations using density function theory is called the *ab-initio* molecular dynamics (AIMD) simulation [34]. Comparing these two, the former can approximately treat a large system (up to tens of millions of atoms) for long time steps (nanoseconds); while the latter can more accurately simulate a small system of hundreds of atoms, and is better at describing reactions involving the change of atomic bonding types.

In principle, the ab-initio molecular dynamics (AIMD) is an extension of the classical molecular dynamics (MD), which includes the first principle-derived potentials for the calculation of the inter-particle forces and the total energy of a system, see [34, 35].

2.2.4.1. Born-Oppenheimer molecular dynamics

In the AIMD scheme, the total potential energy $V_{tot}(\mathbf{R}^N)$ of a system has the same physical meaning as the Kohn-Sham energy (E^{KS}) by means of the Born-Oppenheimer (BO) approximation. Since the Kohn-Sham energy depends only upon nuclear positions and the defined hyper-surface for nuclear motions, the Lagrange functional of the so-called Born-Oppenheimer molecular dynamics (BOMD) has the following formula [35]

$$L_{BO}(\boldsymbol{R}^N, \dot{\boldsymbol{R}}^N) = \sum_{I=1}^{N_I} \tfrac{1}{2} M_I \, \dot{\boldsymbol{R}}_I^2 - \min_{\{\varphi_i\}} E^{KS}[\{\varphi_i\}; \boldsymbol{R}^N] + \sum_{i,j} \Lambda_{ij}[<\varphi_i \,|\, \varphi_j> - \delta_{ij}]. \quad (4)$$

where $\{\Lambda_{ij}\}$ are the Lagrange-multipliers. So the minimization of Kohn-Sham energy, E^{KS}, is the constraint to the orthogonal basis-orbitals $\{\phi_{i,j}\}$. And then, Eq. (3) for many-particles can be modified by means of Eq. (4) as follows:

$$M_I \, \ddot{\boldsymbol{R}}_I = -\frac{\partial}{\partial \boldsymbol{R}_I}\left[\min_{\{\varphi_i\}} E^{KS}[\{\varphi_i\}; \boldsymbol{R}^N]\right] + \sum_{i,j} \Lambda_{ij} \frac{\partial}{\partial \boldsymbol{R}_I}[<\varphi_i \,|\, \varphi_j> - \delta_{ij}], \quad (5)$$

and

$$H_e^{KS} \varphi_i = \sum_j \Lambda_{ij} \varphi_j. \quad (6)$$

where H_e^{KS} denotes an effective Kohn-Sham Hamiltonian operator for a single-electron. Based upon Eqs. (5–6), a *classical* molecular dynamics program can be turned into the BOMD program by replacing the energy and the force routines with the corresponding ones from a quantum chemistry methodology [35]. The accuracy of the force calculations used in the BOMD depends linearly on the accuracy of minimizing the Kohn-Sham energy E^{KS}. To realize this, an algorithm of the conjugate gradient (CG) minimization was effectively applied to such calculations, see [34].

2.2.4.2. *Car-Parrinello molecular dynamics*

The basic idea of the Car-Parrinello molecular dynamics (CPMD) can be seen as exploring a time-scale separation of "fast-electronic" and "slow-nuclear" motions by transforming it into a dynamical framework of "classical-mechanical and adiabatic energies" [36]. In classical mechanics, forces on nuclei are obtained through the derivative of a Lagrange functional with respect to nuclear positions. The CPMD scheme suggested that, a functional derivative with respect to particle orbitals, which were interpreted as classical fields, yielded the forces on those suitable particle orbitals given by the Lagrange functional. Thus, the CPMD scheme postulates the following Lagrange functional

$$L_{CP}[\boldsymbol{R}^N, \dot{\boldsymbol{R}}^N; \{\varphi_i\}, \{\varphi_j\}] = \sum_{I=1}^{N_I} \tfrac{1}{2} M_I \dot{\boldsymbol{R}}_I^2 + \sum_{i=1}^{occ} \tfrac{1}{2} \mu < \dot{\varphi}_i \mid \dot{\varphi}_i > - \\ - E^{KS}[\{\varphi_i\}; \boldsymbol{R}^N] + \sum_{i,j} \Lambda_{ij} [< \varphi_i \mid \varphi_j > - \delta_{ij}] \quad . \tag{7}$$

Like in classical mechanics, nuclear positions, electronic orbitals and the Car-Parrinello's equations of the motion are modified as the following formulae corresponding to the Newton's equations [35]

$$M_I \ddot{\boldsymbol{R}}_I = -\tfrac{\partial}{\partial \boldsymbol{R}_I} \left[E^{KS}[\{\varphi_i\}; \boldsymbol{R}^N] \right] + \sum_{i,j} \Lambda_{ij} \tfrac{\partial}{\partial \boldsymbol{R}_I} [< \varphi_i \mid \varphi_j > - \delta_{ij}] , \tag{8}$$

and

$$\mu \ddot{\varphi}_i = -H_e^{KS} \varphi_i + \sum_j \Lambda_{ij} \varphi_j , \tag{9}$$

where μ is a "fictitious mass" or "inertia parameter" assigned to the orbital degrees of the freedom. The unit of this "mass parameter" μ is the energy multiplies the time square for reasons of dimensionality. Note that constraints within E^{KS} would lead to "force constraints" in Eq. (8). In practice, these constraints would depend upon both the Kohn-Sham orbitals and the nuclear positions through an overlap matrix of basis-functions—the simulated annealing to search for coefficients of the basis-orbitals that minimize electronic energy. In this terminology, a scheme of "low electronic temperature" or "cold electrons" in Eq. (9) is close to its instantaneous minimum energy: $\min_{\{\varphi_i\}} E^{KS}\left[\{\varphi_i\}; R^N\right]$, that is, in this case, it is close to the exact Born-Oppenheimer hyper-surface. Therefore, a ground-state wave function optimized for the initial configuration of nuclei would stay close to its ground state during time evolution if the system is kept at a sufficiently low temperature. A SHAKE algorithm is used to impose constraints on the system to ensure that the KS orbitals remain orthonormal [34]. In practice, it was found that the Car-Parrinello scheme was unnecessary for electronic configurations to be minimized in a coefficient space at each MD time step even though this manner may give errors in force calculations on nuclei. However, it was rather strange to indicate that, the errors of force calculations on nuclei were canceled by the associated errors in depicting electronic motions. According to this, the Car-Parrinello scheme may tolerate force errors on nuclear motions in the AIMD simulation.

In addition, the last term on the right-hand side of Eqs. (5) and (8) would vanish if basis-orbitals $\{\phi_i\}$ are independent of nuclear positions $\{\boldsymbol{R}^N\}$. Comparing the BOMD to the CPMD schemes, the former can be viewed as an alternative to the latter, which may give very accurate coefficients of basis-orbitals through minimizing electronic energy at each MD time step.

http://dx.doi.org/10.5772/64904

2.3. An implemented software: VASP

The Vienna Ab-initio Simulation Package (VASP) is a commercial software under the framework of density functional theory (DFT) code that is developed and distributed by a research group at The University of Vienna, Austria [37]. VASP uses a plane wave (PW) basis-sets (orbitals) to expand the single-electron Kohn-Sham wave function and the pseudopotentials for describing electron-ion interactions. Three separate approximations to the exchange-correlation (XC) energy are employed. The local density approximation (LDA) and the generalized-gradient approximation (GGA) are in two flavors: the PW91 functional and the PBE functional [14–17]. In addition to these functionals, several additional routines have been implemented to allow for estimation of transitional state energies and geometries (NEB, metadynamics [38]), vibrational properties, analyses of various electronic structure aspects, etc. For more details regarding the VASP code, see [37].

3. Density functional study of adhesion at the metal/ceramic interfaces

A fundamental quantity to influence mechanical properties of a material interface is defined as adhesion energy: W_{sep} (also commonly referred to as a "work of separation"), breaking interfacial bonds and reversibly separating the interface into two free surfaces, but neglecting diffusion and plastic deformation [39–41]. Experimental observations on this dissipative process indicated that, due to plastic deformation, the W_{sep} was proportional to, but smaller than, the work done to cleave the interface, see [42]. Thus formally, the W_{sep} is defined in terms of surface and interface energies of respective bulk materials, or to the difference in total energy between an interface system and two isolated surface slabs:

$$W_{\text{sep}} = \sigma_1 + \sigma_2 - \gamma_{12} = \frac{1}{A}(E_1^{\text{tot}} + E_2^{\text{tot}} - E_{12}^{\text{tot}}). \tag{10}$$

Here σ_{i} is the surface energy of slab-i, γ_{12} is the interface energy, $E_{\text{i}}^{\text{tot}}$ is the total energy of slab-i, E_{12}^{tot} is the total energy of a system (slab-1 + slab-2) with an interface, and A is the whole interface-area.

3.1. Results and discussions

Available analytical models for predicting the W_{sep} were still generally limited to liquid-metal and oxide interfaces, and had to rely on some simple empirical correlations that described either the free energy of formation for oxide and liquid-metal, or the mixing enthalpies for respective oxide elements in metals [42]. Unfortunately, many of these models were not applicable to those systems comprised of non-oxidized ceramics interface. Furthermore, these models may only provide qualitative information for trends in adhesion and their generality. Even within the class of metal/oxide interfaces, it was still debatable because many of them only have those interfaces using the α-Al_2O_3 (Alumina) as oxides [43].

In recognition of limitations of analytic models, a number of *ab-initio* studies on metal/ceramic adhesion using density functional theory (DFT) have appeared during past two decades. These studies captured atomic-scale features at the interfacial areas and can reveal invaluable information using electronic structures. For example, one DFT research focused on the interfacial adhesion at several Al-Ceramic couples: Al/α-Al_2O_3, Al/WC, Al/VC, Al/VN, Al/CrN and Al/TiN [44–47]. Among which, large alloy slabs obeying a unit stacking sequence along x- and y-directions, respectively, but up to 15 atomic layers along z-direction were adopted in constructing calculation geometries, so as to accurately evaluate the surface energies and avoid the non-convergence problem via the slab thickness. Superscripts on polar surface systems, α-$Al_2O_3(0001)^O$, $WC(0001)^W$, $WC(0001)^C$, etc., indicated surface terminations. In addition, each non-stoichiometric slab must be constructed to allow for the identical termination of its both surfaces. Regarding their relevant geometric parameters, see [47]. In this research, although Eq. (10) defined the W_{sep} in terms of both surface and interface energies, these quantities do not necessarily play an equal role, that is, since interface energy γ is relatively small in those metal-ceramic systems, the strength of interfacial bonding relevant to the W_{sep} would depend to a large extent upon the reactivity of individual surfaces as reflected by their surface energies, see **Table 1** [47].

Surface	**No. of layers**	**σ (J·m^{-2})**
Al(100)	5	0.89
Al(110)	4	1.05
Al(111)	5	0.81
α-$Al_2O_3(0001)^{Al}$	15	1.59
α-$Al_2O_3(0001)^o$	13	7.64
$WC(0001)^w$	9	3.66
$WC(0001)^c$	9	5.92
$WC(11\bar{2}0)$	4	3.88
VN(100)	7	0.95
VC(100)	7	1.28
CrN(100)	7	0.74
TiN(100)	7	1.25

Superscripts give termination of those surfaces cleaved along a polar plane.

Table 1. Calculated surface energies σ_i and slab-i dimensions (no. of layers) used for the W_{sep} calculations.

If considering atomic relaxations into the calculation, at an Al/Alumina interface, both magnitude and rank ordering of adhesion energies for different stacking sequences are found to change to the unrelaxed results determined by a "universal binding energy relation (UBER, in this framework, both charge density and binding energy of surface via inter-particle separation were empirically described as exponential decaying functions)" [48], see **Figure 1(a)**. Regarding the optimized interfacial stacking sequences, through analyzing electronic structures (charge density contours), it is found that, at the polar Al/α-Al_2O_3 and the polar Al/WC interfacial areas (more dangling bonds on isolated surfaces), metal atoms prefer to the sites which continue the natural stacking sequence in ceramic bulk across the interfacial area into the Al-slab host, see the right picture in **Figure 1(b)**; while for non-polar Al/α-$Al_2O_3^{Al}$, Al/VC, Al/VN, Al/CrN, Al/TiN and Al/WC systems, Al atoms prefer to the sits above metalloid atoms (O, C and N), see the left picture in **Figure 1(b)**.

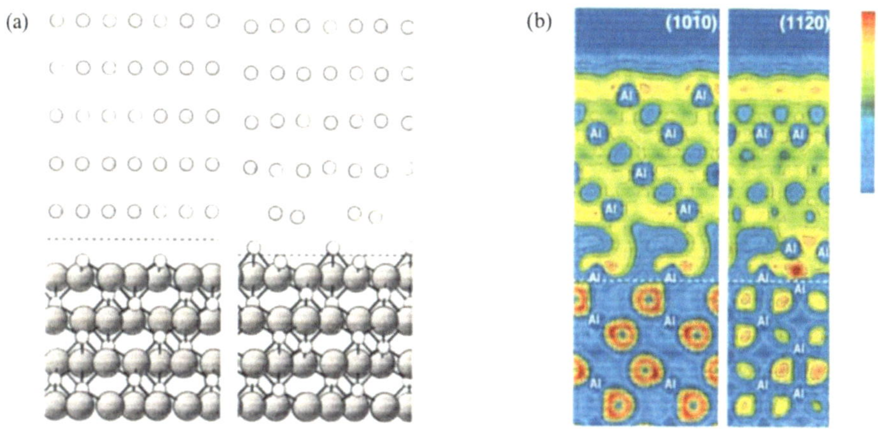

Figure 1. (a) (Left) An optimized geometry for a fcc-Al/α-Al_2O_3 interface determined by the UBER calculations; (right) the final relaxed structure: Small spheres are Al atoms, and large spheres are O atoms. The view direction is along $[1\bar{2}10]$. (b) Side-view of electronic structures at an interface: Two slices through the electron localization function (ELF) for an fcc-Al/α-Al_2O_3 interface taken along $(10\bar{1}0)$ and $(11\bar{2}0)$ planes, indicating four of hexagonal close-packed oxygen layers in oxide (bottom) and all five atomic layers from an Al slab (top).

3.1.1. *Effect of surface geometry to the W_{sep}*

The W_{sep} calculations are listed in **Table 2** [47], where readers may find out obvious differences in the W_{sep} values between polar and non-polar geometries; among which, three polar surfaces behave the largest W_{sep} values from 4.08 $J\cdot m^{-2}$ for the Al/WC^W up to 9.73 $J\cdot m^{-2}$ for the Al/α-$Al_2O_3^O$; for the remaining six non-polar interfaces, five have the W_{sep} values of 2.10 $J\cdot m^{-2}$ or less. Furthermore, values for three nitride-ceramics are among the smallest range and look relatively insensitive to the choice of metallic component, with the W_{sep} values falling within 1.45–1.73 $J\cdot m^{-2}$. Averagely, the W_{sep} values in polar systems are about 4.60 $J\cdot m^{-2}$ larger than in non-polar systems, consistent with their larger surface energies as shown in **Table 1**. It was also noted that, the predicted W_{sep} value of 1.06 $J\cdot m^{-2}$ for the Al/α-$Al_2O_3^{Al}$ system was in favorable agreement with the experimental value of 1.13 $J\cdot m^{-2}$ in [49].

Interface	Orientation	Polarity	Strain (%)	γ (J·m^{-2})	W_{sep} (J·m^{-2})
Al/α-$Al_2O_3^{Al}$	$(111)[\bar{1}10]_{Al} \parallel (0001)[10\bar{1}0]_{Al_2O_3}$	NP	4.9	1.34	1.06
Al/α-$Al_2O_3^{o}$	$(111)[\bar{1}10]_{Al} \parallel (0001)[10\bar{1}0]_{Al_2O_3}$	P	4.9	−1.28	9.73
Al/WC^{w}	$(111)[\bar{1}10]_{Al} \parallel (0001)[11\bar{2}0]_{WC}$	P	2.2	0.39	4.08
Al/WC^{c}	$(111)[\bar{1}10]_{Al} \parallel (0001)[11\bar{2}0]_{WC}$	P	2.2	0.72	6.01
Al/WC	$(110)[1\bar{1}0]_{Al} \parallel (11\bar{2}0)[001]_{WC}$	NP	0.4	1.79	3.14
Al/VN	$(100)[001]_{Al} \parallel (100)[001]_{VN}$	NP	2.3	0.11	1.73
Al/VC	$(100)[001]_{Al} \parallel (100)[001]_{VC}$	NP	3.2	0.03	2.14
Al/CrN	$(100)[001]_{Al} \parallel (100)[001]_{CrN}$	NP	0.5	0.18	1.45
Al/TiN	$(100)[001]_{Al} \parallel (100)[001]_{TiN}$	NP	5.3	0.62	1.52

Terminations of polar ceramic surfaces are indicated with a superscript. The W_{sep} values correspond to the optimized stacking sequences.

Table 2. Interfacial orientation relationship, polarity (P = polar, NP = nonpolar), strain, interface energy (γ), and W_{sep}.

3.1.2. *Effect of surface energies to the W_{sep}*

To further verify whether the W_{sep} is mainly determined by surface energies or not, one Figure (see **Figure 1** in [47]) plots the W_{sep} versus the $[\sigma_{Al}+\sigma_{Ceramic}]$. A dashed-line with a unit slope is added to indicate data points lying in an idealized case: $\gamma = 0$. In this figure, the horizontal deviation of a given data point from the "$\gamma = 0$ – line" corresponds to the γ size. For three polar interfaces in the top region, data are bracketed with horizontal error bars, giving the range of possible σ_i listed in **Table 1**. The W_{sep} values in the whole range indicate a roughly linear trend with respect to the σ_i, all but one of data points fall to the line-right. One system falling to the line-left with a negative γ value and involving an α-$Al_2O_3^O$ surface, shows the largest surface energy and the W_{sep} value. This implies that the possibility of the fracture be within the Al-bulk rather than at the Al-Ceramic junction (interfacial area) due to a tensile stress. It also indicates that such an interface geometry is very firmly constructed.

With the exception of TiN, the rocksalt-structured carbides and nitrides (CrN, VN and VC) exhibit the largest dependence of the W_{sep} on their surface energies: These three systems have the smallest interface energies, only 0.03–0.18 J·m^{-2}, and their W_{sep} values crowd in a nearly linear fashion (close to the $\gamma = 0$ – line) along the bottom left of the figure. Presumably, small γ values at these interfaces can be explained by similarities between bulk and interface bonding. However, since both of bulk components exhibit some extent to metallic bonding, for the Al/TiN system—a singularity with respect to other rocksalt ceramics, the γ value contributes larger, 0.62 J·m^{-2}, which could be due to the relatively large interfacial strain (misfit of slab geometry) of 5.3%.

Correspondingly, **Table 2** also illustrates that those W_{sep} values involving non-polar α-Al_2O_3 and WC (11$\bar{2}$0) systems are not solely depicted by their surface energies (i.e., the γ effects should be more important). For two α-Al_2O_3 interfaces, as with TiN, a possible explanation for this behavior could be the significant strain (4.9 and 5.3%). In addition, a main effect of ionic bonding in bulk α-Al_2O_3 differs substantially from that in Al-bulk. On the contrary for the WC system, the strain is very low (0.4%), and thus differences of bonding types in respective bulks are less appeared (WC is metallic). So it implements an incoherent, misfit geometry characterized by irregular bonding across the interface.

3.2. Conclusion

Comprehensively, the DFT calculations indicated that the surface energy played a dominant role in determining the adhesion energy (the W_{sep}) for systems described above, interface energy was generally much smaller in comparison. In most cases, particularly for those polar systems involving cubic oxides, carbides and nitrides (adhesives) at interfaces, the W_{sep} was described largely in terms of surface energy alone. Exceptions to this rule were mainly for systems with relatively large interfacial strain or incoherent interface bonding.

4. Ab-initio molecular dynamics study at an Al/Fe_2O_3 interface

In this section, without any loss of generality, *ab-initio* molecular dynamics (AIMD) based upon density functional theory (DFT-GGA) is employed to discuss the adhesive transfer of materials at an interface combining a clean Al(111) and a hematite [α-Fe_2O_3(0001)] slabs together, to obtain invaluable information on the dynamical adhesion. As a result, the compression of two slabs may form an amorphous alumina (Al_2O_3)-texture, which welds two slabs together at the interfacial area through a thermite reaction. Subsequently, a tensile process results in the removal of amorphous alumina-texture from the melting Al-slab. The overall reaction is highly exothermic to increase the temperature of the system.

4.1. Modeling geometry and simulation procedures

An AIMD study focusing on a clean Al(111) slab here involves a super-cell geometry with a periodic 4-layer Al(111)-slab (16 ions per layer) in the x- and the y-directions [50]. This orthorhombic super-cell has three orientated vectors: $a[1\bar{1}0] = 2\sqrt{2}a_{\mathrm{Al}}$ at the x-axis; $b[11\bar{2}] = \sqrt{6}a_{\mathrm{Al}}$ at the y-axis; and $c[111] = 40$ Å at the z-axis along with a vacuum distance of 24.0 Å in the z-direction, and one bottom layer fixed along the z-direction. Here $a_{\mathrm{Al}} = 4.053$ Å [51]. Similarly, simulations for a widely studied α-Fe_2O_3(0001) slab with oxygen-termination (more reactive than iron-termination to the Al-surface) are conducted using a periodic seven-layer rectangle slab with two oriented vectors: $\sqrt{5}a_{\mathrm{Fe_2O_3}}$ at the x-axis and $\frac{\sqrt{15}}{2}a_{\mathrm{Fe_2O_3}}$ at the y-axis, and a height of 8.40 Å vacuum space with free boundary along the z-direction (total of 36

Fe and 54 O ions). Here $a_{Fe_2O_3}$ = 5.031 Å [52]. Comparing these two slabs, the Al-slab size is slightly larger than the Fe_2O_3-slab along *x*- and *y*-directions, respectively. So a super-cell fitting for the Al-slab is selected as the final frame to contain both of slabs in the simulation. In this super-cell, misfit percentages of each of two slab lengths along *x*- and *y*-directions are calculated at very small values: 1.9 and 1.9%, respectively (see **Figure 2**).

The simulation time step was selected as 0.001 ps (i.e., 1 ps = 10^{-12} s). The whole system was held at constant energy of AIMD simulation (no temperature control) when two slabs interacted. At the beginning of each simulation, top three layers of the α-Fe_2O_3-slab were constrained to move down along the *z*-direction at a constant velocity, V_Z = – 4.00 Å/ps, corresponding to a displacement d_Z = – 0.004 Å per time step. Since this speed, 400 m/s, was typical of real processing conditions, such shock loading may exceed the yield strength of the Al-slab, and the deformation of top layers in the Al-slab can be reasonably observed during contact simulation (**Figure 2**).

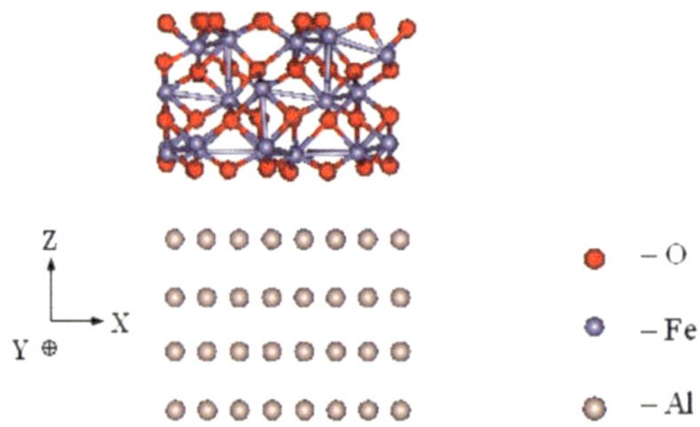

Figure 2. Side-view of geometry at the Al(111)/α-Fe_2O_3(0001) interface.

Before carrying out the compressive simulation, a favorable interface geometry between two slabs is optimized using the energy minimization in the DFT. The result indicates that the final spacing between two slabs would be about 1.88 Å to form Al–O bonds at the interfacial area, which is in very agreement with those observed in experiments [53] (see **Figure 3**). **Table 3** also lists some relevant bond lengths [50].

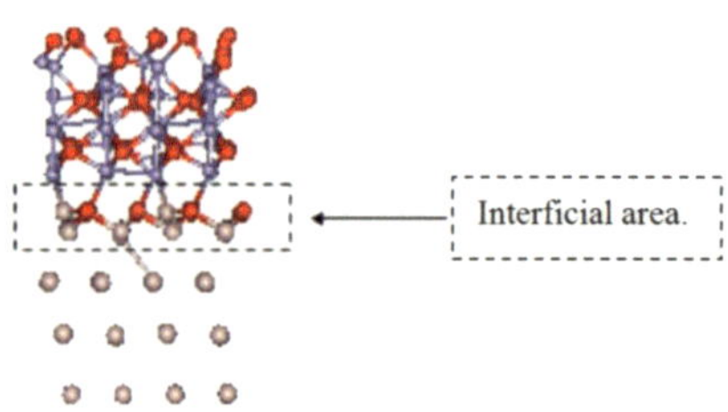

Figure 3. The final Al(111)/α-Fe_2O_3(0001) interfacial geometry optimized by the DFT.

http://dx.doi.org/10.5772/64904

Bond pair	Al–O		Fe–O	
Bond length (Å)	Calculation	Experiment [a]	Calculation	Experiment [b]
	1.86–1.93	1.86–1.97	1.82–1.84	1.83–1.85

[a] Ref. [55].
[b] Ref. [56].

Table 3. Relevant bond lengths at the interfacial area [unit: Å].

Therefore in the following AIMD simulation, the initial spacing between two slabs was selected as 2.50 Å, which was far enough to avoid the significant forming of chemical bonds in the R–L region (the interfacial area) when the simulation started. As two slabs moved closer, chemical bonds would begin forming in the spacing, and the interfacial reactions may occur. In **Figure 4(a)**, since top three layers (outside the R–L region) in the α-Fe_2O_3(0001)-slab may move down at a constant velocity (V_Z = − 4.0 Å/ps) as a rigid block, its final downward displacement is chosen as 0.62 Å, so as to keep the final spacing between two slabs around an optimized value: 1.88 Å.

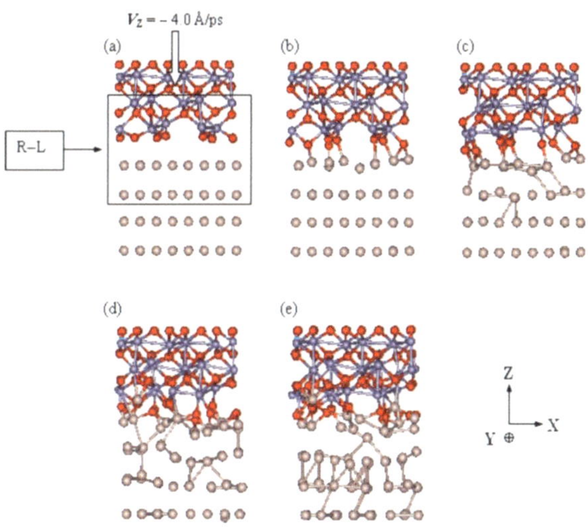

Figure 4. Main sequence of compressing the Al(111)/α-Fe_2O_3(0001) interface without lubricants.

4.2. Results and discussion

4.2.1. Compressive process

Figure 4 shows the compressing sequence of the Al(111) and the α-Fe_2O_3(0001) slabs. A fixed velocity, V_Z = − 4.00Å/ps, is applied onto top three layers in the α-Fe_2O_3(0001)-slab along the *z*-direction during the simulation [see **Figure 4(a)**].

In this figure, during the first 80 time steps, the α-Fe_2O_3(0001)-slab is moving downward the Al(111)-slab without forming Al–O bonds, see **Figure 4(a)**; around the 80th time step, some Al–O bonds begin forming (with Al–O bond length about 1.86~1.93 Å) [see **Figure 4(b)**]; around the 140th time step, most of Al–O bonds in the R–L region are formed [see **Figure 4(c)**]. Local temperature in this region increases to about 870 K. During this evolution, an amorphous Al_2O_3-texture is gradually formed due to the further reaction of oxygen-ions with the Al-surface; around the 390th time step, this reaction continues at the uppermost point [see **Figure 4(d)**]. Local temperature in the R–L region continues rising to about 1130 K; by the 620th time step, some new layers made up of amorphous alumina-texture are completely constructed in the R–L region [see **Figure 4(e)**]. Local temperature in this region increases to about 1200 K.

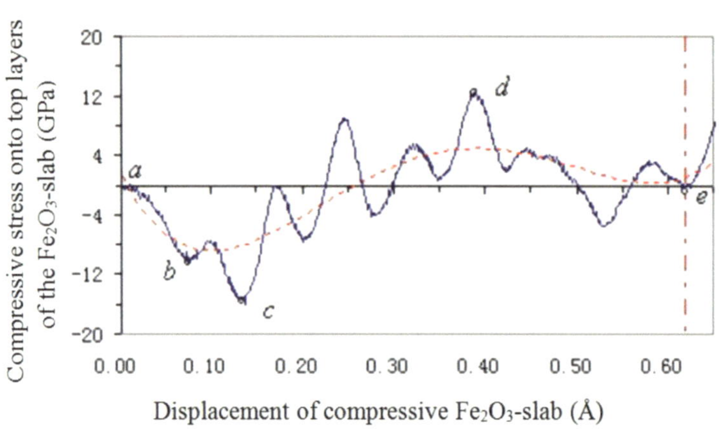

Figure 5. The stress perpendicular to the interfacial area as two slabs compress together.

Figure 5 indicates the distribution of the stress normal to the interfacial area as two slabs compress together. Initially, the stress becomes negative (attractive between two slabs) as Al–O bonds begin forming in the R–L region (points *a–c*); when two slabs are moving closer, the stress tends to the positive values (repulsive between two slabs) because most of Al and O atoms at the interfacial area are forced to come closely, leading to a nearly formation of amorphous alumina-texture with thermal (exothermic) expansion; after two slabs totally react, the stress falls down to a near zero point (point *e*) from the highest positive one (point *d*) as an optimal bond spacing between two slabs is reached; and then, the stress increases again (after point *e*) when two slabs continue compressing. This figure also represents a polynomial (dashed line) fit to the stress curve, which may clearly regress the real stress trend through its oscillation. Please note that the stress trend just depicts the starting stage of a whole P-h curve as stated in [54]. However, since two slabs here are too thin and too small in dimension, and at too short time steps throughout the whole AIMD simulation, it is hardly to rough out the whole P-h curve fulfilled by *classical* molecular dynamics on very large size models at very long time steps. Even if, this stress curve may still provide some invaluable information at the initial stage of nanoindentation process.

Figure 6 shows the number change of Al–O and Fe–O bonds versus normal distance at the interfacial area in **Figure 4(a)** during the compressive process. In this figure, Al–O bonds in the R–L region increase substantially while Fe–O bonds in this region decrease gradually. The reason is that, when two slabs compress together, the dangling bonds on Al and O atoms at the interfacial area would bond together. As the compression continues, more Al–O bonds are forming as some Fe–O bonds are breaking. Nevertheless, only a very porous interface is formed during two slabs compression, which is far from the ideally compressed status. Therefore, the thermite reaction at the interfacial area just partially occurs, only a few Fe–O bonds are breaking in this region.

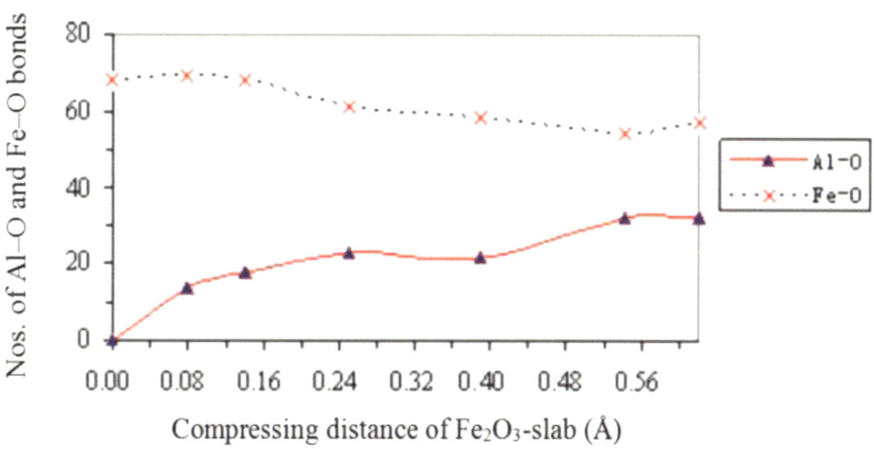

Figure 6. The number change of Al–O and Fe–O bonds in the R–L region during slab compression.

4.2.2. *Tensile process*

After the compression, an effect of a tensile strain acting on the compressed α-Fe_2O_3(0001)/Al(111) slabs is immediately conducted: Top three layers of the α-Fe_2O_3(0001)-slab in **Figure 4(e)** is moving up as a rigid block at a constant velocity, V_Z = + 4.00 Å/ps, reversing the downward direction of previous compression.

Figure 7 shows the stretching sequence of this process: (i) when top three layers of the α-Fe_2O_3(0001)-slab move up about 1.30 Å from the starting point in **Figure 4(e)**, the real tensile process at the interfacial area begins (part *a*). (ii) Then next, as two slabs separate, further deformation at the interfacial area and thermite reaction will continue, more atomic mixing in this area oxidizes the Al-surface (part *b*). (iii) During transitional stages (parts *c* and *d*), the whole interfacial area begins breaking apart, mainly due to the breaking of some more Fe–O and Al–Al bonds. Along with a small increase of Al–O bonds, amorphous alumina-texture is continuously formed, and an adhesive metal transfer begins. (iv) Finally, surface on the Al-slab is totally broken: an adhesive metal transfer has completely occurred, amorphous alumina-texture is removed from the remnant surface of the Al-slab, and bonds onto the bottom of the α-Fe_2O_3(0001)-slab, indicating the final apart step of adhesive transfer of materials. In addition, a fresh disordered Al-surface is exposed on the residual Al-slab after the tensile process finishes (part *e*).

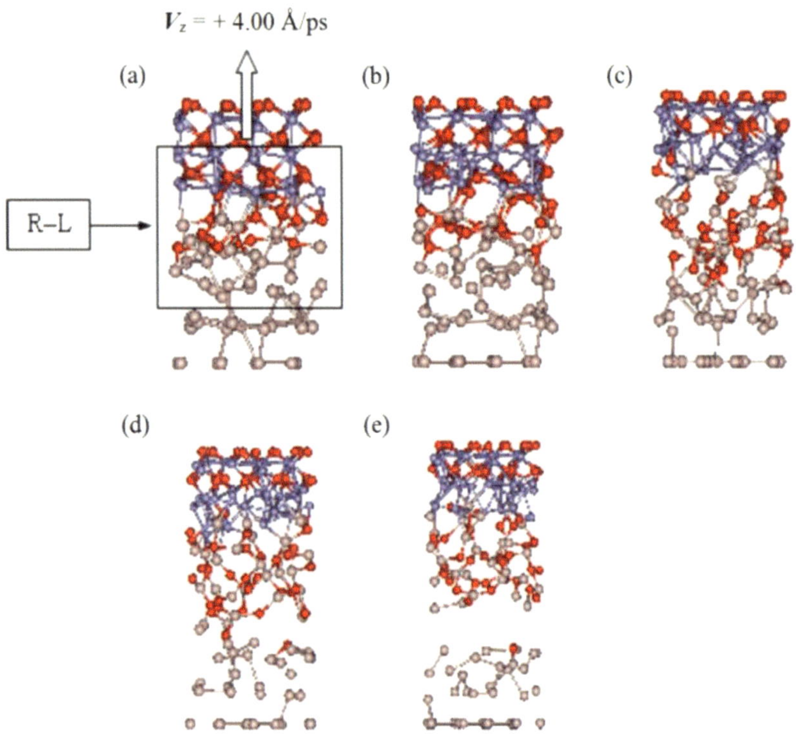

Figure 7. The main sequence of stretching the Al(111)/α-Fe_2O_3(0001) slabs.

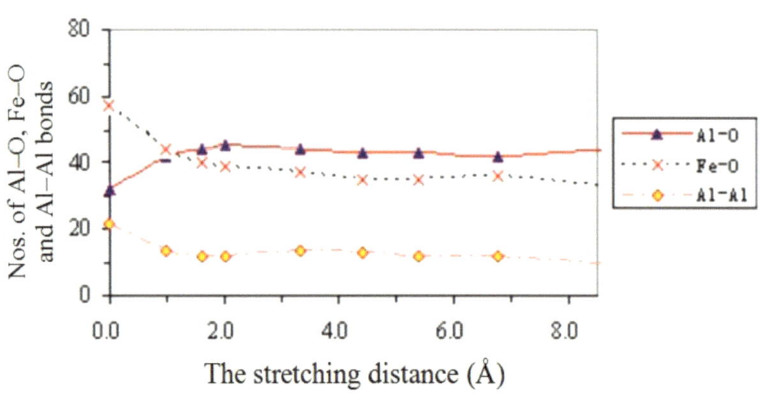

Figure 8. The number change of Al–O, Fe–O and Al–Al bonds in the new R–L region during the slab tension.

Figure 8 shows the number change of Al–O, Fe–O and Al–Al bonds in a new larger R–L region as shown in **Figure 7(a)** during the tensile process. In this figure, at initial stage, a few more Al–O bonds are forming and a few more Fe–O bonds are breaking, which represents the continuous thermite reaction in the R-L region as the Fe_2O_3-slab is being pulled upward. Also, there is a breaking of approximately 10–12 Al–Al bonds in top two layers of the Al-slab. When the Fe_2O_3-slab is pulled upward about 8.0 Å, there is a slight increase in Al–O bonds but a slight decrease in Fe–O bonds. In this case, the interfacial area is totally separated. By the end of the tensile process, the Fe–O coordination is approximately 1:1, while the Al–O one is approximately 2:3. Thus, this reaction could be approximately $2Al + 3Fe_2O_3 \rightarrow 6FeO + Al_2O_3$, instead of a more favorable thermite reaction: $2Al + 3FeO \rightarrow 2Al_2O_3 + 3Fe$, which is mainly due to very limited reaction time. Local temperature in the residual Al-slab is about 810 K, while that in the removed slab ($Fe_2O_3/FeO+Al_2O_3$) is about 3030 K, implying that such a soft slab may bond onto a hard-roller surface as a protective thin film resistant to the elevated temperature.

Comprehensively, the AIMD simulations reveal that the tensile stress at the Al(111)/α-Fe_2O_3(0001) interface may remove the Al_2O_3 fragments from the Al-slab surface after a steel roller (covered with an iron oxide) contacts the slab under an external strain load, which could indicate a typical process of the adhesive transfer of materials at interfaces.

4.3. Conclusion

The AIMD simulations of adhesive transfer of materials at an interfacial area without lubricants were performed for compressive and tensile processes between the Al(111) and the α-Fe_2O_3(0001) slabs. The conclusions can be drawn: (1) Contact of the Fe_2O_3-slab with the Al-slab may result in a strong exothermic reaction to form an amorphous alumina-texture, which welded two slabs together. (2) Separation of two slabs may have the local temperature at the interfacial area increased significantly, leading to the melting of the Al-surface and easy removal of amorphous Al_2O_3-texture from the Al-slab surface. (3) The initial stage of the P-h curve can be drawn throughout the AIMD simulation to provide the invaluable information, which was hardly observed throughout *classical* molecular dynamics simulation. (4) During the AIMD simulation, the whole reaction process at the interfacial area was $2Al + 3Fe_2O_3 \rightarrow 6FeO + Al_2O_3$, that is, due to the very short simulation time, such a thermite reaction was incomplete.

5. Summary

In this chapter, atomistic simulations of adhesive behaviors at several Al/Ceramic/Iron oxide interfaces are comprehensively discussed using density functional theory and *ab-initio* molecular dynamics, to illuminate a reasonable way for people's further understanding to this fascinating field. Some main points in the discussion include, without any loss of generality:

1. Main phenomenon in the Al/Ceramic adhesion indicated: In most cases, surface energy would play a dominant role in determining adhesion energy (W_{sep}) as interface energy was

generally much smaller in comparison if ignoring effects of large stain and incoherent bonding at the interfacial area between two contact slabs, particularly for those polar systems involving cubic oxides, carbides and nitrides to form very solid thin films (adhesives) at interfaces. Apparently, the more the W_{sep} depended upon the surface energy, the more steady and tenacious the adhesives (thin films) bind the interfaces. Therefore, knowledge of surface energies alone could serve as a reasonable starting point for estimating interfacial strength.

2. Common phenomena in the adhesive transfer of materials (phase transformation of indented materials) included: (a) Contact of the Fe_2O_3-slab with the Al-slab may result in a strong exothermic reaction to form an amorphous alumina-texture, which welded two slabs together. (b) Separation of two slabs led to the melting of the Al-surface and easy removal of amorphous Al_2O_3-texture from the Al-slab surface. This texture may bond onto a hard-roller surface as a protective thin film (soft adhesives) resistant to the elevated temperature.

Acknowledgements

This work was supported by the National Science Foundation (Grant no.: DMR9619353), USA; the Hebei Province Science and Technology Support Program (Grant no.: 15961006D), the Start-up Fund of Doctoral Research at North China Institute of Aerospace Engineering (Grant no.: BKY201405), and the Opening Fund of State Key Laboratory of Nonlinear Mechanics at Institute of Mechanics in Chinese Academy of Sciences (Grant no.: LNM201503), China.

Author details

Jun Zhong

Address all correspondence to: settings83@hotmail.com

School of Materials Engineering, North China Institute of Aerospace Engineering, Langfang, China

References

[1] Doerner MF, Nix WD. A method for interpreting the data from depth-sensing indentation instruments. Journal of Materials Research. 1986; 1: 601-609. DOI: http://dx.doi.org/10.1557/JMR.1986.0601

[2] Oliver WC, Pharr GM. An improved technique for determining hardness and elastic modulus using load and displacement sensing indentation experiments. Journal of Materials Research. 1992; 7: 1564-1583. DOI: http://dx.doi.org/10.1557/JMR.1992.1564

[3] Gerberich WW, Nelson JC, Lilleodden1 ET, Anderson E, Wyrobek JT. Indentation induced dislocation nucleation: The initial yield point. Acta Materialia. 1992; 44: 3585-3598. DOI: 10.1016/1359-6454(96)00010-9

[4] Gouldstone A, Koh HJ, Zeng KY. Discrete and continuous deformation during nano-indentation of thin films. Acta Materialia. 2000; 48: 2277-2295. DOI: 10.1016/S1359-6454(00)00009-4

[5] Page TF, Oliver WC, McHargue CJ. The deformation behavior of ceramic crystals subjected to very low load (nano) indentations. Journal of Materials Research. 1992; 7: 450-473. DOI: http://dx.doi.org/10.1557/JMR.1992.0450

[6] Pollock HM, Blau PJ. Friction, Lubrication, and Wear Technology-ASM Metals Handbook. Vol.18. ASM International; 1992. 419p. DOI: 10.1007/s11191-012-9502-4

[7] Shenoy VB, Miller R, Tadmor EB, Phillips R, Ortiz M. Quasi-continuum models of interfacial structure and deformation. Physical Review Letters. 1998; 80: 742-745. DOI: http://dx.doi.org/10.1103/PhysRevLett.80.742

[8] Ortiz M, Cuitino AM, Knap J, Koslowski M. Mixed atomistic-continuum models of material behavior: the art of transcending atomistics and informing continua. MRS Bulletin. 2001; 26: 216-221. DOI: http://dx.doi.org/10.1557/mrs2001.45

[9] Born M, Oppenheimer R. Zur Quantentheorie der Molekeln (On the Quantum Theory of Molecules). Annalen der Physik (in Germany). 1927; 389: 457-484. DOI: 10.1002/andp.19273892002

[10] Hohenberg P, Kohn W. Inhomogeneous electron gas. Physical Review. 1964; 136: B864-871. DOI: 10.1103/PhysRev.136.B864

[11] Kohn W, Sham LJ. Self-consistent equations including exchange and correlation effects. Physical Review. 1965; 140: A1133-1138. DOI: 10.1103/PhysRev.140.A1133

[12] Kresse G, Joubert D. From ultrasoft pseudopotentials to the projector augmented-wave method. Physical Review B. 1999; 59: 1758-1775. DOI: http://dx.doi.org/10.1103/PhysRevB.59.1758

[13] Jones RO, Gunnarsson O. The density functional formalism, its applications and prospects. Reviews of Modern Physics. 1989; 61: 689-746. DOI: http://dx.doi.org/10.1103/RevModPhys.61.689

[14] Perdew, JP. Accurate density functional for the energy: real-space cutoff of the gradient expansion for the exchange hole. Physical Review Letters. 1985; 55: 1665-1668. DOI: http://dx.doi.org/10.1103/PhysRevLett.55.1665

[15] Perdew JP, Chevary JA, Fiolhais C. Atoms, molecules, solids, and surfaces: applications of the generalized gradient approximation for exchange and correlation. Physical Review B. 1992; 46: 6671-6687. DOI: http://dx.doi.org/10.1103/PhysRevB.46.6671

[16] Perdew, JP. Generalized gradient approximations for exchange and correlation: a look backward and forward. Physica B. 1991; 172: 1-6. DOI: 10.1016/0921-4526(91)90409-8

[17] Perdew JP, Burke K, Ernzerhof M. ERRATA: generalized gradient approximation made simple. Physical Review Letters. 1996; 77: 3865-3868. DOI: http://dx.doi.org/10.1103/PhysRevLett.77.3865]. Physical Review Letters. 1997; 78: 1396, DOI: http://dx.doi.org/10.1103/PhysRevLett.78.1396

[18] Chadi DL, Cohen ML. Special points in the Brillouin-zone. Physical Review B. 1973; 8: 5747-5753. DOI: http://dx.doi.org/10.1103/PhysRevB.8.5747.

[19] Monkhorst HJ, Pack JD. Special points for Brillouin-zone integrations. Physical Review B. 1976; 13: 5188-5192. DOI: http://dx.doi.org/10.1103/PhysRevB.13.5188.

[20] Bachelet GB, Hamann DR, Schlüter M. Pseudopotentials that work: from H to Pu. Physical Review B. 1982; 26: 4199-4228. DOI: http://dx.doi.org/10.1103/PhysRevB.26.4199

[21] Vanderbilt D. Soft self-consistent pseudopotentials in a generalized eigenvalue formalism. Physical Review B. 1990; 41: 7892(R)-7895. DOI: http://dx.doi.org/10.1103/PhysRevB.41.7892

[22] Blöchl PE. Projector augmented-wave method. Physical Review B. 1994; 50: 17953-17979. DOI: http://dx.doi.org/10.1103/PhysRevB.50.17953

[23] Kresse G, Hafner J. Norm-conserving and ultrasoft pseudopotentials for first-row and transition elements. Journal of Physics: Condensed Matter. 1994; 6: 8245-8258. DOI: 10.1088/0953-8984/6/40/015

[24] Alder BJ, Wainwright TE. Phase transition for a hard sphere system. Journal of Chemical Physics. 1957; 27: 1208. DOI: http://dx.doi.org/10.1063/1.1743957; *ibid*. Studies in Molecular Dynamics. I. General Method. Journal of Chemical Physics. 1959; 31: 459-466, DOI: http://dx.doi.org/10.1063/1.1730376

[25] Verlet L. Computer "experiments" on classical fluids. I. Thermodynamical properties of Lennard-Jones molecules. Physical Review. 1967; 159: 98-103, DOI: http://dx.doi.org/10.1103/PhysRev.159.98

[26] Beeman D. Some multistep methods for use in molecular dynamics calculations. Journal of Computational Physics. 1976; 20: 130-139. DOI: 10.1016/0021-9991(76)90059-0

[27] Gear CW. Numerical Initial Value Problems in Ordinary Differential Equations. 1st ed. Prentice-Hall, Upper Saddle River; 1971. 253p. ISBN-13: 978-0136266068

[28] Heermann DW. Computer Simulation Method in Theoretical Physics. 2nd ed. Springer Verlag, Berlin; 1990. 148p. ISBN-13: 978-3540522102

[29] Nosé S. A molecular dynamics method for simulations in the canonical ensemble. Molecular Physics. 1984; 52: 255-268. DOI: 10.1080/00268978400101201

[30] Anderson HC. Molecular dynamics simulations at constant pressure and/or temperature. Journal of Chemical Physics. 1980; 72: 2384-2393. DOI: http://dx.doi.org/10.1063/1.439486

[31] Parrinello M, Rahman A. Polymorphic transitions in single crystals: a new molecular dynamics method. Journal of Applied Physics. 1981; 52: 7182-7190. DOI: http://dx.doi.org/10.1063/1.328693

[32] Hoover, WG. Computational Statistical Mechanics. 1st ed. Elsevier Science, Oxford; 1991. 330p. ISBN-13: 978-0444564238

[33] Kaplan IG. Intermolecular Interactions-Physical Picture, Computational methods and Model Potentials. 1st ed. Wiley; 2006. 380p. ISBN-13: 978-0470863329

[34] Payne MC, Allan DC, Arias TA, Joannopoulos JD. Iterative minimization techniques for ab initio total-energy calculations: molecular dynamics and conjugate gradients. Reviews of Modern Physics. 1992; 64: 1045-1098. DOI: http://dx.doi.org/10.1103/RevModPhys.64.1045

[35] Marx D, Hütter J. Ab Initio Molecular Dynamics: Basic Theory and Advanced Methods. 1st ed. Cambridge; 2009. 577p. ISBN-13: 978-1107663534

[36] Car R, Parrinello M. Unified approach for molecular dynamics and density-functional theory. Physical Review Letters. 1985; 55: 2471-2474. DOI: http://dx.doi.org/10.1103/PhysRevLett.55.2471

[37] VASP the GUIDE [Internet]. 2016. Available from: http://cms.mpi.univie.ac.at/vasp/.

[38] Henkelman G, Jónsson H. Improved tangent estimate in the nudged elastic band method for finding minimum energy paths and saddle points. Journal of Chemical Physics. 2000; 113: 9978-9985. DOI: http://dx.doi.org/10.1063/1.1323224

[39] Finnis, MW. The theory of metal-ceramic interfaces. Journal of Physics: Condensed Matter. 1996; 8: 5811-5836. DOI: 10.1088/0953-8984/8/32/003

[40] Humenik M, Kingery WD. Metal-ceramic interactions: III, surface tension and wettability of metal-ceramic systems. Journal of the American Ceramic Society. 1954, 37: 18-23. DOI: DOI: 10.1111/j.1151-2916.1954.tb13972.x

[41] Howe JM. Bonding, structure, and properties of metal/ceramic interfaces: Part 1 Chemical bonding, chemical reaction, and interfacial structure. International Materials Reviews. 1993; 38: 233-256. DOI: http://www.maneyonline.com/doi/abs/10.1179/imr.1993.38.5.233

[42] Eustathopoulos N, Chatain D, Coudurier L. Wetting and interfacial chemistry in liquid metal-ceramic systems. Materials Science and Engineering A. 1991; 135: 83-88, DOI: 10.1016/0921-5093(91)90541-T

[43] Barrera RG, Duke CB. Dielectric continuum theory of the electronic structure of interfaces. Physical Review B. 1976; 13: 4477-4489. DOI: http://dx.doi.org/10.1103/PhysRevB.13.4477

[44] Siegel DJ, Hector LG Jr, Adams JB. Adhesion, atomic structure, and bonding at the Al(111)/α–Al_2O_3(0001) interface: a first principles study. Physical Review B. 2002; 65: 085415. DOI: http://dx.doi.org/10.1103/PhysRevB.65.085415

[45] Siegel DJ, Hector LG Jr, Adams JB. Adhesion, stability, and bonding at metal/metal-carbide interfaces: Al/WC. Surface Science. 2002; 498: 321-336. DOI: 10.1016/S0039-6028(01)01811-8

[46] Siegel DJ, Hector LG Jr, Adams JB. First-principles study of metal–carbide/nitride adhesion: Al/VC vs. Al/VN. Acta Materialia. 2002; 50: 619-631. DOI: 10.1016/S1359-6454(01)00361-5

[47] Siegel DJ, Hector LG Jr, Adams JB. Ab initio study of Al-ceramic interfacial adhesion. Physical Review B. 2003; 67: 092105. DOI: http://dx.doi.org/10.1103/PhysRevB.67.092105

[48] Banerjea A, Smith R. Origins of the universal binding-energy relation. Physical Review B. 1988; 37: 6632-6645. DOI: http://dx.doi.org/10.1103/PhysRevB.37.6632

[49] Lipkin DM, Israelachvili JN, Clarke DR. Estimating the metal-ceramic van der Waals adhesion energy. Philosophical Magazine A. 1997; 76: 715-728. DOI: 10.1080/01418619708214205

[50] Zhong J, Adams JB. Adhesive metal transfer at the Al(1 1 1)/α-Fe_2O_3(0001) interface: a study with ab initio molecular dynamics. Modelling and Simulation in Materials Science and Engineering. 2008; 16: 085001. DOI: 10.1088/0965-0393/16/8/085001.

[51] Zhong J, Adams JB. Adsorption and decomposition pathways of vinyl phosphonic and ethanoic acids on the Al(111) surface: a density functional analysis. Journal of Physical Chemistry C. 2007; 111: 7366-7375. DOI: 10.1021/jp0667487

[52] Yamamoto N. The shift of the spin flip temperature of α-Fe_2O_3 fine particles. Journal of the Physical Society of Japan. 1968; 24: 23-28. DOI: http://dx.doi.org/10.1143/JPSJ.24.23

[53] Coast R, Pikus M, Henriksen NP, Nitowski AG. A vibrational spectroscopic comparison of vinyl-triethoxysilane and vinyl-phosphonic acid adsorbed on oxidized aluminum. Journal of Adhesion Science and Technology. 1996; 10: 101-121. DOI: 10.1163/156856196X00805

[54] Szlufarska I. Atomistic simulations of nanoindentation. Materials Today. 2006; 9: 42-50. DOI: 10.1016/S1369-7021(06)71496-1

[55] Eng PJ, Trainor TP, Brown Jr GE, Waychunas GA, Newville M, Sutton SR, Rivers ML. Structure of the Hydrated α-Al2O3(0001) Surface. Science. 2000; 288: 1029–1033. DOI: 10.1126/science.288.5468.1029.

[56] Hay MT, Geib SJ. Tetra-butyl-ammonium isobutyl silsesquioxane monochloroferrate(III). Acta Crystallographica Section E. 2007; E63: m445-m446. DOI: 10.1107/S1600536807000360.